Annals of Mathematics Studies

Number 111

COMBINATORIAL GROUP THEORY AND TOPOLOGY

EDITED BY

S. M. GERSTEN

AND

JOHN R. STALLINGS

PRINCETON UNIVERSITY PRESS

PRINCETON, NEW JERSEY

1987

Copyright © 1987 by Princeton University Press
ALL RIGHTS RESERVED

The Annals of Mathematics Studies are edited by
William Browder, Robert P. Langlands, John Milnor, and Elias M. Stein
Corresponding editors:
Stefan Hildebrandt, H. Blaine Lawson, Louis Nirenberg, and David Vogan

Clothbound editions of Princeton University Press books are printed on acid-free paper, and binding materials are chosen for strength and durability. Paperbacks, while satisfactory for personal collections, are not usually suitable for library rebinding

ISBN 0-691-08409-2 (cloth)
ISBN 0-691-08410-6 (paper)

Printed in the United States of America
by Princeton University Press, 41 William Street
Princeton, New Jersey

☆

Library of Congress Cataloging in Publication data will
be found on the last printed page of this book

CONTENTS

PREFACE — vii

I
Combinatorial Group Theory

PROBLEMS IN COMBINATORIAL GROUP THEORY — 3
  by Roger Lyndon

POINCARÉ DUALITY GROUPS OF DIMENSION TWO ARE SURFACE GROUPS — 35
  by Beno Eckmann

HOW TO GENERALIZE ONE-RELATOR GROUP THEORY — 53
  by James Howie

GRAPHICAL THEORY OF AUTOMORPHISMS OF FREE GROUPS — 79
  by John R. Stallings

PEAK REDUCTION AND AUTOMORPHISMS OF FREE GROUPS AND FREE PRODUCTS — 107
  by Donald J. Collins

NONSINGULAR EQUATIONS OF SMALL WEIGHT OVER GROUPS — 121
  by S. M. Gersten

A GRAPH-THEORETIC LEMMA AND GROUP-EMBEDDINGS — 145
  by John R. Stallings

THE TODD-COXETER PROCESS, USING GRAPHS — 157
  by John R. Stallings and A. Royce Wolf

A SUBGROUP THEOREM FOR PREGROUPS — 163
  by Frank Rimlinger

GROUPS WITH A RATIONAL CROSS-SECTION — 175
  by Robert H. Gilman

ON THE RATIONAL GROWTH OF VIRTUALLY NILPOTENT GROUPS — 185
  by Max Benson

SJOGREN'S THEOREM FOR DIMENSION SUBGROUPS — THE METABELIAN CASE — 197
  by Narain Gupta

ON GROUP PRESENTATIONS, COPRODUCTS AND INVERSES by Robert Craggs and James Howie	213
ON COMPLEXES DOMINATED BY A TWO-COMPLEX by John G. Ratcliffe	221
SUBCOMPLEXES OF TWO-COMPLEXES AND PROJECTIVE CROSSED MODULES by Michael Dyer	255
LENGTH FUNCTIONS OF GROUP ACTIONS ON Λ-TREES by Roger Alperin and Hyman Bass	265

II
Very Low Dimensional Topology

RESIDUAL FINITENESS FOR 3-MANIFOLDS by John Hempel	379
THE NIELSEN-THURSTON THEORY OF SURFACE AUTOMORPHISMS by Steven A. Bleiler	397
WHITEHEAD GROUPS OF CERTAIN HYPERBOLIC MANIFOLDS, II by A. J. Nicas and C. W. Stark	415
A CHARACTERIZATION OF FINITE SUBGROUPS OF THE MAPPING-CLASS GROUP by Jane Gilman	433
A SEQUENCE OF PSEUDO-ANOSOV DIFFEOMORPHISMS L. Neuwirth and N. Patterson	443
DEHN'S ALGORITHM REVISITED, WITH APPLICATIONS TO SIMPLE CURVES ON SURFACES by Joan S. Birman and Caroline Series	451
PATHS OF GEODESICS AND GEOMETRIC INTERSECTION NUMBERS: I by Marshall Cohen and Martin Lustig	479
PATHS OF GEODESICS AND GEOMETRIC INTERSECTION NUMBERS: II by Martin Lustig	501
SELECTED PROBLEMS by S. M. Gersten	545

PREFACE

Combinatorial Group Theory lives in the fertile region between pure group theory and pure topology. The interplay between the logical precision of algebra and the intuitive depths of geometry gives it charm and strength. Lyndon suggests in his article that perhaps there is no exact definition of the subject. Certainly it includes the study of equations over groups (which tends to involve the study of geometric diagrams), much of 3-manifold theory (the Poincaré Conjecture is equivalent to a group-theoretic question), group actions on geometric-combinatorial objects such as trees, the theory of surface automorphisms, and many other developments. As the Walrus said, the time has come to talk of many things.

At Alta Lodge in the spectacular Wasatch Mountains of Utah, on July 15-18, 1984, sixty of us gathered for an intense conference. Roger Lyndon's opening address, in the tradition of Hilbert, offered a score of open problems to guide the field in the future. In the style of the Séminaire Bourbaki, six speakers were assigned topics for expository talks; these were the origins of the papers in this book by Alperin and Bass, Bleiler, Eckmann, Hempel, and Howie. In addition, there were about twenty-five shorter talks on current research, from which we selected the remaining articles here.

Our goal was to produce a book full of ideas, understandable to ourselves and to students, that will open up the future development of this field. We consciously tried to reach a large class of readers, including good graduate students with backgrounds in group theory and topology. We are grateful for the cooperation of the authors who have helped us in pursuing this goal.

It is a pleasure to acknowledge the assistance given us by Roger Lyndon, with whom we frequently consulted in the planning of the conference. In tribute to his work in the field and his insight into directions for future research, his article has a prominent place in this volume.

We are greatly indebted to James Howie and Geoffrey Mess for frequent consultations. In addition, we wish to thank some thirty referees, who by tradition must remain anonymous, for their valuable aid.

Special thanks are due to Marty Jones of Alta Lodge and to Ann Reed for helping make the conference run smoothly. Finally we acknowledge with thanks and appreciation the financial support of the University of Utah and the National Science Foundation.

STEVE GERSTEN
JOHN STALLINGS

Combinatorial Group Theory
and Topology

PROBLEMS IN COMBINATORIAL GROUP THEORY
Roger Lyndon

0. *Introduction*

Steve Gersten asked me to give a talk like Hilbert gave in Paris in 1900. I said I'd be happy to, but pointed out that I'm no Hilbert. Merzlyakov [117] came to my rescue with the following suggestion: "Rather than waiting [for a new Hilbert] group theorists have come to the more prosaic idea of a Present-day collective Hilbert." So I have appealed to some of you for amendments to a provisional outline, for which I thank you, and I hope this exonerates me of any charge of arrogance in presenting this biased survey.

What is Combinatorial Group Theory? This term acquired official status as the title of the book of Magnus-Karrass-Solitar. The first sentence of the History of Combinatorial Group Theory by Chandler-Magnus [28] says: "Combinatorial Group Theory may be characterized as the theory of groups which are given by generators and relations...." (But compare Magnus-Karrass-Solitar with Coxeter-Moser, Generators and Relations for Discrete Groups.) This hardly does justice to the goals or methods of the subject. On other occasions Chandler-Magnus speak of "group theory with the exception of Lie groups and of group representations and linear groups," but this seems to both include and exclude too much.

Maybe we could define Combinatorial Group Theory to be Very Low Dimensional Topology. In fact, group theorists, like others, will attack whatever problems interest them with whatever tools they have at hand, so perhaps Combinatorial Group Theory is just a state of mind. Maybe it is distinguished by a reluctance to make the great supposedly simplifying assumptions of Commutativity, Linearity, and either Finiteness or Continuity, or, put more positively, a relish for the combinatorial core of a problem. Altogether, it seems fruitless to try to define an elephant we have barely touched.

I intended to talk about the history of the subject, but Chandler-Magnus have let me off that hook. I do not need to tell anyone that the major source and strength of Combinatorial Group Theory has been Topology, wherein we tentatively include Discontinuous Groups, from Poincaré on, with an assist from the more or less abstract or axiomatic side, beginning with Cayley, through the influence of Finite Groups and of problems from Logic. These influences will be evident, and, if my discussion appears biased against topology, it is because as a non-topologist I am reluctant or unable to tell you what you know better than I. Altogether, though, I was agreeably surprised by the homogeneity of the subject, if somewhat inconvenienced by the many interconnections among the various approaches and problems in the subject.

My account is also biased for the most part toward rather recent work, and to pathways to the future more than monuments to the past. The space devoted to a subject should not be taken as a measure of the importance attached to it. Likewise, the mention of a name or citation of a paper is often more to draw attention to it than to bestow an honor. References are illustrative or suggestive, and should usually be followed by 'and others.' I have not cited papers that are well known or easily accessible (with a few exceptions), in particular papers more than about five years old, or listed in the bibliography of Lyndon-Schupp. Incomplete references indicate lack of knowledge.

Despite the unity of the subject, as a first step toward linearization I have divided it into seven unequal and gerrymandered sections. It took some cutting and pasting to bring my list of problems to twenty, a modest three fewer than Hilbert. Even so, many are really 'problem areas,' of the classical form: 'What can be said about ....?.'

1. Properties of free groups, equations in and over groups

The most 'abstract' and 'axiomatic' of our problems is a folklore problem of Alfred Tarski.

PROBLEM 1. *Do all nonabelian free groups have the same elementary theory?*

The elementary theory of a class of groups is the set of all sentences in first order logic (with symbols for equality and group composition but, emphatically, excluding set theory) that are true in all groups of the class. By contrast, among free *abelian* groups those of rank at most 2 are distinguished by the property that there exist elements a and b such that, for all x, one of x, xa, xb, xab is the square y^2 of some element y. See V. Dyson [47].

In this connection, R. Vaught posed the following test problem: In a free group, does $x^2 y^2 = z^2$ always imply $xy = yx$? This was proved by Lyndon and generalized extensively. The method was close to that used by H. Zieschang in studying automorphisms of surface groups and Fuchsian groups.

A much earlier theorem of Frobenius (1895) deals with the number of solutions of the equation $x^n = 1$ in a finite group; for a survey of this problem see H. Finkelstein [57]. See also Finkelstein-Mandelberg [58].

Equations *over* groups entered Combinatorial Group Theory in a paper of B.H. Neumann (1943), where he showed that, for a positive integer n and an element g of a group G, the equation $x^n = g$ has as many solutions as desired in some group H containing G. Higman-Neumann-Neumann extensions are a generalization of the fact that the equation $t^{-1} g_1 t = g_2$, for g_1 and g_2 in G, has a solution t in a group H containing G if and only if g_1 and g_2 have the same order.

The central problem in this area is the Kervaire-Lauderbach problem, which we state as follows.

PROBLEM 2. *If G has a presentation G = (X:R) and H = (X∪t:R∪w) is obtained by adding one new generator and one defining relation, when does the inclusion X → X ∪ t induce an injection of G into H ?*

In simpler language, for $g_1, \cdots, g_n$ in G, when does an equation $w(g_1, \cdots, g_n, t) = 1$ have a solution t in some group H containing G?

We have seen that $t^{-1} g_1 t g_2^{-1} = 1$ has no solution if g_1 and g_2 have different orders. Gerstenhaber-Rothaus showed that if the sum of the exponents on t in w is not 0, and if G can be embedded in a compact connected Lie group, then a solution always exists. Rothaus [144] later improved the condition on G to local residual finiteness. The sufficient condition of local indicability has been studied by J. Howie [94]. The group $G = (a_1, \cdots, a_4 : a_{i+1}^{-1} a_i a_{i+1} = a_i^2$, i modulo 4) of G. Higman satisfies none of the known sufficient conditions for the solvability of an equation (with non-zero exponent sum) over G. Without the exponent condition, probing by Lyndon [113] suggests that the problem is difficult even for G a finite cyclic group. See also Brodskii [19, 20] and Short [152].

The result of Gerstenhaber-Rothaus raises the following question.

PROBLEM 2a. *If the sum of the exponents on t in w is not 0, does the equation w = 1 always have a solution?*

So much for equations *over* groups. Equations *in* groups are exemplified by the Vaught problem. This prompted Lyndon to ask about solutions of an equation $w(a_1, \cdots, a_n, t_1, \cdots, t_m) = 1$ *in* a free group G with basis $a_1, \cdots, a_n$. In the case m = 1 of one unknown he obtained a set of words containing parameters as exponents which, subject to conditions on the parameters, give precisely the set of solutions. For example, $t^{-1} a_1 t a_1^{-1} = 1$ has exactly the solutions $t = a_1^n$ for all integers n. This was substantially improved and extended, but without definitive result. Recently Makanin [115] has given an algorithm that associates with an equation w = 1, as above, an integer N such that, if any solution exists, there exists one with total length of the t_i at most N. This settles the question of existence of solutions and provides an algorithm for finding one if it exists, but leaves open the following problem.

PROBLEM 3. *Given an equation over a free group, find an algebraic description of the set of all solutions.*

See Howie [93, 95, 97], Ozhigov [130].

A special case of this problem, for free groups and other groups, is the Substitution Problem (for free groups, the Endomorphism Problem): Does an equation $w(t_1,\cdots,t_m) = g$ have a solution? Wicks [169] showed that an element g of a free group is a commutator if and only if, relative to any basis, g can be written in the form $g = abca^{-1}b^{-1}c^{-1}$ *without cancellation*. This result has been extended substantially by C. C. Edmunds [51, 52] and L. P. Comerford, Jr. [34, 35] and, using topological methods, by M. Culler [37] and by Goldstein-Turner [73, 74]. Related methods have been used by M. Scharlemann [146] and P. E. Schupp [148]; see also E. Rips [140], and P. Hill and S. J. Pride [90].

PROBLEM 4. *Let $w(a_1,\cdots,a_n)$ be a word in the free group F with basis $a_1,\cdots,a_n$. Is there an algorithm which, given g in F, decides if there exist $t_1,\cdots,t_n$ in F such that $w(t_1,\cdots,t_n) = g$?*

The Substitution Problem has been studied extensively for finite simple groups and for various classical infinite groups. See Finkelstein [57], Finkelstein-Mandelberg [58], Lyndon [112], Mycielski [122], and also Ehrenfeucht-Fajtlowicz-Malitz-Mycielski [53].

2. *Automorphisms of groups*

The general linear group GL(n, R) is the group of automorphisms of the free R-module of rank n. The automorphism group GL(2, Z) of the free Z module, or free abelian group, A_2 of rank 2, and also, in particular, the modular group $PSL(2, \mathbf{Z})$, have been much studied. But it is a giant stride to the study of the automorphism group $\text{Aut } F_n$ of the free nonabelian group F_n of rank n.

Nielsen used automorphisms of free groups to great advantage, and obtained a finite presentation for Aut F. J. McCool recovered this presentation by different methods, which enabled him to obtain finite presentations for the stabilizers of finite sets of elements. His method is based on that used by Whitehead to decide whether two finite sequences of elements of a free group are equivalent under some automorphism. A related problem of Whitehead, to decide whether two finitely generated subgroups are equivalent under an automorphism, has been solved recently by Gersten [68], using new methods. By the same methods he has proved a conjecture of P. Scott that the subgroup of elements fixed by an automorphism of a finitely generated free group is finitely generated. J. L. Dyer and Scott had shown earlier that the set of fixed points of a finite subgroup of Aut F is a free factor of F. There has been some study of the structure of a single automorphism, inducing the dream, or nightmare, of a 'Jordan structure theorem.'

Considerable work on the structure of Aut F, related mainly to its action on various naturally arising characteristic subgroups and their quotients, has been done by S. Andreadakis and by S. Bachmuth-H. Mochizuki.

PROBLEM 5. *Determine the structure of* Aut F, *of its subgroups, especially its finite subgroups, and its quotient groups, as well as the structure of individual automorphisms.*

If N is a normal subgroup of a free group F and $\text{Aut}_N F$ is its stabilizer, then $\text{Aut}_N F$ induces a group of automorphisms of $G = F/N$. Nielsen showed that for the usual presentation of a surface group G the group $\text{Aut}_N F$ maps onto Aut G. G. Rosenberger [143] and H. Zieschang have studied this 'lifting problem,' especially for Fuchsian groups. See also S. J. Pride and A. D. Vella [137].

PROBLEM 6. *If* $G = F/N$, F *free, what subgroups of* Aut G *are images of subgroups of* Aut F ?

The mapping class group M of a surface can be identified essentially with the group of outer automorphisms (automorphisms modulo inner automorphisms) of its fundamental group G. The results of J. McCool give M (or, directly, Aut G) as the fundamental group of a finite complex, which, unfortunately, exhibits too many natural symmetries to make it accessible to present day computation. W. Thurston (see A. Hatcher-W. Thurston [88]) has given a quite different method for obtaining a finite presentation of the mapping class groups, and, for the orientable case, B. Wajnryb [168] has used this method to obtain presentations that are reasonably concise but not entirely perspicuous.

PROBLEM 7. *Obtain finite presentations for the mapping class groups that are at once usably concise and yet in which both the generators and the relations have fairly obvious geometrical meanings.*

For example, Hatcher-Thurston say: "... all relations follow from relations supported in certain subsurfaces, finite in number, of genus at most 2."

The following is an obvious addendum.

PROBLEM 7a. *The same for all Fuchsian groups.*

3. *Morphisms of trees*

The prefix 'auto' is omitted as a salute to the recent work of Gersten [62, 64-72] and Stallings [158, 159] on morphisms in the category of graphs.

A basic paper of J. Tits [164] initiates the study of the group Aut T of automorphisms of a tree T. He shows that, if Aut T leaves invariant no proper subtree and no end of T, then the subgroup G generated by all stabilizers of branch points is a simple group, while the quotient (Aut T)/G is a free product of groups of order 2 and infinite cyclic groups. This frequently cited work deserves to be extended.

A great deal is coming to be known about certain very special but very remarkable groups of automorphisms of trees, introduced by N. Gupta and S. Sidki [84, 85, 86, 153, 154]. If A is an 'alphabet' with a prime number p of letters, then the monoid $T = A^*$ of all finite words, ordered by left divisibility, is a tree T. Generalizing a construction of R I. Grigorchuk [79], Gupta and Sidki show that certain easily described 2-generator subgroups G of Aut T have remarkable properties: they are 'Burnside groups,' that is, infinite 2-generator p-groups (with elements of unbounded order); they contain isomorphically all finite p-groups; they are residually finite, and all their proper quotient groups are finite. These groups are remarkable not only for these properties, but also because they are quite 'concrete' and accessible to detailed study.

PROBLEM 8. *Study the structure of the automorphism groups of trees and of their subgroups.*

For example, do these groups have Sylow subgroups?

Certain naturally arising instances of groups acting on trees encountered by J-P. Serre appear to have led to the Bass-Serre theory of graphs of groups, with their associated groups acting on trees. This method has become a standard tool in the study of infinite groups, especially as obtained by amalgamated product and HNN-extension, and with regard to subgroup theorems. Earlier, Lyndon, in seeking to unify cancellation arguments based on Nielsen transformations in the proofs of the Nielsen-Schreier and Kurosh Subgroup Theorems and of the Grushko-Neumann Theorem, introduced axiomatically characterized length functions on groups. I. M. Chiswell showed that this theory, for integer valued functions, is essentially equivalent to the Bass-Serre theory of groups acting on trees. J. W. Morgan and P. B. Shalen [120], following work of R. C. Alperin and K. N. Moss [4] on real valued length functions, have played off these two theories to obtain new proofs of two theorems of Thurston.

In connection with the Subgroup Theorems, we note that Rosenberger, Zieschang, and Karrass-Solitar all obtained refinements of the basic Subgroup Theorem of Hanna Neumann for amalgamated products. In particular, Karrass and Solitar introduced tree products, which agree with a special case of the Bass-Serre graph products. They also introduced polygonal products: a group G is generated by vertex groups, with the edge groups amalgamated. (This group G differs in a small but significant way from the corresponding Bass-Serre group; the Karrass-Solitar definition is natural in the context of presentations of certain geometrically constructed groups, while that of Bass-Serre is natural for a graph of complexes.) Polygonal products were motivated by the recognition that the Picard group $PSL(2, Z[i])$ can be obtained from four very small groups at the vertices of a square by amalgamating subgroups associated with the sides of the square. See, for example, B. Fine [55, 56] and A. Brunner, M. L. Frame, Y. W. Lee, N. J. Wielenberg [24]. Square products are studied also in a paper of D. Ž. Djokovič [42], which, although failing of its main objective, contains an extensive study of groups acting on cubic trees. See also Djokovič-G. L. Miller [43]. A. Brunner, Y. W. Lee, and N. J. Wielenberg [25] have used polygonal products to obtain elegant descriptions of various 3-dimensional Euclidean groups. M. W. Davis [40] has used a similar construction in a more abstract context to obtain generalized Coxeter groups by sewing together infinitely many copies of an abstract polytope by identifications at vertices according to specified orthogonal groups; in this way he obtains aspherical manifolds of dimension $n \geq 4$ not covered by Euclidean space. There are also various abstractly defined generalizations of symmetric groups, braid groups, and Coxeter groups; K. I. Appel and P. E. Schupp [5, 6] have used small cancellation theory to study certain such groups.

PROBLEM 9. *The various constructions mentioned above appear to open the way to a more comprehensive theory of the structure of infinite groups.*

We mention a curious result concerning the question of accessibility by D. E. Muller and P. E. Schupp [121], that a finitely generated group is virtually free if and only if it has a context-free word problem and is accessible.

See also W. Dicks [41], J-C. Hausmann [89], A. Karrass-A. Pietrowski-D. Solitar [108], Stallings [158], and M. D. Tretkoff [165].

4. *Burnside groups*

W. Burnside asked if every finitely generated torsion group is finite. This is true for groups of certain small exponents, for linear groups, and for analogous Lie rings. In 1960 Golod proved the existence of what we shall call *Burnside groups*, finitely generated infinite torsion groups. In 1964 P. S. Novikov and S. I. Adian showed that $B(m,p)$, the free group on $m \geq 2$ generators in the variety of groups of prime exponent p, is infinite for sufficiently large p; see [1]. We have mentioned that Grigorchuk [79] gave a very simple construction for an infinite 3-generator 2-group, and that N. D. Gupta and S. Sidki extended this construction to obtain infinite 2-generator p-groups for all primes p. Before Grigorchuk, S. V. Aleshin [3] had constructed groups like Grigorchuk's in terms of automata, which Y. I. Merzlyakov [118] showed to be essentially equivalent to those of Grigorchuk.

A. Y. Olshanskii [124, 125, 126, 128, 129] used small cancellation theory to construct infinite 2-generator groups, with solvable conjugacy problem, in which all nontrivial proper subgroups are isomorphic cyclic groups, all infinite cyclic or all of order p, for some prime $p > 10^{75}$. Olshanskii's groups are Tarski Monsters, groups all of whose proper subgroups have smaller cardinality. S. Shelah [151], also using small cancellation theory, had constructed earlier an uncountable Tarski Monster. E. Rips (see [140]) has obtained results, not wholly published, parallel to results of Olshanskii, using similar methods.

The groups B(m,p) of Novikov-Adian are universal in the sense of being as large as possible given the number m of generators and the exponent p, and thus they have a simple definition, but at the expense of being complicated to work with. Despite this Adian [1] has obtained considerable detailed information about them. The groups of Olshanskii are more economical, as images of the Novikov-Adian groups, but their definition is much more complicated, their presentation being dictated naturally by the problem at hand; from this it is easier to establish their crucial properties. The groups of Gupta-Sidki, although not of finite exponent, combine the virtues of being very easily defined and relatively easy to work with.

The Burnside problem may be viewed as a test problem, of more interest in terms of methods than of results. Nonetheless, the natural generality of the Novikov-Adian groups together with Adian's detailed results on them, the fact that Olshanskii's groups of exponent p are infinite finitely generated simple groups, and the fact that the Gupta-Sidki groups arise as groups of automorphisms of trees and have interesting properties, all lead to the conclusion that Burnside groups are of considerable intrinsic interest. (Perhaps there was a time when finite p-groups were considered as of interest only as Sylow groups.)

PROBLEM 10. *Develop a general theory of Burnside groups, including a detailed study of certain particular such groups.*

Despite enormous work, mainly for very small and very large prime values of n, the following problem is far from solved.

PROBLEM 10a. *For which pairs m and n is the group B(m,n) finite?*

Another natural problem seems to be folklore, and little more.

PROBLEM 10b. *Is every finitely presented torsion group finite?*

I have found no reference to this problem in print. Gupta has pointed out that all those who have constructed infinite Burnside groups have been at pains to point out that their groups are not finitely presented. Scott has told me that he and, independently, W. Jaco, had recognized that the existence of a finitely presented infinite torsion group would follow from the existence of a 3-manifold whose nonabelian fundamental group had an infinitely generated center.

For a survey of Burnside groups see Gupta [82].

5. Generators and relations

From the point of view of presentations, the simplest groups after free groups are the one-relator groups. These groups have been studied extensively, partly because the surface groups are one-relator groups, and partly because they share some of the accessibility of free groups while exhibiting considerable individual complexity. Magnus, giving credit to Dehn's topological insight, solved the word problem for one-relator groups, and obtained other results by the same method, notably the Freiheitssatz. Although his method looks rather easy, and perhaps even obvious from the algebraic side, in retrospect, it has yet to be incorporated into a general theory in an entirely satisfactory way, especially from the topological point of view. In connection with the Freiheitssatz, see Lyndon, and B. Baumslag-Pride, and J. Howie [93]. It is striking that, despite the solution of the conjugacy problem for a fair variety of groups, the conjugacy problem for one-relator groups remains unsettled fifty years after Magnus' solution of the word problem.

PROBLEM 11. *Is the conjugacy problem solvable for all one-relator groups?*

Note that B.B. Newman has solved this problem for all one-relator groups with torsion.

Recently A. Juhász [103, 104, 105, 106] has developed new methods in small cancellation theory that solve the conjugacy problem for certain groups, including a new special class of one-relator groups. Other recent work, especially of S. J. Pride [133, 134, 135, 136] and of Pride together with P. Hill [90] and with P. Hill and A. D. Vella [91], has extended small cancellation theory to derive considerable information about the structure of one-relator groups and their subgroups. For example, Pride has shown that a one-relator group with torsion can have only finitely many isomorphism types of nonfree two-generator subgroups. See also G. Baumslag [9] and H. D. Hurwitz [102].

PROBLEM 11a. *Determine further the structure of one-relator groups.*

Although small cancellation theory seems now to be the best tool for studying one-relator groups and, especially, their conjugacy problem, if it turns out that there are one-relator groups with unsolvable conjugacy problem, it would seem to require very considerable ingenuity on the side of logic to construct one.

A second question arising simply in terms of generators, initially without reference to relations, is that of the rate of growth on a group. If a group G is generated by a finite set X, let g(n) be the number of elements of G represented by words in X of length at most n. J. Milnor, who introduced these ideas in connection with differential geometry, asked if the asymptotic rate of growth of g(n), which does not depend on the choice of X, is always either polynomial (as for a free abelian group) or exponential (as for a nonabelian free group), and, with J. A. Wolf, established this conjecture for finitely generated solvable groups; for these M. Gromov [80] (see also A. J. Wilkie-L. van den Dries [170]) showed that the growth is polynomial just in case the group contains a nilpotent group of finite index. Recently Grigorchuk has shown that one of his groups is neither polynomial nor exponential.

J. W. Cannon [27] showed that for certain Fuchsian groups and Coxeter groups, with a natural choice of generating set X, the growth generating function $G(z) = \Sigma g(n) z^n$ is a rational function, and characterized its zeros and poles. M. Benson [14] showed that $G(z)$ is rational for finite extensions of free abelian groups of finite rank. See also M. Grayson [77]. Grigorchuk [78] considered a function $H(z) = \Sigma h(n) z^n$ for H a subgroup of a free group F, where $h(n)$ is now the number of elements of H of minimal length n relative to a basis X for F, and showed that $H(z)$ is a rational function. For H normal and $G = F/H$, he obtains the spectral radius r_G of a random walk on G in terms of the radius of convergence of $H(z)$, and proves the following. To given $\varepsilon > 0$ and integers M and N, there corresponds an integer L with the property that, if G has a presentation with M generators and $n \leq N$ relators, all of length at least L, which satisfies a standard small cancellation condition, then $|r_G - \sqrt{2M-1}/M| < \varepsilon$, and hence G is not amenable. Adian uses these ideas to show that his infinite groups B(m,p) are not amenable. C. Series [149, 150] uses similar methods to study the distribution of the set of limits of random walks on Fuchsian groups. See also H. Bass [8], W. J. Floyd [59], and P. Wagreich [167].

PROBLEM 12. *There is clearly much to be done in determining the possible growth functions of groups and in relating them to properties of groups.*

A similarity between the formula for the growth function of a free product and that for the characteristic suggests a connection between these two concepts, which is borne out in some of Cannon's examples. It appears that further understanding is needed of the dependence of the detailed structure of the growth function on the choice of generating set, and, with it, on properties of the set of relations.

A long standing question is that of the significance of the deficiency of a presentation, the excess of the number of relations over the number of generators. Deficiency of a finite group has been much studied. The deficiency can be viewed as a truncation of a Poincaré series of a resolution of the presentation, and seems related to the Euler-Poincaré characteristic. See B. Baumslag-S. J. Pride [10, 11], K. S. Brown [21, 22] (and the review of [21] by K. W. Gruenberg [81]), I. M. Chiswell [29], M. Edjvet [49], J. G. Ratcliffe [138], N. S. Romanovskii [142] and R. Stöhr [160]. We pose only a rather vague problem.

PROBLEM 13. *Extend and relate the theories of the deficiency, the rate of growth, and the Euler-Poincaré characteristic. In particular, what influence does the deficiency have on the structure of an infinite group?*

This seems an appropriate point to speak of certain problems of a directly topological origin which have led to equivalent or related problems that can be stated in purely group theoretic terms. I do not feel it is my part to do more than mention these problems.

PROBLEM 14. *The Poincaré conjecture.*

See G. A. Swarup [161]. Among the many problems in group theory stimulated by the Poincaré conjecture, that of J. J. Andrews and M. L. Curtis is possibly the best known.

PROBLEM 15. *Let the trivial group have a balanced presentation $(X:R)$, where $|X| = |R| < \infty$. Can this presentation be reduced to a trivial presentation by a succession of transformations of the following kinds:*
1. *Nielsen transformations of X;*
2. *Nielsen transformations of R;*
3. *Replacing an element of R by a conjugate;*
4. *Tietze transformations introducing (or deleting) a new element x of X together with a new relator r in R defining x?*

See R. Craggs [36], W. Metzler [119], and C. P. Rourke [145].

A geometrically finite group G is one possessing a finite $K(G,1)$, and its geometric dimension is then the least dimension of such a $K(G,1)$.

Geometrically finite groups, along with many related concepts and problems, are discussed in G. Baumslag-E. Dyer-A. Heller [12]; in connection with a related problem of Baumslag-Dyer-Heller see T. Bortnik [18]. The equality of geometric dimension and cohomological dimension has been proved with one exception.

PROBLEM 16 (The Eilenberg Problem). *If G has cohomological dimen- 2, must it also have geometric dimension 2 ?*

We remark that Baumslag-Dyer-Heller examine algebraically closed groups, using, as does S. D. Brodskii [19], a definition that is natural but which differs from the original definition given by W. R. Scott.

A third problem from topology with many group theoretic connections is Whitehead's asphericity problem.

PROBLEM 17. *Is every subcomplex of an aspherical 2-complex aspherical?*

This problem has received considerable attention recently. See R. Brown-J. Huebschmann [23], I. M. Chiswell-D. J. Collins-Huebschmann [30], Collins-Huebschmann [31], M. Gutierrez-J. G. Ratcliffe [87], J. Howie [96], Huebschmann [99, 100, 101], and A. J. Sieradski [155].

The problems above touch on the large area of questions of a homological nature, of which we mention only one more. That is the study of Poincaré duality groups; for this see R. Bieri [2] and, especially, the talk by B. Eckmann.

We cannot resist introducing small cancellation theory with a quotation from J. Nielsen (1912), see [123].

"While in the solution of the generation problem one arranges elements in sequence in all possible ways — one could speak of a linear problem — one represents the [relators] in the group graph by polygons and in the treatment of the identity problem one puts these together in a map. This latter problem has, therefore, in an expression of Dehn's, 'one more dimension' than the first. Whether this is a real or only an apparent difference in level ... remains undecided."

Small cancellation theory has become a useful and ubiquitous tool. Applied small cancellation theory is exemplified by recent work of Juhász, Olshanskii, Pride, and Rips, who have also, as well as, for example, Do Long Van [44], J. Howie-S. J. Pride [98], and J. Perraud [132], greatly extended the methods of pure small cancellation theory.

The theory, as developed so far, could be described as follows. From 'local' hypotheses, on 'small' subcomplexes of the (2-dimensional) Cayley complex (Gruppenbild) of a presentation (initially, on the stars of faces), one derives properties of ' large' subcomplexes (singular discs, spheres, annuli, tori, ...) or of the entire complex. There is an evident analogy, or connection, with corresponding differential-integral concepts, for example, in small cancellation theory the formulas relating 'arc length,' curvature,' and 'area.' (Two papers in the 40's by C. Blanc [16] and F. Fiala [54], developing and applying complex function theory on planar graphs do not seem to have been pursued further. There is more recent work by P. Cartier on harmonic analysis on trees, and also work on harmonic analysis on free groups.) Thus, as well as Dehn's original application of small cancellation theory in the hyperbolic plane to logical decision problems, from which the theory arose, and of the more or less obvious connections with topology (for example, asphericity) and with purely group theoretic 'structure problems' (for example, Pride's work on subgroups), there are connections of an analytic nature, concerning rate of growth, random walks, analysis on groups, etc.

PROBLEM 18. *Extend small cancellation theory, especially in accordance with the connection or analogy with analysis, and unify the special extensions and applications noted above.*

6. *Geometric groups*

Much of combinatorial group theory arose from the study of Fuchsian groups, if one includes surface groups, and there is no need to mention the role of hyperbolic groups in the study of higher dimensional manifolds. In the study of Fuchsian groups, especially surface groups, there seems to be a long tradition, beginning with Dehn and Magnus, and with Nielsen, and continuing through the work of Reidemeister and Zieschang and, in an extreme form, two papers of A. H. M. Hoare, A. Karrass, and D. Solitar on subgroups of Fuchsian group, which could be described, somewhat invidiously, as a more or less conscious attempt to free the subject from analysis. This tendency is not entirely a matter of prejudice, in view of the increasing difficulty of applying analytic methods in higher dimensions and in more abstract situations, notably to 'geometric' groups defined combinatorially without any direct reference to an analytic structure, as mentioned in Section 3.

To a non-analytic mind it seems a great mystery that combinatorial and analytic methods often lead to the same results; it happens often that objects definable combinatorially are realizable analytically. A modest example is the combinatorially regular tessellations of the n-sphere and n-space for $n \geq 2$, and a more impressive one is the definition by Zieschang, E. Vogt, and H. D. Coldewey of Fuchsian groups as automorphism groups of planar 2-complexes. So perhaps we should not reject analysis, but seek to assimilate it.

A striking example of such assimilation is J. Cannon's work based in part on the fact that, if the Cayley complex for a discontinuous hyperbolic group is realized in the natural way in hyperbolic space, then the concepts of combinatorial (word) and metric hyperbolic geodesic agree about as well as could be hoped for. This provides a new approach to small cancellation theory as well as opening up other vistas.

Something along the same line appears, with a different purpose, in the papers of C. Series. A geometrically combinatorial approach appears in a number of papers by A. L. Edmonds, J. E. Ewing, and R. S. Kulkarni (e.g. [50]) in which they study, for example, torsion free subgroups of finite index in Fuchsian groups. Somewhat in the same spirit is the rather extensive use by a number of workers of coset graphs in the study of Riemann surfaces and of subgroups and quotient groups of Fuchsian groups and other groups. See, for example, recent work of J. L. Brenner-R. C. Lyndon and of W. W. Stothers.

PROBLEM 19. *Develop a unified combinatorial theory of a suitably comprehensive class of 'geometric' groups.*

See also Floyd-Hoare-Lyndon [60] and C. L. and M. D. Tretkoff [165].

7. *Algebraic representations of groups*

We are now far distant from logic and topology and approaching the forbidden vastnesses of representation theory, Lie theory, and group rings. But, despite the interdiction by Chandler-Magnus, we must mention Magnus' representation of free groups in associative rings, Lie rings, and matrix rings. The first and central example here is Magnus' representation of a free group F in the power series completion A of a free associative ring, carrying with it a representation of the descending central quotients of F by the dimension modules of the free Lie ring associated with A. We refrain from mentioning all but one of the many extensions of these ideas.

The connection between the commutator structure of a group, especially a p-group, and Lie theory, although not as perfect as for continuous groups, has been highly developed, especially by P. Hall and his followers. Closely related to the Magnus representation is the free differential calculus of R. H. Fox, originating in knot theory, but with a homological interpretation along the lines of the work of Reidemeister.

Any group G is naturally embedded in its integral group ring ZG. Let Δ be the fundamental ideal, the kernel of the augmentation map $ZG \to Z$. The n-th dimension subgroup $D_n(G)$ of G is then defined to be $G \cap (1+\Delta^n)$, the group of all g in G such that $g \equiv 1$ modulo Δ^n. Magnus (1935) showed that if G is a free group, then $D_n(G) = \gamma_n G$, the n-th term of the descending central series of G; from $\cap \Delta^n = 0$ it follows that $\cap \gamma_n G = 1$.

It was conjectured that $D_n(G) = \gamma_n G$ for all groups G and all positive integers n, and this conjecture is easily reduced to the case that G is a finite p-group. However, Rips (1972) exhibited a 2-group G with $\gamma_4 G = 1$ but $D_4(G) \neq 1$. This very special counterexample, using the only even prime, seems to leave the door open for a possible rehabilitation of the problem. See I. B. S. Passi [131], J. A. Sjogren [156], K-I. Tahara [162, 163], and the talk by Gupta.

Fox (1953) introduced the groups $F(n,R) = F \cap (1 + J\Delta^n)$, where R is a normal subgroup of the free group F and J is the kernel of the natural map $ZF \to Z(F/R)$. The dimension subgroups of $G = F/R$ are represented analogously by the groups $D(n,R) = F \cap (1 + J + \Delta^n)$. By the result of Magnus, $F(n,F) = D_{n+1}(F) = \gamma_{n+1} F$, and Magnus showed also that $F(1,R) = \gamma_2 R$. Enright showed that $F(2,R) = \gamma_2(R \cap \gamma_2 F)\gamma_3 R$, and Gupta and Passi derived properties of $F/F(n,R)$ from a matrix representation. For these results see the survey article by Gupta, A problem of R. H. Fox, Canad. Math. Bull. *24* (1981), 129-136.

PROBLEM 20. *Obtain intrinsic descriptions of the Fox subgroups* $F(n,R)$ *and the dimension subgroups* $D(n,R)$ *in terms of the commutator structure of the free group* F *and its normal subgroup* R.

Acknowledgement. The author gratefully acknowledges partial support of the National Science Foundation.

DEPARTMENT OF MATHEMATICS
UNIVERSITY OF MICHIGAN
ANN ARBOR MICHIGAN 48109-1003

REFERENCES

[1] Adian, S. I. The Burnside problem and identities in groups. Nauka, Moscow 1975; Springer 1978.

[2] _____. Random walks on free periodic groups. Izv. Akad. Nauk 46 (1982); Math. USSR Izv. 21 (1983), 425-434.

[3] Aleshin, S. V. Finite automata and the Burnside problem for periodic groups. Mat. Zametki 11 (1972), 319-328; Math. Notes 11 (1972), 199-203.

[4] Alperin, R. C., Moss, K. N. Complete trees for groups with a real-valued length function. J. London Math. Soc., to appear.

[5] Appel, K. I. On Artin groups and Coxeter groups of large type. Contemporary Math. 33 (1984), 50-78.

[6] Appel, K. I., Schupp, P. E. Artin groups and infinite Coxeter groups. Invent. Math. 72 (1983), 201-220.

[7] Bass, H. Euler characteristics and characters of discrete groups. Invent. Math. 35 (1976), 155-196.

[8] _____. H. Growth of finitely generated groups. Proc. 11th Braz. Math. Coll. I. Inst. Mat. Pur. Apl., Rio de Janeiro 1978, 103-115.

[9] Baumslag, G. Some problems on one-relator groups. Proc. 2nd Internat. Conf. Theory of Groups, Canberra 1973. Springer Lecture Notes 372 (1974), 75-81.

[10] Baumslag, B., Pride, S. J. Groups with two more generators than relators. J. London Math. Soc. 17 (1978), 425-426.

[11] _____. Groups with one more generator than relators. Math. Z. 167 (1979), 279-281.

[12] Baumslag, G., Dyer, E., Heller, A. The topology of discrete groups. J. Pure Appl. Alg. 16 (1980), 1-47.

[13] Beardon, A. F. The Geometry of Discrete Groups. Springer 1983.

[14] Benson, M. Growth series of finite extensions of Z^n are rational. Invent. Math. 73 (1983), 251-269.

[15] Bieri, R. On groups of cohomological dimension 2. Topology and Algebra. Monograph Enseign. Math. 26 (1978), 55-62.

[16] Blanc, C. Une interprétation élémentaire des théorèmes fondamentaux de M. Nevanlinna. Comm. Math. Helv. 12 (1940), 153-163.

[17] _____. Les réseaux Riemanniens. Comm. Math. Helv. 13 (1941), 54-67.

[18] Bortnik, T. Two-generator two-relator acyclic group. Bull. Acad. Polon. Sci. Math. 28 (1980), 433-436.

[19] Brodskii, S, D. Equations over groups and groups with a single defining relation. Uspekhi Math. Nauk 35 (1980), 183; Russ. Math. Surveys 35 (1980), 165.

[20] _____. Equations over groups and groups with one defining relation. Sib. Mat. Zh. 28 (1984), 84-103.

[21] Brown, K. S. Complete Euler characteristics and fixed-point theory. J. Pure Appl. Alg. 24 (1982), 103-121.

[22] _____. Cohomology of Groups. Springer Graduate Texts 87 (1982). (Review by K. W. Gruenberg, Bull. Amer. Math. Soc. 11 (1984), 244-246.)

[23] Brown, R., Huebschmann, J. Identities among relations. Proc. Conf. on Topol. in Low Dim., Bangor 1978. London Math. Soc. Lecture Notes 48 (1982), 153-202.

[24] Brunner, A. M., Frame, M. L., Lee, Y. W., Wielenberg, N. J. Classifying torsion-free subgroups of the Picard group. Trans. Amer. Math. Soc. 282 (1984), 205-235.

[25] Brunner, A. M., Lee, Y. W., Wielenberg, N. J. Polyhedral groups and graph amalgamation products, Topology and Its Applications 20 (1985), 289-304.

[26] Cannon, J. W. The combinatorial structure of cocompact discrete hyperbolic groups. Geom. Dedicata 16 (1984), 123-148.

[27] _____. The growth of the closed surface groups and the compact hyperbolic Coxeter groups. Preprint.

[28] Chandler, B., Magnus, W. The History of Combinatorial Group Theory. A Case Study in the History of Ideas. Springer 1982.

[29] Chiswell, I. M. Euler characteristics of groups. Math. Z. 147 (1976), 1-11.

[30] Chiswell, I. M., Collins, D. J., Huebschmann, J. Aspherical group presentations. Math. Z. 178 (1981), 1-36.

[31] Collins, D. J., Huebschmann, J. Spherical diagrams and identities among relations. Math. Ann. 261 (1981), 155-183.

[32] Collins, D. J., Zieschang, H. Rescuing the Whitehead method for free products. I. Math. Z. 185 (1984), 487-504; II. The algorithm.

[33] _____. On the Whitehead method in free products. Contemporary Math. 33 (1984), 141-158.

[34] Comerford, L. P., Jr. Quadratic equations over free groups and free products. J. Algebra 68 (1980), 75-86.

[35] Comerford, L. P., Jr., Edmonds, C. C. Quadratic parametric equations over free groups. Contemporary Math. 33 (1984), 159-196.

[36] Craggs, R. Freely reducing group readings for 2-complexes in 4-manifolds, Preprint.

[37] Culler, M. Using surfaces to solve equations in groups. Topology 20(1981), 113-145.

[38] _____. Finite groups of outer automorphisms of a free group. Contemporary Math. 33(1984), 197-207.

[39] Culler, M., Vogtmann, K. Moduli of graphs and automorphisms of free groups. Invent. Math. 84(1986), 91-119.

[40] Davis, M. W. Groups generated by reflections and aspherical manifolds not covered by Euclidean space. Annals Math. 117(1982), 293-324.

[41] Dicks, W. Groups, trees and projective modules. Springer Lecture Notes 790(1980).

[42] Djokoviĉ, D. Ž. Another example of a finitely presented infinite simple group. J. Algebra 69(1980), 261-269.

[43] Djokoviĉ, D. Ž., Miller, G. L. Regular groups of automorphisms of cubic graphs. J. Comb. Theory B29(1980), 195-230.

[44] Do Long Van. The word and conjugacy problem for a class of groups with nonhomogeneous conditions of small cancellation. Arch. Math. 41(1983), 481-490.

[45] Dyer, E., Vasquez, A. T. Some small aspherical spaces. J. Austral. Math. Soc. 16(1973), 332-352.

[46] Dyer, J. L. A remark on automorphism groups. Contemporary Math. 33(1984), 208-211.

[47] Dyson (Huber-Dyson), V. An inductive theory for free products of groups. Algebra Universalis 9(1979), 35-44.

[48] Eckmann, B., Linnell, P. Poincaré duality groups of dimension two, II. Comm. Math. Helv. 58(1983), 111-114.

[49] Edjvet, M. Groups with balanced presentations. Arch. der Math. 42(1984), 311-313.

[50] Edmonds, A. L., Ewing, J. E., Kulkarni, R. S. Torsion free subgroups of Fuchsion groups and tessellations of surfaces. Bull. Amer. Math. Soc. 86(1982), 456-458; Invent. Math. 69(1982), 331-346.

[51] Edmunds, C. C. On the endomorphism problem for free groups II. Proc. London Math. Soc. 38(1979), 153-168.

[52] _____. A condition equivalent to the solvability of the endomorphism problem for free groups. Proc. Amer. Math. Soc. 76(1979), 23-24.

[53] Ehrenfeucht, A., Fajtlowicz, S., Malitz, J., Mycielski, J. Some problems on the universality of words in groups. Algebra Universalis 11(1980), 261-263.

[54] Fiala, F. Sur les polyèdres à faces triangulaires. Comm. Math. Helv. *19*(1946), 83-90.

[55] Fine, B. Fuchsian subgroups of the Picard group. Canad. J. Math. *28*(1976), 481-485.

[56] _____, The HNN and generalized free product structure of certain linear groups. Proc. Amer. Math. Soc. *81* (1976), 413-416.

[57] Finkelstein, H. Solving equations in groups: a survey of Frobenius' theorem. Per. Math. Hung. *9*(1978), 187-204.

[58] Finkelstein, H., Mandelberg, K. I. On solutions of "equations in symmetric groups." J. Comb. Theory A *25*(1978), 142-152.

[59] Floyd, W. J. Group completions and limit sets of Kleinian groups. Invent. Math. *57*(1980), 205-218.

[60] Floyd, W. J., Hoare, A. H. M., Lyndon, R. C. The word problem for geometrically finite groups. Geom. Dedicata *20*(1986), 201-207.

[61] Gersten, S. M. Intersections of finitely generated subgroups of free groups and resolutions of graphs. Invent. Math. *71* (1983), 567-592.

[62] _____. On fixed points of automorphisms of finitely generated free groups. Bull. Amer. Math. Soc. *8*(1983), 451-454.

[63] _____. Conservative groups, indicable groups, and a theorem of Howie's. J. Pure Appl. Alg. *29*(1983), 59-74.

[64] _____. Solution of equations over ω-nilpotent groups. Rocky Mountain J. Math., to appear.

[65] _____. Geometric automorphisms of finitely generated free groups are rare. Proc. Amer. Math. Soc. *89*(1983), 27-31.

[66] _____. On fixed points of certain automorphisms of free groups. Proc. London Math. Soc. *48*(1984), 72-90.

[67] _____. Fixed points of automorphisms of free groups. Adv. Math., to appear.

[68] _____. On Whitehead's algorithm. Bull. Amer. Math. Soc. *10* (1984), 281-284.

[69] _____. A presentation of the special automorphism group of a free group. J. Pure Appl. Math. *33*(1984), 269-279.

[70] _____. On fixed points of certain automorphisms of free groups, addendum. Proc. London Math. Soc. *48*(1984), 340-342.

[71] _____. Geometry of Automorphisms of Free Groups. Cambridge Univ. Press, to appear.

[72] _____. Dynamics of positive automorphisms, preprint.

[73] Goldstein, R. Z., Turner, E. C. Applications of topological graph theory to group theory. Math. Z. *165* (1978), 1-10.

[74] _____. Solving quadratic equations in groups. Preprint, SUNY Albany, 1982.

[75] _____. Automorphisms of free groups and their fixed points. Invent. Math. *78* (1984), 1-12.

[76] Gottlieb, D. H. Transfers, centers, and group cohomology. Proc. Amer. Math. Soc. *89* (1983), 157-162.

[77] Grayson, M. Thesis, Princeton 1983.

[78] Grigorchuk, R. I. Symmetric random walks on discrete groups. Chapter 7, Multicomponent Random Systems, Nauka, Moscow 1978; Dekker, 1980.

[79] _____. On the Burnside problem for periodic groups. Funk. Anal. Priložen *14* (1980), 53-54; Funct. Anal. Appl. *14* (1980), 41-43.

[80] Gromov, M. Groups of polynomial growth and expanding maps. Publ. Math. IHES *53* (1981).

[81] Gruenberg, K. W. Review of Brown, K. S., Cohomology of Groups. Bull. Amer. Math. Soc. *11* (1984), 244-246.

[82] Gupta, N. D. Burnside Groups and Related Topics. Univ. of Manitoba, 1976.

[83] _____. Fox subgroups of free groups. J. Pure Appl. Alg. *11* (1977), 1-7; II. Contemporary Math. *33* (1984), 223-231.

[84] Gupta, N., Sidki, S. On the Burnside problem for periodic groups. Math. Z. *182* (1983), 385-388.

[85] _____. Extensions of groups by tree automorphisms. Contemporary Math. *33* (1984), 232-246.

[86] _____. Some infinite p-groups. Algebra i Logika *22* (1983), 584-589.

[87] Gutierrez, M. A., Ratcliffe, J. G. On the second homotopy group. Quart. J. Math. *32* (1981), 45-86.

[88] Hatcher, A., Thurston, W. A presentation for the mapping class group of a closed orientable surface. Topology *19* (198), 221-237.

[89] Hausmann, J-C. Sur l'usage de critères pour reconnaitre un groupe libre, un produit amalgamé ou une HNN-extension. Enseign. Math. *27* (1981), 221-242.

[90] Hill, P., Pride, S. J. Commutators, generators and conjugacy equations in groups. Arch. der Math. *44* (1985), 1-14.

[91] Hill, P., Pride, S. J., Vella, A. D. Subgroups of small cancellation groups. J. Reine Angew. Math. *349* (1984), 24-54.

[92] Hoare, A. H. M. Nielsen methods in groups with a length function. Math. Scand. *48* (1981), 153-164.

[93] Howie, J. On pairs of 2-complexes and systems of equations over groups. J. Reine Angew. Math. *324* (1981), 165-174.

[94] ———. On locally indicable groups. Math. Z. *180* (1982), 445-461.

[95] ———. The solution of length three equations over groups. Proc. Edinburgh Math. Soc. *26* (1983), 89-96.

[96] ———. Some remarks on a problem of J. H. C. Whitehead. Topology

[97] ———. Spherical diagrams and equations over groups. Math. Proc. Comb. Phil. Soc. *96* (1984), 257-270.

[98] Howie, J., Pride, S. J. A spelling theorem for staggered generalized 2-complexes, with applications. Invent. Math. *76* (1984), 55-74.

[99] Huebschmann, J. Cohomology theory of aspherical groups and of small cancellation groups. J. Pure Appl. Alg. *14* (1979), 137-143.

[100] ———. The homotopy type of a combinatorially aspherical presentation. Math. Z. *173* (1980), 163-169.

[101] ———. Aspherical 2-complexes and an unsettled problem of J. H. C. Whitehead. Math. Ann. *258* (1981), 17-37.

[102] Hurwitz, H. D. A survey of the conjugacy problem. Contemporary Math. *33* (1984), 278-298.

[103] Juhász, A. The solution of the conjugacy problem for certain one-relator groups I. Israel J. Math.

[104] ———. The solution of the word and conjugacy problem for certain groups. Israel J. Math.

[105] ———. The solution of the word and conjugacy problems for Engel groups.

[106] ———. Weakening small cancellation hypothesis and application to groups with commutator type relators. Conf. on Infinite Group Theory. Iraklion, Crete, 1984.

[107] Kalia, R. M., Rosenberger, G. Automorphisms of the Fuchsian groups of type $(0; 2,2,2,q; 0)$. Comm. Alg. *6* (1978), 115-129.

[108] Karrass, A., Pietrowski, A., Solitar, D. The subgroups of graph amalgamation products, Preprint.

[109] Losey, G., Losey, N. The augmentation quotients of the groups of order 2^4. Contemporary Math. *33* (1984), 412-435.

[110] Lubotzky, A. Combinatorial group theory for pro-p groups. J. Pure Appl. Alg. 25 (1982), 311-325.

[111] _____. Group presentations, p-adic analytic groups, and lattices in $SL_2(C)$. Annals Math. 118 (1983), 115-130.

[112] Lyndon, R. C. Equations in groups. Bol. Soc. Bras. Mat. 11 (1980), 79-102.

[113] _____. Equations over cyclic groups. Laboratoire I. T. P., Paris VII, Publ. (1981), 81-41.

[114] Macbeath, A. M. Residual nilpotency of Fuchsian groups. Illinois J. Math. 28 (1984), 299-311.

[115] Makanin, G. S. Equations in a free group. Izv. Akad. Nauk 46 (1982); Math. USSR Izv. 21 (1983), 483-546.

[116] McCool, J. A characterization of periodic automorphisms of a free group. Trans. Amer. Math. Soc. 260 (198), 309-318.

[117] Merzlyakov, Y. I. The group theory problems of the Kourovka Notebook — progress from the sixth to the seventh symposium. Uspekhi Mat. Nauk 37 (1982), 147-170; Russ. Math. Surveys 37 (1982), 165-191.

[118] _____. On infinite finitely generated periodic groups. Dokl. Akad. Nauk 268 (1983), 803-805; Soviet Math. Dokl. 27 (1983), 169-172.

[119] Metzler, W. On the Andrews-Curtis conjecture and related problems. In: Comb. Methods in Topol. and Alg. Geom., Amer. Math. Soc. 1985.

[120] Morgan, J.W., Shalen, P.B. Valuations, trees, and degenerations of hyperbolic structures, I, Annals Math. 120 (1984), 401-476; II, to appear.

[121] Muller, D. E., Schupp, P. E. Context-free languages, groups, the theory of ends, second-order logic, tiling problems, cellular automata, and vector addition systems. Bull. Amer. Math. Soc. 4 (1981), 331-334.

[122] Mycielski, J. Can one solve equations in groups? Amer. Math. Monthly 84 (1977), 723-726.

[123] Nielsen, J. On calculation with noncommutative factors and its application to group theory. Math. Sci. 6 (1981), 73-85.

[124] Olshanskii, A. Y. Infinite groups with cyclic subgroups. Dokl. Akad. Nauk 245 (1979); Soviet Math. 20 (1979), 343-289.

[125] _____. An infinite simple Noetherian group without torsion. Izv. Akad. Nauk 43 (1979), 1328-1393; Math. USSR Izv. 15 (1980), 531-588.

[126] Olshanskii, A. Y. An infinite group with subgroups of prime orders. Izv. Akad. Nauk *44* (1980), 309-321.

[127] ———. On the question of an invariant mean on a group. Uspekhi Mat. Nauk *35* (1980), 199-200.

[128] ———. Groups of bounded period with subgroups of prime order. Algebra i Logika *21* (1982), 553-618.

[129] ———. On the Novikov-Adian theorem. Math. Sb. *118* (1982), 203-235.

[130] Ozhigov, Y. I. Equations with two unknowns in a free group. Dokl. Akad. Nauk *268* (1983), 809-813.

[131] Passi, I. B. S. The associated graded ring of a group ring. Bull. London Math. Soc. *10* (1978), 241-255.

[132] Perraud, J. Sur le problème des mots des quotients de groupes et produits libres. Bull. Soc. Math. France *108* (1980), 285.

[133] Pride, S. J. The isomorphism problem for two-generator one-relator groups with torsion is solvable. Trans. Amer. Math. Soc. *227* (1977), 109-139.

[134] ———. One-relator quotients of free products. Math. Proc. Camb. Phil. Soc. *88* (1980), 233-243.

[135] ———. Subgroups of small cancellation groups: a survey. Groups St. Andrews 1981. London Math. Soc. Lecture Notes *71* (1982), 298-302.

[136] ———. Small cancellation conditions satisfied by one-relator groups. Math. Z. *184* (1983), 283-286.

[137] Pride, S. J., Vella, A. D. On the hopficity and related properties of some two-generator groups. In: Groups St. Andrews 1981, London Math. Soc. Lecture Notes *71* (1982), 303-312.

[138] Ratcliffe, J. G. Euler characteristics of 3-manifold groups and discrete subgroups of SL(2,C). Notices Amer. Math. Soc. 1983, 789-20-92.

[139] Richardson, J. S., Rubinstein, J. H. Hyperbolic manifolds from regular polyhedra, Preprint.

[140] Rips, E. Commutator equations in free groups. Israel J. Math. *39* (1981), 326-340.

[141] ———. Generalized small cancellation theory and applications, I. Israel J. Math. *41* (1982), 1-146.

[142] Romanovskii, N. S. Free subgroups of finitely presented groups. Algebra i Logika *16* (1977), 88-97.

[143] Rosenberger, G. Automorphismen ebener diskontinuierlicher Gruppen. Riemann Surf. Rel. Topics. Princeton, 1980, 439-455.

[144] Rothaus, O. S. On the nontriviality of some group extensions given by generators and relators. Ann. Math. *106* (1977), 559-612.

[145] Rourke, C. P. Presentations and the trivial group. Proc. 2nd Sussex Conf. Springer Lecture Notes *722* (1977), 134-143.

[146] Scharlemann, M. Certain free subgroups of SL(2,R): a geometric view. Lin. Multilin. Alg. *7* (1979), 177-191.

[147] Scharlemann, M., Squier, C. Automorphisms of the free group of rank two without finite orbits. Contemporary Math. *20* (1983), 341-346.

[148] Schupp, P. E. Quadratic equations in groups, cancellation diagrams on compact surfaces and automorphisms of surface groups. Word Problems II, North Holland 1980, 347-371.

[149] Series, C. The infinite word problem and limit sets in Fuchsian groups. Ergod. Theory and Dynam. Systems *1* (1981), 337-360.

[150] _____. Martin boundaries of random walks on Fuchsian groups. Israel J. Math. *44* (1983), 221-242.

[151] Shelah, S. On a problem of Kurosh, Jónsson groups, and applications. Word Problems II, North Holland 1980, 373-394.

[152] Short, H. B. Topological methods in group theory: the adjunction problem. Thesis, Warwick, 1981.

[153] Sidki, S. On a 2-generated infinite 3-group. The presentation problem.

[154] _____. On a 2-generated infinite 3-group. Subgroups and automorphisms.

[155] Sieradski, A. J. Framed links for Peiffer identities. Math. Z. *175* (1980), 125-137.

[156] Sjogren, J. A. Dimension and lower central series. J. Pure Appl. Alg. *14* (1979), 175-194.

[157] Squier, C. Fixed points and finite orbits of free group automorphisms, Preprint.

[158] Stallings, J. R. Topologically unrealizable automorphisms of free groups. Proc. Amer. Math. Soc. *84* (1982), 21-24.

[159] _____. Topology of finite graphs. Invent. Math. *71* (1983), 551-565.

[160] Stöhr, R. Groups with one more generator than relators. Math. Z. *182* (1983), 45-47.

[161] Swarup, G. A. Two reductions of the Poincaré conjecture. Bull Amer. Math. Soc. *1* (1979), 774-777.

[162] Tahara, K-I. The fourth dimension subgroups and polynomial maps, II. Nagoya Math. J. *69* (1978), 1-7.

[163] ———. Augmentation quotients and dimension subgroups of semidirect products. Math. Proc. Camb. Phil. Soc. *91* (1982), 39-49.

[164] Tits, J. Sur le groupe des automorphismes d'un arbre. Essays on Topology and Related Topics. Mémoires dédiés à Georges de Rham. Springer 1970, 188-211.

[165] Tretkoff, C. L., Tretkoff, M. D. Combinatorial group theory, Riemann surfaces and differential equations. Contemporary Math. *33* (1984).

[166] Tretkoff, M. D. A topological approach to the theory of groups acting on trees. J. Pure Appl. Alg. *16* (1980), 323-333.

[167] Wagreich, P. The growth function of a discrete group. Proc. Conf. Alg. Varieties with Group Actions, Vancouver, 1982.

[168] Wajnryb, B. A simple presentation for the mapping class group of an orientable surface. Israel J. Math. *45* (1983), 157-174.

[169] Wicks, M. J. Commutators in free products. J. London Math. Soc. *37* (1962), 433-444.

[170] Wilkie, A. J., van den Dries, L. Effective bounds for groups of linear growth. Archiv der Math. *286* (1984), 339-349.

Added in proof, February 1, 1985. A letter from G. S. Makanin and R. I. Grigorchuk, containing many valuable comments on the manuscript of this paper, was received too late to incorporate in the text. They propose the following very reasonable additions to the list of problems: (1) (Tarski) Decidability of the elementary theory of free groups; (2) Finiteness of the Burnside groups $B(m,2^n)$; (3) The conjugacy and equivalence problems for tame knots.

They point out that Problem 4 as stated is solved by the cited result of Makanin. This problem should perhaps be construed as asking for a simpler and more explicit description of the set of all solutions of an equation of the special form $w(x_1, \cdots, x_n) = a$.

They have provided the following four important additional references; note that the first two of these papers contain extensive bibliographies.

Adian, S. I. Studies on the Burnside problem and related questions. Trudy Mat. Akad. Nauk SSSR *168*(1984), 171-196.

Adian, S. I., Makanin, G. S. Studies on algorithmic questions in algebra. Trudy Mat. Akad. Nauk. SSSR *168*(1984).

Grigorchuk, R. I. On Milnor's problem of group growth. Dokl. Akad. Nauk SSSR *271* (1983); Soviet Math. Dokl. *28*(1983), 23-26.

———. Powers of growth of finitely generated groups and the theory of invariant averages. Izv. Akad. Nauk SSSR *48*(1984).

Suščanskii, V. I. Periodic p-groups of permutations and the unrestricted Burnside problem. Dokl. Akad. Nauk SSSR *247*(1979).

Added in proof, February 20, 1985. Cameron Gordon has pointed out that the solution to the equivalence problem for knots is contained in the paper by G. Hemion, "On the classification of homeomorphisms of 2-manifolds and the classification of 3-manifolds," Acta Math. *142*(1979), 123-155.

POINCARÉ DUALITY GROUPS OF DIMENSION TWO ARE SURFACE GROUPS

Beno Eckmann

1. Introduction

We first explain the terms appearing in the title, and add some general comments.

1.1. A *surface group* G is a group isomorphic to the fundamental group $\pi_1(\Sigma_g)$ of a closed surface Σ_g, orientable or not, of genus $g \geq 1$. Such a group admits a presentation

$$G = <x_1,y_1,\cdots,x_g,y_g | [x_1,y_1]\cdots[x_g,y_g] = 1>$$

in the orientable case,

$$G\ <z_0,z_1,\cdots,z_g | z_0^2 z_1^2 \cdots z_g^2 = 1>$$

in the non-orientable case.

We will also use the concept of a *surface group-pair* $(G; \{S_0, S_1, \cdots, S_m\})$: it consists of the (free) fundamental group of a closed surface of genus $g \geq 0$ with $m+1$ disks removed, $m \geq 0$ (but ≥ 1 if the surface is a sphere), together with the $m+1$ infinite cyclic subgroups generated by the boundary circles of these disks. Surface group-pairs have presentations

$$G = <t_1,t_2,\cdots,t_m,x_1,y_1,\cdots,x_g,y_g>,$$

$$S_0 = <t_1 t_2 \cdots t_m [x_1,y_1]\cdots[x_g,y_g]>,\ S_j = <t_j> \text{ for } j = 1,\cdots,m$$

in the orientable case, $m+g > 0$;

$$G = \langle t_1, t_2, \cdots, t_m, z_0, z_1, \cdots, z_g \rangle$$
$$S_0 = \langle t_1 t_2 \cdots t_m \, z_0^2 z_1^2 \cdots z_g^2 \rangle, \quad S_j = \langle t_j \rangle \text{ for } j = 1, \cdots, m$$

in the non-orientable case, $m \geq 0$, $g \geq 0$.

The "lowest" cases of surface group-pairs are

$$G = \langle t_1 \rangle, \quad S_0 = \langle t_1 \rangle, \quad S_1 = \langle t_1 \rangle$$

and

$$G = \langle z_0 \rangle, \quad S_0 = \langle z_0^2 \rangle.$$

1.2. A *Poincaré duality group* G *of dimension* n, in short a PD^n group, is a group fulfilling Poincaré duality in (co-) homology for all coefficient ZG-modules A with respect to the formal dimension n and to a certain G-action on the additive group of integers $\tilde{Z}$:

$$H^i(G;A) \cong H_{n-i}(G; \tilde{Z} \otimes A), \quad i \in Z.$$

Here $\tilde{Z} \otimes A$ is a ZG-module by diagonal action, and the isomorphisms are natural in the ZG-modules A. As we will see the dimension n and the ZG-module structure of $\tilde{Z}$ are determined by G (the PD^n-group being orientable or non-orientable according to whether the action is trivial or not).

The definition above is, of course, analogous to the Poincaré duality valid for all closed n-dimensional manifolds X, the coefficients A being $\pi_1(X)$-modules. If X is aspherical then the (co-) homology of X is isomorphic to that of $G = \pi_1(X)$ so that, in that case, G is a PD^n-group. It is not known whether in general the converse is true; i.e., whether a PD^n-group is necessarily isomorphic to the fundamental group of a closed n-dimensional aspherical manifold.

Since the universal covering of the surface Σ_g, $g \geq 1$, is R^2 the surface Σ_g is aspherical. Thus the surface groups in 1.1 above are PD^2-groups. The *Theorem* formulated in the title states that the converse is true, thus solving the problem in the case $n = 2$. The proof of that *Theorem* has been achieved in several steps contained in a series of papers by the present author and various collaborators; these papers were written partly with other objectives in mind. The conference organizers have asked the author to present a survey, as complete as possible, of that proof. We do so and use the opportunity to simplify some of the arguments.

1.3. There are, in fact, different techniques and methods involved in that proof. They belong roughly speaking to the following three areas of ideas:

(I)   Homological algebra, homology of groups.

(II) Structure and splitting theorems for groups (Stallings-Bass-Serre and others).

(III) Ranks of projective modules and Euler characteristic (Hattori-Stallings-Bass-Kaplansky).

The papers leading to or containing the steps of the proof are as follows. In the field (I) by Robert Bieri and the author [2], [4], [5]; in (II) by Heinz Müller and the author [12], [14]; in (III) by Peter Linnell and the author [10], [11].

1.4. For compact manifolds-with-boundary (∂-manifolds), in particular for the closed surfaces with disks removed, one has the well-known "relative" Poincaré duality. The surface group-pairs listed in 1.1 above fulfill such a relative Poincaré duality, of dimension 2. To formulate this in a precise way, relative (co-) homology groups for pairs of groups $(G; \{S_0, S_1, \cdots, S_m\})$ have to be considered, cf. Section 2.3 below; this yields the concept of a PD^n-pair of groups. An important step in the proof of the *Theorem* will be to show that all PD^2-pairs of groups are surface group-pairs ("Relative Theorem," Section 3.2).

1.5. We mention here two corollaries of the *Theorem*.

COROLLARY 1.1 (cf. [12], [11]). *All Poincaré-2-complexes are homotopy equivalent to closed surfaces (of genus ≥ 0).*

COROLLARY 1.2. *Let G be a torsion-free group containing a surface group S as subgroup of finite index; then G is also a surface group.*

Indeed, a homological argument (see Section 2.2) shows that G is a PD^2-group. This corollary is a special case of the "Nielsen realization conjecture" proved by Kerckhoff [Annals of Math. 117 (1983), 235-265]. The special case above was established by Eckmann-Müller [12] before Kerckhoff's proof, and before our Theorem on PD^2-groups had been completely settled.

1.6. This survey is organized as follows. Section 2 contains general preliminaries on duality and relative duality groups. In Section 3 we assume that the PD^2-group fulfills a certain "splitting" property and show that the *Theorem* can then be reduced to the Relative Theorem; both that reduction and the proof of the Relative Theorem are given on the basis of general splitting arguments explained in Section 4. In Section 5 we show, by quite different methods involving ranks of finitely generated projective modules, that any PD^2-group fulfills the splitting assumption.

2. *Duality groups*

2.1. The group G is called a *duality group of dimension* $n > 0$ with respect to a dualizing ZG-module C, in short a D^n group, if one has isomorphisms
$$H^i(G;A) \cong H_{n-i}(G; C \otimes A)$$

for all $i \in Z$ and all ZG-modules A; they are assumed to be natural in A, and $C \otimes A$ is endowed with the diagonal G-action. If $C = Z$ as an Abelian group, one has Poincaré duality (cf. 1.2), i.e., G is a PD^n-group.

From the definition it follows that $H^i(G;A)$ commutes with direct limits in A. This is possible only if G admits a projective resolution

$\cdots \to P_i \to P_{i-1} \to \cdots \to P_0 \twoheadrightarrow Z$ with all P_i finitely generated over ZG (by the Bieri-Eckmann-Brown finiteness criterion, see [3] and [7]). Furthermore one easily checks that $C \otimes ZG \cong C_0 \otimes ZG$ (where C_0 is the Abelian group underlying C) is an induced module, and thus $H^i(G;ZG) = 0$ for all $i \neq n$; as for $H^n(G;ZG)$, it is isomorphic to $H_0(G;C \otimes ZG) = C$, and this is a (right) ZG-module isomorphism. The dualizing module C is thus determined by G; it is easily seen to be torsion-free as an Abelian group.

The cohomology dimension $cd\, G$ is clearly $\leq n$, and by the above it is $= n$; hence the integer n is also determined by G, and G admits a finitely generated projective resolution of finite length (equal to n); such groups are said to be of type FP.

Summarizing we see that a D^n-group G fulfills

(1) G is of type FP.

(2) $H^i(G;ZG) = 0$ for $i \neq n$ and $H^n(G;ZG)$ is torsion-free,

(3) $cd\, G = n$.

It has been proved by Bieri-Eckmann [2] that, conversely, a group fulfilling (1) and (2) is a D^n-group with dualizing module $C = H^n(G;ZG)$.

We note that $cd\, G = n$ implies that G is torsion-free.

2.2. *As an application, let* G *be torsion-free and* S *a subgroup of finite index.* By Serre's theorem (see e.g. [8], p. 190) $cd\, G = cd\, S = n$. Clearly G admits a finitely generated free resolution over ZG if and only if S does over ZS; hence G is of type FP if and only if S is. Moreover $H^i(S;ZS) \cong H^i(G; \text{Hom}_S(ZG,ZS)) \cong H^i(G; ZS \otimes_S ZG) = H^i(G;ZG)$, and it follows that G fulfills (1) and (2) above if and only if S does, with the "same" dualizing module C.

Thus G is a D^n-group if and only if S is a D^n-group, and the dualizing modules are isomorphic as Abelian groups. In particular, G *is a* PD^n-*group if and only if* S *is.*

1) REMARKS: *Dimension 1*. G is a D^1-group if and only if it is finitely generated free. It is a PD^1-group if and only if it is infinite cyclic.

2) *Subgroups of infinite index*. For PD^2-groups G, Strebel [17] has proved, by homological methods, that for a subgroup S of infinite index in G one has $cd\, S \leq n-1$. For $n=2$ it follows that S is a free group; this is the PD^2-analogue of a fact well known for surface groups.

2.3. We briefly recall *relative duality* for group pairs (for details see Bieri-Eckmann [4], [5]). A group pair $(G,\underline{S})$ consists of a group G and a family $\underline{S} = \{S_j, j \in I\}$ of subgroups, not necessarily distinct. For any subgroup $S \subset G$ one writes ZG/S for the G-module whose underlying Abelian group is freely generated by the cosets xS, with G-action by left multiplication. Relative (co-) homology is defined by means of the "augmentation kernel" $\Delta = \ker\{\oplus_j ZG/S_j \xrightarrow{\varepsilon} Z\}$ where $\varepsilon(xS_j) = 1$ for all $x \in G$ and $j \in I$:

$$H_i(G,\underline{S};A) = H_{i-1}(G;\Delta \otimes A),$$

$$H^i(G,\underline{S};A) = H^{i-1}(G;\mathrm{Hom}(\Delta,A)),$$

A being a G-module, $\otimes$ and Hom equipped with diagonal G-action.

A duality pair of dimension n with dualizing module C is a pair $(G,\underline{S})$ fulfilling

$$H^i(G;A) \cong H_{n-i}(G,\underline{S};C \otimes A)$$

and

$$H^i(G,\underline{S};A) \cong H_{n-i}(G;C \otimes A).$$

In the (orientable) Poincaré duality case, $C = Z$, each of these isomorphisms implies the other one and the first one becomes

$$H^i(G;A) \cong H_{n-i-1}(G;\Delta \otimes A);$$

i.e., $(G,\underline{S})$ is a PD^n-pair of groups if and only if G is a D^{n-1}-group with dualizing module Δ. Relative exact sequences show that $\underline{S}$ must be a finite family of PD^{n-1}-groups $S_0, S_1, \cdots, S_m$.

In particular, if $(G,\underline{S})$ is a PD^2-pair then G is finitely generated free and $\underline{S}$ is a finite family of infinite cyclic groups. The duality isomorphisms yield $H^2(G,\underline{S};ZG) = Z$, $H^1(G,\underline{S};ZG) = 0$, $H^1(G;ZG) = \Delta$. As shown in [4] the only important homological property is $H^2(G,\underline{S};ZG) = Z$; indeed, if G is finitely generated free and $\underline{S}$ a finite family of infinite cyclic groups it characterizes PD^2-pairs.

3. PD^2-groups splitting over a finitely generated subgroup

3.1. A group G is said to split over the subgroup H if either (α) G is a non-trivial amalgamated free product $G = G_1 *_H G_2$, or (β) G is an HNN-extension $G = G_1 *_{H,p}$. Two cases are of special importance: 1) H is finitely generated, 2) H is finite. We recall that Stallings' structure theorem [15], [16] for finitely generated groups tells that 2) holds if and only if $H^1(G;ZG) \neq 0$.

PROPOSITION 3.1. *The assertion of the Theorem holds if the PD^2-group G splits over a finitely generated group L.*

The proof of Proposition 3.1 makes strong use of the "simultaneous splitting theorem" (in short: SST) for groups and subgroups, established by Heinz Müller [14]. The SST is a refinement of the relative version (Swan [18], Swarup [19]) of Stallings' structure theorem; it deals with splittings over finite subgroups H, and since in our case G is torsion-free only $H = 1$ will occur. A short outline of SST and its application in the present context will be given in Section 4 below.

The application of SST to PD^2-groups (α) $G = G_1 *_L G_2$, or (β) $G = G_1 *_{L,p}$, with L finitely generated, is as follows. Since $H^1(G;ZG) = 0$, L is $\neq 1$. The index of L in G is infinite; by Strebel's theorem [17] one has $cd\, L \leq 1$, and hence L is (finitely generated) free. We only describe the case (α), the case (β) being similar.

If in $G = G_1 *_L G_2$ the rank of L is > 1 one has, by virtue of SST (see Section 4.4, A)), splittings

$$G_1 = H_1 * H_2, \quad L = L_1 * L_2 \text{ with } L_i \subset H_i, \quad i = 1, 2,$$

or

$$G_1 = H * <q> = H *_{1,q}, \quad L = L_1 * qL_2^{-1}q \text{ with } L_1, L_2 \subset H.$$

The first possibility yields

$$G = H_1 *_{L_1} (H_2 *_{L_2} G_2);$$

if $L_1 \neq H_1$ then G splits over L_1, and if $L_1 = H_1$ then G splits over L_2. Thus G splits over a subgroup whose rank is less than that of L. The second possibility yields

$$G = (H *_{L_1} G_2) *_{L_2, q^{-1}},$$

so G splits over L_2, again of rank less than that of L.

Thus we are always reduced to the case where G splits over an infinite cyclic subgroup C as (α) $G = G_1 *_C G_2$, or (β) $G = G_1 *_{C,p}$.

Since C is a PD^1-group we can apply the general homological arguments of [4], Theorem 8.1 and 8.3. It follows that in the case (α) the group pairs (G_1, C) and (G_2, C) and in the case (β) the group pair $(G_1, \{C, p^{-1}Cp\})$ are PD^2-pairs. Now the *Relative Theorem* below tells that these pairs are surface group-pairs (see 1.1) corresponding to closed surfaces with one disk, or two disks respectively, removed.

In (α), $G = G_1 *_C G_2$ is the fundamental group of the closed surface obtained by identifying the boundary circles; in (β), $G = G_1 *_{C,p}$ is the fundamental group of the closed surface obtained by joining the two boundary circles by a tube.

3.2. RELATIVE THEOREM. *Any PD^2-pair $(G; \{S_0, \cdots, S_m\})$ is a surface group-pair.*

Again the proof uses SST, in addition to the properties of PD^2-pairs mentioned in 2.3. We proceed by induction on the rank of the finitely generated free group G.

If that rank is 1, i.e., G infinite cyclic $C = <c>$, one has $H^1(G;ZG) = Z$ since $G = C$ is a PD^1-group, and $H^1(G;ZG) = \Delta$ since G is a duality group of dimension 1 with dualizing module Δ. The exact sequence

$$\Delta \rightarrowtail \bigoplus_j ZG/S_j \twoheadrightarrow Z$$

then yields $\bigoplus_j ZG/S_j = Z \oplus Z$; this is possible only if $\underline{S} = \{C,C\}$ or $\underline{S} = \{<c^2>\}$. Thus the pair $(G,\underline{S})$ is either $(C,\{C,C\})$ or $(C = <c>, \{<c^2>\})$; i.e., we obtain precisely the lowest orientable case $g = 0$, $m = 1$, or the lowest non-orientable case $g = 0$, $m = 0$ of the presentation list of surface group-pairs in 1.1.

If the rank of G is >1, SST (see Section 4.4, B)) yields splittings (α) $G = G_1 * G_2$ with $S_0 = <g_1 g_2>$, $1 \neq g_j \epsilon G_j$, while the S_j for $j > 0$ are (conjugate to) subgroups of G_1 and G_2, say $S_1,\cdots,S_k \subset G_1$, $S_{k+1},\cdots,S_m \subset G_2$; or ($\beta$) $G = G_1 *<p> = G_1 *_{1,p}$ with $S_0 = <pq_1 p^{-1} g_2>$, $g_1, g_2 \epsilon G_1$, while $S_1,\cdots,S_m$ are (conjugate to) subgroups of G_1. We will deal with (α) only, the case (β) being similar. We can write

$$G = (G_1 *<g_2>) *_{<g_2>} G_2 .$$

Then $S_0 \subset G_1 *<g_2>$. The pair $(G_2; \{<g_2>, S_{k+1},\cdots,S_m\})$ is a PD^2-pair by Theorem 8.1 of [4]; note that the case $<g_2> = G_2$ needs a special argument, and the pair must then be $(<g_2>,\{<g_2>,<g_2>\})$, cf. [4] p. 517. Similarly the pair $(G_1,\{<g_1>,S_1,\cdots,S_k\})$ is a PD^2-pair. By induction they are surface group-pairs, and one easily checks that so is $(G;\{S_0,S_1,\cdots,S_m\})$.

4. Simultaneous splitting of groups and subgroups

4.1. We consider throughout this section a finitely generated group G which splits over a finite subgroup K as $G = G_1 *_K G_2$ or $G = G_1 *_{K,p}$, i.e., with $H^1(G;ZG) \neq 0$, cf. Section 3.1. Given a finite family $\{S_1,\cdots,S_m\}$ of subgroups of G let $N(G;S_1,\cdots,S_m)$ be the intersection of the kernels of the restriction maps $res_j : H^1(G;ZG) \to H^1(S_j;ZG)$, $j = 1,\cdots,m$. By the relative version of the structure theorem (Swan [18], Swarup [19]) $N(G;S_1,\cdots,S_m) \neq 0$ if and only if there is a splitting of G such that the S_j are conjugate to subgroups of G_1 or G_2; this will be the case in the following and only such splittings will be considered.

Let T be a further subgroup of G, and assume that T is finitely generated. $\{x_\nu\}$ denoting a set of coset representatives of G mod T, we consider the restriction map

$$H^1(G;ZG) \xrightarrow{res} H^1(T;ZG) \cong \bigoplus_\nu H^1(T;ZT)x_\nu.$$

The minimal number of non-zero components of $res(c)$ for all $0 \neq c \in N(G;S_1,\cdots,S_m)$ is called the *weight* $n(T)$ of T with respect to G (and to $S_1,\cdots,S_m$). Note that $n(T) = 0$ if and only if $N(G;T,S_1,\cdots,S_m) \neq 0$; i.e., if there is a splitting of G with $T \subset G_1$ or G_2.

4.2. The simultaneous splitting theorem (SST) established by H. Müller [14] (actually for more general G and T) concerns the case $n(T) > 0$. It can be formulated roughly as follows, its full content being in fact more complicated.

There is a tree Γ on which G acts with finite edge-stabilizers and with proper subgroups of G as vertex-stabilizers such that Γ/G has one edge (and that each S_j, $j = 1,\cdots,m$ stabilizes a vertex of Γ); and Γ contains a subtree Γ_T invariant under T with Γ_T/T having at most $n(T)$ edges.

EXAMPLES. 1) Let G be torsion-free, and $n(T) = 1$. Then one has the following possibilities:

(1) $G = G_1 * G_2$, $T = T_1 * T_2$, $T_1 \subset G_1$, $T_2 \subset G_2$.

(2) $G = G_1 *<p> = G_1 *_{1,p}$, $T = T_1 * pT_2 p^{-1}$, $T_1, T_2 \subset G_1$.

(3) $G = <p>$, $T = <p>$, $S_1 = \cdots = S_m = 1$ or $m = 0$.

2) Let G be torsion-free, T infinite cyclic, and $n(T) = 2$. Then one has the following possibilities (cf. [14])

(1) $G = G_1 * G_2$, $T = <g_1 g_2>$, $1 \neq g_i \in G_i$, $i = 1, 2$.

(2) $G = G_1 *<p> = G_1 *_{1,p}$, $T = <pg_1 p^{-1} g_2>$, $g_1, g_2 \in G_1$.

(3) $G = <p>$, $T = <p^2>$, $S_1 = \cdots = S_m = 1$ or $m = 0$.

4.3. We restrict ourselves to some remarks concerning the proof of SST. We write Z_2 for $Z/2Z$ and use $Z_2 G$, $Z_2 T$ instead of ZG, ZT; one easily checks that this yields the same weight $n(T)$. Then $H^1(G; Z_2 G)$ can be interpreted as group of all subsets of G which are "almost invariant" under translation without being G or $\emptyset$; almost invariant means invariant except for finite sets. Namely, writing $\overline{Z_2 G}$ for $\text{Hom}(ZG, Z_2) = \prod_{x \in G} Z_2 x$ and εG for $\overline{Z_2 G}/Z_2 G$ (arbitrary modulo finite subsets of G), the exact sequence

$$H^0(G; \overline{ZG}) \to H^0(G; \varepsilon G) \to H^1(G; ZG) \to H^1(G; \overline{Z_2 G}) = 0$$

yields

$$(\varepsilon G)^G / (Z_2 G)^G \cong H^1(G; Z_2 G).$$

The restriction map $H^1(G; Z_2 G) \xrightarrow{\text{res}} \oplus_\nu H^1(T; ZT) x_\nu$ is then given by $U \mapsto U \cap T x_\nu$ for all ν where U is a non-trivial almost invariant set.

The components $U \cap Tx_\nu$ are almost T-invariant. By the techniques of Dunwoody [9] and Swarup [19] these almost invariant sets yield the theorem.

4.4. There remains to compute the weights of those subgroups which occur in the proofs of Proposition 3.1 (subgroup L) and of the "Relative Theorem" in 3.2 (subgroup S_0).

A) We again restrict ourselves to the case (a) in the proof of Proposition 3.1. Thus G is a PD^2-group with $G = G_1 *_L G_2$, L free of rank > 1. We claim that $n(L)$ with respect to $(G_1; \emptyset)$ or to $(G_2; \emptyset)$ is equal to 1. By example 1) in 4.2 this yields the required simultaneous splittings of G_1 and L.

To prove the claim we consider the exact Mayer-Vietoris sequence

$$\cdots \to 0 \to H^1(G_1;ZG) \oplus H^1(G_2;ZG) \xrightarrow{(\mathrm{res}_1 - \mathrm{res}_2)} H^1(L;ZG) \xrightarrow{\delta} H^2(G;ZG) \to \cdots$$

and note that the weight is not 0 since res_1 and res_2 are injective. $H^1(L;ZL)$ is free Abelian of infinite rank since L is free of rank > 1. Thus the restriction of δ to $H^1(L;ZL)$ cannot be injective, $H^2(G;ZG)$ being $= Z$; i.e., the intersection $\mathrm{im}(\mathrm{res}_1, -\mathrm{res}_2) \cap H^1(L;ZL)$ is $\neq 0$. On the other hand, if both $n(L)$ with respect to $(G_1, \emptyset)$ and to $(G_2, \emptyset)$ are > 1, the image $\mathrm{res}_1(c_1) - \mathrm{res}_2(c_2)$ of $0 \neq (c_1, c_2) \in H^1(G_1;ZG) \oplus H^1(G_2;ZG)$ cannot lie in $H^1(L;ZL) \subset H^1(L;ZG)$; this is seen by looking at the lengths of elements with respect to coset representatives of $G \bmod G_1$ and G_2. Thus $n(T)$ with respect to, say, $(G_1, \emptyset)$ is $=1$, and we obtain the simultaneous splitting.

B) In 3.2, $(G,\underline{S})$ is a PD^2-pair, $\underline{S} = \{S_0, S_1, \cdots, S_m\}$ with $m \geq 0$ and all S_j infinite cyclic, and we have to consider the case where the free group G is of rank > 1. The claim is that $n(S_0)$ with respect to $(G, \{S_1, \cdots, S_m\})$ is $= 2$. By example 2) in 4.3 this yields the required splittings.

The exact relative cohomology sequence of G mod $\underline{S}$ is

$$0 \longrightarrow (G,\underline{S};ZG) \longrightarrow H^1(G;ZG) \xrightarrow{r} \bigoplus_{j=0}^{m} H^1(S_j;ZG) \xrightarrow{\delta} H^2(G,\underline{S};ZG) \longrightarrow 0,$$

where r denotes the map with components res_j, $j = 0,1,\cdots,m$. The PD^2-pair properties tell that the first term is 0, the last isomorphic to Z. If S_0 (or any proper subset) is omitted from the family $\underline{S}$, the last term becomes 0 and the first term must be $\neq 0$ (cf. [4], Section 11); i.e., the intersection $N = N(G;S_1,\cdots,S_m)$ of the ker res_j, $j = 1,\cdots,m$ is non-zero. The weight $n(S_0)$ is the minimal number of components in $H^1(S_0;ZG) = \oplus_\nu H^1(S_0;ZS_0)x_\nu \cong \oplus_\nu Zx_\nu$ of $\mathrm{res}_0(c)$ for all $0 \neq c \in N$; note that ker $\mathrm{res}_0 \cap N = 0$.

Now $r(N) = (\mathrm{res}_0(N),0,\cdots,0) = (H^1(S_0;ZG),0,\cdots,0) \cap \ker \delta$. One easily checks that δ restricted to any summand Zx_ν of $H^1(S_0;ZG)$ is bijective. So obviously the minimum number of components of elements $\neq 0$ in $\mathrm{res}_0(N)$ is 2, which proves the claim.

5. The first Betti number of a PD^2-group

5.1. In order to conclude the proof of the Main *Theorem* we have to show that a PD^2-group, without any further assumption, splits over a finitely generated subgroup. This is guaranteed by the following proposition whose proof will be given in 5.2 and 5.3 below. Recall that PD^n-groups are of type FP so that Betti numbers $\beta_i(G) = \mathrm{rank}\, H_i(G;Z)$ are defined.

PROPOSITION 5.1. *If G is a PD^2-group then $\beta_1(G) > 0$.*

From this it follows that $H_1(G;Z)$, the abelianized group G, contains at least one infinite cyclic summand C, and thus G admits a factor group $\cong C$. By a result of Bieri-Strebel ([6], Theorem A) this implies that G splits as $G = G_1 *_{L,p}$ over a finitely generated subgroup L (as shown in [6], this holds for any group G of type FP_2, i.e., admitting a projective resolution which is finitely generated in dimensions

≤ 2, and having an infinite cyclic factor group). The splitting is constructed explicitly, the generator p being any element projecting onto a generator of C.

5.2. For the proof of Proposition 5.1 we use the homological Euler characteristic $\chi(G) = \beta_0(G) - \beta_1(G) + \beta_2(G)$ of G. By the Euler-Poincaré formula $\chi(G)$ is equal to the alternating sum of the ranks of the free Abelian groups $Z \otimes_G P_i$ for any FP-resolution over ZG

$$0 \to P_2 \to P_1 \to P_0 \twoheadrightarrow Z .$$

We recall that in the orientable case $\beta_0(G) = \beta_2(G) = 1$ and $\beta_1(G) =$ even; in the non-orientable case $\beta_0(G) = 1$, $\beta_2(G) = 0$ (the proofs are the same as for closed surfaces). Thus in both cases the claim of Proposition 5.1 is $\chi(G) \leq 0$. If G is a non-orientable PD^2-group, let G_0 be the orientable subgroup of index 2. By the multiplicative property of the Euler characteristic (valid for groups of type FP, cf. [8], §IX.6) $\chi(G_0) = 2\chi(G)$; hence $\chi(G_0) \leq 0$ implies $\chi(G) \leq 0$, and we are reduced to the orientable case.

G being an orientable PD^2-group we choose a resolution

(1) $$0 \longrightarrow P \longrightarrow ZG^d \xrightarrow{\delta} ZG \longrightarrow Z$$

with P finitely generated projective over ZG. Applying $\text{Hom}_G(-, ZG)$ to (1) we get the sequence

(2) $$0 \longleftarrow P^* \xleftarrow{\delta} ZG^d \longleftarrow ZG \longleftarrow 0$$

where $P^* = \text{Hom}_G(P, ZG)$ is finitely generated projective. Since $H^i(G; ZG) = 0$ for $i \neq 2$ and $H^2(G; ZG) = Z$ we obtain the exact sequence

(3) $$Z \longleftarrow P^* \xleftarrow{\delta} ZG^d \longleftarrow ZG \longleftarrow 0 ,$$

which is another FP-resolution for G; for $\chi(G)$ it yields

$$\chi(G) = \text{rank}(Z \otimes_G P^*) - d + 1$$

whence

(4) $$\beta_1(G) = 2 - \chi(G) = 1 + d - \text{rank}(Z \otimes_G P^*).$$

Comparing (1) and (3) we see that $P^*/\delta ZG^d \cong ZG/\partial ZG^d$, and therefore $P^* \oplus \partial ZG^d \cong ZG \oplus \delta ZG^d$. One then has a surjection $ZG^{d+1} \twoheadrightarrow P^* \oplus \partial ZG^d$, and since $\partial ZG^d \neq 0$ a surjection $ZG^{d+1} \twoheadrightarrow P^*$ with non-zero kernel N, i.e.,

(5) $$ZG^{d+1} \cong P^* \oplus N.$$

N is finitely generated projective, and $\text{rank}(Z \otimes_G P^*) + \text{rank}(Z \otimes_G N) = d + 1$. From (4) we obtain

(6) $$\beta_1(G) = \text{rank}(Z \otimes_G N).$$

5.3. It is not known in general whether, for a group G and a non-zero finitely generated projective ZG-module N, the free Abelian group $Z \otimes_G N$ is non-zero. It has, however, proved useful to compare $\text{rank}(Z \otimes_G N)$ with another rank concept (for f.g. projectives) which we propose to call the *Kaplansky rank* $\kappa(N)$. It is defined as follows: Let $N \oplus M$ be finitely generated free, and ϕ the idempotent endomorphism of $N \oplus M$ which is 1_N on N and 0 on M. The trace of ϕ (in a basis of $N \oplus M$) is an element of ZG; its coefficient of $1 \in G$ does not depend on the choice of M and of the basis, and this is $\kappa(N)$. A theorem of Kaplansky [13] states that if $N \neq 0$ then $\kappa(N) > 0$.

For free modules N one clearly has $\kappa(N) = \text{rank}_{ZG} N = \text{rank}(Z \otimes_G N)$; but for projective N the middle term is not defined and one does not know in general whether $\kappa(N) = \text{rank}(Z \otimes_G N)$.

In the case of the PD^2-group G and the projective module N in Section 5.2 let us first assume that P and hence P^* is free (e.g., G is the fundamental group of a finite CW-complex). Then (5) immediately implies $\kappa(N) = \text{rank}(Z \otimes_G N)$, and by Kaplansky's theorem we get from (6) that $\beta_1(G) > 0$.

In the general case, P^* projective, this does not work. However, an interesting criterion of Bass [1] tells that if, for some projective module, the two ranks are not equal then G contains a subgroup H isomorphic to the additive group $Z\left[\frac{1}{p}\right]$ for some prime p. If the index of H in G is finite H would be a PD^2-group — which it is not; if the index is infinite, this would imply, by Strebel's theorem [17], that H is free — which it is not. Thus the two ranks coincide, and we again get from (6) that $\beta_1(G) > 0$.

BENO ECKMANN
MATHEMATIK
ETH-ZENTRUM
CH-8092 ZÜRICH
SWITZERLAND

REFERENCES

[1] Bass, H., Euler characteristic and characters of discrete groups. Inventiones Math. 35 (1976), 155-196.

[2] Bieri, R., and Eckmann, B., Groups with homological cuality generalizing Poincaré duality. Inventiones Math. 20 (1973), 103-124.

[3] _____, Finiteness properties of duality groups. Comment. Math. Helv. 49 (1974), 460-478.

[4] _____, Relative homology and Poincaré duality for group pairs. J. of Pure and Applied Algebra 13 (1978), 277-319.

[5] _____, Two-dimensional Poincaré duality groups and pairs, in: Homological Group Theory, London Math. Soc. Lecture Notes 36 (1979), 225-230.

[6] Bieri, R., and Strebel, R., Almost finitely presented soluble groups. Comment. Math. Helv. 53 (1978), 258-278.

[7] Brown, K. S., Homological criteria for finiteness. Comment. Math. Helv. 50 (1975), 129-135.

[8] _____, Cohomology of Groups. Springer Verlag New York 1982.

[9] Dunwoody, M. J., Accessibility and groups of cohomological dimension one. Proc. of the London Math. Soc. 38 (1979), 193-215.

[10] Eckmann, B., and Linnell, P., Groupes à dualité de Poincaré de dimension 2. C.R. Acad. Sci. Paris 295 (1982), Série I, 417-418.

[11] Eckmann, B., and Linnell, P., Poincaré duality groups of dimension two, II. Comment. Math. Helv. 58 (1983), 111-114.

[12] Eckmann, B., and Müller, H., Poincaré duality groups of dimension two. Comment. Math. Helv. 55 (1980), 510-520.

[13] Montgomery, S., Left and right inverses in group algebras. Bull. Amer. Math. Soc. 75 (1969), 539-540.

[14] Müller, H., Decomposition theorems for group pairs. Math. Zeitschrift 176 (1981), 223-246.

[15] Stallings, J. R., On torsion-free groups with infinitely many ends. Ann. of Math. 88 (1968), 312-334.

[16] ———, Group theory and three-dimensional manifolds. Yale Math. Monographs 4, Yale Univ. Press 1971.

[17] Strebel, R., A remark on subgroups of infinite index in Poincaré duality groups. Comment. Math. Helv. 52 (1977), 317-324.

[18] Swan, R. G., Groups of cohomological dimension one. J. Algebra 12 (1969), 585-601.

[19] Swarup, G. A., Relative version of a theorem of Stallings. J. of Pure and Applied Algebra 11 (1977), 75-82.

HOW TO GENERALIZE ONE-RELATOR GROUP THEORY

James Howie[1]

The primary purpose of this paper is to survey a number of recent results in combinatorial group theory. The main theme is the one indicated in the title: that of trying to generalize the rich theory of one-relator groups to a more general construction known as the *one-relator* (or *anomalous*) *product*. There are also a number of subplots. For example, I want to illustrate the use of a geometric technique from 3-manifold topology—that of (*Papakyriakopoulos*)—*towers* in combinatorial group theory. The relevance of one-relator theory here is that the classical one-relator technique initiated by Magnus [38] is just the tower technique in disguise. I also want to discuss equations over groups and quasi-varieties. Last, but not least, I want to give some publicity to the class of locally indicable groups, which I believe to be of fundamental importance in low-dimensional topology.

Our story begins in 1980, with a 1-page announcement [6] by Sergei Brodskii of the following three strong results.

THEOREM 1. *Torsion-free 1-relator groups are locally indicable.*

THEOREM 2. *Any non-degenerate equation over a locally indicable group has a solution in some overgroup.*

THEOREM 3. *There exists a nontrivial quasivariety $\widetilde{\mathcal{Q}}$ of groups, closed under extensions, such that every group in $\widetilde{\mathcal{Q}}$ can be embedded in an equationally closed[2] group in $\widetilde{\mathcal{Q}}$.*

---

[1] Supported by an SERC Advanced Fellowship.

[2] The Russian term is "algebraicheski zamknuty," which means literally "algebraically closed," and is usually translated as such (see [6] and [32], p. 2, Question 2). Unfortunately there already exists a distinct notion of "algebraically closed group" [57, 43] in English. I am trying to avoid confusion between these two notions.

By the time [6] appeared, I had independently discovered Theorems 1 and 2 [21, 22], and Hamish Short had independently discovered Theorem 2 [58]. Brodskii's proofs were published only recently [5]. One interesting factor is that his arguments are purely algebraic, while Short and I use (different) geometric methods. B. Baumslag [2] later rediscovered Brodskii's proof of Theorem 2, which essentially follows Magnus' proof of the Freiheitssatz [38].

The paper is organized as follows. In sections 1-3 below I shall briefly discuss, respectively: locally indicable groups; one-relator products and equations over groups; and quasivarieties. In section 4 I shall introduce the notion of a tower of 2-complexes, and sketch a proof of the classical Freiheitssatz to illustrate their use. In section 5 I shall list a number of results, and in section 6 I shall discuss some open problems.

Almost all the results mentioned in this paper have appeared (or will soon appear) elsewhere, so I will for the most part omit proofs. Lemmas 2 and 3 in section 4 and Theorem 16 in section 5 are new, and I will indicate how to prove them.

This is an expanded version of my lecture at the conference. I have deliberately included more material here than I could present in a single lecture, in the hope of broadening the scope of the article. Much of this material was contained (in a very unpolished form, and complete with gory details of proofs) in a course of lectures given at the University of Glasgow in 1982. I am grateful to the audience there for their fortitude. In particular I am grateful to Steve Pride for reading and commenting on a draft version of the paper. I am also grateful to Steve Gersten for further helpful comments.

1. *What are locally indicable groups, and why are they interesting?*

A group G is *indicable* (or *Z-indicable*) if $H^1(G,Z) = \text{Hom}(G,Z) \neq 0$. In other words, if there exists an epimorphism $G \to Z$ (called an *indexing* *function*). A group is *locally indicable* if every nontrivial, finitely generated subgroup is indicable. Thus, for example, every locally indicable group is torsion free. These groups first appeared in Higman's thesis [20] (see also [19, 50]) on group rings.

THEOREM 4 [19]. *Let R be an integral domain and G a locally indicable group. Then the group algebra RG has no zero divisors, no idempotents other than 0 and 1, and no units other than those of the form ug (u a unit in R, $g \in G$).*

Higman's results have subsequently been extended to larger classes of groups. (See for example [47] for details.) One example is that Theorem 4 holds whenever G is *right orderable*. In other words G admits a total order relation $<$ which is invariant under right multiplication $x < y \Longrightarrow xz < yz$.

THEOREM 5 [7]. *Locally indicable groups are right orderable.*

Here are two further characterizations of locally indicable groups. In his paper [1] on Whitehead's Conjecture [67] that any subcomplex K of an aspherical 2-complex L is aspherical, Adams introduced the class of *conservative* (or Z-*conservative*) groups. A group G is *conservative* if, whenever G acts freely and cellularly on a 2-complex X such that $H_2(X/G, Z) = 0$, then $H_2(X, Z) = 0$. Strebel [61, 62] rediscovered this concept in a cohomological form: say that G is a $\mathcal{D}$ group, or $G \in \mathcal{D} = \mathcal{D}(Z)$ if, whenever $f : P \to Q$ is a ZG-homomorphism between projective ZG-modules, with $\operatorname{Ker}(1 \otimes f) = 0$, then $\operatorname{Ker} f = 0$, where $1 \otimes f : Z \otimes_G P \to Z \otimes_G Q$.

THEOREM 6 [13, 29]. *G is locally indicable $\iff$ G is conservative $\iff$ $G \in \mathcal{D}$.*

More generally, let p be a prime number, and consider the classes of locally p-indicable groups, p-conservative groups, and $\mathcal{D}_p$, obtained by substituting Z_p for Z in the definitions. Then these three classes coincide.

From the topological point of view, locally indicable groups are of interest for two reasons. The first is the connection with Whitehead's Conjecture arising from Adams' work [1] mentioned above.

COROLLARY. *Let K be a connected 2-complex such that $H_2(K) = 0$ and $\pi_1(K)$ is locally indicable. Then K is aspherical.*

This is exploited, for example, in [27], where special cases of Whitehead's Conjecture are proved by showing that the fundamental group is locally indicable.

The second interesting point about locally indicable groups for the topologist is the fact that "most" 3-manifold groups are locally indicable.

THEOREM 7. *Let M be an orientable 3-manifold. Then either $\pi_1(M)$ is locally indicable, or $M = M_1 \# M_2$ with M_1 a Q-homology 3-sphere.*

Theorem 7 has appeared in the literature only recently [8, 22, 30], but some forms of it were certainly known to various people prior to that. The proof is an easy exercise, given the Compact Submanifold Theorem [56] and the Sphere Theorem [46]. In [30], Theorem 7 is applied to give an algebraic proof that any band-connected sum of nontrivial knots is nontrivial.

We shall see in subsequent sections that locally indicable groups are also of interest in connection with equations over groups, one-relator products, and quasivarieties.

2. *One-relator products and equations over groups*

Let $r \in A*B$ be a cyclically reduced word of length at least 2 in the free product $A*B$ of two groups A and B, and let $N(r)$ denote the smallest normal subgroup of $A*B$ containing r. Then the group $G = (A*B)/N(r)$ is called a *one-relator product* of A and B. We call A and B the *factors* and r the *relator*. (Brodskii [5] refers to this construction as an *anomalous product*, and to r as the *anomaly*.)

If A and B are free groups in the above, then G is just a one-relator group in the usual sense. Conversely, every (infinite) one-relator group is a one-relator product of two free groups. Thus one-relator products are natural generalizations of one-relator groups, and it is natural to try to extend theorems about one-relator groups to apply to other one-relator products. A good test case is the Freiheitssatz of Magnus [38] (see also [37], Chapters II.5, II.6). This says that a one-relator product of free groups contains each factor as a (naturally embedded) subgroup: the natural maps $A \to G$, $B \to G$ are injective. For more general one-relator products, we will say that *the Freiheitssatz holds* (in a given situation) if $A \to G$, $B \to G$ are injective. Unfortunately, the Freiheitssatz does not always hold.

EXAMPLE. Let A, B be simple groups, and let $a \in A$, $b \in B$ be elements of distinct finite orders. Then $(A * B)/N(ab) = \{1\}$.

This example clearly represents the worst possible pathology, from the point of view of one-relator products. Since the Freiheitssatz is fundamental to one-relator group theory, we cannot hope to pursue our program of generalization without imposing some conditions to avoid such examples. One possibility is a condition on the relator.

THEOREM 8 [16]. *The Freiheitssatz holds whenever* $r = s^m$ *for some* $m > 4$.

We shall be concentrating, however, on the other possibility, namely imposing restrictions on the factors. The following "folklore" conjecture is one plausible example.

CONJECTURE. *The Freiheitssatz holds for torsion free factors.*

At present this is merely a conjecture. None of the known methods of attack apply to torsion free groups in full generality. To actually obtain results, a stronger restriction is needed. There are generalizations of the Freiheitssatz in [3, 15, 35, 48], but the following is the most general form to date.

THEOREM 2′ [6, 21, 58]. *The Freiheitssatz holds for locally indicable factors.*

It turns out that the condition of local indicability is just the right one to make proofs work—not just of the Freiheitssatz, but also, as we will see later, of other results from one-relator group theory. These results, in fact, can all be extended to constructions more general than one-relator products, namely to the fundamental groups of *staggered generalized 2-complexes* in the sense of [28]. I shall not go into details here, but these groups all have the form $(*_i A_i)/N\{r_j\}$, where the A_i are to be assumed locally indicable, and the relators r_j are "staggered." This is a natural generalization of staggered presentations, and also of *staggered 2-complexes*, which will be defined in section 4 below. An interesting special case is the *tree anomalous product* of Brodskii [5]. Here the factors A_i correspond to the vertices of a tree whose edges correspond to the relators r_j, and each r_j involves only the two factors corresponding to the endpoints of the corresponding edge.

As the numeration suggests, Theorems 2 and 2′ are closely related. To explain this relationship, I must now discuss equations over groups. Consider a one-relator product $G = (A*B)/N(r)$ in which $B = \langle t \rangle$ is infinite cyclic. We can regard r as a "polynomial" $r(t)$ in the variable t, with coefficients from the group A. The polynomial equation $r(t) = 1$ has a solution in A if and only if the natural map $A \to G$ is *split injective*. If this map is merely *injective*, then we can regard G as an overgroup of A in which the equation has a solution. Moreover, this solution is *universal* in the obvious sense: if H is any overgroup of A in which the equation $r(t) = 1$ has a solution h, then the inclusion $A \to H$ factors uniquely through $A \to G$ in such a way that $t \in G$ is sent to $h \in H$. (Hence, in particular, $A \to G$ is injective.) We can think of G as being formed from A by "adjoining a root" of the polynomial $r(t)$. These ideas are due originally to B. H. Neumann [42].

Call a polynomial $r = r(t)$ *nondegenerate* if r does not belong to any conjugate of A, and call an equation $r(t) = 1$ *nondegenerate* if r is nondegenerate. Now Theorem 2´ clearly implies Theorem 2 (put $B = <t>$). Conversely, Theorem 2 also implies Theorem 2´: the "twisted" embedding $A * B = A * (tBt^{-1}) \hookrightarrow (A * B) * <t>$ sends any cyclically reduced word of length at least 2 in $(A * B)$ to a nondegenerate polynomial.

We can also consider the idea of a system

$$\Sigma : r_1(t_1, \cdots, t_n) = \cdots = r_m(t_1, \cdots, t_n) = 1$$

of several equations in several unknowns, over a group A. It is then appropriate to study the group $G = (A * <t_1, \cdots, t_n>)/N\{r_1, \cdots, r_m\}$, and the natural map $A \to G$. A group A is called *algebraically closed* [43, 57] (or *existentially closed*) if $A \neq 1$ and any finite system Σ of equations over A, having a solution in some overgroup, has a solution in A. (In other words, whenever the map $A \to G$ is injective in the above, then it is split injective.) A group A is called *equationally closed* if every (single) nondegenerate equation over A has a solution in A. (In other words the map $A \to (A * <t>)/N(r)$ is always split injective.) It is not too difficult to see that *equationally closed* groups are torsion free, while *algebraically closed* groups contain elements of all finite orders, so that *no* group falls into both classes (cf. footnote 2).

To any *finite* system Σ of equations over a group, we can naturally associate an integer matrix $M(\Sigma) = (\mu_{ij})$, where μ_{ij} is the algebraic sum of the exponents of t_i appearing in the word r_j. If rank $M(\Sigma)$ is equal to the number of equations in Σ, then we call Σ *(linearly) independent* or *nonsingular*. If this remains true working modulo some prime p, then Σ is called p-*independent*.

THEOREM 9 [14]. *Every independent system of equations over a compact, connected Lie group* A *has a solution in* A.

COROLLARY 1 [14]. *Every independent system of equations over a finite group has a solution in some finite overgroup.*

COROLLARY 2 [49]. *Every independent system of equations over a locally residually finite group has a solution in some overgroup.*

THEOREM 10 [13]. *Every p-independent system of equations over a locally p-indicable group has a solution in some overgroup.*

COROLLARY [21]. *Every independent system of equations over a locally indicable group has a solution in some overgroup.*

There are also numerous results about equations over groups based on restrictions on the equations [11, 23, 33, 42, 52, 53, 54] (see also [37], Chapter I.6, and the survey article [36] of Lyndon). A number of other results and conjectures related to these problems are discussed in [25, 60].

3. *Quasivarieties*

Varieties of groups have been extensively studied (see for example [44]). A *variety* is a class of groups defined by some collection of *laws* or *identities*, for example the variety of metabelian groups is defined by the single identity $[[a,b],[c,d]] = 1$. Similarly a *quasivariety* is a class of groups defined by a collection of *quasi-identities* (in which implication signs may appear). Thus for example the class $\mathcal{TF}$ of torsion-free groups is a quasivariety defined by the collection

$$g^n = 1 \Longrightarrow g = 1 \quad (n = 2,3,4,\cdots)$$

of quasi-identities.

All the above can be made precise using the language of predicate calculus. For details see [39], Chapter V. Alternatively, a class A of groups is a quasivariety if and only if it contains the trivial group and is closed under subgroups, direct products and ultraproducts. The only nontrivial notion here is that of an ultraproduct. The reader is referred to [39], Chapters IV-V for a comprehensive introduction, but here is a brief sketch.

Let I be a set. An *ultrafilter* U on I is a subset of the power set of I which satisfies:

(i) $X, Y \in U \Longrightarrow X \cap Y \in U$;
(ii) $X \in U, X \subset Y \Longrightarrow Y \in U$;
(iii) $X \subset I \Longrightarrow$ *precisely* one of $X, I-X$ belongs to U.

Obvious examples are the *principal* ultrafilters $U_x = \{X \subset I; x \in X\}$ for $x \in I$, but these are not interesting from our point of view. If I is finite, then all ultrafilters on I are principal, but otherwise examples of nonprincipal ultrafilters can be found using Zorn's Lemma.

Now let $\{G_i\}$ be a family of groups indexed by I. The *ultraproduct* of the G_i defined by U is the group $G = (\Pi_i G_i)/\sim$, where $\sim$ is the equivalence relation

$$(g_i) \sim (h_i) \text{ if and only if } (i \in I : g_i = h_i) \in U .$$

For example $G \cong G_x$ whenever U is the principal ultrafilter U_x. (This is why principal ultrafilters are uninteresting.) It is an easy exercise to check that the classes $\mathcal{D}, \mathcal{D}_p, \mathcal{RO}, \mathcal{TF}$ of locally indicable, locally p-indicable, right orderable and torsion-free groups, respectively, are closed under the formation of ultraproducts (and hence quasivarieties). It is not too difficult to write down explicit sets of quasi-identities defining $\mathcal{D}, \mathcal{D}_p$ and (as we have seen) $\mathcal{TF}$, but I do not know of an explicit set of quasi-identities defining $\mathcal{RO}$.

Ultraproducts can be very powerful. An interesting application is the proof of van den Dries and Wilkie [65] of Gromov's Theorem on groups of polynomial growth.

A class $\mathcal{C}$ of groups is *idempotent* or *extension-closed* if $\mathcal{C} \cdot \mathcal{C} = \mathcal{C}$, in other words if any extension of an $\mathcal{C}$-group by an $\mathcal{C}$-group is an $\mathcal{C}$-group. Brodskii [5] constructs a quasivariety $\widetilde{\mathcal{C}}$ of groups satisfying:

(i) $Z \in \widetilde{\mathcal{C}}$;
(ii) $\widetilde{\mathcal{C}} \cdot \widetilde{\mathcal{C}} = \widetilde{\mathcal{C}}$;
(iii) $\widetilde{\mathcal{C}}$ is minimal with respect to (i) and (ii).

(Indeed properties (i)-(iii) uniquely determine $\widetilde{\mathcal{A}}$.) He then proves the following result.

THEOREM 3 [5, 6]. *The quasivariety $\widetilde{\mathcal{A}}$ is nontrivial and closed under extensions, and every group in $\widetilde{\mathcal{A}}$ can be embedded in an equationally closed group in $\widetilde{\mathcal{A}}$.*

This answers, in an extremely strong way, a question of Bokut' [32], Question 2, p. 2: does there exist a nontrivial, equationally closed group?

It would be more satisfactory to have an explicit description of the quasivariety $\widetilde{\mathcal{A}}$. There is a chain of inclusions

$$\widetilde{\mathcal{A}} \subset \mathcal{D} \subset \mathcal{RO} \subset \mathcal{TF}$$

of which the first follows from property (iii) of $\widetilde{\mathcal{A}}$ and the second from Theorem 5. Only the third inclusion is known to be strict ([47], Lemma 3.3, p. 606), so $\mathcal{D}$ is a plausible candidate for $\widetilde{\mathcal{A}}$ (as also is $\mathcal{RO}$). Some evidence in this direction is given by Theorem 3´ in section 5 below.

Here is a mod p version. Replace Z by Z_p in property (i) of $\widetilde{\mathcal{A}}$. Then Brodskii's construction works equally well to produce a quasivariety $\widetilde{\mathcal{A}}_p$ with $\widetilde{\mathcal{A}}_p \subset \mathcal{D}_p$.

4. Towers and the Magnus argument

There is a "standard" method of proving results about one-relator groups, which was first exploited by W. Magnus in his fundamental paper [38]. There have since been various refinements (see for example [40]), but the basic idea remains the same. Briefly, and omitting a few technical tricks, it is as follows. Given a one-relator group presentation $G = \langle x_1, \cdots, x_n; r \rangle$, choose an epimorphism $\phi: G \to Z$, and consider the presentation of $H = \text{Ker } \phi$ obtained by the Reidemeister-Schreier rewriting process. This is *staggered* in the sense that it has the form $\langle y_i (i \in Z), z_j (j \in I); r_i (i \in Z) \rangle$ for some index set I, and there are constants $m < M$ such that each r_i is a word in $\{y_{i+m}, \cdots, y_{i+M}, z_j (j \in I)\}$, properly involving y_{i+m} and y_{i+M}. Furthermore, things can always be arranged so that each r_i is strictly shorter than r.

Now, provided the problem under consideration can be "lifted" from G to H and then "restricted" to a finite subpresentation, and provided it behaves nicely with respect to amalgamated free products, we can proceed as follows. Restate the desired theorem for finite (subpresentations of) staggered presentations, and argue by double induction on the number of relations and the maximal relator length.

Geometrically, what is happening is this. Initially we have a 2-complex with a single 2-cell. We lift to an infinite cyclic cover and restrict to a finite subcomplex of that cover. Then, by some suitable inductive argument, we reduce to the case of a subcomplex containing a single 2-cell. This process makes the attaching map of the unique 2-cell "simpler" in some sense than in the initial case, so after finitely many repetitions we arrive at a situation which is so simple that there are no more infinite cyclic covers. With luck, our problem will have a trivial solution in this case, and from that we will be able to deduce a solution in the general case.

The type of construction described above—cover, subspace, cover, etc.—is very familiar to people who work with 3-manifolds. It is called a *tower*. Such things were first used in 3-manifolds by Papakyriakopoulos [46] to prove Dehn's Lemma and the Sphere Theorem (see also [18], Chapter 4). They are still being used today (for example [12], Theorem 2.1). It is also possible to use towers of spaces other than 3-manifolds [13, 21, 22, 24, 26, 64].

Formally, a *tower* is a map

$$g = i_0 \circ p_1 \circ i_1 \circ \cdots \circ p_h \circ i_h : K' \to K$$

between connected CW-complexes (simplicial complexes, smooth or PL manifolds, $\cdots$), such that each i_j is an inclusion map, and each p_j is a covering projection. It is often useful to restrict the p_j to belong to some particular class of coverings, depending on the problem under consideration. For the purposes of this paper, we will restrict the p_j to be *infinite cyclic* coverings. In other words regular and connected, with infinite cyclic covering transformation group.

A commutative triangle

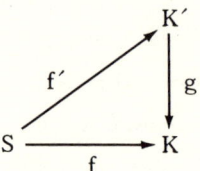

of connected complexes, in which g is a tower, is called a *tower lifting* of f. It is *proper* if g is not an isomorphism, and it is *maximal* if f' does not have a proper tower lifting.

LEMMA 1. *Let S be a finite CW-complex (compact manifold,$\cdots$). Then any "reasonable" map $f: S \to K$ has a maximal tower lifting, unique up to unique isomorphism.*

I have been deliberately vague about the hypothesis on f in Lemma 1. See [21, 24] for more precise formulations. "Reasonable" can be taken to mean simplicial when dealing with simplicial complexes, PL when dealing with PL manifolds, etc. The important point is that any map can be replaced, up to homotopy, by a "reasonable" map, and this will not in general alter the nature of the problem we are interested in.

The existence part of Lemma 1 is proved using a cell-counting argument. (See [21], Lemma 3.1.) Uniqueness is proved by staring hard at the pullback of two maximal tower liftings.

Now we introduce the idea of a *staggered* 2-complex. A 2-complex K is *staggered* if all its 2-cells are attached by cyclically reduced paths of positive length, and if the sets of 1-cells and of 2-cells are equipped with linear orderings which are compatible in the following sense: whenever $\alpha < \beta$ are 2-cells, then $(\min \alpha) < (\min \beta)$ and $(\max \alpha) < (\max \beta)$.

Here min a (resp. max a) denotes the least (greatest) 1-cell involved in the attaching map for a. This is closely related to the notion of staggered presentation mentioned above, and also to that of a staggered generalized 2-complex [28].

LEMMA 2. *Let K be a staggered 2-complex, and* $g : K' \to K$ *a tower. Then K' is staggered.*

The proof easily reduces to the case where g is a covering. Fix a generator γ of the (infinite cyclic) covering transformation group. Define linear orderings on the 1- and 2-cells by $a < \beta$ if either $g(a) < g(\beta)$ or $\beta = \gamma^n(a)$ for some $n > 0$. These are easily seen to be compatible.

LEMMA 3. *Let K' be a finite staggered 2-complex with at least one 2-cell, such that* $H^1(K') = 0$, *and suppose the greatest 2-cell a of K' is not attached by a proper power (in* $\pi_1(K'^{(1)})$). *Then K' collapses across a, with free edge* max a.

Proof. Note first that, if some 2-cell γ of K' is attached by a proper power s^m say, then replacing γ by a 2-cell attached along s will not change $H^1(K')$. Nor will this procedure affect the staggering of K'. Hence we may assume that *no* 2-cell of K' is attached by a proper power.

We argue by induction on the number of 2-cells in K', which by hypothesis is at least 1. If there is only one 2-cell, then the first Betti number of the 1-skeleton $K'^{(1)}$ is at most 1, since $H^1(K') = 0$. On the other hand, since the attaching map P of the 2-cell is a cyclically reduced path of positive length, and not a proper power, it follows that $K'^{(1)}$ cannot be a tree, and P is the unique simple closed path in $K'^{(1)}$, whence the result.

For the inductive step, consider the Mayer-Vietoris sequence

$$\cdots \to H^1(K') \to H^1(K'-a) \oplus H^1(D^2) \to H^1(S^1) \to \cdots$$

associated to the adjunction of a. From this we see that $H^1(K'-a)$ is at most cyclic. If the subcomplex $K'' = K' - \{a, \max a\}$ is connected, then $H^1(K'-a) = H^1(K'') \oplus Z$, so $H^1(K'') = 0$. Otherwise K'' has two components K_1 and K_2 say, and $H^1(K'-a) = H^1(K_1) \oplus H^1(K_2)$. Without loss of generality we may assume $H^1(K_1) = 0$. Also, in this case K_1 cannot be a tree, since a is attached by a cyclically reduced closed path which meets K_1.

Now apply the inductive hypothesis either to K'' or to K_1, but with the staggering *opposite* to that inherited from K' (that is, the orderings of cells is opposite). Then the complex in question collapses across its least 2-cell β say (in the original ordering), with free edge $\min \beta$. But a does not involve $\min \beta$ since $\beta < a$, so K' also collapses across β with free edge $\min \beta$. Let $L = K' - \{\beta, \min \beta\}$ be the result of this collapse.

Then the inductive hypothesis applies to L, so L collapses across a with free edge $\max a$. But β does not involve $\max a$ since $\beta < a$. Hence $K' = L \cup \{\beta, \min \beta\}$ also collapses across a with free edge $\max a$.

COROLLARY 3.1. *Let K' be a finite staggered 2-complex, with $H^1(K') = 0$, in which no 2-cell is attached by a proper power. Then K' is collapsible.*

COROLLARY 3.2. *Let K' be a finite staggered 2-complex, with $H^1(K') = 0$. Then K' has the homotopy type of a wedge of pseudo-projective planes.*

(A *pseudo-projective plane* is the mapping cone of some self-map of S^1 of nonzero degree.)

Lemmas 1-3 and the two corollaries, together with the observation that any 2-complex with a single 2-cell (and nontrivial attaching map) is staggered, form the basis for applying towers to one-relator theory. Roughly, one restates the problem under consideration in terms of a "reasonable" map $f : S \to K$, where S and K are 2-complexes, S is finite, and K

has a single 2-cell. Apply Lemma 1 to obtain a maximal tower lifting, and then Lemma 2 and 3 (or one of the corollaries) to show that the problem can be solved at the top of the tower. By way of illustration, here is a sketch proof of Magnus' Freiheitssatz.

THEOREM 11 [38]. *Let* $G = <x_1,\cdots,x_n;r>$ *where* r *properly involves* x_n. *Then* $\{x_1,\cdots,x_{n-1}\}$ *freely generate a free subgroup of* G.

Proof. There is no loss of generality in supposing that r is not a proper power: if $r = s^m$, then replace r by s. Let K be the 2-complex model of the stated presentation, and let Γ be the subgraph of its 1-skeleton $K^{(1)}$ consisting of the unique 0-cell and the 1-cells $x_1,\cdots,x_{n-1}$. We must show that $\pi_1\Gamma \to \pi_1 K$ is injective. If not, then there is a map $f: D^2 \to K$ with $f(S^1) \subset \Gamma$ representing a nontrivial element of $\pi_1(\Gamma)$. Make f "reasonable," and let

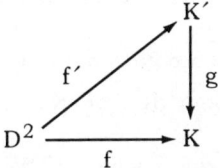

be a maximal tower lifting. Then K' is staggered, and indeed the proof of Lemma 2 shows that we may assume $g(\max \alpha) = x_n$ for each 2-cell α of K'. Also K' is finite (for otherwise f' would factor through an inclusion, contradicting maximality), and $H^1(K') = 0$ (for otherwise f' would lift over an infinite cyclic cover, contradicting maximality). Finally, no 2-cell in K' is attached by a proper power. Repeated applications of Lemma 3 show that K' collapses to a 1-complex Γ' (necessarily a tree), such that $g(y) = x_n$ for each 1-cell y in $K' - \Gamma'$. In particular $f'(S^1) \subset g^{-1}(\Gamma) \subset \Gamma'$, so $f(S^1)$ is nullhomotopic in $g^{-1}(\Gamma)$, and $f(S^1)$ is nullhomotopic in Γ. Contradiction.

5. Results

Many other theorems about one-relator groups can be proved using towers in a manner similar to the proof of Theorem 11 above. These include Brodskii's Theorem 1, that torsion free one-relator groups are locally indicable. The argument is identical, except that the disc D^2 is replaced by some finite 2-complex S with $H^1(S) = 0$ (so that $f_*(\pi_1(S))$ is a finitely generated, nonindicable subgroup of $\pi_1(K)$).

A large class of genuinely new results can be obtained by a relative version of the tower method. I shall omit the details, except to say that we are now considering pairs (L,K) of 2-complexes, with $L = K \cup e^1 \cup e^2$ connected, and such that the attaching map for e^2 properly involves e^1. On the face of it, there are two distinct cases, depending as K is connected or not, but there is in fact no essential difference between these two cases. To obtain positive results, we also have to assume that (each component of) K has locally indicable fundamental group.

THEOREM 2″. *With the above notation and conventions, $\pi_1(K_1) \to \pi_1(L)$ is injective for each component K_1 of K.*

Theorem 2″ is just Theorem 2 when K is connected, and Theorem 2′ otherwise. We can also obtain the following generalization of Theorem 1.

THEOREM 12 [22]. *A one-relator product of locally indicable groups is locally indicable, provided the relator is not a proper power (in the free product).*

The following remark is due to S. D. Brodskii. When solving an equation $r(t) = 1$, in which $r = s^m$ is a proper power, it is enough to solve the equation $s(t) = 1$. Thus by Theorems 2 and 12, each nondegenerate equation over a locally indicable group actually has a solution in some *locally indicable* overgroup. By a standard argument we now have the following embedding theorem.

THEOREM 3′ [22]. *Every locally indicable group can be embedded in an equationally closed locally indicable group.*

The next result generalizes a theorem of Weinbaum [66].

THEOREM 13 [22]. *In a one-relator product of locally indicable groups, no proper subword of the relator represents the identity element.*

The Spelling Theorem of B. B. Newman [45] says that, in a one-relator group with relator $r = s^m$, any nonempty word which represents the identity element must contain a (cyclic) subword of r or r^{-1} longer than s^{m-1}. More precise versions are due to Gurevich [17], Schupp [55] and Pride [48]. These results all have appropriate generalizations to one-relator products of locally indicable groups [28]. The details are somewhat technical, but the spirit is conveyed by the following (slightly vague) statement.

THEOREM 14 [28]. *Let $G = (A * B)/N(r)$, where A and B are locally indicable and $r = s^m$. Let w be a nonempty word in $A * B$ which lies in $N(r)$. Then either w is conjugate to r or r^{-1}, or w contains two almost-disjoint cyclic subwords, each of which is a cyclic subword of r or r^{-1} longer than s^{m-1}.*

COROLLARY. *If $m > 1$ and A and B have solvable word problem in Theorem 14, then G also has solvable word problem.*

The above Corollary has been independently obtained by Brodskii and Mazurovskii (unpublished). They have also solved the word problem for G in the case $m = 1$, but under the stronger hypothesis that A and B be *effectively locally indicable* (there exists an algorithm which, given a finite set of generators for a subgroup H, will decide whether or not $H = \{1\}$, and if not will exhibit an epimorphism $H \to Z$).

THEOREM 15 [26]. *Let $G = (A * B)/N(r)$, where A and B are locally indicable and $r = s^m$ where s is not a proper power. Let K_A, K_B, K_C*

be Eilenberg-MacLane complexes $K(A,1)$, $K(B,1)$, $K(C,1)$ respectively, where C is cyclic of order m. Let $K_G = (K_A \vee K_B) \cup_S K_C$, where $S = S^1 \to K_A \vee K_B$ represents s and $S = S^1 \to K_C$ represents a generator. Then K_G is a $K(G,1)$-complex.

Theorem 15 is a generalization of a theorem of Dyer and Vasquez [10] for one-relator groups, which can be thought of as a geometric version of Lyndon's Identity Theorem [34]. The usual (algebraic) version of the Identity Theorem is in terms of the *relation module* $N(r)^{ab} = N(r)/[N(r), N(r)]$.

COROLLARY 1. *Let G, A, B, r, s be as in Theorem 15. Then $N(r)^{ab}$ is isomorphic, as a ZG-module, to the cyclic module $ZG/(1-s)ZG$, generated by* $r \cdot [N(r), N(r)]$.

Either interpretation allows one to compute the (co)homology of G in high dimensions.

COROLLARY 2. *There are natural isomorphisms, for each* $q > 2$:

$$H^q(G; -) = H^q(A; -) \oplus H^q(B; -) \oplus H^q(C; -)$$

$$H_q(G; -) = H_q(A; -) \oplus H_q(B; -) \oplus H_q(C; -) .$$

Combining Corollary 2 with a result of Serre [31], we see that the conjugates of $C = gp(s)$ are precisely the maximal finite subgroups of G; and that these conjugates are disjoint in the strong sense that $C \cap gCg^{-1} \neq \{1\} \Longrightarrow g \in C$.

A similar type of argument (involving computation of H^2 as well as of H^q, $q > 2$) shows that any subgroup of the form $S = A \cap gBg^{-1}$ ($g \in G$) or $S = A \cap gAg^{-1}$ ($g \in G - A$) or $S = B \cap gBg^{-1}$ ($g \in G - B$) has cohomological dimension at most 1, and hence [59, 63] is free. One can then show by induction on the length of r in the standard way that S has rank at most 1. In the case where the relator is a proper power, the situation is much simpler. It follows without too much difficulty from the

Spelling Theorem of [28] (see Theorem 14) that $S = \{1\}$. Putting these facts together, we have the following result.

THEOREM 16. *In the above notation, the group* S *is cyclic. If* r *is a proper power then* $S = \{1\}$.

In the ordinary one-relator group case, these statements are due to Bagherzadeh [4] and Newman [45] respectively.

6. *Open problems*

I'll finish with some "homework" for the interested reader. On the whole these are difficult unsolved problems, but it may be that there is hope of partial progress in some of them.

6.1. Does the Freiheitssatz extend to some class $\mathcal{C}$ of groups strictly greater than $\mathcal{D}$? Ideally, one would like to prove it for all torsion free factors ($\mathcal{C} = \mathcal{TF}$). Failing that, what about $\mathcal{C} = \mathcal{D}_p \cap \mathcal{TF}$ or $\mathcal{C} = \mathcal{D}_p \cap \mathcal{D}_q$ for distinct primes p and q?

The Freiheitssatz fails for $\mathcal{D}_p$ in general because of the occurrence of torsion. For example, put $p = 2$, $A = \langle a; a^4 \rangle$, $B = \langle b \rangle$, $r = aba^2b^{-1}$. The classes $\bigcap_{p \in \pi} \mathcal{D}_p$ are distinct, for *finite* sets of primes π, and are contained in $\mathcal{TF}$ whenever $|\pi| > 1$ [29].

6.2. Can Theorem 8 be improved so that "$m > 4$" can be replaced by "$m > 3$" (or "$m > 2$" or "$m > 1$")?

6.3. Is the word problem for a one-relator product of two locally indicable groups solvable under some condition on the factors weaker than "effectively locally indicable" (see section 5)?

Effectively locally indicable clearly implies a solution to the word problem (for the factors), but this is a necessary condition in any case, since the factors embed in the whole group by the Freiheitssatz.

Does there exist a locally indicable group, with solvable word problem, which is not effectively locally indicable?

6.4. Let $r \in A * <t>$ be such that $(A *<t>)/N(r) = \{1\}$. Is $A = \{1\}$? This is the so-called Kervaire (or Laudenbach) conjecture. It is clear from abelian considerations that A must be perfect, and that t must occur in r with exponent sum ± 1. It suffices to consider the case when A is an infinite simple group.

6.5. Let Σ be a system of n equations in n unknowns over a group A. Does Σ have a solution in some overgroup of A whenever $\det M(\Sigma) \neq 0$? Whenever $\det M(\Sigma) \neq 0 \mod p$ for some fixed prime p? Whenever $\det M(\Sigma) = \pm 1$?

The third (and weakest) of these would be sufficient to prove the Kervaire conjecture 6.4. See [25, 60] for other related problems.

6.6. Let $M_1 \subset M_2$ be a tame embedding of (smooth or PL) 3-manifolds, such that $H_2(M_2, M_1; \mathbf{Z}) = 0$ (resp. $H_2(M_2, M_1; \mathbf{Z}_p) = 0$ for some fixed prime p; resp. $H_2(M_2, M_1; \mathbf{Z}_p) = 0$ for all primes p). Let S be a compact orientable surface and $f: S \to M_2$ a (smooth or PL) map such that $f(\delta S) \subset M_1$. Then the homology condition ensures that there is a compact orientable surface T, a homeomorphism $h: \delta S \to \delta T$, and a (smooth or PL) map $g: T \to M_1$, such that

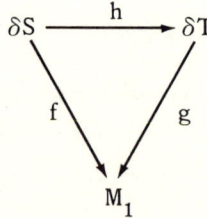

commutes. Can T be chosen such that genus $T \leq$ genus S?

Unrelated as these seem to the questions above, Stallings [60] shows that each implies an affirmative answer to the corresponding question in

6.5, and hence also to 6.4. He also shows that the answer is yes whenever f is an embedding, and that in fact g may then also be taken to be an embedding. Scharlemann [51] shows that the answer is yes whenever f restricts to an embedding of δS into δM_1. Stallings' result also has a form for nonorientable surfaces, valid whenever $H_2(M_2, M_1; Z_2) = 0$.

6.7. Consider the chain $\widetilde{\mathcal{Q}} \subset \mathcal{D} \subset \mathcal{RO}$ of quasivarieties, discussed in section 3. Which of these inequalities are strict?

The similarity between Theorems 3 and 3´ provide some evidence in support of the conjecture that $\widetilde{\mathcal{Q}} = \mathcal{D}$, but this remains merely a conjecture, and it is not clear how one would go about proving it. Brodskii informs me that he has spent some time trying unsuccessfully to prove it. As for the inequality $\mathcal{D} \subset \mathcal{RO}$, this is presumably strict, but to prove that would involve exhibiting a finitely generated, right orderable group with finite abelianization. Any suggestions?

6.8. A mod p version of 6.7. Is the inclusion $\widetilde{\mathcal{Q}}_p \subset \mathcal{D}_p$ strict?

Again, the conjectured answer is no, but how to prove it?

6.9. Is there a Cohen-Lyndon Theorem for one-relator products of locally indicable groups? In other words, let A and B be locally indicable groups, and let $r = s^m \in A*B$ be a cyclically reduced word of length at least 2, where m is a nonnegative integer and s is not a proper power. Is there a set T of left coset representatives for $<s> \cdot N(r)$ in $A*B$ such that $N(r)$ is free with basis $\{trt^{-1}; t \in T\}$?

The case when A and B are free is due to Cohen and Lyndon [9]. In the general case $N(r)$ is free by the Freiheitssatz (Theorem 2´) and the Kurosh Theorem. Furthermore *any* choice of T will give the correct answer modulo $[N(r), N(r)]$ by the Identity Theorem (Corollary 1 to Theorem 15).

6.10. What about two-relator groups?

Clearly the picture is fairly complicated. Two-relator groups can be a long way short of being locally indicable. Indeed many 3-manifold groups can be presented on 2 generators and 2 relators, but are not indicable (see for example [41]). On the other hand, some two-relator presentations behave almost like one-relator presentations after "passing to an infinite cyclic cover." Groups with such presentations are locally indicable, provided neither relator is a proper power (modulo the other).

EXAMPLE [27]. Let F be the free group with basis $\{x,y,z\}$, and let $f \in \text{End}(F)$ be defined by $x \mapsto x^2yx^{-1}y^{-1}$, $y \mapsto y^2zy^{-1}z^{-1}$, $z \mapsto z^2xz^{-1}x^{-1}$. Then for each $n \geq 0$ the group

$$G_n = <x,y,z; f^n(x), f^n(y)>$$

is locally indicable.

Here is another way of stating Theorem 1: *A one-relator group is locally indicable if and only if each of its one-generator subgroups is locally indicable.* This suggests a conjecture.

CONJECTURE. *A two-relator group is locally indicable if and only if each of its two-generator subgroups is locally indicable.*

Of course, one could generalize this conjecture by replacing 2 by n. I would like to be able to prove something like this for arbitrary n, but so far even the case $n = 2$ is elusive.

6.11. Let K be a subcomplex of a contractible 2-complex L. Is $\pi_1(K)$ locally indicable?

This implies the Whitehead Conjecture (cf. Corollary to Theorem 6).

6.12. Let I denote the augmentation ideal in the group ring $\mathbb{Z}G$. A $\mathbb{Z}G$-module M is *perfect* if $IM = M$. Do there exist nonzero, finitely generated, perfect projective $\mathbb{Z}G$-modules?

From the definition of $\mathcal{D}$-groups, it is clear that the answer is no whenever $G \in \mathcal{D}$ (or even $G \in \mathcal{D}_p$ for some p). Indeed, the answer is no unless G has a nontrivial, finitely generated, perfect subgroup [29]. On the other hand, if G does contain such a subgroup, then there exists an *infinitely generated* perfect projective $\mathbb{Z}G$-module [68].

JAMES HOWIE
DEPARTMENT OF MATHEMATICS
UNIVERSITY OF GLASGOW
UNIVERSITY GARDENS
GLASGOW G12 8QW
U. K.

REFERENCES

[1] J. F. Adams, A new proof of a theorem of W. H. Cockcroft, J. London Math. Soc. 49 (1955), 482-488.

[2] B. Baumslag, Free products of locally indicable groups with a single relator, Bull Austral. Math. Soc. 29 (1984), 401-404.

[3] B. Baumslag and S. J. Pride, An extension of the Freiheitssatz, Math. Proc. Camb. Phil. Soc. 89 (1981), 35-41.

[4] G. H. Bagherzadeh, Commutativity in one-relator groups, J. London Math. Soc. 13 (1976), 459-471.

[5] S. D. Brodskii, Equations over groups and groups with a single defining relator (Russian), Sibirskii Mat. Zh. 25, 2 (1984), 84-103.

[6] ———, Equations over groups and groups with a single defining relator, Uspehi Mat. Nauk. 35, 4 (1980), 183. (Russian Math. Surveys 35, 4 (1980), 165.)

[7] R. G. Burns and V. W. D. Hale, A note on group rings of certain torsion-free groups, Can. Math. Bull. 15 (1972), 441-445.

[8] T. Cochran, Ribbon knots in S^4, J. London Math. Soc. 28 (1983), 563-576.

[9] D. E. Cohen and R. C. Lyndon, Free bases for normal subgroups of free groups, Trans. Amer. Math. Soc. 108 (1963), 528-537.

[10] E. Dyer and A. T. Vasquez, Some small aspherical spaces, J. Austral. Math. Soc. 16 (1973), 332-352.

[11] M. H. Freedman, Remarks on the solution of first degree equations in groups, Lect. Notes in Math. 664 (1978), 87-93.

[12] M. H. Freedman, J. Hass and P. Scott, Least area incompressible surfaces in 3-manifolds, Invent. Math. 71 (1983), 609-642.

[13] S. Gersten, Conservative groups, indicability and a conjecture of Howie, J. Pure Appl. Alg. 29 (1983), 59-74.

[14] M. Gerstenhaber and O.S. Rothaus, The solution of sets of equations in groups, Proc. Nat. Acad. Sci. U.S.A. 48 (1962), 1531-1533.

[15] D. Gildenhuys, A generalization of Lyndon's Theorem on the cohomology of one-relator groups, Can. J. Math. 28 (1976), 473-480.

[16] F. Gonzalez-Acuña and H. Short, Knot surgery and primeness, Math. Proc. Comb. Phil. Soc. 99 (1986), 89-102.

[17] G.A. Gurevich, On the conjugacy problem for groups with one defining relator, Dokl. Akad. Nauk. SSSR 207, 1 (1972), 18-20 (Soviet Math. Dokl. 13 (1972), 1436-1439).

[18] J. Hempel, 3-manifolds, Ann. of Math. Studies 86, Princeton 1976.

[19] G. Higman, The units of group rings, Proc. London Math. Soc. 46 (1940), 231-248.

[20] _____, Units in group rings, D. Phil. Thesis, University of Oxford 1940.

[21] J. Howie, On pairs of 2-complexes and systems of equations over groups, J. Reine Angew. Math. 324 (1981), 165-174.

[22] _____, On locally indicable groups, Math. Z. 180 (1982), 445-461.

[23] _____, The solution of length three equations over groups, Proc. Edinburgh Math. Soc. 26 (1983), 89-96.

[24] _____, Some remarks on a problem of J.H.C. Whitehead, Topology 22 (1983), 475-485.

[25] _____, Spherical diagrams and equations over groups, Math. Proc. Camb. Phil. Soc. 96 (1984), 255-268.

[26] _____, Cohomology of one-relator products of locally indicable groups, J. London Math. Soc. 30 (1984), 419-430.

[27] _____, On the asphericity of ribbon disc complements, Trans. Amer. Math. Soc. 289 (1985), 281-302.

[28] J. Howie and S.J. Pride, Spelling theorems for staggered generalized 2-complexes, with applications, Invent. Math. 76 (1984), 55-74.

[29] J. Howie and H.R. Schneebeli, Homological and topological properties of locally indicable groups, Manuscripta Math. 44 (1983), 71-93.

[30] J. Howie and H. Short, The band sum problem, J. London Math. Soc. 31 (1985), 571-576.

[31] J. Huebschmann, Cohomology theory of aspherical groups and small cancellation groups, J. Pure Appl. Alg. 14 (1979), 137-143.

[32] L. Ya. Leifman and D. L. Johnson (eds.) The Kourovka Notebook. Unsolved problems in group theory, AMS Translations (2) 121, 1983.

[33] F. Levin, Solutions of equations over groups, Bull. Amer. Math. Soc. 68 (1962), 603-604.

[34] R. C. Lyndon, Cohomology theory of groups with a single defining relation, Ann. of Math. 52 (1950), 650-665.

[35] _____, On the Freiheitssatz, J. London Math. Soc. 5 (1972), 95-101.

[36] _____, Equations in groups, Bol. Soc. Bras. Math. 11 (1980), 79-102.

[37] R. C. Lyndon and P. E. Schupp, Combinatorial Group Theory, Springer-Verlag 1977.

[38] W. Magnus, Ueber diskontinuierliche Gruppen mit einer definierenden Relation (Der Freiheitssatz), J. Reine Angew. Math. 163 (1930), 141-165.

[39] A. I. Mal'cev, Algebraic Systems, Springer-Verlag 1973.

[40] J. McCool and P. E. Schupp, On one-relator groups and HNN extensions, J. Austral. Math. Soc. 16 (1973), 249-256.

[41] J. Milnor, On the three-dimensional Brieskorn manifolds M(p,q,r), Ann. of Math. Studies 84 (1975), 175-225.

[42] B. H. Neumann, Adjunction of elements to groups, J. London Math. Soc. 18 (1943), 4-11.

[43] _____, A note on algebraically closed groups, J. London Math. Soc. 27 (1952), 227-242.

[44] H. Neumann, Varieties of Groups, Springer-Verlag 1967.

[45] B. B. Newman, Some results on one-relator groups, Bull. Amer. Math. Soc. 74 (1968), 568-571.

[46] C. D. Papakyriakopoulos, On Dehn's Lemma and the asphericity of knots, Ann. of Math. 66 (1957), 1-26.

[47] D. Passman, The Algebraic Structure of Group Rings, Wiley 1977.

[48] S. J. Pride, One-relator quotients of free products, Math. Proc. Camb. Phil. Soc. 88 (1980), 233-243.

[49] O. S. Rothaus, On the nontriviality of some group extensions given by generators and relators, Ann. of Math. 106 (1977), 599-612.

[50] R. Sandling, Graham Higman's thesis "Units in group rings," Lect. Notes in Math. 882 (1981), 93-116.

[51] M. Scharlemann, 3-manifolds with $H_2(A,\delta A) = 0$ and a conjecture of Stallings, Lect. Notes in Math. 1144 (1985), 138-145.

[52] H. Schiek, Adjunktionsproblem und inkompressible Relationen I, Math. Ann. 146 (1962), 314-320; II, Math. Ann. 161 (1965), 163-170.

[53] _____, Das Adjunktionsproblem der Gruppentheorie, Math. Ann. 147 (1962), 158-165.

[54] H. K. Schuff, Ueber Wurzeln von Gruppenpolynomen, Math. Ann. 124 (1952), 294-297.

[55] P. E. Schupp, A strengthened Freiheitssatz, Math. Ann. 221 (1976), 73-80.

[56] P. Scott, Compact submanifolds of 3-manifolds, J. London Math. Soc. 7 (1973), 246-250.

[57] W. R. Scott, Algebraically closed groups, Proc. Amer. Math. Soc. 2 (1951), 118-121.

[58] H. Short, Topological methods in group theory: the adjunction problem, Ph.D. Thesis, University of Warwick 1984.

[59] J. Stallings, On torsion free groups with infinitely many ends, Ann. of Math. 88 (1968), 312-334.

[60] _____, Surfaces in three-manifolds and non-singular equations in groups, Math. Z. 184 (1983), 1-17.

[61] R. Strebel, Die Reihe der Derivierten von E-Grippen, Diss. 5148, ETH Zuerich 1973.

[62] _____, Homological methods applied to the derived series of groups, Comment. Math. Helv. 49 (1974), 302-332.

[63] R. G. Swan, Groups of cohomological dimension one, J. Alg. 12 (1969), 585-610.

[64] C. B. Thomas, Splitting theorems for certain PD^3-groups, Math. Z. 186 (1984), 201-209.

[65] L. van den Dries and A. J. Wilkie, Gromov's theorem on groups of polynomial growth and elementary logic, J. Alg. 89 (1984), 349-374.

[66] C. M. Weinbaum, On relators and diagrams for groups with a single defining relator, Illinois J. Math. 16 (1972), 308-322.

[67] J. H. C. Whitehead, On adding relations to homotopy groups, Ann. of Math. 42 (1941), 409-428.

[68] J. M. Whitehead, Projective modules and their trace ideals, Comm. Alg. 8 (1980), 1873-1901.

GRAPHICAL THEORY OF AUTOMORPHISMS OF FREE GROUPS

John R. Stallings*

Abstract. The equalizer of two monomorphisms of finitely generated free groups is finitely generated. This generalizes Gersten's Theorem that the fixed subgroup of an automorphism of a finitely generated free group is finitely generated. The proof is an exposition of Gersten's ideas, using graph-theory. Techniques are developed which may have other uses, "dyads," "folds," "ladder-attachments." The paper contains a historical sketch and a list of related unsolved problems, and also an appendix relating dyads to Heegaard diagrams of 3-manifolds.

Introduction

In the fall of 1982, S. M. Gersten [G3] proved his fixed-point theorem:

GERSTEN'S THEOREM. *If* $a : F \to F$ *is an automorphism of a finitely generated free group, then*

$$\text{Fix}(a) = \{w \in F | a(w) = w\}$$

is finitely generated.

This has incited a number of authors to offer their own variations and improvements (such as Cooper [Co], Goldstein and Turner [G-T], and Hoare [Hoa]). Indeed, I have my own variation, which I want to explain in this paper. Because of curious facts about mathematical publication, it is likely that many of Gersten's successors may get their works published before his paper is printed. However, much of the idea has been published already [G1, G2].

---

*Partly supported by NSF Grant DMS 83-03283

The history of this matter goes back at least to Jaco and Shalen [J-S], who proved that if α is induced by a surface homeomorphism, then Fix (α) has rank bounded by the rank of F, and to Dyer and Scott [D-S] who proved the same bound for periodic automorphisms. In [St1] I described a class of "PV" automorphisms which were not approached by either of these cases. Using methods tailored to each case, Squier [Sq] showed that several PV automorphisms have trivial fixed subgroup. Gersten [G1] developed a theory of CMT ("Change of Maximal Tree") automorphisms, which includes many PV examples, and proved that for a CMT $\alpha: F \to F$, the rank of Fix (α) is bounded by the rank of F; in the general case [G3] this bound was not proved.

The general study of automorphisms of free groups has been studied by many people, including Nielsen [Ni], Whitehead [Wh1, Wh2], Rapaport [Ra], and McCool [Mc]. A very recent development (Culler and Vogtmann [C-V] and Gersten [G4]) has discovered the homological fact that, for F finitely generated, Aut (F) contains a subgroup of finite index which has a finite Eilenberg-Mac Lane space.

There are many viewpoints from which to see the subject, such as geometry, analysis, combinatorics. The great advance of Whitehead, using a technique of manipulating surfaces in 3-manifolds, was explained to the more combinatorially minded by Rapaport, for instance. Gersten proved his fixed point theorem using graph theory; Cooper's proof uses analytic topology.

I prefer graph theory because of its logical simplicity and its adaptability to algorithmic methods, and because one can draw pictures of interesting examples.

The theorem I shall describe is this:

THEOREM. *Let* $\alpha, \beta : F \to A$ *be two injective homomorphisms, where* F *and* A *are free groups and* F *is finitely generated. Then their equalizer*

$$\mathrm{Eq}(\alpha, \beta) = \{w \in F \mid \alpha(w) = \beta(w)\}$$

is finitely generated.

This implies Gersten's result, by taking $A = F$ and β the identity. I wonder whether some simple argument which has eluded me can get this out of Gersten's Theorem. My proof is in fact a very close imitation of Gersten's. Here is a list of the main differences: My "dyads" are more restrictive than the Gersten analogue (Hoare [Hoa] has developed a similar concept, with extra conditions, something like my "DQI dyads"; he calls this the "W-graph"); I substitute the attachment of "ladders" for Gersten's "big fibre product"; I introduce the notion of "quasi-immersive" to deal with the possibility that α and β may not be surjective; and I describe the significant idea of "path-surgery" (Gersten's axiom G4) by formal lemmas, 5.6 and 5.11.

1. Unsolved problems

Before getting into the real mathematics, I would like to discuss some questions and conjectures.

P1. Let $\alpha: F \to F$ be an automorphism of a finitely generated free group. Is it always true that $\text{rank}(\text{Fix}(\alpha)) \leq \text{rank}(F)$?

This is so when α is CMT, periodic, or realizable by a surface autohomeomorphism, by the works previously cited. Gersten's method does provide a bound on $\text{rank}(\text{Fix}(\alpha))$ depending on the nature of α, but this can be arbitrarily large for a fixed F.

P2. A PV automorphism $\alpha: F \to F$ is one whose abelianization is such that all its eigenvalues but one have absolute value less than 1. If the rank of F is at least 3, then it is easy to see that $\text{Fix}(\alpha) \subset F_3$, the third term in the lower central series. Is it always true that $\text{Fix}(\alpha) = \{1\}$?

It is possible to find automorphisms β (e.g., a free product of certain inner automorphisms), such that $\{1\} \neq \text{Fix}(\beta) \subset F_3$, but Marc Culler pointed out to me that this only seems to happen in very special circumstances. The general questions here are: What subgroups S of F can be of the form $\text{Fix}(\beta)$? If $\text{Fix}(\beta)$ has some special property, such as $\text{Fix}(\beta) \subset F_3$, what does this imply about β?

P3. The set of all automorphisms of all free groups of finite rank with generators in the list $\{x_1, x_2, \cdots\}$ is effectively enumerable. Thus one can ask: Is $\text{rank}(\text{Fix}(\alpha))$ a recursive function of α? Because of Gersten's Theorem (and this problem was suggested by its proof), this depends on a question like this: Let $|S|$ be the sum of the lengths of a minimal generating set of S. Is $|\text{Fix}(\alpha)|$ bounded by a recursive function of α? I suspect the answer to these questions may be "No," because there are relatively simple examples of α for which $|\text{Fix}(\alpha)|$ is unreasonably large (see section 8).

Another piece of evidence that the answer to these questions may be "No" is the recursive unsolvability of Post's "correspondence decision problem" [Po]. This problem is: Given free, finitely generated semigroups P and Q, to decide for pairs of homomorphisms $\alpha, \beta : P \to Q$, whether the equalizer $\text{Eq}(\alpha, \beta)$ is trivial or not. Can this undecidability theorem be extended to the world of free *groups*?

P4. Let $\{a_1, a_2, \cdots\}$ be an infinite set of automorphisms of the finitely generated free group F. Is the intersection

$$\bigcap \text{Fix}(a_n)$$

finitely generated?

For a finitely generated group of automorphisms this follows from Gersten's and Howson's [How] Theorems. One might conjecture that this intersection has rank bounded by $\text{rank}(F)$. This question derives from that originally asked by Dyer and Scott.

P5. Questions analogous to the above can also be asked about the equalizers of injective homomorphisms.

P6. If F is finitely generated and free, then the direct product $F \times F$ has many strange subgroups. For examples, see Mihailova [Mi] and

Lyndon-Schupp [L-S], p. 193. Questions about the fixed subgroup of α and about $\text{Eq}(\alpha,\beta)$ can be translated into questions about the intersection of certain subgroups of $F \times F$ with the diagonal subgroup $\Delta(F)$. One conjecture is:

If A and B are finitely *presented* subgroups of $F \times F$, then is $A \cap B$ finitely *generated*? In particular, if A is finitely presented, then is $A \cap \Delta(F)$ finitely generated? If the answer is "Yes," then if we only assume that α is injective and make no assumption on β, it would follow that $\text{Eq}(\alpha,\beta)$ is finitely generated.

2. Résumé of graph-theory

2.1. It is convenient to have a category of graphs somewhat more flexible than that of Serre-Bass [Se], so that it allows collapsing of edges. Gersten describes this as follows:

A *graph* $\Gamma = \{X, \iota,{}^{-1}\}$ consists of a set X and two functions $X \to X$ satisfying the two axioms:

$$\forall x \in X, \quad \iota(\iota(x)) = \iota(x) \text{ and } (x^{-1})^{-1} = x .$$
$$\forall x \in X, \quad \iota(x) = x \text{ if and only if } x^{-1} = x .$$

A *map of graphs* $f : \Gamma \to \Delta$ is then a function compatible with this structure.

We define the *vertices* (= degenerate edges):

$$V(\Gamma) = \{x \in X \mid \iota(x) = x\}$$

and *non-degenerate edges*:

$$E(\Gamma) = X - V(\Gamma) .$$

A map $f : \Gamma \to \Delta$ is *non-degenerate* if $f(E(\Gamma)) \subset E(\Delta)$.

We think, for $x \in X$, of x^{-1} as the *reverse* of x, and $\iota(x)$ as the *initial vertex* of x; and we define the *terminal vertex* of x as $\tau(x) = \iota(x^{-1})$.

The *degenerate set* of a map $f:\Gamma \to \Delta$ is

$$Df = f^{-1}(V(\Delta)).$$

This is a subgraph which contains $V(\Gamma)$.

2.2. We can define *paths*, *fundamental groupoid* $\pi(\Gamma)$, and *fundamental group* based at $v_0 \in V(\Gamma)$, $\pi_1(\Gamma, v_0)$, without any difficulty.

Thus, a *path* p in Γ is an n-tuple $p = x_1 x_2 \cdots x_n$ of elements $x_i \in X$ satisfying

$$\forall\, i \in [1, n-1], \tau(x_i) = \iota(x_{i+1}).$$

We define the *initial* and *terminal* vertices of p by

$$\iota(p) = \iota(x_1) \text{ and } \tau(p) = \tau(x_n)$$

and the *length* $|p| = n$. [A path of length 0 is supposed to determine a single vertex.]

We call $p = x_1 \cdots x_n$ *non-degenerate* if

$$\forall\, i,\, x_i \in E(\Gamma).$$

If p is an arbitrary path, by striking out degenerate edges, we get a non-degenerate path that will be denoted $\#(p)$. By striking out, in any order, "round-trips" xx^{-1}, in $\#(p)$, we obtain a unique path denoted $\rho(p)$ which is *reduced*, that is, contains no round-trips and no degenerate edges.

The set of all paths q having the reduced path $\rho(p)$ is written $[p]$ and called the *homotopy-class* of p. Then $\pi(\Gamma)$ is made up of homotopy-classes of paths, and $\pi_1(\Gamma, v_0)$ consists of those homotopy-classes starting and ending at v_0.

2.3. The notions of *immersion* and *fold* are essential [St2]. An *immersion* $f:\Gamma \to \Delta$ is a *non-degenerate* map such that, for all $v \in V(\Gamma)$, the set

$$\text{St}_\Gamma(v) = \{e \in E(\Gamma) | \iota(e) = v\}$$

is mapped by f *injectively* into $\text{St}_\Delta(f(v))$.

A *fold* under a map f is a pair of edges $e_1, e_2 \in E(\Gamma)$ such that $e_1 \neq e_2$, $\iota(e_1) = \iota(e_2)$, $f(e_1) = f(e_2)$. If there is such a fold, then we can construct a graph

$$\Gamma' = \Gamma/(e_1 = e_2)$$

in which $\tau(e_1)$ is identified to $\tau(e_2)$, e_1 to e_2, and e_1^{-1} to e_2^{-1}. The map f then factors through the *fold map*

$$\Gamma \to \Gamma'$$

and a map $f' : \Gamma' \to \Delta$.

Three important properties are

1. An immersion $f : \Gamma \to \Delta$ is injective on π_1.
2. A fold map $\epsilon : \Gamma \to \Gamma/(e_1 = e_2)$ is both surjective itself, and is surjective on π_1.
3. A non-degenerate map f is an immersion if and only if it admits no fold.

2.4. Here are some technical definitions about paths. We say of paths p and q in Γ, that p is a *left segment* of q, and write $p \leq q$, if:

$$\iota(p) = \iota(q); \; p = x_1 \cdots x_n, \; q = y_1 \cdots y_k; \; n \leq k; \text{ and } \forall i \in [1,n], \; x_i = y_i.$$

If p and q are paths with the same initial vertex, we define $p \wedge q$ to be the unique *maximal common left segment* of p and q.

Note these facts:

If $f : \Gamma \to \Delta$ is a map of graphs and p and q are paths in Γ with the same initial vertex, then

$$f(p \wedge q) \leq f(p) \wedge f(q).$$

If, additionally, f is an immersion, then $f(p \wedge q) = f(p) \wedge f(q)$.
In any case, $p \leq q$ implies $\#(p) \leq \#(q)$.

3. Dyads

3.1. Suppose $f : F \to A$, $g : F \to B$ are homomorphisms, where F, A, B are free groups and F is finitely generated. We can describe this, in terms of maps of graphs, in the following way.

Choose bases of these free groups. Let Γ and Δ be 1-vertex graphs with

$$\pi_1(\Gamma) = A \text{ and } \pi_1(\Delta) = B,$$

the edges of Γ and Δ corresponding to the chosen bases of A and B.

Let x be a basis element of F. Then $f(x)$ and $g(x)$ are words in the bases of A and B. Let C_x be a basepointed circle subdivided into two arcs E_{Ax} and E_{Bx}, which are further subdivided into as many edges as needed. Then $f(x)$ gives a recipe for mapping E_{Ax} into Γ, and $g(x)$ a recipe for mapping E_{Bx} into Δ. We define

$$\alpha_x : C_x \to \Gamma, \ \beta_x : C_x \to \Delta$$

so that α_x does $f(x)$ on E_{Ax} and is degenerate on E_{Bx}, and β_x is degenerate on E_{Ax} and does $g(x)$ on E_{Bx}.

Taking the union of the C_x along their basepoints, we obtain a graph Θ, and maps (the unions of $\{\alpha_x\}$ and $\{\beta_x\}$)

$$\alpha : \Theta \to \Gamma, \ \beta : \Theta \to \Delta.$$

We can identify $\pi_1(\Theta, \text{basepoint})$ with F, and see that α and β induce the homomorphisms on π_1 that are f and g, respectively.

3.2. Now, (α, β) may not be as simple or convenient as possible. One obvious point is that, for example, α may not be an immersion on the degenerate set $D\beta$, consisting of the arcs E_{Ax} and vertices. We can then factor α through a series of folds and thus obtain a map $\alpha' : \Theta' \to \Gamma$, that is an immersion on the image of $D\beta$; the identifications only occur in $D\beta$ and so β induces a map $\beta' : \Theta' \to \Delta$ as well.

3.3. This discussion leads us to some definitions:

Let Γ and Δ be 1-vertex graphs. A (Γ,Δ)-*dyad* consists of a connected, finite graph Θ and a pair of maps $\alpha : \Theta \to \Gamma$, $\beta : \Theta \to \Delta$ such that $D\alpha$ and $D\beta$ are complementary, that is:

$$\Theta = D\alpha \cup D\beta$$

$$D\alpha \cap D\beta = V(\Theta).$$

The *minor complexity* of the dyad (α, β, Θ) is defined to be the number of edges of θ.

A *fold* of a dyad (α, β, Θ) consists of a dyad $(\alpha', \beta', \Theta')$ and a graph-map $\phi : \Theta \to \Theta'$ which is an identification of a pair of edges e_1, e_2 having the same initial vertex and having the properties that $\alpha(e_1) = \alpha(e_2)$ and $\beta(e_1) = \beta(e_2)$. Then α' and β' are the unique graph-maps such that $\alpha'\phi = \alpha$ and $\beta'\phi = \beta$.

We call (α, β, Θ) *degenerate-immersive* or "DI," if no fold is possible; that is, if $\alpha|D\beta$ and $\beta|D\alpha$ are immersions.

A fold reduces the minor complexity and is surjective on π_1. Thus we obtain:

3.4. LEMMA. *Let (α, β, Θ) be a (Γ,Δ)-dyad. Then there is a (Γ,Δ)-dyad $(\alpha', \beta', \Theta')$ and a graph-map $\phi : \Theta \to \Theta'$ such that*

(i) $\alpha'\phi = \alpha$, $\beta'\phi = \beta$. *--In other words, ϕ is a map of dyads.*

(ii) *ϕ is the composition of a finite number of folds.*

(iii) *$(\alpha', \beta', \Theta')$ is degenerate-immersive.*

(And it follows from (ii) that:)

(iv) *ϕ is surjective.*

(v) *$\phi_* : \pi_1(\Theta) \to \pi_1(\Theta')$ is surjective.*

4. *Quasi-immersion*

4.1. A (Γ,Δ)-dyad (α, β, Θ) is said to be symmetric to the (Δ,Γ)-dyad (β, α, Θ). In what follows, all definitions and theorems are to be interpreted symmetrically. For example, "quasi-fold" is defined in terms of the ordering (α, β) and is supposed to include the symmetric possibility.

The *major complexity* of a (Γ, Δ)-dyad is the sum of the number of components of $D\alpha$ and the number of components of $D\beta$.

4.2. PROPOSITION. *A fold of a dyad does not increase the major complexity.*

Proof. What a fold does is to perform a fold on one of $D\alpha$, $D\beta$ and to identify a pair of vertices, which may already be equal, of the other. If $D\alpha$ is folded, its number of components stays the same, and the number of components of $D\beta$ stays the same or decreases by one. □

4.3. A *quasi-fold* of a dyad (α, β, Θ) is the following (or its symmetric counterpart): A triple (e_1, e_2, p) where e_1 and e_2 are non-degenerate edges of $D\beta$ such that $\alpha(e_1) = \alpha(e_2)$; where p is a non-degenerate path in $D\alpha$ from $\iota(e_1)$ to $\iota(e_2)$; such that $\tau(e_1)$ and $\tau(e_2)$ are in different components of $D\alpha$.

A dyad is said to be *quasi-immersive* or "QI" if it has no quasi-folds. If a dyad is both DI and QI we call it "DQI."

4.4. A *ladder* $L = L(a,p)$ in (Γ, Δ) is the following, or its symmetric counterpart:

L is a (Γ,Δ)-dyad; a is a non-degenerate edge of Γ, and p is a non-degenerate path in Δ of length $n > 0$.

The underlying graph of L consists of two disjoint arcs A_1, A_2 subdivided into $n = |p|$ edges; these are the *sides* of the ladder; and of n+1 edges $\varepsilon_0, \varepsilon_1, \cdots, \varepsilon_n$ connecting corresponding vertices in A_1 and A_2, with initial vertices in A_1; these are the *rungs* of the ladder.

Here are the two graph maps $\alpha, \beta : L \to \Gamma, \Delta$. α is degenerate on both A_1 and A_2 and maps each ε_i to a. β maps each of A_1 and A_2 by the path p and is degenerate on the ε_i. We can describe this pictorially (Figure 1). To describe a dyad in pictures, we draw a picture of a directed graph, and label each edge with a pair of labels, the first describing the map α, the second describing the map β; the symbol 0 denotes the unique vertex of Γ or Δ; an edge labelled "c0" then belongs to $D\beta$ and an edge labelled "0z" belongs to $D\alpha$.

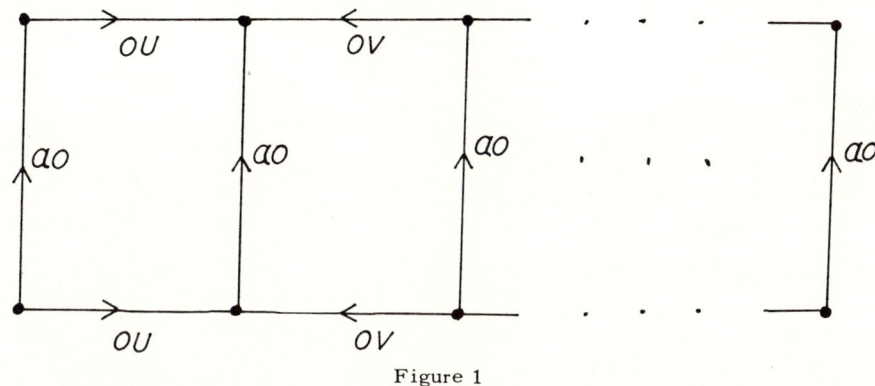

Figure 1

Note that both α and β map $\pi_1(L)$, with any basepoint, to $\{1\}$, that is, trivially.

4.5. If the (Γ, Δ)-dyad (α, β, Θ) admits a quasi-fold (e_1, e_2, p), we construct the ladder $L(\alpha(e_1), \beta(p))$ and *attach* it to (α, β, Θ), by identifying the side A_1 to Θ along the path p and identifying the extreme rungs ε_0 and ε_n to e_1 and e_2 respectively. This results in a new dyad $(\alpha', \beta', \Theta')$. We can imagine this attachment done in stages; first identify the corner vertex $\iota(\varepsilon_0)$ to $\iota(p)$; then fold, in succession, the edges of A_1 onto the edges of p; and then fold the extreme rungs onto e_1 and e_2.

4.6. PROPOSITION. *The attachment of a ladder L to a dyad (α, β, Θ) along a quasi-fold reduces the major complexity.*

Proof. $D\beta$ changes by adding the rungs of the ladder and the vertices of A_2; since one vertex of each rung is in $D\beta$ to begin with, the number of components of $D\beta$ does not change. But $D\alpha$ is changed by joining the points $\tau(e_1)$ and $\tau(e_2)$ by the arc A_2, thus reducing the number of components of $D\alpha$ by 1. □

4.7. LEMMA. *If (α, β, Θ) is any (Γ, Δ)-dyad, then there is a (Γ, Δ)-dyad $(\alpha', \beta', \Theta')$ and a map*

$$\phi : \Theta \to \Theta'$$

such that
1. *$\alpha'\phi = \alpha$, $\beta'\phi = \beta$.*
2. *ϕ is the composition of a finite number of folds and ladder attachments.*
3. *$(\alpha', \beta', \Theta')$ is DQI.*

Proof. If the dyad is not QI, a quasi-fold may involve a path p of length 0; in this case, the dyad may be folded to reduce the major complexity; if the path p has length greater than 0, a ladder may be attached to reduce the major complexity. We can then reduce the minor complexity by folding, and this does not, by 4.2, increase the major complexity. Thus, a double induction establishes that we eventually reach the point that there are neither quasi-folds nor folds. □

5. The common-kernel condition

5.1. PROPOSITION. *Let $\phi : T \to T'$, $\alpha : T \to C$, $\beta : T \to D$, $\alpha' : T' \to C$, $\beta' : T' \to D$ be group-homomorphisms such that $\alpha = \alpha'\phi$, $\beta = \beta'\phi$. Suppose $\ker \alpha = \ker \beta$. Then*
   (i) *If ϕ is surjective, then $\ker \alpha' = \ker \beta'$.*
   (ii) *If $T' = T * K$, a free product, $\phi : T \to T * K$ is the inclusion, and $\alpha'(K) = \{1\} = \beta'(K)$, then $\ker \alpha' = \ker \beta'$.*

Proof. Easy. □

5.2. Let (α, β, Θ) be a (Γ, Δ)-dyad. We say that this dyad satisfies the *common-kernel* condition if $\ker \alpha = \ker \beta$ on π_1 with some choice of basepoint in Θ. Since Θ is connected, this is independent of the choice of basepoint.

5.3. LEMMA. *If (α, β, Θ) is a (Γ, Δ)-dyad with common kernel, then (by 4.7) there is a DQI-dyad $(\alpha', \beta', \Theta')$ and a map of dyads $\phi : \Theta \to \Theta'$ which is the composition of a finite series of folds and ladder-attachments. The end result $(\alpha', \beta', \Theta')$ then satisfies the common-kernel condition.*

Proof. A fold is surjective on π_1, and thus by 5.1 (i) preserves the common-kernel condition. A ladder-attachment can be done, by 4.5, in two stages; the first, wedging a ladder on at a corner, preserves the common-kernel condition by 5.1 (ii); the second is a series of folds. □

5.4. PROPOSITION. *If (α, β, Θ) is a DI (Γ, Δ)-dyad with common kernel, then $D\alpha$ and $D\beta$ are forests.*

Proof. Let γ be a reduced closed path in $D\alpha$. Since $\alpha(\gamma)$ is a completely degenerate path, $[\alpha(\gamma)] = 1$ in $\pi_1(\Gamma)$. By the DI condition, $\beta|D\alpha$ is an immersion, and so $\beta(\gamma)$ is a reduced path in Δ; by the common kernel assumption, $[\beta(\gamma)] = 1$ in $\pi_1(\Delta)$. Therefore $\beta(\gamma)$ has length 0, and so γ has length 0. □

5.5. Let (α, β, Θ) be a DQI (Γ, Δ)-dyad with common kernel. We call a pair of edges e_1, e_2 of Θ *parallel* if the following conditions, or the symmetric analogue, hold:

e_1 and e_2 belong to $D\beta$.
$\alpha(e_1) = \alpha(e_2)$.
$\iota(e_1)$ and $\iota(e_2)$ belong to the same component of $D\alpha$.

5.6. PROPOSITION. *Let (α, β, Θ) be a DQI (Γ, Δ)-dyad with common kernel. Suppose that e_1 and e_2 are parallel edges of Θ. Then there is a ladder $L(\varepsilon, \beta(p))$ and a map of $L(\varepsilon, \beta(p))$ into (α, β, Θ), such that e_1 and e_2 are the images of the extreme rungs of the ladder. The reduced path p in $D\alpha$ is uniquely determined.*

Proof. Suppose $e_1, e_2 \in D\beta$, and $a(e_1) = a(e_2) = \varepsilon$; there is, by 5.4 and the definition of parallel, a unique reduced path p from $\iota(e_1)$ to $\iota(e_2)$ in Da. Since the dyad is QI, there is also a unique reduced path q from $\tau(e_1)$ to $\tau(e_2)$ in Da. Then $[a(e_1 q e_2^{-1} p^{-1})] = [\varepsilon \varepsilon^{-1}] = 1$. By the common kernel condition, $[\beta(e_1 q e_2^{-1} p^{-1})] = [\beta(q)\beta(p)^{-1}] = 1$. By DI, $\beta(q)$ and $\beta(p)$ are reduced paths in Δ; hence they are equal. In particular, q and p have the same length. Let $q = f_1 q'$ and $p = f_2 p'$, where f_1 and f_2 are edges of Da. We have $\beta(f_1) = \beta(f_2)$ and $\iota(f_1)$ is joined to $\iota(f_2)$ by e_1^{-1} in $D\beta$. Therefore, by the symmetric version of the above argument, $\tau(f_1)$ is joined to $\tau(f_2)$ by an edge e_3^{-1} in $D\beta$ such that $a(e_3) = a(e_1)$. By induction on $|p|$, we see that e_3 and e_2 are extreme rungs of a ladder mapped into Θ; by adjoining the extra square $\{e_1, e_3, f_1, f_2\}$ we get e_1 and e_2 to be the extreme rungs of the desired ladder. □

5.7. REMARK. The above, 5.6, if the form of Gersten's "path-surgery" axiom G4 which is appropriate in this situation.

5.8. In a (Γ, Δ)-dyad (a, β, Θ), a non-degenerate path p is called (a, β)-reduced if $\#a(p)$ and $\#\beta(p)$ are reduced paths in Γ and Δ. (Recall that $\#q$ is got from q by striking out degenerate edges.)

Pseudo-homotopy is the equivalence relation $\simeq$ on paths in Θ generated by two operations:

(a) Homotopy: $pee^{-1}q \simeq pq$, and $pvq \simeq pq$ if v is a vertex.

(b) Replacing a segment joining diagonal points of a ladder by the other segment; that is:

$$pefq \simeq pf'e'q$$

when $f'e'f^{-1}e^{-1}$ is a closed path, $e, e' \in D\beta$, $f, f' \in Da$, $a(e) = a(e')$, $\beta(f) = \beta(f')$.

5.9. PROPOSITION. *If* $p \simeq q$, *then* $[a(p)] = [a(q)]$, $[\beta(p)] = [\beta(q)]$, *and p and q have the same endpoints.*

Proof. Elementary. □

5.10. PROPOSITION. *Let (α,β,Θ) be a DQI (Γ,Δ)-dyad with common kernel. For every path p in Θ, there exists an (α,β)-reduced path p' such that $p \simeq p'$; and in this case*

$$|p'| = |[\alpha(p)]| + |[\beta(p)]| .$$

Proof. Choose $p' \simeq p$ so that $|p'|$ is minimal. If p' were not (α,β)-reduced, it would have a segment s of the following kind or its symmetry:

$$p' = p_1 s p_2 , \quad s = e_1^{-1} p_3 e_2$$

where e_1, e_2 are parallel edges in $D\beta$, joined by a path p_3 in $D\alpha$. Since $|p'|$ is minimal, p_3 is a reduced path. By 5.6, there is a ladder in Θ connecting these edges, and so there is a path p_4 in $D\alpha$ from $\tau(e_1)$ to $\tau(e_2)$ such that $\beta(p_3) = \beta(p_4)$. Then:

$$p'' = p_1 p_4 p_2 \simeq p'$$

and p'' would have shorter length. □

5.11. THEOREM. *Let (α,β,Θ) be a DQI (Γ,Δ)-dyad with common kernel. Let p and q be (α,β)-reduced paths in Θ having the same initial vertex. Consider paths p', q' which are (α,β)-reduced, such that $p \wedge q \leq p' \wedge q'$ and $p' \simeq p$ and $q' \simeq q$. Of these, choose a pair p', q' such that $|p' \wedge q'|$ is maximal. Then*

$$\#\alpha(p' \wedge q') = \#\alpha(p') \wedge \#\alpha(q') ,$$

and

$$\#\beta(p' \wedge q') = \#\beta(p') \wedge \#\beta(q') .$$

Proof. Note that $|p' \wedge q'|$ is bounded by $|p'| = |[\alpha(p)]| + |[\beta(p)]|$. So such a pair with $|p' \wedge q'|$ maximal exists. Let $r = p' \wedge q'$. Clearly

$$\#\alpha(r) \leq \#\alpha(p') \wedge \#\alpha(q') .$$

Suppose $\#a(r) < \#a(p') \wedge \#a(q')$. Let e_1 be the first edge of $D\beta$ in p' after r, and e_2 the first edge of $D\beta$ in q' after r. The situation is that

$$p' = rs_1 e_1 t_1, \quad q' = rs_2 e_2 t_2$$

where s_1 and s_2 are paths in $D\alpha$, and $a(e_1) = a(e_2)$. The picture is:

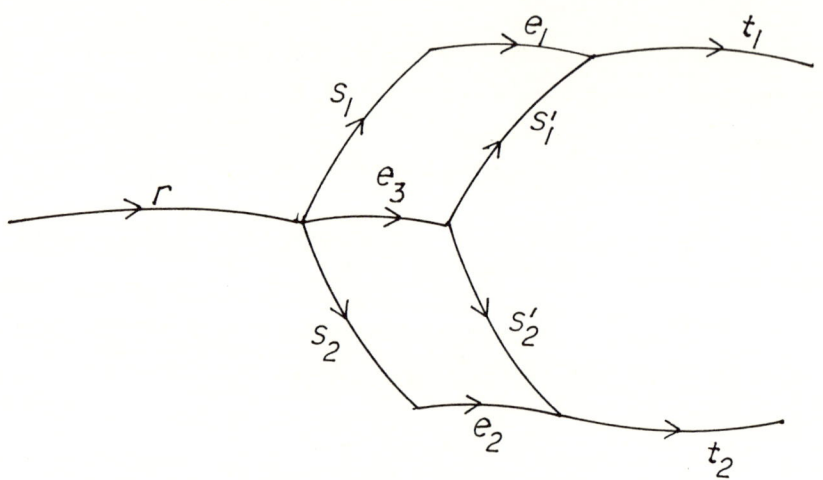

Figure 2

By 5.6, e_1 and e_2 are joined by a ladder $L(a(e_1), \beta(s_1^{-1} s_2))$ in (a, β, Θ). Thus, as in the picture, there is an edge e_3 in $D\beta$ with $a(e_3) = a(e_1)$, and paths s_1', s_2' in $D\alpha$ with $\beta(s_1) = \beta(s_1')$ and $\beta(s_2) = \beta(s_2')$. Then

$$p'' = re_3 s_1' t_1, \quad \text{and} \quad q'' = re_3 s_2' t_2$$

are pseudo-homotopic to p' and q', but they have $re_3 \leqq p'' \wedge q''$ and so $|p' \wedge q'|$ was not maximal. □

5.12. COROLLARY. *Let (a, β, Θ) be a $\Gamma, \Delta)$-dyad which is DQI with common kernel. Let p and q be paths in Θ such that p and q have the same initial vertex and such that $[a(p)] = [a(q)]$, $[\beta(p)] = [\beta(q)]$. Then $p \simeq q$. In particular, p and q have the same terminal vertex.*

Proof. Apply 5.11 to find (α,β)-reduced p´, q´ such that p $\simeq$ p´, and q $\simeq$ q´, and the conclusions of 5.11 hold. Since p´, q´ are (α,β)-reduced the paths #a(p´), etc., are reduced paths in Γ and Δ. Thus the assumption that the homotopy classes of the images of p and q under α and β are equal implies

$$\#\alpha(p´\wedge q´) = \#\alpha(p´) = \#\alpha(q´)$$

$$\#\beta(p´\wedge q´) = \#\beta(p´) = \#\beta(q´) .$$

We can thus compute the lengths of p´, q´, and p´$\wedge$q´, and find that they are all the same. Thus p´$\wedge$q´ = p´ = q´. □

5.13. REMARK. This implies that the kernel of the homomorphism

$$\alpha_* \times \beta_* : \pi_1(\Theta) \to \pi_1(\Gamma) \times \pi_1(\Delta)$$

is normally generated by ladders. This is not surprising, since this image is isomorphic to the image of α, which is finitely generated and free, and thus finitely related. The "relations" are given by ladder-squares. The hope that this could be arranged only under the assumption that the image of $\alpha_* \times \beta_*$ is finitely related (without assuming the common kernel condition) is the basis for problem P6.

6. *Coincidences*

We shall now discuss the case of (Γ,Γ)-dyads. That is, $\Delta = \Gamma$. In this case, we can compare $\alpha(p)$ and $\beta(p)$ and therefore discuss the equalizer on π_1.

6.1. Let $\alpha : T \to C$, $\beta : T \to C$ be homomorphisms of groups. We define the *image of the equalizer*:

$$\text{IEq}(\alpha,\beta) = \{x \in C | \exists y \in T \ni \alpha(y) = \beta(y) = x\} .$$

6.2. PROPOSITION. Let $\phi : T \to T´$, $\alpha : T \to C$, $\beta : T \to C$, $\alpha´ : T´ \to C$, $\beta´ : T´ \to C$ be group-homomorphisms such that $\alpha = \alpha´\phi$, $\beta = \beta´\phi$. Then:

(i) *If ϕ is surjective, then* $\mathrm{IEq}(\alpha',\beta') = \mathrm{IEq}(\alpha,\beta)$.

(ii) *If* $T' = T * K$, $\phi : T \to T * K$ *is the inclusion, and* $\alpha'(K) = \{1\} = \beta'(K)$, *then* $\mathrm{IEq}(\alpha',\beta') = \mathrm{IEQ}(\alpha,\beta)$.

Proof. Easy. □

6.3. LEMMA. *If* (α,β,Θ) *is a* (Γ,Γ)-*dyad with common kernel, then there is a DQI-dyad* (α',β',Θ') *and a map of dyads* $\phi : \Theta \to \Theta'$ *which is the composition of a finite series of folds and ladder-attachments. The end result* (α',β',Θ') *satisfies the common-kernel condition, and*

$$\mathrm{IEq}(\alpha'_*, \beta'_*) = \mathrm{IEq}(\alpha_*, \beta_*) .$$

[This is to be interpreted by choosing a basepoint in Θ whose image in Θ' is to be the basepoint of Θ'. Then α_*, etc., are the homomorphisms on π_1 relative to these basepoints.]

Proof. This is a restatement of 5.3 with the additional conclusion about IEq, which follows from 6.2 since folds and ladder-attachments satisfy on π_1 the conditions of 6.2 (i) and (ii). □

6.4. Suppose that (α,β,Θ) is a DQI (Γ,Γ)-dyad with common kernel.

An (α,β)-reduced path p in Θ will be called *invariant* (or a *coincidence*) if
$$\#\alpha(p) = \#\beta(p) .$$

Let e be an edge of Θ. There may or may not be an invariant path p whose first edge is e. [REMARK: As Gersten shows, there is an efficient and direct mechanical procedure to find such a path p if it exists. This procedure may not terminate if such a path does not exist.] If there is an invariant path p whose first edge is e, then out of all such, choose one whose length is minimal, and call it m(e). The plan now is to invent a graph made up of $\{m(e)\}$ which determines, up to pseudo-homotopy, all the invariant paths.

6.5. Let (α, β, Θ) be a DQI (Γ, Γ')-dyad with common kernel, and choose, as above, for each edge e of Θ for which it is possible, a minimal invariant path m(e). Define the *coincidence graph and map*, (Ξ, σ), as follows:

First, we take disjoint arcs $\{A(e)\}$, where A(e) is subdivided into $|m(e)|$ edges, and the set of vertices $V(\Theta)$; identify the initial and terminal vertices of A(e) to $\iota(m(e))$ and $\tau(m(e))$, respectively. This is the graph Ξ.

Then, we define $\sigma: \Xi \to \Theta$ as follows: On $V(\Theta)$, let σ be the identity. On A(e), let σ define the path m(e).

6.6. THEOREM. *Let (α, β, Θ) be a DQI (Γ, Γ')-dyad with common kernel, and let a choice of minimal invariant paths $\{m(e)\}$ be made, with which the coincidence graph and map (Ξ, σ) are formed. Then*:

(i) *For every path p of Ξ whose initial and terminal vertices are in $V(\Theta)$*,
$$[\alpha(\sigma(p))] = [\beta(\sigma(p))] .$$

(ii) *For every path q of Θ, if $[\alpha(q)] = [\beta(q)]$, then there exists a path p of Ξ whose endpoints are those of q in $V(\Theta)$, such that*

$$\#\sigma(p) \simeq \#q .$$

Proof. (i) Since p has its endpoints in $V(\Theta)$, it is homotopic to a product of paths, each crossing some A(e) once, and so $\sigma(p)$ is a product of paths of the form m(e) or $m(e)^{-1}$. Since the latter are invariant, (i) follows.

(ii) We can suppose, by 5.10, that q is non-degenerate, and is (α, β)-reduced. The proof is then by induction on $|q|$, being clear if $|q| = 0$. Let e be the first edge of q. Apply 5.11 to the pair of paths m(e), q. We obtain paths m(e)', q'; these both start with e since

$$e \leqq m(e) \wedge q \leqq m(e)' \wedge q' .$$

We have $m(e) \simeq m(e)'$, $q \simeq q'$; all four of these are (α,β)-reduced; and the $\wedge$-equalities in the conclusion of 5.11 hold. Because $m(e)'$ and q' are invariant, we can deduce the single equality:

$$\#\alpha(m(e)' \wedge q') = \#\beta(m(e)' \wedge q') .$$

That is, $m(e)' \wedge q'$ is invariant and starts with e. Since $|m(e)'| = |m(e)|$ is minimal among such invariant paths, we conclude that $m(e)' \wedge q' = m(e)'$. Thus,

$$q' = m(e)'q'' \simeq m(e)q''$$

and, clearly, q'' is an invariant path itself, with $|q''| < |q'| = |q|$. Thus, inductively, $q'' \simeq \sigma(p'')$, and then $m(e)q''$ is pseudo-homotopic to the σ-image of p'' preceded by a path crossing $A(e)$. □

7. The Main Theorem

7.1. THEOREM. *Let* F, C *be free groups, with* F *finitely generated. Let* $\alpha_*, \beta_* : F \to C$ *be homomorphisms with* $\ker \alpha_* = \ker \beta_*$. *Then* $\mathrm{IEq}(\alpha_*, \beta_*)$ *is finitely generated.*

Proof. The discussion in 3.1 shows that we can realize α_*, β_* as the induced π_1 maps in a (Γ,Γ)-dyad (α, β, Θ). We can perform folds and ladder attachments, reducing the major and minor complexities. We start out with the common-kernel hypothesis, and this is preserved, and IEq is left unchanged, by 6.3. Eventually then, we end up with a DQI-dyad $(\alpha', \beta', \Theta')$. We started with Θ finite, since F is supposed to be finitely generated, and therefore, after a finite number of improvements, Θ' is still finite. We now construct, by a choice of minimal invariant paths, a coincidence graph $\sigma : \Xi \to \Theta'$, by 6.5. We see that Ξ is the union of a finite number of vertices and finitely subdivided arcs, and so is finite. Now, by 6.6, we can conclude that $\mathrm{IEq}(\alpha'_*, \beta'_*)$, which is the same as $\mathrm{IEq}(\alpha_*, \beta_*)$, is exactly

$$\alpha_* \sigma_* \pi_1(\Xi) [= \beta_* \sigma_* \pi_1(\Xi)] .$$

This is finitely generated because Ξ is finite. □

[This proof requires keeping track of the basepoint. It would seem that the finitely many subgroups of C, up to conjugacy, described by the components of Ξ have some undetermined significance.]

7.2. COROLLARY. *If F is a finitely generated free group, and C is a free group, and $\alpha,\beta: F \to C$ are monomorphisms, then $\mathrm{Eq}(\alpha,\beta)$ is finitely generated.*

Proof. Note that $\ker \alpha = \ker \beta = \{1\}$, and $\mathrm{IEq}(\alpha,\beta) = \alpha(\mathrm{Eq}(\alpha,\beta))$ is isomorphic to $\mathrm{Eq}(\alpha,\beta)$. □

7.3. REMARK. 7.2 implies 7.1 by an easy argument, left to the reader. But 7.1 is the form of the result which is proved by this graph-theory. The monomorphism condition deteriorates into the common kernel condition because of the need to attach ladders.

8. *Two examples*

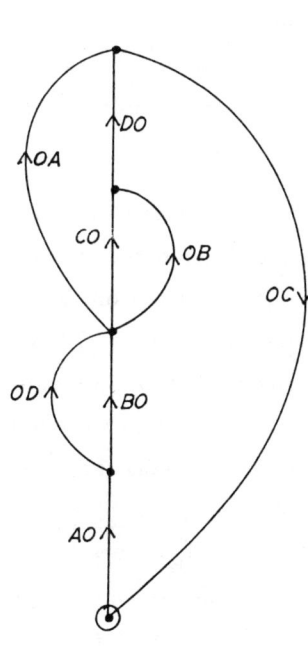

Figure 3

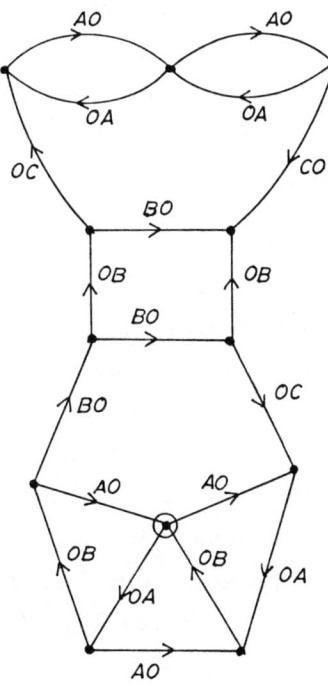

Figure 4

These are DQI (Γ,Γ')-dyads, with the pictorial description discussed in 4.4. The circled vertices are the basepoints. Figure 4 is a general case, in which various types of ladders can be made out. Figure 3 is a picture of one of Gersten's "CMT" automorphisms. What is notable about Figure 3 is that there is a minimal invariant path starting at the basepoint (it happens to be a closed path); it has length 100. Thus, long after you want to quit, if you keep at it, you may find that a minimal invariant path exists.

9. Appendix: Heegaard diagrams

9.1. The notions of "dyad," "fold," and "degenerate-immersive" can be used to explain some of the work of Volodin, Kuznetsov, and Fomenko [V-K-F]. They developed a technique for simplifying Heegaard diagrams of 3-manifolds. They conjectured that an unsimplifiable diagram of the 3-sphere was the standard one, and got a computer to verify the conjecture on 1,000,000 randomly constructed diagrams. Viro and Kobelskii [V-K] then drew a picture of a genus three counterexample. This is an instance in mathematics of the common "scientific" phenomenon of a survey of an unexpectedly biased population (cf. [La]). Nevertheless, the technique of Volodin *et al.*, is worth further study.

9.2. A *Heegaard diagram* of genus n, (T,A,B), consists of a closed, connected, oriented 2-manifold T of genus n, and two sets, A and B, each consisting of n oriented disjoint simple closed curves on T, such that T – A and T – B are connected (and therefore of genus 0), and such that A and B intersect transversely. Given (T,A,B), we can construct H_A by attaching n 2-cells to T along A, and then attaching a 3-cell to the result; H_A is a handlebody, the result of taking the 3-cell and identifying n pairs of disjoint 2-cells on its boundary. Similarly, there is the handlebody H_B. The union of H_A and H_B along their common boundary T is a closed, connected, orientable 3-manifold M; and every such M is obtained from some Heegaard diagram in this way.

We obtain the same 3-manifold M if certain changes are made on A and B. Thus, we can change A by an isotopy to minimize the points of intersection of A∩B. If some component of T − A∪B is not simply connected, then an analysis shows that M can be decomposed into a non-trivial connected sum, each summand having a Heegaard diagram of smaller genus (or, in the extreme case, we have $M = S^1 \times S^2$ and genus = 1). Once we have made these immediately simplifying changes, we shall have what we call a *simple* Heegaard diagram: One having all components of T − A∪B simply connected, and having the cardinality of A∩B minimal within the isotopy class of A on T.

9.3. A subspace A of a topological space X is said to be *bicollared* in X, if it has a neighborhood homeomorphic to $(-1,+1) \times A$ in such a way that A corresponds to $0 \times A$.

In a Heegaard diagram (T,A,B) we see that A is bicollared in T, and B is bicollared in T, and A∪B − A∩B is bicollared in T − A∩B.

9.4. Suppose A is bicollared in X. We *construct a graph* $\Gamma = \Gamma(X,A)$ as follows:

V(Γ) is the set of connected components of X − A.

E(Γ) contains a reverse pair of edges for each component of A. We think of $e \in E(\Gamma)$ as a "normal orientation" to the corresponding component of A, and thus connects one side of that component to the other side. The components of X − A containing these sides constitute $\iota(e)$ and $\tau(e)$.

9.6. Let (T,A,B) be a Heegaard diagram of genus n. We can construct $\Gamma = \Gamma(T,A)$, $\Delta = \Gamma(T,B)$, and $\Theta = \Gamma(T-A\cap B, A\cup B - A\cap B)$. There are maps of graphs in the Gersten sense, $\alpha: \Theta \to \Gamma$, $\beta: \Theta \to \Delta$.

Both Γ and Δ are 1-vertex graphs, each with n reverse pairs of edges. An edge of Θ corresponding to a component of A − A∩B (resp., B − A∩B) belongs to Dβ (resp., Dα). Thus (α, β, Θ) is a (Γ, Δ)-dyad.

(A similar construction works for any space X and pair of bicollared subspaces intersecting transversely.)

Now, there is an embedding of the topological realization of the associated graph Θ into T. Surrounding each point of intersection of A and B there is the image of a ladder-square. Each component of A and of B gives rise to a ladder that goes around in a circle.

If (T,A,B) is a simple Heegaard diagram, then $D\alpha$ and $D\beta$ are deformation retracts of T – A and T – B, respectively.

9.7. What does it mean for the dyad (α, β, Θ) associated to the Heegaard diagram (T,A,B) to admit a fold? It is the following, or its symmetric analogue:

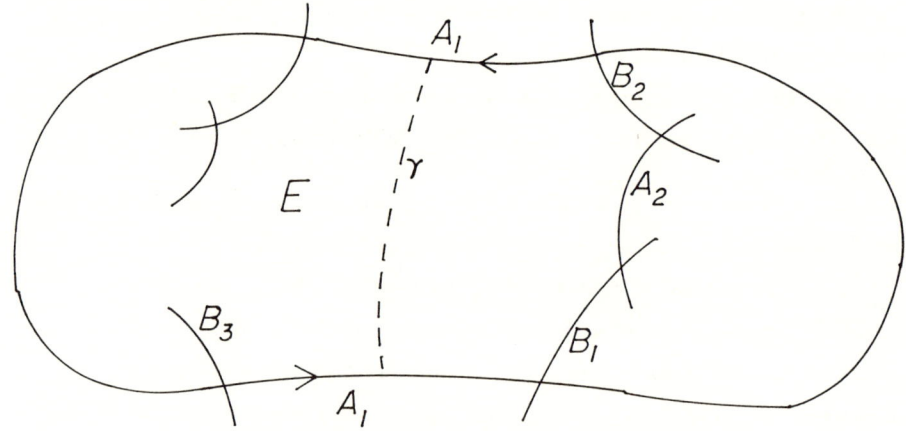

Figure 5

There is a component E of T – A∪B on whose boundary there are two components of A – A∩B which belong to the same component A_1 of A, and which are oriented coherently in the boundary of E. Volodin et al., term the arc γ which is drawn in Figure 5, a "wave."

Ordinarily, in higher dimensions, a fold would be topologically realized by joining up these two pieces of Bd E by a tube along the arc γ in the picture. However, because of the fact that T is 2-dimensional, this complicates the picture by replacing one component A_1 of A by two curves A_{11} and A_{12}. Thus, the attempt to simplify, in fact complicates; this is a common experience in 3-manifold labor.

9.8. *However*, Volodin *et al.*, note that one can simply throw away one of A_{11} or A_{12}. This produces a new Heegaard diagram (T,A',B) of the same 3-manifold, with fewer points of intersection $A' \cap B$. On the associated dyad this produces a change that is rather different from simply a fold; in particular, the graph Γ can be imagined to change, changing the free basis of $\pi_1(\Gamma) = \pi_1(H_A)$.

The choice of which of A_{11} or A_{12} to omit is made like this: Let $A'' = A - A_1$; then $T - A''$ is a surface of genus 1, and $A_1 \neq 0$ in $H_1(T-A'', \mathrm{Bd}(T-A''))$. Homologically, $A_1 = A_{11} + A_{12}$, and therefore at least one of A_{11} or A_{12} is $\neq 0$ in $H_1(T-A'', \mathrm{Bd}(T-A''))$; throw the other one away. This choice insures the condition that $T - A'$ be connected and thus be of genus 0.

9.9. The conjecture of Volodin *et al.*, which was disproved by the Viro-Kobelskii example, was that if (T,A,B) is a simple Heegaard diagram of the 3-sphere whose associated dyad admits no fold, then it is the obvious diagram of genus 0. In other words, given any Heegaard diagram of S^3 in which the number of components of $A \cap B$ has been minimized, then either $T - A \cup B$ has a non-simply-connected piece, or else there is a "wave."

If that conjecture had been true, it would have given some hope that the Poincare Conjecture might be attacked from that angle. There are other kinds of manipulations on Heegaard diagrams; the dyad picture gives an abstract idea of what is happening to the fundamental groups. It might be worthwhile to pursue this further to find some clarification of algorithmic problems in 3-manifolds, as well as to quest for the solution to the Poincaré Conjecture.

Perhaps the major technical snag is that a path-surgery result, such as 5.6, is lacking. The dyads that come up do not have the common-kernel condition.

DEPARTMENT OF MATHEMATICS
UNIVERSITY OF CALIFORNIA
BERKELEY, CALIFORNIA 94720

REFERENCES

[Co] D. Cooper, "Automorphisms of Free Groups Have F. G. Fixed Point Sets," Preprint Princeton U. (1983).

[C-V] M. Culler, K. Vogtmann, "Moduli of Graphs and Automorphisms of Free Groups," *Inv. math* (to appear).

[D-S] J. L. Dyer, G. P. Scott, "Periodic Automorphisms of Free Groups," *Comm. Alg.* 3 (1975), 195-201.

[G1] S. M. Gersten, "On Fixed Points of Certain Automorphisms of Free Groups," *Proc. London Math. Soc.* 48 (1984), 72-90, "Addendum," 49 (1984), 340-342.

[G2] _____, "On Fixed Points of Automorphisms of Finitely Generated Free Groups," *Bull. Amer. Math. Soc.* 8 (1983), 451-454.

[G3] _____, "Fixed Points of Automorphisms of Free Groups," *Adv. in Math.* (to appear).

[G4] _____, *Topology of the Automorphism Group of a Free Group*, (to appear).

[G-T] R. Z. Goldstein, E. C. Turner, "Automorphisms of Free Groups and Their Fixed Points," *Inv. math.* 78 (1984), 1-12.

[Hoa] A. H. M. Hoare, "On Automorphisms of Free Groups" (to appear).

[How] A. G. Howson, "On the Intersection of Finitely Generated Free Groups," *J. London Math. Soc.* 29 (1954), 428-434.

[J-S] W. Jaco, P. B. Shalen, "Surface Homeomorphisms and Periodicity," *Topology* 16 (1977), 347-367.

[La] S. Lang, *The File*, Springer-Verlag (1981).

[L-S] R. C. Lyndon, P. E. Schupp, *Combinatorial Group Theory*, Springer-Verlag (1977).

[Mc] J. McCool, "Some Finitely Presented Subgroups of the Automorphism Group of a Free Group," *J. Algebra* 35 (1975), 205-213.

[Mi] K. A. Mihailova, "The Occurrence Problem for Direct Products of Groups," *Doklady Akad. Nauk SSSR* 119 (1958), 1103-1105.

[Ni] J. Nielsen, "Die Isomorphismengruppe der Freien Gruppen," *Math. Ann.* 91 (1924), 169-209.

[Po] E. L. Post, "A Variant of a Recursively Unsolvable Problem," *Bull. Amer. Math. Soc.* 52 (1946), 264-268.

[Ra]   E. S. Rapaport, "On Free Groups and their Automorphisms," *Acta Math.* 99 (1958), 139-163.

[Se]   J-P. Serre, *Arbres, Amalgames* SL_2, Astérisque 46, Soc. Math. de France (1977).

[Sq]   C. Squier (private communication).

[St1]   J. R. Stallings, "Topologically Unrealizable Automorphisms of Free Groups," *Proc. Amer. Math. Soc.* 84 (1982), 21-24.

[St2]   _____, "Topology of Finite Graphs," *Inv. math.* 71 (1983), 551-565.

[V-K]   O. Ya. Viro, V. L. Kobelskii, "The Volodin-Kuznetsov-Fomenko Hypothesis on Heegaard Diagrams of the 3-Sphere is False," *Uspehi Mat. Nauk* 32:5 (197) (1977), 175-176.

[V-K-F] I. A. Volodin, V. E. Kuznetsov, A. T. Fomenko, "On the Problem of Algorithmic Discrimination of the Standard Three-Dimensional Sphere," *Uspehi Mat. Nauk* 29:5 (179) (1974), 71-168. English translation: *Russian Math. Surveys* 29:5, London Math. Soc. (1974), 71-172.

[Wh1]   J. H. C. Whitehead, "On Certain Sets of Elements in a Free Group," *Proc. London Math. Soc.* 41 (1936), 48-56.

[Wh2]   _____, "On Equivalent Sets of Elements in a Free Group," *Ann. of Math.* 37 (1936), 782-800.

PEAK REDUCTION AND AUTOMORPHISMS OF FREE GROUPS AND FREE PRODUCTS

Donald J. Collins

"Every mountain and hill shall be made low, the crooked made straight and the rough places smooth." (Isaiah 40:4).

Abstract. In a well known paper published in 1936, J. H. C. Whitehead gave an algorithm to decide whether two elements of a free group of finite rank are equivalent under an automorphism. An algebraic proof of Whitehead's result was later given by P. J. Higgins and R. C. Lyndon and this was the basis for three important papers by J. McCool on presentations of groups of automorphisms of a free group. The purpose of this article is to describe to what extent these results can be carried over to arbitrary free products.

§1. *Introduction*

In a well known paper [14] published in 1936, J. H. C. Whitehead gave an algorithm to decide whether two elements of a free group of finite rank are equivalent under an automorphism. An algebraic proof of Whitehead's result was later given by P. J. Higgins and R. C. Lyndon [9] and this was the basis for three important papers [10, 11, 12] by J. McCool on presentations of groups of automorphisms of a free group. The purpose of the present article is to outline to what extent these results can be carried over to arbitrary free products. For the most part the article is a summary of [3] and [4], where complete details are given, but the author has also sought to give a clear and easily understood description of the essential features of the method, introduced by Whitehead, which has been christened "Peak Reduction." Much of what is new or recent in what follows has been obtained by the author and Heiner Zieschang in the course of a harmonious and stimulating collaboration.

We begin with a little notation and terminology. Let $G = *_{i \in I} G_i$ be the free product of finitely many indecomposable factors. Some factors may be infinite cyclic and we separate these out, writing $G = (*_{i \in J} G_i) * F$ where F is the free group, on some basis S, given by the infinite cyclic factors. A *letter* of G is a non-trivial element of G_i, $i \in J$ or an element of SUS^{-1}. The length of a word w of G, written $|w|$, is the number of letters it contains. For technical reasons we deal almost entirely with *cyclic* words, i.e. reduced cyclically ordered strings of letters. Clearly Aut G acts on the set of all cyclic words.

The key to Whitehead's method is the choice of a set Ω of generators for Aut G which, in terms to be explained below, is

(1.1)   large enough to allow "peak reduction";

(1.2)   small enough to be "effectively finite."

Firstly, by a *peak* (over Ω) is meant a quintuple (u,w,v,σ,τ) where $u, w, v,$ are cyclic words, σ, τ are elements of Ω and

(a) $u\sigma = w$, $w\tau = v$,

(b) $|u| \leq |w| \geq |v|$,

(c) $|u| < |w|$ or $|w| > |v|$.

Plotting length upwards, we can visualize a peak as one of:

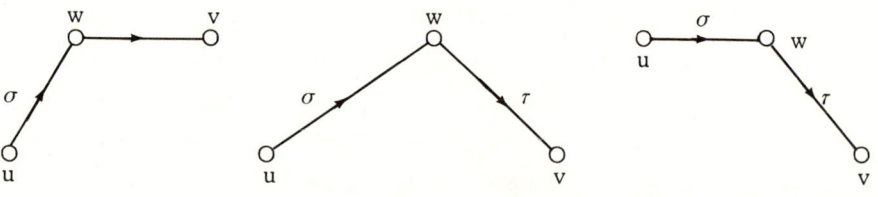

We say such a peak is *reducible* (over Ω) if

$$\sigma\tau = \rho_1 \rho_2 \cdots \rho_r, \rho_i \in \Omega \text{ with } |u\rho_1 \cdots \rho_j| < |w|, \; j = 1, 2, \cdots, r-1.$$

PEAK REDUCTION

Reducibility can be visualized as

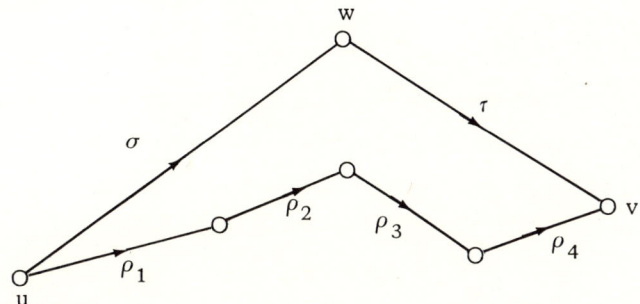

We say Ω *allows peak reduction* if every peak is reducible. A cyclic word u and $\rho_1, \rho_2, \cdots, \rho_r \in \Omega$ form a *valley* if, for $0 \leq s \leq t \leq r$,

(1.4) (a) $|u\rho_1 \cdots \rho_{j-1}| > |u\rho_1 \cdots \rho_j|$, $1 \leq j \leq s$

(b) $|u\rho_1 \cdots \rho_{k-1}| = |u\rho_1 \cdots \rho_k|$, $s+1 \leq k \leq t$

(c) $|u\rho_1 \cdots \rho_{l-1}| < |u\rho_1 \cdots \rho_l|$, $t+1 \leq l \leq r$

(interpreting e.g. $s = 0$ to mean that (a) is vacuous). A valley can be visualized as:

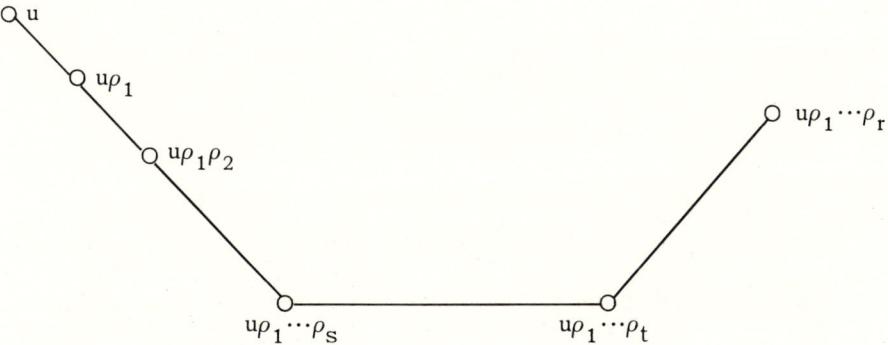

The easy part of Whitehead's argument is contained in the following two propositions.

1.5 PROPOSITION. *Suppose Ω allows peak reduction. Then for any cyclic word u of G and any $a \in$ Aut G, there exist $\rho_1, \rho_2, \cdots, \rho_r \in \Omega$ such that $u, \rho_1, \cdots, \rho_r$ form a valley.*

Proof. Since Ω generates Aut G, $a = \sigma_1 \sigma_2 \cdots \sigma_s$, with $\sigma_k \in \Omega$, $k = 1, 2, \cdots, s$. In general plotting the lengths of $u\sigma_1 \cdots \sigma_k$, $0 \leq k \leq s$ yields a "mountain range."

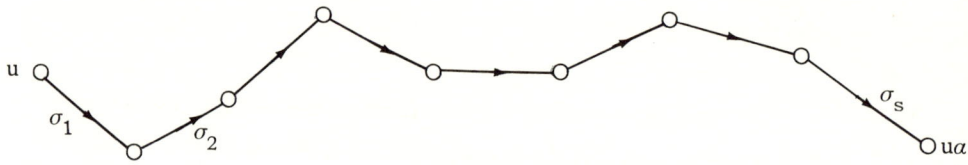

Repeated reduction of peaks produces the desired valley.

The second proposition illustrates an instance of how finiteness, referred to in (1.2), is significant.

1.6 PROPOSITION. (i) *Suppose*

    (a) Ω *allows peak reduction,*

    (b) Ω *is finite.*

Then one can decide if an element of G is of minimal length in its automorphism class.

    (ii) *If, in addition,*

    (c)   *G has only finitely many elements of any given length*

then one can decide if any two elements of G lie in the same automorphism class.

Proof. By making use of inner automorphisms, one sees immediately that it suffices to deal with cyclic rather than linear words.

(i) If a cyclic word u is not minimal, it follows from Proposition 1.5 that there exists $\rho \in \Omega$ with $|u\rho| < |u|$. Since Ω is finite this can be discovered by simple exhaustion of cases.

(ii) Repeated use of (i) and Proposition 1.5 reduce the problem to words u and v of the same length which are of minimal length in their automorphism classes.

Let Γ be the graph whose vertices consist of all words of the same length as u and v, two vertices being joined by an edge if there exists $\rho \in \Omega$ carrying one to the other. By the two finiteness assumptions (b) and (c), Γ is finite. By Proposition 1.5 we have to decide if u and v lie in the same path component, which can be settled by simple inspection.

REMARK. Condition (c) in (1.6) is satisfied only when the factors of G are infinite cyclic or finite. So at least in this case the tacit assumption that we can effectively evaluate the action of an automorphism is clearly justified.

§2. *The choice of* Ω

We firstly examine the case when $G = F$ is just a free group. The most usual generating set for Aut F consists of

(2.1) the automorphisms which permute or invert elements of the basis S;

(2.2) the Nielsen automorphisms, i.e., those where for some fixed $x \in S$ and fixed $y \in S \cup S^{-1}$.

$$x \to xy$$
$$s \to s, \qquad s \neq x.$$

However this set of generators does not allow peak reduction as can be seen by taking $S = \{x,y,z\}$ and considering the peak

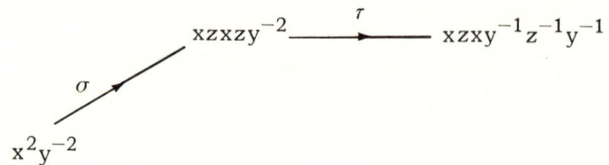

where $\sigma : x \to xz$ and $\tau : y \to yz$.

The way to achieve peak reduction is to enlarge the generating set so as to include products such as $\sigma\tau$. The above peak is then trivially reducible. Specifically Ω is chosen to consist of the automorphisms (2.1) and (2.3) the *Whitehead automorphisms*, i.e., those automorphisms such that, for some fixed $y \in S \cup S^{-1}$ and any $s \in S$, $s\sigma$ is one of s, sy, $y^{-1}s$, $y^{-1}sy$. Clearly Ω is finite and Whitehead was able to show that Ω allows peak reduction.

Now let $G = \underset{i \in I}{*} G_i = (\underset{i \in J}{*} G_i) * F$ be an arbitrary free product of finitely many indecomposable factors G_i. According to D. I. Fouxe-Rabinowitsch [6. 7] a generating set for Aut G can be formed from

(2.4) the *permutation automorphisms*, i.e. automorphisms which permute, via fixed isomorphisms, any factors that are isomorphic to one another;

(2.5) the *factor automorphisms*, i.e., those of form $\underset{i \in I}{*} \phi_i$, where $\phi_i \in \text{Aut } G_i$;

(2.6) those automorphisms in which the only non-trivial action is that either some factor G_i that is not infinite cyclic is conjugated by a letter $y \notin G_i$, or some element x of the basis S of F is postmultiplied by a letter $y \neq x^{\pm 1}$: in symbols, either

$$G_i \to y^{-1} G_i y, \text{ some } i \in J, y \notin G_i,$$

or

$$x \to xy, \text{ some } x \in S, y \neq x^{\pm 1}.$$

The automorphisms (2.6) are the analogues of the Nielsen automorphisms (2.2). So if the example of an irreducible peak given above is to be avoided then the generating set must be extended to include

(2.7) the *Whitehead automorphisms*, i.e., those automorphisms σ for which there is some fixed letter y (called the *operative letter*) and

(a)   for *any* $i \in J$ either $G_i \xrightarrow{\sigma} y^{-1} G_i y$

$$\text{or} \quad G_i \xrightarrow{\sigma} G_i,$$

(b)   for *any* $s \in S$, $s\sigma$ is one of s, sy, $y^{-1}s$, $y^{-1}sy$.

(In (a) σ is understood to operate pointwise and if $y \in G_k$, $k \in J$, then σ leaves G_k fixed.)

An unsuccessful attempt to establish peak reduction with this choice of Ω is recorded in [2]. The sticking point occurs for peaks involving two Whitehead automorphisms whose operative letters come from the same factor G_k (not infinite cyclic) and examples of irreducible peaks are given there. The resolution of the difficulty is achieved by further enlarging Ω so that it contains products of Whitehead automorphisms all of whose operative letters lie in the same factor G_k (not infinite cyclic). We refer to these as *multiple Whitehead automorphisms*.

2.8. PROPOSITION. *Let Ω consist of all permutation, factor and multiple Whitehead automorphisms. Then Ω allows peak reduction.*

Proposition 2.8 is the principal result of [3]. The proof parallels the Higgins-Lyndon argument for Whitehead's original theorem by using a formula which gives the change in length that occurs when a multiple Whitehead automorphism is applied to a cyclic word. We give a brief outline of the argument and need the following notation. Purely for simplicity of exposition, assume no factor is infinite cyclic. If σ is a Whitehead automorphism with operative letter $x \in G_k$, write

$$A = \{G_i : G_i \xrightarrow{\sigma} x^{-1} G_i x\} \cup \{G_k\}.$$

Then σ is uniquely defined by the pair (A,x) and we shall denote σ by (A,x). It is easy to check that

(2.9) if $A_1 \cap A_2 = \{G_k\}$, then $(A_1,x)(A_2,x) = (A_1 \cup A_2, x)$

and

(2.10) $(A,x_1)(A,x_2) = (A, x_2 x_1)$, where $x_1, x_2 \in G_k$.

It follows, after a little calculation, that any multiple Whitehead automorphism can be written as a product $\rho = (A_1, x_1)(A_2, x_2) \cdots (A_r, x_r)$, where $x_j \in G_k$, $j = 1, 2, \cdots, r$, $A_p \cap A_q = \{G_k\}$, $1 \leq p, q \leq r$ and, hence $r \leq |I|$. We call $A = A_1 \cup A_2 \cup \cdots \cup A_r$ the *domain* of ρ. The reduction of a peak (u, w, v, σ, τ) is achieved as follows:

(2.11) If, say, σ is factor or permutation, then $\sigma \tau = (\sigma \tau \sigma^{-1}) \sigma$, $\sigma \tau \sigma^{-1}$ is multiple Whitehead and $|u \sigma \tau \sigma^{-1}| - |u| = |w \tau| - |w|$.

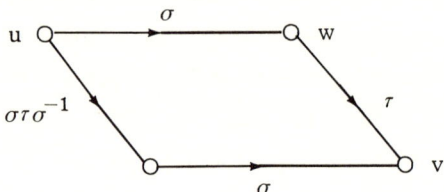

So we may suppose both σ and τ are multiple Whitehead.

(2.12) If the operative letters of σ and τ lie in the same factor, then $\sigma \tau$ is also multiple Whitehead.

(2.13) If σ and τ have disjoint domains, then $\sigma \tau = \tau \sigma$ and $|u \tau| - |u| = |w \tau| - |w|$; this is illustrated by

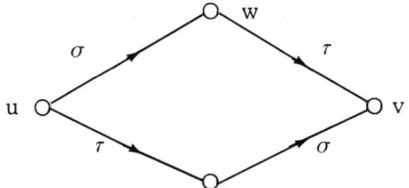

The argument for this is an instance of the well-known diamond principle — roughly, if two operations have disjoint domains of application, the order in which they are performed is irrelevant.

(2.14) Let σ and τ have non-disjoint domains A and B, respectively. Let σ have operative letters in G_k, τ have operative letters in G_l, and suppose that $G_k \not\in B$, $G_l \not\in A$. In particular, $G_k \neq G_l$. Using the relations (2.9), (2.10) one can write $\sigma = \sigma_1 \sigma_2$ where σ_1 has domain $(A \cap B) \cup \{G_k\}$ and σ_2 has domain $A-B$. With an appeal to symmetry, since $(v, w, u, \tau^{-1}, \sigma^{-1})$ is also a peak, one can show by detailed computation that $|u\sigma_1| < |w|$. Since σ_2 and τ have disjoint domains, (2.1) gives the reduction $\sigma\tau = \sigma_1 \tau \sigma_2$:

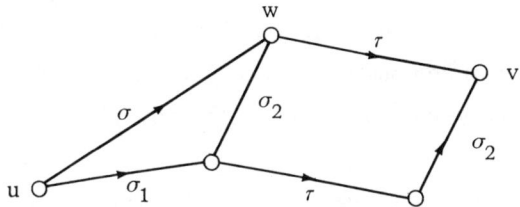

(2.15) The final step is the use of "complementation." Let L be the set of all factors of G. If $\sigma = (C, z)$ with $z \in G_k$, its *complement* is $\sigma' = ((L-C) \cup \{G_k\}, z^{-1})$. Trivially, σ and σ' have the same effect on any cyclic word u. The remaining cases where, in the notation of (2.14), $G_k \in B$ or $G_l \in A$ are quickly reduced to (2.13) and (2.14) by replacing a term in the factorization of σ or τ by its complement.

We must now try to clarify the notion of being "effectively finite," referred to in (1.2). If all the factors of G that are not infinite cyclic are finite, then, by the remarks after (2.9) and (2.10), Ω is finite and no further comment is required. To obtain the analogue of (i) of Proposition (1.6) when Ω is not finite, one needs the following, which is proved in [3], and is not especially difficult.

(2.16) PROPOSITION. *Let* u *be a cyclic word and suppose there exists* $\sigma \in \Omega$ *such that* $|u\sigma| < |u|$. *Then there is a finite subset* Ω_0 *of* Ω, *which is (uniformly) effectively calculable from* u *and contains an element* ρ *such that* $|u\rho| < |u|$.

To see why the proposition holds, recall that any multiple Whitehead automorphism can be written in the form

$$\sigma = (A_1, x_1)(A_2, x_2) \cdots (A_r, x_r)$$

where $r \leq |I|$. The fact that Ω may be infinite stems from the fact that there may be infinitely many possible choices for $x_1, x_2, \cdots, x_r \in G_k$. However, if $|u\sigma| < |u|$, then more letters from the factor G_k are cancelled than are introduced by the action of σ. Thus, it is reasonable to expect that $x_1, x_2, \cdots, x_r$ can be constructed from those letters of u that lie in G_k. In what should be obvious notation, the example

$$u = x_1 a_1 x a_1 (x_1^{-1} x_2) a_2 x b \xrightarrow{(A_1, x_1)(A_2, x_2)} a_1 (x_1 x x_1^{-1}) a_1 a_2 (x_2 x) b$$

shows that it is by no means the case that $x_1, x_2, \cdots, x_r$ are all actual letters of u.

The final step, in generalizing Whitehead's result, is to decide when two cyclic words which are of minimal length within their respective automorphism classes are equivalent. To indicate, albeit very sketchily, how this is done, we need some further notation. Let Σ denote the group of all permutation automorphisms and Φ the group of all factor automorphisms (see (2.4) and (2.5)).

For cyclic words u and v , which are of minimal length write
u ∼ v if there exists $\tau \in \langle \Phi, \Sigma \rangle$ such that $u\tau = v$. We take Γ to be
the graph whose vertices are the equivalence classes under ∼ , two
vertices being joined by an edge if there is a multiple Whitehead automorphism carrying a member of one vertex to a member of the other. The
crucial step is given by:

(2.17) PROPOSITION. (i) *The path component in* Γ *of a vertex* [u]
has its diameter bounded in terms of |u| *and* |I|.

(ii) *If no factor of* G *is infinite cyclic the path
component of any vertex* [u] *of* Γ *is finite and bounded in terms of* |u|
and |I|.

When the conclusion of 2.17 (ii) is valid, the effect is as follows. To
decide if two minimal words u and v lie in the same automorphism
class, compute a representative of each vertex of the path component of
[u] in Γ. Then test whether there exists $\tau \in \langle \Sigma, \Phi \rangle$ carrying one of
these representatives into v . Carrying out this last step requires, of
course, some unavoidable assumptions about the factors G_i and their
automorphism groups.

When infinite cyclic factors occur, path components in Γ are no
longer finite. However, it turns out that the boundedness of the diameter
of a path component still gives enough control to squeeze out an algorithm.

It is difficult to say anything very useful about the proof of Proposition (2.17) without going into a lot of detail and we content ourselves with
two very brief, and only approximately correct, comments.

(2.18) At the minimal level, a multiple Whitehead automorphism "frequently"
simply permutes the letters of the cyclic word on which it operates. So
any "long" path in Γ must contain a loop.

(2.19) When no factor of G is infinite cyclic, an argument somewhat
like that for Proposition (2.16) shows that for any minimal cyclic word u,

there are only finitely many multiple Whitehead automorphisms σ such that $|u\sigma| = |u|$, and so every vertex of Γ has finite degree.

§3. Applications

The principal applications of Whitehead's result are by McCool who was able:

(3.1) to give a finite presentation for the full automorphism group of a free group F of finite rank, thereby confirming the correctness of the presentation found by J. Nielsen [13] (see Note 3.3),

(3.2) to give a finite presentation for the stabilizer of an m-tuple of elements of F — and thereby show that mapping class groups of surfaces are finitely presented.

The application of Proposition 2.8, paralleling (3.1) to obtain a presentation of the full automorphism group Aut G of a free product G has been carried out by N. Gilbert [8] and confirms the correctness of the presentation given by D. I. Fouxe-Rabinowitsch [6, 7] (see Note 3.3). The analogue of (3.2) is carried out in [12] but only for the case of a single word when no factor is infinite cyclic. The full analogue of (3.2) remains open, there being technical difficulties not yet understood in the general case.

The argument for (3.2), and its analogue for a free product, is roughly as follows. Let w be a fixed cyclic word of minimal length. Let Δ be the 2-complex whose vertices are minimal words in the automorphism class of w, whose edges are labelled by automorphisms in Ω that carry one endpoint to the other, and whose faces correspond to any relations among the edge labels which are instances of the equalities used in reducing peaks, all such relations being of bounded length. Reading off labels on closed paths based at w defines an isomorphism from $\pi_1(\Delta, w)$ to $\text{Stab}_{\text{Out } G}(w)$.

When $G = F$ is free of finite rank, Δ is finite and hence $\pi_1(\Delta,w)$ has a finite presentation. When no factors are infinite cyclic, Proposition (2.17) (ii) shows, roughly, that Δ has a "finite cross-section" from which the whole of Δ can be obtained by "translating" along edges labelled by permutation and factor automorphisms. With suitable assumptions on the factors and their automorphism groups, the finite presentability of $\pi_1(\Delta,w)$ follows.

(3.3) NOTE. The question of checking Nielsen's presentation for Aut F, where F is free of finite rank, is discussed by B. Chandler and W. Magnus in [1]. Nielsen's argument is complicated and difficult to follow and the same is true of Fouxe-Rabinowitsch's argument in [7], based as it is on Nielsen's. (The first paper [6] by Fouxe-Rabinowitsch, which deals with the case when no infinite cyclic factors are present, is more straightforward.) It therefore seems worthwhile to provide an alternative derivation of Fouxe-Rabinowitsch's results.

SCHOOL OF MATHEMATICAL SCIENCES
QUEEN MARY COLLEGE
MILE END ROAD
LONDON E1 4NS
U. K.

REFERENCES

[1] B. Chandler, W. Magnus; A history of combinatorial group theory: a case study in the history of ideas, Springer, Berlin-Heidelberg, New York, 1982.

[2] D. J. Collins, H. Zieschang; On the Whitehead method in free products, to appear in Contemporary Mathematics.

[3] ———; Rescuing the Whitehead method for free products, I: Peak Reduction, Math. Z. 185 (1984), 487-504.

[4] ———; Rescuing the Whitehead method for free products, II: The algorithm, Math. Z., 186 (1984), 335-361.

[5] ———; A presentation for the stabiliser of an element in a free product, to appear (Journal of Algebra).

[6] D. I. Fouxe-Rabinowitsch; Über die Automorphismengruppen der freien Produkte I, Mat. Sbornik 8 (1940), 265-276.

[7] D. I. Fouxe-Rabinowitsch; Über die Automorphismengruppen der freien Produkte II, Mat. Sbornik, 9 (1941), 183-220.

[8] N. Gilbert; Ph.D thesis Queen Mary College, London 1985.

[9] P. J. Higgins, R. C. Lyndon; Equivalence of elements under automorphisms of a free group, J. Lond. Math. Soc., 8 (1974), 254-258.

[10] J. McCool; A presentation for the automorphism group of a free group of finite rank, J. Lond. Math. Soc., 8 (1974), 259-266.

[11] ————; On Nielsen's presentation of the automorphism group of a free group, J. Lond. Math. Soc., 10 (1975), 265-270.

[12] ————; Some finitely presented subgroups of the automorphism group of a free group, J. Algebra, 35 (1975), 205-213.

[13] J. Nielsen; Die Isomorphismengruppe der freien Gruppen, Math. Ann., 91 (1924), 169-209.

[14] J. H. C. Whitehead; On equivalent sets of elements in a free group, Ann. of Math., 37 (1936), 728-800.

NONSINGULAR EQUATIONS OF SMALL WEIGHT OVER GROUPS

S. M. Gersten

Abstract. This paper presents two new results on the Kervaire conjecture and formulates a new conjecture on conjugacy classes in a group. The conjecture is shown to be equivalent to a conjecture of Stallings.

Introduction. The Kervaire problem for groups, as it is understood today [4], asks whether a nonsingular system of n equations in n unknowns over a group G has a solution in an overgroup of G. This is the case if G is locally residually finite [3] or locally indicable [4] (or more generally locally p-indicable for some prime number p [2]). Also, Howie has shown that a single equation, involving at most three occurrences of variables, has a solution in an overgroup [5].

In this note, we show that the Kervaire problem has an affirmative answer in general if and only if it has an affirmative answer for nonsingular systems of equations, each of which involves at most three occurrences of variables ("weight at most 3"). Then we show in §2 that any nonsingular system, each equation of which has at most two occurrences of variables ("weight at most 2"), has a solution in an overgroup. In fact, we prove a much stronger result in §2. We show that any spherical diagram for the relative two-complex defined by these equations has precisely two vertices and that the vertex labels are conjugate classes of inverse elements (Theorem 1). There are easier proofs of the Kervaire conjecture for these systems (as Howie has pointed out to us). The length of §2 is justified by this stronger result of inverse conjugacy of vertex labels (indeed we offer two proofs of this fact).

Next we use the result of Theorem 1 to motivate a much stronger conjecture, our 'reciprocity law' (§3) which implies the Kervaire conjecture and many other group theoretic properties as well. The conjectural reciprocity law is then shown in §4 to be equivalent to Stallings' conjecture B [8] as well as equivalent to a purely group theoretic property, which we call "Conjecture G." Since this latter conjecture is perhaps the easiest to state of all three, we state it here. Let G be a group and let $G \to H$ be the homomorphism induced by inclusion, where $H = G * <t_1, t_2, \cdots, t_n> / \ll w_1, w_2, \cdots, w_n \gg$. Here the group $<t_1, t_2, \cdots, t_n>$ is freely generated by $t_1, t_2, \cdots, t_n$ and the determinant of the matrix of exponent sums of the variables t_i in the elements w_j is assumed to be different from zero (such a map $G \to H$ is called a "Kervaire extension"). Let $g_1, g_2, \cdots, g_r \in G$ and assume that $1 \in \prod_{i=1}^{r} g_i^H$, where g_i^H denotes the conjugacy class of the image of g_i in H. Then we conjecture that $1 \in \prod_{i=1}^{r} g_i^G$ (the Kervaire conjecture is the special case $r = 1$).

Finally we show that F. Levin's argument [6] can be strengthened to show that the reciprocity law is valid for a single equation in one variable of weight n and exponent sum n (so all occurrences of the variable occur with positive exponent).

I want to thank John Stallings for suggesting that Theorem 1 below may be related to a theorem of Baumslag and Taylor. This suggestion proved to be the case. I also want to thank Jim Howie for his careful reading of the manuscript. Howie observed the change in focus that occurs in this paper in §2 from the Kervaire conjecture to the reciprocity law, which was gradually taking place in my mind as I wrote it. He suggested that I inform the reader of this change of emphasis at the start, as I have done here, so the reader will be spared the pains I had to endure in discovering the reciprocity law.

§1. Weight

Let G be a group and let F be the free group freely generated by elements $t_1, t_2, \cdots, t_n$. An equation over G is simply an element $w \in G * F$. We may write w uniquely in the form

$$w = g_1 t_{i_1}^{\varepsilon_1} g_2 t_{i_2}^{\varepsilon_2} \cdots g_r t_{i_r}^{\varepsilon_r} g_{r+1}$$

where $g_i \in G$, $\varepsilon_i \neq 0$, and where $t_{i_j} \neq t_{i_{j+1}}$ if $g_i = 1$ (so no cancellation can occur). We define the weight of w to be $\sum_{i=1}^{r} |\varepsilon_i|$. Informally, the weight of w is the number of $t_i^{\pm 1}$ occurring in the reduced expression for w.

We say that an n-tuple of elements $(h_1, h_2, \cdots, h_n)$ in an overgroup H of G is a solution of the equation w if w is in the kernel of the composite homomorphism

$$G * F \xrightarrow{\subset} H * F \xrightarrow{1 * \gamma} H,$$

where $\gamma: F \to H$ is given by $\gamma(t_i) = h_i$.

The homomorphism $G * F \xrightarrow{\eta_i} Z$ sending G to 0 and t_j to δ_{ij} (Kronecker's delta function) is called the exponent sum in the variable t_i. The system of n-equations $w_1, w_2, \cdots, w_n \in G * F$ is called nonsingular if $\det(\eta_i(w_j)) \neq 0$. Thus the Kervaire conjecture states that there is a simultaneous solution of the nonsingular system $w_1, w_2, \cdots, w_n$ in an overgroup of G.

PROPOSITION 1. *The Kervaire conjecture is valid if and only if it is valid for nonsingular systems of equations of weight at most three.*

The argument is analogous to the familiar procedure of converting a differential equation of high order into a system of first order equations. We shall illustrate the procedure in two examples. Observe that in each case the determinant of the associated system of equations of weight ≤ 3 is the same as that of the original system.

EXAMPLE 1. $w = t^3 a t^{-1} b t^{-1} c$, where $a, b, c \in G$. We define $t_1 = t$, $t_2 = t^2 a t^{-1} b t^{-1} c$, $t_3 = t a t^{-1} b t^{-1} c$, $t_4 = t^{-1} b t^{-1} c$, $t_5 = t^{-1} c$. The associated system of equations is

$$\begin{cases} w_1 = t_1 t_2 \\ w_2 = t_2 \cdot (t_1 t_3)^{-1} \\ w_3 = t_3 \cdot (t_1 a t_4)^{-1} \\ w_4 = t_4 \cdot (t_1^{-1} b t_5)^{-1} \\ w_5 = t_5 \cdot (t_1^{-1} c)^{-1} \end{cases}$$

Observe that the original equation w has weight 5, the number of equations of the associated system, and determinant is unchanged. The 5-tuple of elements $(h_1, h_2, h_3, h_4, h_5)$ is a solution of the second system in an overgroup only if h_1 is a solution of the original equation w.

EXAMPLE 2.
$$w_1 = x^2 a y b x c$$
$$w_2 = x^{-2} d y e$$

(a, b, c, d, e in G; x, y indeterminates). Define $t_1 = x$, $t_2 = y$, $t_3 = xaybxc$, $t_4 = ybxc$, $t_5 = xc$, $t_6 = x^{-1} dye$, and $t_7 = ye$. The associated system of equations is

$$\begin{cases} w_1 = t_1 t_3 \\ w_2 = t_3 \cdot (t_1 a t_4)^{-1} \\ w_3 = t_4 \cdot (t_2 b t_5)^{-1} \\ w_4 = t_5 \cdot (t_1 c)^{-1} \\ w_5 = t_1^{-1} t_6 \\ w_6 = t_6 \cdot (t_1^{-1} d t_7)^{-1} \\ w_7 = t_7 \cdot (t_2 e)^{-1}. \end{cases}$$

The sum of the weights of the original system is equal to the number of equations in the associated system. The determinants are equal. A seven-tuple $(h_1,h_2,h_3,h_4,h_5,h_6,h_7)$ in an overgroup is a solution of the associated system only if the pair (h_1,h_2) is a solution of the original system.

The argument for the general case, which we omit, shows in fact the following refinement of Proposition 1.

PROPOSITION 2. *Let* $w_1,w_2,\cdots,w_n$ *be a system of equations over* G *of positive weights* $\gamma_1,\gamma_2,\cdots,\gamma_n$ *respectively. Then there is an associated system of* $\gamma = \sum_{i=1}^{n} \gamma_i$ *equations in* γ *unknowns, each of weight* ≤ 3, *having the same determinant as the original system and such that* $(h_1,h_2,\cdots,h_\gamma)$ *is a solution of the associated system in an overgroup of* G *only if* $(h_1,h_2,\cdots,h_n)$ *is a solution of the original system.*

§2. *Systems of equations of weight* 2

In considering nonsingular systems of equations of weight at most two, we may eliminate variables occurring in equations of weight one to obtain a new system with the same determinant and consisting of equations of weight two. Hence we shall consider a nonsingular system of equations $w_1,w_2,\cdots,w_n$ over a group G where each w_i is of weight exactly 2 in the indeterminates $t_1,t_2,\cdots,t_n$.

Let K be a CW complex with one vertex where $\pi_1(K) = G$, and let L be obtained from K by attaching n 1-cells trivially, so that $\pi_1(K \cup L^{(1)}) = G * F$, where F is freely generated by $t_1,t_2,\cdots,t_n$ (thus the oriented 1-cells of L−K are identified with the elements t_i in $\pi_1(K \cup L^{(1)})$), and then attaching n 2-cells by the maps w_i, $1 \leq i \leq n$. Call these 2-cells a_i, $1 \leq i \leq n$, so a_i is attached by word $w_i \in G * F$. The Kervaire conjecture states that the homomorphism $\pi_1(K) \to \pi_1(L)$ induced by inclusion K ∪ L is injective. From the exact homotopy sequence, we see that this is equivalent to showing that the map $\pi_2(L) \to \pi_2(L,K)$ is surjective. An element of $\pi_2(L,K)$ is represented by a map $f:(D^2,S^1) \to (L,K)$. We shall make this map transverse regular to a tamely imbedded subgraph Γ of L we now describe.

Choose one point P_i in the interior of the 2-cell a_i of L. Since the attaching map $g_i:S^1 \to K \cup L^{(1)}$ for a_i is w_i, where w_i has weight two, if we fix an interior point Q_j in the interior of the 1-cell of L corresponding to t_j, then we may assume the map g_i is transverse regular to $\bigcup_{j=1}^{n} Q_j$. Thus $g_i^{-1}(\cup Q_j)$ consists of precisely two points P_i' and P_i'' in ∂a_i.

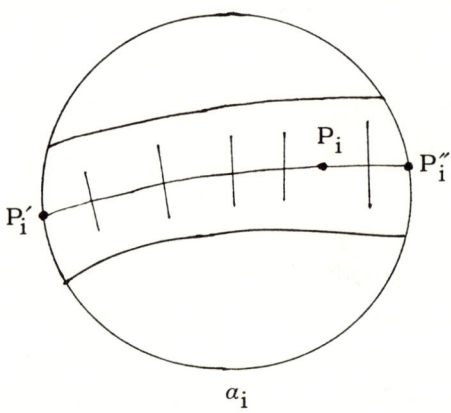

Join P_i', P_i'', and P_i by a properly imbedded PL interval J_i in a_i and let Γ be the image of $\bigcup_{i=1}^{n} J_i$ in L. Since Γ has a regular neighborhood in L, we may make $f:(D^2,S^1) \to (L,K)$ transverse regular to Γ, and adjust the notation so this is done. Then $f^{-1}(\Gamma)$ is a 1-manifold properly imbedded in D^2. Since $f(S^1) \subset K$ and $K \cap \Gamma = \phi$, it follows that $\Delta = f^{-1}(\Gamma)$ is a disjoint union of circles in the interior of D^2 and that $f^{-1}(\bigcup_{1 \le i \le n} P_i)$ is a finite union of points subdividing these circles. We consider the points of $f^{-1}(\cup P_i)$ to be the vertices of the graph Δ.

Observe that $f(D^2 - \Delta) \subseteq L - \Gamma$ and the latter space deformation retracts onto K. Hence to each corner of each complementary region of $D^2 - \Delta$ we may associate a label in $\pi_1(K)$. Reading counterclockwise around one side of a component of Δ, we may assign the product of the labels at the corners encountered. Choose an innermost circle Δ_0 of Δ in D^2. The product of the interior corner labels around Δ_0 is trivial in $\pi_1(K)$. It suffices to prove that the product of the exterior corner labels around Δ_0 is also trivial in $\pi_1(K)$. For if this is the case, we can modify f on an ε-neighborhood of the closed disc bounded by Δ_0 to obtain a new map

$$f_1 : (D^2, S^1) \to (L, K)$$

with $f_1|S^1 = f|S^1$ and with f_1 transverse regular to Γ and with $f_1^{-1}(\Gamma)$ possessing one fewer connected component than Δ. An obvious induction will then show that $f|S^1 : S^1 \to K$ is null homotopic.

Thus we may assume that $\Delta = \Delta_0 = f^{-1}(\Gamma)$, a circle properly imbedded in D^2 with subdivision points $f^{-1}(\bigcup_{i=1}^{n} P_i)$. We embed D^2 in S^2 as a hemisphere and adopt the dual point of view. The dual graph X to $\Delta \subset S^2$ has two vertices v and v' (corresponding to the complementary regions of $S^2 - \Delta$) and directed line segments connecting v to v' (corresponding to the intervals of the subdivision of Δ). Each of the vertices v, v' has a label $\mathcal{L}_v, \mathcal{L}_{v'}$ in $\pi_1(K)$ corresponding to the products of interior and exterior labels around Δ. The complementary regions of $S^2 - X$ are di-gons and each di-gon has an interior label, which is a reduced cyclic word in $\pi_1(L)$ corresponding to one of the words $w_i^{\pm 1}$.

THEOREM 1. $\mathcal{L}_{v'} = \mathcal{L}_v^{-1}$ as conjugacy classes in $\pi_1(K) = G$.

EXAMPLE.
$$\begin{cases} w_1 = t_1 a t_2 b \\ w_2 = t_1^{-1} c t_2 d \end{cases}$$

$(a, b, c, d \in G)$.

We draw below one possible configuration for $X \subset S^2$ (the exterior of the graph X corresponds to the region at ∞) along with corner labels:

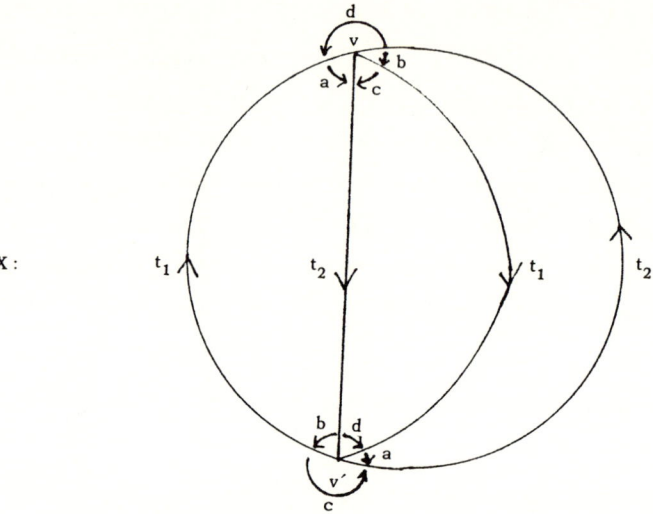

The labels $\mathcal{L}_v$, $\mathcal{L}_{v'}$ are the cyclic words $ac^{-1}b^{-1}d$ and $ca^{-1}d^{-1}b$ respectively. Observe that as cyclic words $\mathcal{L}_{v'} = \mathcal{L}_v^{-1}$ in this example.

REMARK 1. Observe that Theorem 1, when interpreted for the graph Δ dual to X, proves the Kervaire conjecture for nonsingular systems of equations each of weight two. For if $\mathcal{L}_v = 1$, then it follows that $\mathcal{L}_{v'} = 1$, and we've already observed that this is the Kervaire conjecture.

REMARK 2. It's important to observe that Theorem 1 is a combinatorial result. To emphasize this point, we reformulate it as follows. We continue to assume we are given a nonsingular system of equations $w_1, w_2, \cdots, w_n$ over the group G, each of weight two. Assume we are given a tamely imbedded directed graph X in S^2 with two vertices, v and v´, whose directed edges are labeled by the variables of our equations $t_1, t_2, \cdots, t_n$ and such that each edge connects v and v´ (thus $S^2 - X$ is a union of di-gons). We assume each corner has a label from G so that the interior label of each di-gon is one of the reduced cyclic words $w_i^{\pm 1}$. We form the vertex labels $\mathcal{L}_v, \mathcal{L}_{v'}$ by reading the corner labels around v, v´ counterclockwise and taking the product of these corner labels.

THEOREM 1′. $\mathcal{L}_{v'} = \mathcal{L}_v^{-1}$ as conjugacy classes of elements of G.

Clearly Theorems 1 and 1′ are equivalent, so we shall prove Theorem 1′.

Call the directed edges of X from v' to v in counterclockwise order from some starting point $e_1, e_2, \cdots, e_r$, $e_{r+1} = e_1$

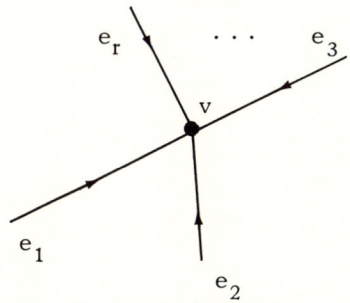

Let $\phi(e_i)$ be the label of e_i, so $\phi(e_i) \in \{t_j^{\pm 1}, 1 \leq j \leq n\}$.

LEMMA 1. *If $\phi(e_1) = \phi(e_s)^{\pm 1}$ for some $1 < s < r$, where $\phi(e_i) \neq \phi(e_j)^{\pm 1}$ for $1 \leq i < j < s$, then $\phi(e_1) = \phi(e_s)^{-1}$.*

Proof. We may reindex the variables t_i so that $\phi(e_i) = t_i$ for $1 \leq i \leq s-1$ and $\phi(e_s) = t_1^{\pm 1}$

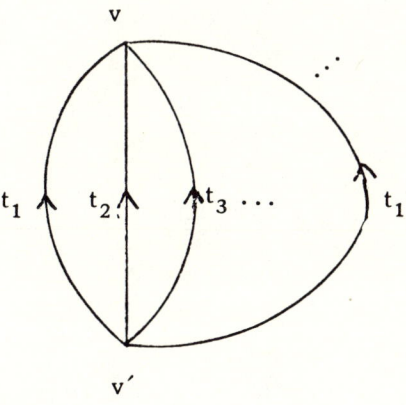
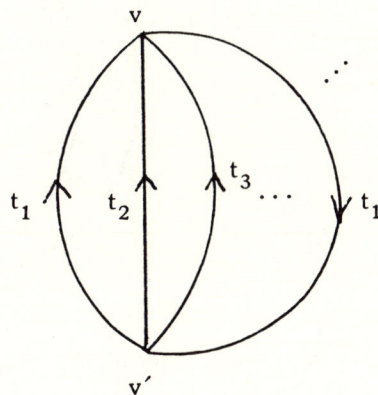

If $\phi(e_s) = t_1$, then the matrix of exponent sums $\eta_i(w_j)$ has an $(s-1) \times (s-1)$ diagonal block of the form below (the entries in the matrix $\eta_i(w_j)$ to the right of this block are zero):

$$\begin{pmatrix} 1 & -1 & 0 & 0 & \cdots & 0 \\ 0 & 1 & -1 & 0 & \cdots & 0 \\ 0 & 0 & 1 & -1 & \cdots & 0 \\ \cdot & & & & & \\ \cdot & & & & & \\ \cdot & & & & & \\ 0 & 0 & \cdots & & 1 & -1 \\ 1 & 0 & \cdots & & 0 & -1 \end{pmatrix}$$

Letting r_i denote the ith row of this matrix, we see that $r_1 + r_2 + \cdots + r_{s-2} = r_{s-1}$. This contradicts the independence of the original system (the reader should check that the alternative conclusion $\phi(e_s) = t_1^{-1}$ is consistent with the nonsingularity hypothesis).

We shall call the configuration of Lemma 1 depicted below

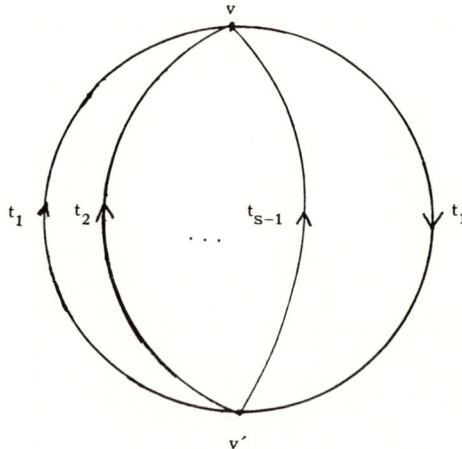

an *atom* (we are ignoring the component of $S^2 - X$ at ∞).

DEFINITION. A 2-vertex configuration of di-gons with internal labels $w_i^{\pm 1}$ is called *reducible* if two adjacent di-gons are opposites of each other. For example, the configuration shown below is reducible:

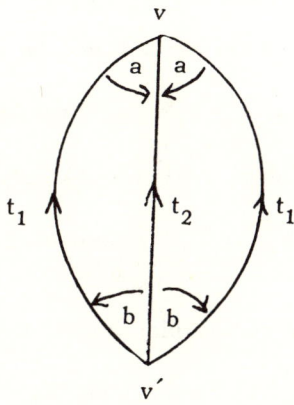

We can always remove pairs of adjacent opposite di-gons to assume our configurations are irreducible.

LEMMA 2. *In the situation of Lemma 1, suppose we are given another irreducible configuration of di-gons with two vertices (ignoring the component at ∞) and interior labels $w_i^{\pm 1}$ whose directed edges from v' to v are $e_1', e_2', \ldots, e_t'$ in order, counterclockwise from e_1'. Suppose that $\phi(e_1') = t_1$, $\phi(e_t') = t_1^{\pm 1}$, but $\phi(e_i') \neq t_1^{\pm 1}$ for $1 < i < t$. Then $s = t$ and either $\phi(e_i) = \phi(e_i')$ for $1 \leq i \leq s$, or $\phi(e_i') = \phi(e_{s+1-i})^{-1}$ for $1 \leq i \leq s$.*

REMARK. The content of Lemma 2 is that the atom of Lemma 1 is unique, up to switching top for bottom (i.e. interchanging v for v') and reading labels inverted in the reverse direction. A musician would recognize this as a retrograde inversion.

Proof. Construct the graph Σ whose vertex set is $\{t_i^{\pm 1}, 1 \leq i \leq n\}$ and whose edge set is the set of *unordered* pairs $\{\{t_i^{\varepsilon_i}, t_j^{\varepsilon_j}\} | \varepsilon_i = \pm 1, t_i^{\varepsilon_i} \neq t_j^{\varepsilon_j}\}$

(we are using "graph" in the sense of Bourbaki [1]). The graph Σ admits an involution which sends a vertex $t_i^{\epsilon_i}$ to $t_i^{-\epsilon_i}$ and sends an edge $\{t_i^{\epsilon_i}, t_j^{\epsilon_j}\}$ to $\{t_i^{-\epsilon_i}, t_j^{-\epsilon_j}\}$. Observe that the only fixed points of the involution are edges $\{t_i, t_i^{-1}\}$, $1 \leq i \leq n$. Let $\overline{\Sigma}$ be the quotient graph of Σ whose edges and vertices are orbits under the involution, and let $\rho = \Sigma \to \overline{\Sigma}$ be the orbit map. If we denote the image of vertex $t_i^{\epsilon_i}$ in $\overline{\Sigma}$ by $\tau_i^{\epsilon_i}$, and the image of edge $\{t_i^{\epsilon_i}, t_j^{\epsilon_j}\}$ by $\{\tau_i^{\epsilon_i}, \tau_j^{\epsilon_j}\}$, then $\tau_i = \tau_i^{-1}$, $1 \leq i \leq n$, and $\{\tau_i^{\epsilon_i}, \tau_j^{\epsilon_j}\} = \{\tau_i^{-\epsilon_i}, \tau_j^{-\epsilon_j}\}$. Note that $\overline{\Sigma}$ possesses loop-like edges $\{\tau_i, \tau_i^{-1}\}$.

An equation w_i of our nonsingular system determines an edge of $\overline{\Sigma}$ as follows. If $w_i = t_{i_1}^{\epsilon_1} a t_{i_2}^{\epsilon_2} b$, we associate to w_i the edge $\{\tau_{i_1}^{\epsilon_1}, \tau_{i_2}^{-\epsilon_2}\}$. Observe that w_i^{-1} and cyclic permutations of the letters of w_i or w_i^{-1} determine the same edge of $\overline{\Sigma}$. Observe also that the equation $t_i a t_i^{-1} b$ is never a member of a nonsingular system, so the construction is well defined (this construction is intimately related to the star graph construction of Whitehead's [10]).

We consider two subgraphs Γ and Γ' of $\overline{\Sigma}$ associated to the configurations $e_1, e_2, \cdots, e_s$ and $e_1', e_2', \cdots, e_t'$ of Lemmas 1 and 2 respectively. Here Γ is the smallest subgraph of $\overline{\Sigma}$ containing edges $\rho(\{\phi(e_i),$ $\phi(e_{i+1})^{-1}\}) = \{\tau_i, \tau_{i+1}^{-1}\}$ $(1 \leq i \leq s-2)$ and $\rho(\{\phi(e_{s-1}), \phi(e_s)^{-1}\} = \{\tau_{s-1}, \tau_1\}$ and Γ' is the smallest subgraph of $\overline{\Sigma}$ containing edges $\rho\{\phi(e_i'), \phi(e_{i+1}')^{-1}\}$, $1 \leq i \leq t-1$. Observe that Γ is a circle with $s-1$ subdivision points. In addition, Γ' has no end vertex, for otherwise we'd have a backtrack and the configuration of Lemma 2 would be reducible. We claim that $\Gamma = \Gamma'$. If not, then the Euler characteristic $\chi(\Gamma \cup \Gamma')$ is negative, so the connected graph $\Gamma \cup \Gamma'$ has more edges than vertices. These edges correspond to certain equations of our original system, so it follows that the determinant of the original system is zero, a contradiction.

Since $\Gamma = \Gamma'$ and since $\phi(e_1) = \phi(e_1') = t_1$, we see that the *oriented* cycle associated to the configuration $e_1', e_2', \cdots, e_t'$ is either directed the same as or oppositely to the *oriented* cycle associated to the original atom $e_1, e_2, \cdots, e_s$. From this observation and from the observation that the cycles are reduced it follows that $s = t$ and either $\phi(e_i') = \phi(e_i)$ for $1 \leq i \leq s$, or $\phi(e_i') = \phi(e_{s+1-i})^{-1}$ for $1 \leq i \leq s$. This completes the proof of Lemma 2.

Proof of Theorem 1'. If we are given a configuration in S^2 of di-gons with two vertices, with internal labels reduced cyclic words $w_i^{\pm 1}$, we may assume the configuration is irreducible. It follows from Lemmas 1 and 2 that the configuration is obtained by juxtaposing an even number of atoms, where adjacent atoms are reverses of each other, top to bottom, and read in opposite directions. It follows that the vertex labels $\mathcal{L}_v$ and $\mathcal{L}_{v'}$ are inverse to each other. This completes the proof of Theorem 1.

REMARK 1. It should be possible to use the uniqueness result of Lemma 2 for atoms to deduce a structure theorem for $\pi_2(L,K)$. Observe that in contrast with the form of the Kervaire conjecture proved in [9] in this volume, $\pi_2(L,K) \neq 0$ in general in our situation.

REMARK 2. Here is an alternative proof of Theorem 1' based on the following result of Baumslag and Taylor (a complete proof of which appears in [8] as Theorem 3.6): let G be a finitely generated free group, $\alpha, \beta \in G$. Suppose that α and β are conjugate mod G_n for all n, where $\{G_n\}$ is the p-lower central series for G. Then α and β are conjugate in G. In our nonsingular system of equations $w_1, w_2, \cdots, w_n$ over G, we may assume that $w_i = a_i t_{i_1}^{\varepsilon_1} b_i t_{i_2}^{\varepsilon_2}$ where elements $a_1, a_2, b_1, b_2, \cdots, a_n, b_n$ *freely* generate G (the general result follows by specializing the coefficients of the equations). The configuration of Theorem 1' of di-gons in S^2 with 2 vertices v and v' may be regarded (by excising tubular neighborhoods of v and v') as a combinatorial map of pairs $F: (S^1 \times I, \partial(S^1 \times I)) \to (L,K)$ for a suitable cellular subdivision of

$S^1 \times I$ consisting of quadrilaterals, two of whose sides lie on $\partial(S^1 \times I)$. If we slit $S^1 \times I$ along one of the interior edges, we obtain a map of a 2-cell into L, from which it follows that $\mathcal{L}_v^{-1}$ and $\mathcal{L}_{v'}$ are conjugate in $\pi_1(L) = G<t_1,t_2,\cdots,t_n>/<<w_1,w_2,\cdots,w_n>> = H$. If p is any prime number not dividing $\det(\eta_i(w_j))$, it follows from the fact that $H_*(L,K;\mathbb{Z}_p)$ = 0 that the inclusion $G \to H$ induces an isomorphism on H_1 and an epimorphism on H_2, with $\mathbb{Z}_p$ coefficients. From Stallings' theorem [7] it follows that $G/G_n \xrightarrow{\cong} H/H_n$ for all terms of the p-lower central series. Thus $\mathcal{L}_v^{-1}$ and $\mathcal{L}_{v'}$ are conjugate mod G_n for all n, and hence conjugate in G, by the result of Baumslag and Taylor quoted above.

§3. *Reciprocity law*

Having proved a version of the Kervaire conjecture, we feel we may indulge in a flight of fancy and propose a conjecture which is stronger than the Kervaire conjecture, which is geometrically motivated, and which is valid in the case of nonsingular equations of weight two.

If G is an arbitrary group, and $A_1, A_2, \cdots, A_n$ are subsets of G, we define $\prod_{i=1}^{n} A_i$ to be $\{a_1 a_2 \cdots a_n \in G | a_i \in A_i\}$. In general this product depends on order of factors.

LEMMA 3. *If A_i is the conjugacy class of the element a_i in the group G, then $\prod_{i=1}^{n} A_i$ is independent of order of factors.*

DEFINITION. A *Howie diagram* for the nonsingular system of equations $w_1, w_2, \cdots, w_n$ over the group G is a pair (C, ϕ) where C is a cellular subdivision of S^2 with oriented 1-skeleton $C^{(1)}$ and where ϕ, the labeling function, associates to each oriented edge (1-cell) of C an indeterminate $t_i^{\pm 1}$, $1 \leq i \leq n$, of our system of equations. One demands $\phi(\bar{e}) = \phi(e)^{-1}$, where $\bar{e}$ is the oppositely oriented 1-cell to e. In addition we demand that ϕ associate to each corner of each 2-cell of C an element of G. One requires in addition one axiom be satisfied:

HD. Reading the labels around any face (2-cell) of C in order clockwise from a suitable starting point gives either w_i or w_i^{-1} in cyclically reduced form.

REMARK. The notion of Howie diagram is abstracted from Howie's treatment of the equation $atbtct^{-1}$ in [5].

Given a Howie diagram (C,ϕ), we may form the vertex label $\mathcal{L}_v$ for any vertex (0-cell) v of C, by reading the corner labels around v in order counterclockwise. The label $\mathcal{L}_v$ is interpreted as a conjugacy class in G, so is independent of starting point.

CONJECTURE (Reciprocity Law). If (C,ϕ) is a Howie diagram for the nonsingular system of equations $w_1, w_2, \cdots, w_n$ over G, then
$$1 \in \prod_{v \in C(0)} \mathcal{L}_v.$$

REMARK 1. Observe that by Lemma 3, the product $\prod_{v \in C(0)} \mathcal{L}_v$ is independent of order of factors.

REMARK 2. The conjecture is true for nonsingular systems of equations of weight two. One reduces the consideration of the general Howie diagram for this case to Theorem 1'.

§4. *Relation with Stallings' conjecture* $B_n(p)$

In [8] Stallings proposed a number of group theoretic conjectures, each of which implies the Kervaire conjecture (one of Stallings' conjectures, $C_n(p)$, has been shown to be false by J. Howie.* This does not affect the conjecture $B_n(p)$ we shall discuss).

We write g^G for the conjugacy class of the element g in the group G. If $a, g_1, g_2, \cdots, g_n \in G$, we say that a is *dependent* on $(g_1, g_2, \cdots, g_n)$

---

*J. Howie, Math Z. *187* (1984), 25-27.

if $a \in \prod_{i=1}^{n} g_i^G$. Observe that this notion is independent of order of factors in the product by Lemma 3. We also write $g^x = xgx^{-1}$ for $x, g \in G$.

STALLINGS' CONJECTURE $B_n(p)$. Let p be a prime number. Let S be a subgroup of the finitely generated free group F with the induced map $H_1(S, Z_p) \to H_1(F, Z_p)$ an isomorphism. Then for all $a, g_1, g_2, \cdots, g_n$ in S, if a is dependent on $(g_1, g_2, \cdots, g_n)$ in F, then a is dependent on $(g_1, g_2, \cdots, g_n)$ in S.

DEFINITION. We let "conjecture B" denote the conjunction of all assertions $B_n(p)$ for all n and all prime numbers p.

DEFINITION. If G is a group, a *Kervaire extension* $G \to H$ is a homomorphism of groups induced by inclusion where $H = G * <t_1, t_2, \cdots, t_n> / \ll w_1, w_2, \cdots, w_n \gg$, with $\det(\eta_i(w_j)) = 0$. Here, the group $<t_1, t_2, \cdots, t_n>$ is freely generated by $t_1, t_2, \cdots, t_n$ and $\ll w_1, w_2, \cdots, w_n \gg$ denotes the normal closure of the equations $w_1, w_2, \cdots, w_n$. The Kervaire conjecture states that the map $G \to H$ is injective for a Kervaire extension.

CONJECTURE G. If $G \to H$ is a Kervaire extension and if $g_1, g_2, \cdots, g_r \in G$, then $1 \in \prod_{i=1}^{r} g_i^H$ implies $1 \in \prod_{i=1}^{r} g_i^G$ (here g_i^H means the conjugacy class of the image of g_i in H).

REMARK. The Kervaire conjecture is the special case of conjecture G for $r = 1$.

We can now state the main results of this section.

PROPOSITION 3. Conjecture G is equivalent to Reciprocity Law of §3.

PROPOSITION 4. Conjecture B is equivalent to conjecture G.

Proof of Proposition 3. We show first that "G" implies the Reciprocity Law. Let (C, ϕ) be a Howie diagram for the nonsingular system of equations $w_1, w_2, \cdots, w_n$ over G and let $G \to H$ be the corresponding Kervaire extension. Let T be a maximal tree in C and let U be a regular neighborhood of T in S^2.

Let (L,K) be a relative 2-complex for the system $w_1, w_2, \cdots, w_n$. Thus $\pi_1(K) = G$ and L is obtained by attaching n 1-cells trivially to K, then attaching n 2-cells by maps corresponding to $w_1, w_2, \cdots, w_n$. The Howie diagram (C, ϕ) may be viewed as a combinatorial map $(S^2 - U_1, \partial U_1) \xrightarrow{f} (L,K)$ where U_1 is a regular neighborhood of $C^{(0)}$ in S^2, so $U_1 \subset U$.

Now applying f to ∂U and reading the resulting homotopy element in the exterior of U gives an element of $\pi_1(L)$ which is 1. But read in the interior of U, $f|\partial U$ is a product of conjugates in H of the vertex labels, i.e.

$$1 \in \prod_{v \in C^{(0)}} L_v^H.$$

Now apply conjecture G to get

$$1 \in \prod_{v \in C^{(0)}} L_v^G = \prod_{v \in C^{(0)}} L_v,$$

which is the reciprocity law.

REMARK. The conjectural reciprocity law can be stated in a more geometric form as follows (it is an easy exercise to prove the equivalence of the two formulations): Let $f : (M^2, \partial M^2) \to (L,K)$ be a combinatorial map [5], where M^2 is a cell structure on a compact oriented 2-manifold of genus zero and (L,K) is a relative 2-complex with $H_2(L,K) = \{0\}$. Assume that $f^{-1}(K) = \partial M$. Then

$$1 \in \prod_{i=1}^{n} f_*([C_i]),$$

where $C_1, C_2, \cdots, C_n$ are the connected components of the boundary ∂M, oriented compatibly with M, and $f_*([C_i])$ is the conjugacy class in $\pi_1(K)$ of the map $S^1 \xrightarrow{\cong} C_i \xrightarrow{f|C_i} K$, where $S^1 \xrightarrow{\cong} C_i$ is an orientation preserving homeomorphism.

We proceed now to the proof that the Reciprocity Law implies conjecture G. Suppose that $g_1, g_2, \cdots, g_n \in G$ with $1 \in \prod_{i=1}^{n} g_i^H$, where $G \to H$ is a Kervaire extension. Construct a map $f: (M^2, \partial M) \to (L, K)$, where M^2 is a disc with holes and (L, K) is a relative 2-complex associated to the Kervaire extension $G \to H$, so that the boundary components of M are mapped by $1, g_1, g_2, \cdots, g_n$. This is possible since $1 \in \prod_{i=1}^{n} g_i^H$. We may assume that f is transverse regular to the centers of the 2-cells of $L - K$. If A is the preimage under f of the centers of these 2-cells, then A is a finite set of points in the interior of M. If we remove an ε-neighborhood U of A from M, then $f|(M-U)$ maps into a subspace of L which deformation retracts onto $K \cup L^{(1)}$. If we assume the composite map $f_1: (M-U) \to K \cup L^{(1)}$ is transverse regular to the centers of the 1-cells of $L-K$, we may take the preimage of these centers under f_1 to get a 1-manifold Γ_1 properly imbedded in $M-U$ and such that $\Gamma_1 \cap \partial M = \phi$. We extend Γ_1 to a graph Γ tamely imbedded in the interior of M by coning over the components of ∂U, so that A is the set of vertices of Γ. The graph Γ has the property that $M - \Gamma$ is mapped into a subset of L which deformation retracts onto K. Using this latter property we can assign labels in $\pi_1(K)$ to the corners of the faces (connected components) of $M - \Gamma$.

Filling in holes, Γ may be considered to be a graph in S^2. Note that a component of $S^2 - \Gamma$ may contain components of ∂M or may contain none at all. If we take an innermost component of Γ, we can reduce the problem to the case where Γ is connected, which we assume is the case from now on.

Let X be a component of $S^2 - \Gamma$. Let Λ_X denote the conjugacy class in $G = \pi_1(K)$ of the product of the corner labels along the boundary of X in order.

LEMMA 4. *Let $C_{X,1}, C_{X,2}, \cdots, C_{X,r}$ be the connected components of ∂M in the interior of X. Then $\Lambda_X \subseteq \prod_{i=1}^{r} f_*[C_{X,i}]$ in G.*

Proof. Since $\Gamma \cap \overset{\circ}{X} = \phi$, the cell X is mapped by f into a subset of L which deformation retracts to K. If we connect the components $C_{X,i}$ by arcs in $\overset{\circ}{X}$ to get a circuit γ that runs around each $C_{X,i}$ precisely once, we can arrange that γ is homotopic in $\overset{\circ}{X} \cap M$ to a circuit δ that runs parallel to ∂X and is in $\overset{\circ}{X}$. From this it follows that

$$\Lambda_X = f_*(\delta) = f_*(\gamma) \subseteq \prod_{i=1}^{r} f_*[C_{X,i}]$$

as required.

Now let Δ be the dual graph to Γ in S^2 and label the edges and corners of Δ using f to get a Howie diagram. If $v_1, v_2, \cdots, v_m$ are the vertices of Δ, let L_{v_i} be the vertex label for v_i (product of the corner labels at v_i in order). The reciprocity law for the Howie diagram Δ says that

$$1 \in \prod_{i=1}^{m} L_{v_i} \quad \text{in } \pi_1(K) = G.$$

But by Lemma 4, each L_{v_i} has an element in common with

$$\prod_j f_*[C_{X_i,j}] \quad \text{in } G,$$

where X_i is the face of $S^2 - \Gamma$ corresponding to v_i and the index set j runs over indices of components $C_{X_i,j}$ of ∂M contained in X_i. It follows that

$$1 \in \prod_C f_*[C] \text{ in } G,$$

where C runs over the (oriented) components of ∂M. But this means $1 \in \prod_{i=1}^{n} g_i^G$, as required. The proof of Proposition 3 is complete.

REMARK. The reciprocity law, in its geometrical form, is valid if ∂M^2 has at most two connected components. This amounts to a restatement of Theorem 1.

Proof of Proposition 4. We assume conjecture B and we let $G \to H$ be a Kervaire extension where $H = G*\langle t_1, t_2, \cdots, t_n \rangle / \langle\langle w_1, w_2, \cdots, w_n \rangle\rangle$ and let p be a prime number not dividing $\det(\eta_i(w_j))$. We may assume that G is finitely generated without loss of generality. Let S be finitely generated and free and let $\phi: S \twoheadrightarrow G$ be a surjective homomorphism. Let $T = \langle t_1, t_2, \cdots, t_n \rangle$ and let $\hat{w}_i \in S*T$ be chosen such that $(\phi*1_T)(\hat{w}_i) = w_i$. Let W be free on symbols $\tilde{w}_1, \tilde{w}_2, \cdots, \tilde{w}_n$ and construct the commutative diagram with exact columns below

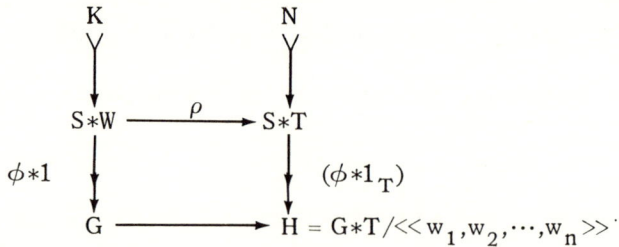

where $\rho|S = 1_S$ and $\rho(\tilde{w}_i) = \hat{w}_i$ and where the map $\phi*1$ maps S by ϕ and maps W trivially. The map ρ is a mod p homology isomorphism, hence injective. One checks that N is the normal closure of K in $S*T$.

Let $x_i \in S$ be such that $\phi(x_i) = g_i$, $1 \leq i \leq r$. If $1 \in \prod_{i=1}^{r} g_i^H$, we deduce from the diagram above that there are elements $k_1, k_2, \cdots, k_s \in K$ such that

$$1 \in \prod_{i=1}^{r} x_i^{S*T} \prod_{i=1}^{s} k_j^{S*T}.$$

By conjecture $B_{r+s}(p)$, we deduce that

$$1 \in \prod_{i=1}^{r} x_i^{S*W} \prod_{j=1}^{s} k_j^{S*W}.$$

Now apply the map $\phi * 1$ to get

$$1 \in \prod_{i=1}^{r} g_i^G,$$

which is conjecture G.

Conversely, assume conjecture G and let $\phi : G \to F$ be a mod p homology isomorphism of finitely generated free groups G and F, where p is a prime number. We can find a commutative diagram

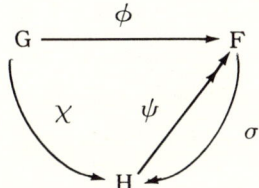

where the map $\chi : G \to H$ is a Kervaire extension, where ψ is surjective, and where $\psi\sigma = 1_F$, $\sigma\phi = \chi$, and $\psi\chi = \phi$. To see this, let $s_1, s_2, \cdots, s_n$ be a free basis for G and let $x_1, x_2, \cdots, x_n$ be a free basis for F. Then there exist free words W_i in n variables such that $W_i(x_1, x_2, \cdots, x_n) = \phi(s_i)$ and $\det \eta_i(W_j) \not\equiv 0 \pmod{p}$. Let

$$H = G<t_1, t_2, \cdots, t_n>/<<W_i(t_1, t_2, \cdots, t_n)s_i^{-1}, 1 \leq i \leq n>>.$$

Define $\chi : G \to H$ to be induced by inclusion and define $\psi : H \twoheadrightarrow F$ by $\psi\chi(g) = \phi(g)$ and $\psi(t_i) = x_i$, $1 \leq i \leq n$. This is possible since χ is necessarily injective. Since F is free, we can define $\sigma: F \to H$ by $\sigma(x_i) = t_i$, $1 \leq i \leq n$. Clearly $\psi\sigma = 1_F$. It remains to check that $\sigma\phi = \chi$. But $\sigma\phi(s_i) = \sigma(W_i(x_1, x_2, \cdots, x_n)) = W_i(t_1, t_2, \cdots, t_n) = \chi(s_i)$, so the claims are established.

Suppose now that $g_1, g_2, \cdots, g_r \in G$ and $1 \in \prod_{i=1}^{r} \phi(g_i)^F$. By applying σ we deduce that $1 \in \prod_{i=1}^{r} \chi(g_i)^H$. But conjecture G now shows that $1 \in \prod_{i=1}^{r} g_i^G$. Thus conjecture B follows. This completes the proof of Proposition 4.

EXAMPLE. Let $F = F(x,y)$ be a free group freely generated by x and y and let S be the subgroup of F generated by xyx^{-1} and yxy^{-1}. Then S is freely generated by these elements and the inclusion $S \to F$ is a Z-homology isomorphism. We claim that $S \to F$ is a Kervaire extension. Observe first that $x^{-1}(xyx^{-1}) \cdot x \cdot (xyx^{-1})^{-1} \cdot x \cdot (yxy^{-1})^{-1} = 1$.

We define a homomorphism $\phi: F(a,b,t) \to F(x,y)$ by $a \to xyx^{-1}$, $b \to yxy^{-1}$, and $t \to x$, where $F(a,b,t)$ is freely generated by a, b, and t. Then ϕ is surjective and contains $t^{-1}ata^{-1}tb^{-1}$ in its kernel. However the element $t^{-1}ata^{-1}tb^{-1}$ is clearly primitive, so the factor group $F(a,b,t)/\ll t^{-1}ata^{-1}tb^{-1} \gg$ is free. Since finitely generated free groups are Hopfian, this implies that ϕ induces an isomorphism $F(a,b,t)/\ll t^{-1}ata^{-1}tb^{-1} \gg \to F(x,y)$. Since the image of $F(a,b)$ under this isomorphism is S, it follows that the inclusion $S \to F$ is a Kervaire extension.

However the general equation in one variable u of weight 3 and exponent sum 1, $w = \alpha\beta u y u^{-1}$, can, by a change of variables, be put in the form $ata^{-1}tb^{-1}t^{-1}$, where $a = \alpha$, $ab^{-1}a^{-1} = \beta^{-1}\gamma\beta$, and $t = u\beta\alpha$. This is of course the equation considered by Howie [5] for which he proved the Kervaire conjecture is valid.

REMARK. It can be shown that the reciprocity law is equivalent to the following extension property of maps. Let M be a compact oriented surface of genus 0 and let (L,K) be a relative 2-complex with K connected and with $H_2(L,K) = 0$.

THEOREM 2. *The reciprocity law holds if and only if every map* $f: \partial M \to K$ *which extends to a map* $f_1: M \to L$ *also extends to a map* $f_2: M \to K$.

We omit the proof. Observe however that the Kervaire conjecture is the special case $M = D^2$.

We shall finish with one result which produces examples of equations for which the reciprocity law holds.

PROPOSITION 5. *The reciprocity law holds for a single equation of weight* $n > 0$ *and exponent sum* n.

Proof. We may assume that the equation is $w = a_1 t a_2 t \cdots a_n t$, where $a_1, a_2, \ldots, a_n \in G$, the coefficient group. Let $H = G{<}t{>}/{\ll}w{\gg}$ and assume that $g_1, g_2, \ldots, g_r \in G$ with $1 = \prod_{i=1}^{r} g_i^{h_i}$, for $h_i \in H$. Let $H_1 = G \wr Z_n = G^n \rtimes Z_n$, the wreath product. By a result of F. Levin's [6], there is a commutative diagram

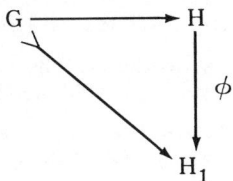

where $G \to H_1$ imbeds G diagonally in $G^n \subset H_1$, $g \mapsto (g, g, \ldots, g)$. Let $k_i = \phi(h_i)$, so we have $1 = \prod_{i=1}^{r} g_i^{k_i}$ in H_1.

LEMMA 5. *If* $x \in G$ *and* $y \in H_1$, *then* x^y *is in* $G^n \subset H_1$, *and* x^y *is of the form* $(x^{y_1}, x^{y_2}, \ldots, x^{y_n})$ *for* $y_1, y_2, \ldots, y_n \in G$.

Proof. Let τ generate Z_n, acting as the shift operator on $G^n \subset H_1$. If we conjugate an element of G^n of the form $(x^{y_1}, x^{y_2}, \cdots, x^{y_n}) = \xi$ $(y_i \in G)$ by $(z_1, z_2, \cdots, z_n) \in G^n$, we obtain $(x^{z_1 y_1}, x^{z_2 y_2}, \cdots, z^{z_n y_n})$. If we conjugate ξ by τ, we obtain $(x^{y_2}, x^{y_3}, \cdots, x^{y_n}, x^{y_1})$. Since G_n and τ generate H_1, the lemma follows.

Returning to the proof of the proposition, we observe that each factor $g_i^{k_i}$ in the expression $1 = \prod_{i=1}^{r} g_i^{k_i}$ has the form x^y of Lemma 5, so $g_i^{k_i}$ lies in G^n and is an n-tuple of conjugates of g_i in G. If we project into the first factor, we obtain $1 \in \prod_{i=1}^{r} g_i^G$, as required.

DEPARTMENT OF MATHEMATICS
UNIVERSITY OF UTAH
SALT LAKE CITY, UTAH 84112

BIBLIOGRAPHY

[1] N. Bourbaki, Groupes et algèbres de Lie, Chapitre IV §2 annex, Hermann 1968.

[2] S. M. Gersten, Conservative groups, indicability and conjecture of Howie, J. Pure Appl. Alg. *29* (1983), 59-74.

[3] M. Gerstenhaber and O. S. Rothaus, The solution of sets of equations in groups, Proc. Nat. Acad. Sci. USA *48* (1962), 1531-1533.

[4] J. Howie, On pairs of 2-complexes and systems of equations over groups, J. reine angew. Math. *324* (1981), 165-174.

[5] ———, The solution of length three equations over groups, Proceedings of the Edinburgh Math. Soc. (1983), *26*, 89-96.

[6] F. Levin, Bull. Amer. Math. Soc. *68* (1962), 603-604.

[7] J. Stallings, Homology and central series of groups, J. Algebra *2* (1965), 170-181.

[8] ———, Surfaces in three manifolds and non-singular equations in groups, Math Z. *184* (1983), 1-17.

[9] ———, these proceedings. A graph-theoretic lemma and group-embeddings.

[10] J. H. C. Whitehead, On certain sets of elements in a free group, Proc. London Math. Soc. *41* (1936), 48-56.

A GRAPH-THEORETIC LEMMA AND GROUP-EMBEDDINGS

John R. Stallings[*]

Abstract. A graph-theoretic lemma is proved, which states, roughly, that in a town in the plane in which all the streets are one-way (and no corner is arranged to be a source or a sink), there is some block around which the traffic circulates. This, and the theory of diagrams of relations, is used to prove a theorem (3.1) about embedding semigroups in groups (similar to a theorem of Adyan's , and a theorem (4.1) about solving a series of equations over a group if a peculiar freeness hypothesis on the equations holds.

1. Graph-theoretic Lemma

Let Γ be a finite directed graph, realized as a subspace of the 2-sphere S^2. "Directed" means that each edge of Γ has an arrow on it.

A *source* of Γ is a vertex such that all adjacent edges point away. A *sink* is a vertex towards which all adjacent edges point.

Orient S^2. A *clockwise region* of $S^2-\Gamma$ is a simply-connected component of $S^2-\Gamma$ whose boundary consists of edges of Γ, all arrows of which go clockwise. There may also be *counter-clockwise regions*.

Any one of these four phenomena will be called a *consistent item* of (S^2,Γ).

1.1. Note that if, in clockwise order around a vertex v, there is no instance of an edge pointing towards v immediately followed by an edge pointing away from v then v is either a source or a sink. Similarly, if, in counter-clockwise order around the boundary of a simply connected region E of $S^2-\Gamma$, there is no instance of an edge of Γ pointed counter-clockwise immediately followed by an edge pointed clockwise, then E is either a clockwise or a counter-clockwise region.

---

[*]Partly supported by NSF Grant DMS 83-03283.

1.2. LEMMA. *Let Γ be a finite directed graph, non-empty, without isolated vertices, embedded in S^2. Then there are at least two consistent items of (S^2, Γ).*

Proof. If Γ has n connected components, then we can produce a graph Γ' which is connected, by adding n–1 edges to Γ. This does not add any new source or sink vertices; and the regions of $S^2-\Gamma'$ having the new edges on their boundaries cannot be clockwise or counter-clockwise, since a new edge is traversed twice, in opposite directions, on the boundary of such a region. This shows that it suffices to consider the case where Γ is connected; in that case each component of $S^2-\Gamma$ is automatically simply-connected.

For each edge e of Γ, consider the edges in clockwise order around the target vertex v of e. Let the next edge after e be e'. If v is the source vertex of e', associate a "red dot" to v. If v is the target vertex of e', associate a "green dot" to the region E of $S^2-\Gamma$ to the left of e. The total number of dots is the number of edges. Let r(v) be the number of red dots associated to v and g(E) be the number of green dots associated to E. Then

$$\chi(S^2) = 2$$
$$= \#(\text{vertices}) + \#(\text{regions}) - \#(\text{edges})$$
$$= \Sigma(1-r(v)) + \Sigma(1-g(E)).$$

Therefore, for at least two items of the form v or E, r(v) = 0 or g(E) = 0. By 1.1 these are consistent items. □

1.3. COROLLARY. *Let Γ be a finite, non-empty, directed graph without isolated vertices, embedded in the plane R^2. Suppose that Γ has neither a source nor a sink. Then there is a bounded, simply-connected component E of $R^2-\Gamma$, such that all the edges of Γ on the boundary of E circulate coherently (clockwise or counter-clockwise).*

Proof. After adding the point at ∞, 1.2 shows there are at least two consistent regions, at most one of which contains ∞. □

1.4. REMARK. This result was suggested to me by conversations with Royce Wolf and Frank Rimlinger. The proof given here is nearly identical with that of Glass [Gl], who extends the equivalent result to other 2-manifolds. Remmers [Re] uses something like this in the theory of semigroups, and various topologists, such as Scharlemann [Sch], in the theory of 3-manifolds. Sieradski [Si], p. 98, proves a more general fact, although the statement of his lemma is weaker than what he proves. There must be other combinatorial theorems about planar graphs along these lines which will have group-theoretic consequences.

2. *Topological set-up*

The following fact is a standard kind of thing in the interface of group-theory and topology (Cf. [Si]). It is proved by interpreting the hypothesis and smoothing things out.

2.1. Let $K_0 \subset K_1 \subset K_2$ be CW-complexes. K_0 is arbitrary. K_1 is assumed to be got from K_0 by attaching 1-cells $\{e_i\}$ along their endpoints; let c_i denote the midpoint of e_i. K_2 is assumed to be got from K_1 by attaching 2-cells $\{D_j\}$ along their boundary curves by a reasonably nice map of the boundary curves into K_1.

Let γ be a closed path in K_1. We suppose that γ is reasonably nice, that is, given by a non-degenerate graph-map from a subdivided circle into the 1-skeleton of K_1. Assume that γ is contractible in K_2.

Then under these conditions: There is a map $f : \Delta \to K_2$, where Δ is a 2-cell, and there is a finite family of mutually disjoint 2-cells $\{\Delta_k\}$, called "relator-cells," in Int Δ, such that:

.0. $f|\text{Bd }\Delta = \gamma$.

.1. $f(\Delta - \cup\{\text{Int }\Delta_k\}) \subset K_1$.

.2. For each k there exists j such that $f|\text{Int }\Delta_k$ is a homeomorphism onto Int D_j.

.3. Let $\Theta = \Delta - \cup\{\text{Int } \Delta_k\}$. Then $f|\Theta$ is transverse to the points $\{c_i\}$; so that $f^{-1}(c_i) = A_i$ is a 1-manifold properly embedded in Θ. Orienting e_i gives A_i a normal orientation, and thus, by virtue of an orientation of Δ, each A_i receives an orientation.

2.2. We call this situation a *diagram* of the fact that γ is null-homotopic in K_2. The *complexity* of such a diagram is the lexicographically ordered pair: (a) the number of cells Δ_k, (b) the number of components of $\cup\{A_i\}$.

2.3. LEMMA. *In a diagram as above, suppose that E is a simply-connected component of $\Theta - \cup\{A_i\}$. As one goes around Bd E, one encounters parts of A_i and arcs on Bd Δ_k. The images under f of the arcs on Bd Δ_k can be interpreted as being elements a_n in the fundamental groupoid $\pi(K_0)$. The product of the elements a_n, in order around Bd E, is then an identity element in $\pi(K_0)$.*

Proof. This happens because E is a 2-cell and the parts of A_i on Bd E go to constant paths. What we really have is something going on in $K_1 - \cup\{c_i\}$, but this deforms canonically to K_0. □

2.4. LEMMA. *Suppose we have a diagram of the fact that γ is null-homotopic in K_2, and which has minimal complexity. Then:*

.1. If D_j is any 2-cell attached by a null-homotopic map of its boundary into K_1, then D_j does not occur as the image of any 2-cell Δ_k of the diagram.

.2. No component of A_i is a simple closed curve.

.3. Let D_1, D_2, D_3 be 2-cells of K_2 attached by boundary maps of the form uv^{-1}, vw^{-1}, and uw^{-1}, where u, v, w are paths with equal initial and equal terminal points in K_1. Then there is no instance of Δ_1 and Δ_2 mapping to D_1 and D_2, and an arc L in some A_i joining points on Bd Δ_1 and on Bd Δ_2 that correspond to exactly the same point of v (that is, v may go across c_i in several spots; the endpoints

of L match up exactly corresponding spots; if this happened, the complexity could be reduced).

.4. Let Δ_1 and Δ_2 both map to D_1. Then there is no arc L of A_i joining points of Δ_1 and Δ_2 that correspond to exactly the same place where $Bd\ D_1$ crosses c_i.

Proof. Figure 1 is a picture of the proof of .3.

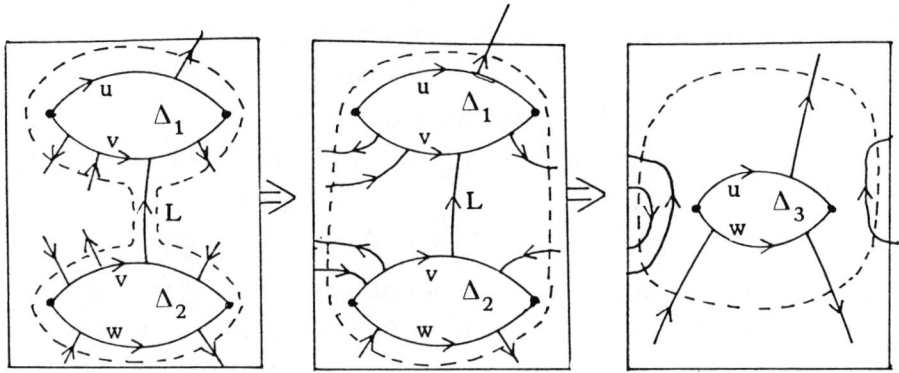

Figure 1

That is, the map f is modified on a 2-cell neighborhood of $\Delta_1 \cup \Delta_2 \cup L$ so as to involve only a single relator-cell Δ_3 which maps to D_3. Note that this reduces both parts of the complexity.

The proofs of .1 and .4 are similar to that of .3. In .1 we eliminate one relator cell which maps trivially. In .4 we eliminate two relator cells which are mirror images of each other along the arc L.

As for .2, if C is a simple closed curve in A_i, it bounds a 2-cell $\Delta' \subset \Delta$. The diagram can be redefined to map Δ' to c_i and then moved slightly to eliminate C and anything in Δ'. □

3. The semigroup embedding Theorem

3.1. THEOREM (Adyan [Ad1, Ad2], Remmers [Re]). *Let $\mathcal{A}$ be an alphabet. Let $\mathcal{U}$ be a set of non-empty words in $\mathcal{A}$ with the property* that no two elements of $\mathcal{U}$ have the same first letter and no two have the same last letter. Let $\mathcal{E} \subset \mathcal{U} \times \mathcal{U}$. Let $\mathcal{P}$ denote the presentation:

$$\{\mathcal{A} : \forall (u,v) \in \mathcal{E}, u = v\}$$

Let $S(\mathcal{P})$ and $G(\mathcal{P})$ be the semigroup and group defined by this presentation. Then the natural map $S(\mathcal{P}) \to G(\mathcal{P})$ is injective.

Proof. First, we can suppose $\mathcal{E}$ is an equivalence relation on $\mathcal{U}$. Suppose that p and q are words in $\mathcal{A}$ such that pq^{-1} is in the normal closure of the elements $\{uv^{-1} | (u,v) \in \mathcal{E}\}$ in the free group $F(\mathcal{A})$. We must show that $p = q$ is true in the semigroup $S(\mathcal{P})$.

The topological set-up is thus: K_0 is a point. K_1 is got by attaching a 1-cell e_α for each letter $\alpha \in \mathcal{A}$. K_2 is got by attaching a 2-cell $D(u,v)$ for each $(u,v) \in \mathcal{E}$, along the attaching map uv^{-1}. Then $\gamma = pq^{-1}$ is null-homotopic in K_2. There is a diagram of this fact, of minimal complexity. In this diagram, the boundary of Δ maps by pq^{-1}, and there are in Δ 2-cells Δ_j mapping to $D(u,v)$, and $\Theta = \Delta - \cup\{\Delta_j\}$ maps to K_1. By 2.4.2, each component of $A_\alpha = f^{-1}(c_\alpha)$ is an oriented arc. Let us call a *circuit* a series of these oriented arcs that connect up the Δ_j in a circular fashion.

The claim is: *There is no circuit.* If there were, it would bound a part of Θ, call it Ψ. We can imagine Ψ to be a graph in Δ by imagining each Δ_j to be a point, these points being joined by some of the oriented arcs in $\cup\{A_\alpha\}$. Now, Ψ has no source or sink vertex; none on the bounding circuit since the arcs there are circularly oriented; and none elsewhere since each Δ_i has edges going out and in, because uv^{-1} contains both positive and negative exponents.

By Corollary 1.3 there is a consistent region within Ψ. In this region E, look at a vertex Δ_i on its boundary; this joins up to a vertex Δ_j on the boundary of E by an arc L in A_α. If $\Delta_i = \Delta_j$ the picture is Figure 2, in which case u and v have the same last letter; thus $u = v$

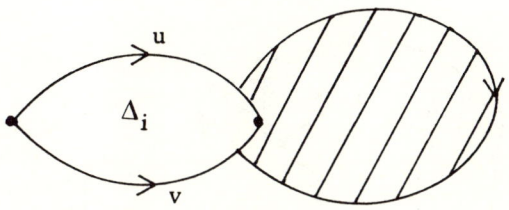

Figure 2

and the complexity could be reduced by 2.4.1 (it is also possible to have a circuit showing the first letters of u and v are equal). If Δ_i is not equal to Δ_j the picture is Figure 3. Here v and v´ have the same last

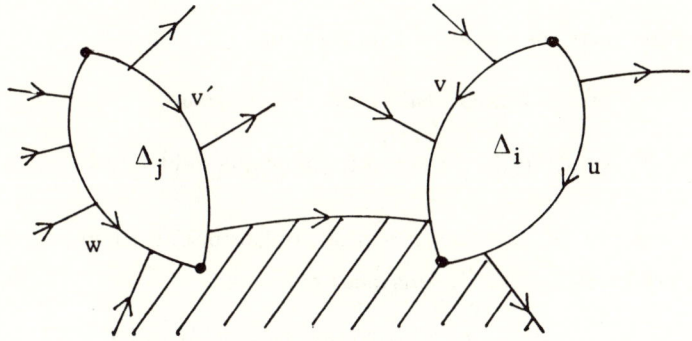

Figure 3

letter and so are equal. Thus, by 2.4.3 the two Δs can be replaced by one whose boundary maps to uw^{-1}. (There is also the dual possibility regarding first letters.)

Since now there is no circuit, we can follow the arrows from Δ_i to Δ_j and so on until we can go no further. Suppose that Δ_1 is where we stop. See Figure 4. Bd Δ_1 maps by uv^{-1}; what we see from this picture is that p must contain a subword u. If we replace u by v at this occurrence, we get p´. Now p = p´ in $S(\mathcal{P})$, and the picture shows that $p´q^{-1}$ has a diagram involving one less Δ_j. An inductive argument then shows p = q in $S(\mathcal{P})$. □

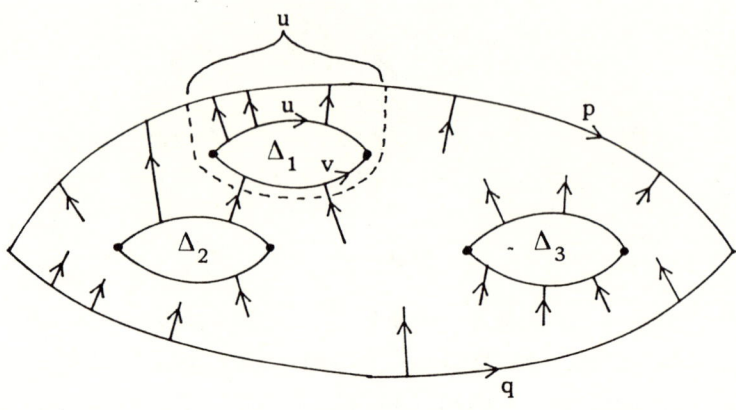

Figure 4

3.2. Here are some instances of this theorem:

$\mathcal{P}_1$ $\{a,b : aba = bab\}$, the trefoil group,

$\mathcal{P}_2 = \{a,b,c,d,e,f : adc = cfb = bea, dae = ebf = fcd\}$

the free product of $\mathbf{Z}$ with the group of the Borromean Rings. It appears that there are many such 3-dimensional examples.

3.3. A close examination of the proof would show the additional fact that, for any presentation $\mathcal{P}$ satisfying the hypotheses of Theorem 3.1, if we reduce the number of relations to the minimum that will give the same group, then the 2-complex of that presentation is aspherical. There is therefore a suggestion of a connection with Papakyriakopoulos's Sphere Theorem.

3.4. REMARK. James Howie and Roger Lyndon pointed out to me that this theorem is an easy consequence of a theorem by Remmers (4.6 in [Re]); the proof given here is closely related to his proof, although the arrangement of ideas is somewhat different. Remmers in fact reproved, using geometric intuition in place of combinatorial arguments, a theorem by Adyan (Theorem 5 in [Ad1], Theorem 3 of Chapter 2 in [Ad2]).

4. The solution of certain equations over a group

4.0. Consider the free product $G*Z$ where Z is infinite cyclic generated by t. Let w be a cyclically reduced element. An occurrence of $\cdots tat^{-1}\cdots$ in w considered cyclically, for $a \in G$, will be called *positive*, and a will be called the *corresponding element* of G. Thus, in $t^{-1}atbt^{-2}ctdte$, there are two positive occurrences, with corresponding elements b and e. An occurrence of $\cdots t^{-1}at\cdots$ will be called *negative*. In the element above, there are two negative occurrences with corresponding elements a and c. Note that positive occurrences in w^{-1} have corresponding elements that are the inverses of the elements corresponding to positive occurrences in w; also note that there are exactly as many negative as positive occurrences in w.

Let $\{w_i\}$ be a set of elements of $G*Z$. We denote by $G_+\{w_i\}$ the subgroup of G generated by all the elements corresponding to positive occurrences in the w_i. Similarly, $G_-\{w_i\}$ is the subgroup generated by the elements corresponding to negative occurrences.

To say that $G_+\{w_i\}$ is "free on the positive occurrences" is equivalent, when $\{w_i\}$ is finite, to the assertion that $G_+\{w_i\}$ is a free group of rank equal to the number of positive occurrences in the list $\{w_i\}$. In particular, no element of G of finite order is an element corresponding to a positive occurrence in w_i, and no elements of G corresponding to different positive occurrences can equal each other.

4.1. THEOREM. *Let $\{w_i\}$ be a set of cyclically reduced elements in the free product $G*Z$ where Z is infinite cyclic generated by t. Suppose that in each w_i both t and t^{-1} occur. Let $G_+ = G_+\{w_i\}$ and $G_- = G_-\{w_i\}$. Suppose that G_+ is the free group with basis the positive occurrences in $\{w_i\}$ and that G_- is free on the negative occurrences. Let N be the normal closure of $\{w_i\}$ in $G*Z$. Then the map $G \to (G*Z)/N$ is injective.*

[*Thus, the equations $\{w_i = 1\}$, considering t as the "unknown," have a simultaneous solution in the "overgroup" $(G*Z)/N$ containing G.*]

Proof. Let K_0 be the Eilenberg-Mac Lane space $K(G,1)$; $K_1 = K_0$ wedged with a circle "t"; K_2 is K_1 with 2-cells $\{D_i\}$ attached along $\{w_i\}$. Let γ be a closed path in K_0 which is null-homotopic in K_2; we want to show γ is null-homotopic in K_0. There is a diagram of minimal complexity of the fact that γ is null-homotopic in K_2.

The claim is that in this diagram there are no 2-cells Δ_k and that A, the inverse image of the midpoint of "t" is empty. We note that the arcs in A do not reach Bd Δ since γ maps Bd Δ into K_0. If we make up a graph Γ whose vertices are the cells Δ_k and whose edges are the components of A, we see Γ embedded in Δ, and note, by the assumption that both t and t^{-1} occur in each w_i, that Γ has no source and no sink. Therefore, by 1.3, there is a simply connected region E of $\Delta-\Gamma$ whose boundary is consistently oriented. If clockwise, we use the assumption on G_+; if counter-clockwise, G_-. By 2.3, the arcs on Δ_k occurring on Bd E multiply together to give 1 in G; these arcs represent basis elements and their inverses in the free group G_+ (or G_-), and so there are two consecutive ones that are inverses of each other. See Figure 5. Those inverses represent exactly the same "occurrence," and thus Δ_1 and Δ_2, by 2.4.4, can be removed from the diagram.

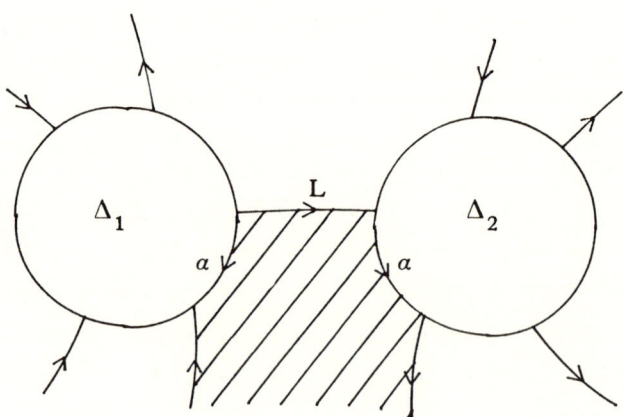

Figure 5

Now we have shown the minimal diagram contains nothing, and so maps Δ into K_0, showing γ to be null-homotopic in K_0. □

4.2. Here is an example: G is the free group on $\{a,b\}$ and $w_1 = a^2 t a^{-3} t^{-1}$, $w_2 = t^2 b t^{-3} b^{-1}$. We have G_+ generated by a^{-3} and b, and G_- generated by a^2 and b^{-1}. The freeness hypothesis is evidently satisfied, and we conclude that $\{a,b\}$ generate a free subgroup of rank 2 in the group

$$\{a,b,t : a^2 t a^{-3} t^{-1}, t^2 b t^{-3} b^{-1}\}.$$

4.3. A closer examination of the proof would show that $\pi_2(K_2, K_0) = 0$ and that K_2 is itself aspherical.

DEPARTMENT OF MATHEMATICS
UNIVERSITY OF CALIFORNIA
BERKELEY, CALIFORNIA 94720

REFERENCES

[Ad1] S. I. Adyan, 'On the Embeddability of Semigroups in Groups,' *Doklady Akad. Nauk SSSR* 133 (1960), 255-257. English translation: *Soviet Mathematics* 1 (1960), Amer. Math. Soc.

[Ad2] ———, 'Defining Relations and Algorithmic Problems in Groups and Semigroups,' *Trudy Mat. Inst. Steklova* LXXXV (1966). English translation by M. Greendlinger: *Proc. Steklov Inst. of Math.* 85 (1967), Amer. Math. Soc.

[Gl] Leon Glass, 'A Combinatorial Analog of the Poincaré Index Theorem,' *J. Comb. Theory* Ser. B 15 (1973), 264-268.

[Re] John H. Remmers, 'On the Geometry of Semigroup Presentations,' *Adv. in Math.* 36 (1980), 283-296.

[Sch] Martin Scharlemann, 'Smooth Spheres in $\mathbf{R}^4$ with Four Critical Points are Standard,' Preprint U. C. Santa Barbara (1984).

[Si] Allan J. Sieradski, 'A Coloring Test for Asphericity,' *Quart J. Math. Oxford* (2) 34 (1983), 97-106.

THE TODD-COXETER PROCESS, USING GRAPHS

John R. Stallings* and A. Royce Wolf

Abstract. The Todd-Coxeter Coset Enumeration Algorithm method is described in terms of graphs, where the notions of immersion and edge-folding apply naturally. This viewpoint eases the conceptualization and simplifies the record-keeping needed. It is easily programmed on a microcomputer where it can be applied to presentations of low complexity.

Suppose you have a finite presentation

$$\mathcal{P} = \{a, b, \cdots, c : r_1, \cdots, r_2\},$$

the r_j being words in the free group on the generators $a, b, \cdots, c$. In order to see if the group of $\mathcal{P}$ is finite, and if so, what it is, the "Todd-Coxeter Coset Enumeration Algorithm" can be applied. (See [L-S], pages 164-166 for a mathematical description, and [C-D-H-W] for a discussion of methods of implementation.) There is a particularly simple interpretation of this in terms of graphs:

1. Let Δ be the graph with one vertex, and one directed edge for each generator. Thus $\pi_1(\Delta)$ is the free group on the generators.

If r_k has length n_k, then r_k gives a recipe for mapping a graph A_k into Δ; A_k is a circle subdivided into n_k edges, with a basepoint. If we take the union of the A_k identifying their basepoints, we obtain a graph Γ_1' and a map $f_1' : \Gamma_1' \to \Delta$.

2. Now, f_1' can be factored through a series of edge-foldings and an immersion $f_1 : \Gamma_1 \to \Delta$. [S]. The images of the fundamental groups of Γ_1'

---

*Partly supported by NSF Grant DMS 83-03283.

and Γ_1 are equal; call this R_1. We are trying to find out something about the quotient of $\pi_1(\Delta)$ by the normal closure of R_1. [The "cosets" in the usual explanation are, at this point, cosets of R_1 in $\pi_1(\Delta)$, some of which correspond to the vertices of Γ_1.]

3. Suppose then that we take two copies of Γ_1 and identify a pair of vertices, one in one copy and one in the other. We obtain $f_2' : \Gamma_2' \to \Delta$; Γ_2' is the "wedge" of two copies of Γ_1, wedged not along their basepoints but along some arbitrarily selected pair of vertices. Now f_2' can be factored through a series of folds and an immersion $f_2 : \Gamma_2 \to \Delta$. The image of fundamental groups is the group R_2 generated by the union of two conjugates of R_1.

4. We can continue this process indefinitely. We could do it in two modes: In the first "automatic" mode, we can make the vertex identification, to get Γ_{n+1}' out of two copies of Γ_n, in some systematic manner. In the second "interactive" mode, we can examine Γ_n and try to choose the identification cleverly.

In either case, the objective is to try to make $f_n : \Gamma_n \to \Delta$ look like a covering projection in larger and larger neighborhoods of the basepoint. Since f_n is an immersion, the major requirement is to get f_n locally surjective on large parts of Γ_n.

Perhaps, eventually, f_n is a covering projection. This can be checked numerically: The immersion f_n is a covering if and only if $\chi(\Gamma_n) = k \cdot \chi(\Delta)$, where "$\chi$" means "Euler characteristic" and k is the number of vertices of Γ_n. In general, $\chi(\Gamma_n) \geq k \cdot \chi(\Delta)$, equality occurring only when f_n is a covering.

If f_n is a covering, that indicates that the image in $\pi_1(\Delta)$ of $\pi_1(\Gamma_n)$, which is the subgroup generated by a number of conjugates of R_1, is a subgroup of finite index.

5. Suppose f_n is a covering projection. Then f_n is a *regular* covering if and only if each generator of $\pi_1(\Delta)$ leads to a covering

translation. This can be checked as follows: Let v_0 be taken as the base vertex of Γ_n; for a generator a, there is a unique edge e of Γ_n starting at v_0 and mapping to the edge a of Δ. Let v_1 be the other vertex of e. Take two copies of Γ_n and identify v_0 in one copy with v_1 in the other copy. We factor the resulting $\Gamma_{n+1}' \to \Delta$ through an immersion $f_{n+1} : \Gamma_{n+1} \to \Delta$. There is a covering translation taking v_0 to v_1 if and only if $\pi_1(\Gamma_{n+1})$ can be identified with $\pi_1(\Gamma_n)$; since $\pi_1(\Gamma_n) \subset \pi_1(\Gamma_{n+1})$ and these are of finite index in $\pi_1(\Delta)$, this condition is equivalent to the numerical condition that $\chi(\Gamma_{n+1}) = \chi(\Gamma_n)$.

If f_n *is* a regular covering, this indicates that we have found the desired group of the presentation $\mathcal{P}$. The number of elements of this group is the number of vertices of Γ_n, and the multiplication table of the group is encoded in the map f_n.

6. Clearly, this is just a translation into graphs of the usual description of the Todd-Coxeter "algorithm." It is not exactly an algorithm, because there is no bound which is a recursive function of the data $\mathcal{P}$ that tells us how many steps we have to go before we can be sure that the group is infinite.

However, this method is easy to implement on a computer, even on a small home-computer toy, and can provide some fascinating group-theoretic play. The limitations are in storage space, which grows exponentially in random examples, and in time. The time-consuming part of the program is in the comparison of edges to find a fold; if you have N edges, you have to make about $N^2/2$ comparisons, which is an order of magnitude more tedious than the rest of the program. Therefore, group-theoretic hackers would be well-advised to make this comparison as short as possible, or else to plan to do something useful and healthful while the computer is cooking.

As an example, our implementation of the "lap computer" NEC|PC-8201A takes 153 seconds to show that the group

$$\{a, b : abab^{-1}a^{-2}b^{-1}, ab^2ab^{-1}a^{-1}b^{-1}\}$$

has 120 elements, and 21 seconds to show the group

$$\{a,b : aba^{-2}b^{-1}a^2, ab^2a^{-1}b^{-3}\}$$

is trivial.

7. REMARK #1. In the general form of the Todd-Coxeter method, there is an additional piece of data, a finite set

$$\Sigma = \{u,v,\cdots,w\}$$

of elements of the free group on the generators. Let G denote the group of the presentation $\mathcal{P}$ and H the subgroup of G generated by Σ. The problem is to determine the set of left cosets G/H.

This can be handled by similar methods. We look for the covering corresponding to the subgroup generated by Σ and the normal closure of the relators. Σ determines, as in (1), a basepointed wedge Ψ of subdivided circles, and a map $g : \Psi \to \Delta$. We determine, using $\mathcal{P}$, $f_n : \Gamma_n \to \Delta$ as before, and at each step look at

$$g \vee f_n : \Psi \vee \Gamma_n \to \Delta .$$

Factor this through an immersion $h_n : \Theta_n \to \Delta$.

If h_n is a covering, then G/H is finite. An operation similar to that in (5) can be done to determine whether G/H has been found; wedge Θ_n with Γ_1 along various basepoints and factor the result through an immersion; if this always reproduces Θ_n, then G/H has been determined. In general, since h_n may not become a regular covering, one must test all vertices of Θ_n, not just those near the basepoint.

8. REMARK #2. The point of this article is to give a *conceptual* description of the Todd-Coxeter method. We hope it is easy to see that this description contains redundancies which could be avoided or arranged more efficiently. [C-D-H-W] discuss various techniques to manage this, for serious application to large problems.

DEPARTMENT OF MATHEMATICS
UNIVERSITY OF CALIFORNIA
BERKELEY, CALIFORNIA 94720

REFERENCES

[C-D-H-W] John J. Cannon, Lucien A. Dimino, George Havas, Jane M. Watson, 'Implementation and Analysis of the Todd-Coxeter Algorithm,' *Math. of Computation* 27 (1973), 463-490.

[L-S] Roger C. Lyndon, Paul E. Schupp, *Combinatorial Group Theory*, Springer-Verlag (1977).

[S] John R. Stallings, 'The Topology of Finite Graphs,' *Invent. math.* 71 (1983), 551-565.

A SUBGROUP THEOREM FOR PREGROUPS

Frank Rimlinger[*]

Abstract. This paper investigates the structure of pregroups, Stallings' generalization of free products with amalgamation. We show that pregroups are related to length functions on groups, and derive a subgroup theorem for the universal group of a pregroup. Our methods may be used to construct HNN groups from more or less arbitrary "template" groups.

Pregroups are a generalization of free products with amalgamation. They were discovered by Stallings [2]. The original axiomatic definition of pregroups was motivated by a desire to generalize a theorem of van der Waerden [3] concerning the structure of free products with amalgamation. We show that the class of pregroups is in natural correspondence with the "groups with length function derived from reduced words." We also prove a more subtle result: that subgroups of groups with pregroup structures inherit a pregroup structure of their own. This result is proved by exploiting the tree-like nature of the Cayley graph of the universal group of a pregroup with colors in the pregroup.

1. *Stallings' theorem about pregroups*

Let P be a set. Let $i: P \to P$ be an involution, denoted $x \to x^{-1}$. Let $1 \in P$ be a distinguished element. Let $D \subset P \times P$ and let $m: D \to P$ be a set map, denoted $(x,y) \to xy$. Suppose that

(P1) for all $x \in P$, $(x,1), (1,x) \in D$ and $x1 = 1x = x$, and

(P2) for all $x \in P$, $(x, x^{-1}), (x^{-1}, x) \in D$ and $xx^{-1} = x^{-1}x = 1$.

---

[*]Sloan Foundation Doctoral Dissertation Fellow.

The sets $(P,i,1,D,m)$, usually referred to as (P,D), form a category C whose morphisms $\phi \in C((P,D),(Q,E))$ are set maps $\phi : P \to Q$ such that $(x,y) \in D$ implies $(\phi(x),\phi(y)) \in E$ and $\phi(xy) = \phi(x)\phi(y)$. There is a forgetful functor from groups to C whose left adjoint $U(P)$ is presented by $(P; \{xy[m(x,y)^{-1}] | (x,y) \in D\})$. Let $F(P)$ be the free group on the set P. Then the map $\iota : P \to U(P)$ is the restriction of the quotient map $F(P) \to U(P)$, and for every C morphism $\phi : P \to G$ whose codomain is a group, there is a unique group homomorphism $h : U(P) \to G$ such that $h\iota = \phi$. $U(P)$ is called the *universal group* of P.

Pregroups arise out of a natural desire to make $\iota : P \to U(P)$ injective. To do this we need more axioms. The point of the category C is to set up all the notation. Thus given an object (P,D), a P-word $(x_1, \cdots, x_n) \in P^n$ of length n represents $x_1 \cdots x_n \in U(P)$. This P-word is *P-reduced* if each adjacent pair $(x_i, x_{i+1}) \notin D$. We also write $(x_1, \cdots, x_n)_D$ if each pair $(x_i, x_{i+1}) \in D$. If $(w,x,y)_D$, both $(w,xy)_D$ and $(wx,y)_D$, and $w(xy) = (wx)y$, then (w,x,y) *associates*. Given P-words $X = (x_1, \cdots, x_n)$ and $A = (a_0 = 1, a_1, \cdots, a_{n-1}, a_n = 1)$ such that (a_{i-1}^{-1}, x_i, a_i) associates for $i = 1, \cdots, n$ define X *interleaved by* A to be the P-word $(x_1 a_1, a_1^{-1} x_2 a_2, \cdots, a_{n-1}^{-1} x_n)$.

We are now ready to define pregroups and state the classical result of Stallings.

DEFINITION (Stallings [2, chap. 3A]). Let (P,D) be an object of C. Suppose that

(P3) for all $w,x,y \in P$, if $(w,x,y)_D$, then $(w,xy)_D$ or $(wx,y)_D$ implies (w,x,y) associates, and

(P4) for all $w,x,y,z \in P$, $(w,x,y,z)_D$ implies $(w,xy)_D$ or $(xy,z)_D$.

Then (P,D) is a *pregroup*.

THEOREM 1 (Stallings [2, p. 20]). *Let (P,D) be a pregroup. Let X and Y be P-reduced representations of an element of $U(P)$. Then X*

and Y have the same length, and there is a P-word A such that Y is X interleaved by A. In particular, $\iota: P \to U(P)$ is injective.

In view of Theorem 1, we may define the P-*length* $l_P(x)$ of an element $x \in U(P)$ to be the length of any P-reduced word representing x. For technical reasons we insist that $l_P(1) = 0$.

2. *A characterization of pregroups*

There is a converse to Theorem 1 which asserts that if a group has a length function derived from reduced words, then it may be realized as the universal group of a pregroup.

THEOREM 2. *Let G be a nontrivial group. Suppose* $P \subset G$, $P = P^{-1}$, *and P generates G. Suppose* $D \subset P \times P$; *and suppose that* $(x,y) \in D$ *implies* $xy \in P$. *Clearly* $(P,D) \in C$. *By abuse of notation, the P-words represent elements of G. Suppose all the P-reduced words representing the same element of G are of the same P-length. Then* (P,D) *is a pregroup and the natural map* $h: U(P) \to G$ *is an isomorphism.*

Proof. We verify the pregroup axioms. First notice that if $x,y \in P$ and $xy \in P$, then we must have $(x,y)_D$; for otherwise we would contradict the invariance of P-length.

(P1) Since P generates G, $P \neq \phi$. Pick any $x \in P$. Since $P = P^{-1}$, $x^{-1} \in P$, so $(x, x^{-1})_D$ or $(x^{-1}, x)_D$ implies $1 = xx^{-1} = x^{-1}x \in P$. If (x, x^{-1}) and (x^{-1}, x) are both P-reduced, then so is (x, x^{-1}, x). But this implies the P-length of x is not well defined, a contradiction. Thus $1 \in P$. Similarly we deduce that $(x,1)_D$ and $(1,x)_D$ for all $x \in P$.

(P2) Suppose $x \in P$, so $x^{-1} \in P$. Then $(x, x^{-1})_D$ and $(x^{-1}, x)_D$ follow from the fact that $1 \in P$ and has P-length zero.

(P3) Suppose $(w, x, y)_D$. Then $(w, xy)_D$ iff $w(xy) \in P$ iff $l_P(wxy) = 1$ iff $(wx, y)_D$.

(P4) Suppose $(w,x,y,z)_D$. Since $wxyz = (wx)(yz)$, the P-length of $wxyz$ is ≤ 2. But if (w,xy) and (xy,z) are both P-reduced, then $wxyz = w(xy)z$ implies $wxyz$ has P-length 3. Hence either $(w,xy)_D$ or $(xy,z)_D$.

Thus (P,D) is a subpregroup of G. Let $h: U(P) \to G$ be induced by the inclusion of (P,D) in G. Let $x \in U(P)$ be represented by the P-reduced word $(x_1, \cdots, x_n)$. If $h(x) = 1$, then it follows that $n = 1$. Since h is injective on P, it follows that $x = 1$. Thus h is an isomorphism since P generates G. ∎

Combining Theorems 1 and 2, we see that if a group has a length function derived from reduced words, then we may pass from one word representing a group element to any other by means of an interleaved product.

3. *A pregroup structure for subgroups of* $U(P)$

Let (P,D) be a pregroup and S a subgroup of $U(P)$. We say that $x \in S$ is *S-irreducible* if there is a P-word $(x_1, \cdots, x_n)$ representing x such that $x_1 \cdots x_i \notin S$ for all i such that $1 < i < n$. In particular, the elements of $S \cap P$ are S-irreducible.

THEOREM 3. *Let (P,D) be a pregroup, S a subgroup of $U(P)$. Let Q be the set of S-irreducible elements of S. Let $E = \{(x,y) \in Q \times Q | xy \in Q\}$. Then (Q,E) is a subpregroup of $U(P)$, and the natural map $h: U(Q) \to U(P)$ maps $U(Q)$ isomorphically onto S.*

Proof. The proof is divided into three parts. In part 1, we set up the Cayley graph Γ of $U(P)$ with colors in P. By collapsing some of the structure of this graph we obtain a tree T. In part 2, we use the fact that any two vertices of T are connected by a unique geodesic in T to study the structure of paths in Γ between two fixed vertices of Γ. This results in a lemma about E-reduced words. In part 3, we apply the lemma to demonstrate that axiom (P4) holds for (Q,E). As the other axioms are easily verified, it follows that (Q,E) is a pregroup. Finally, we show $U(Q) \approx S$.

Part 1. Constructing the graph Γ and the tree T

We shall work in the category of graphs defined by Serre [1, chap. 1.2.1]. We construct the Cayley graph Γ of $U(P)$ with colors in P as follows. The vertices of Γ are $U(P)$ and the positively oriented edges are $E^+ = \{(x,y) \in U(P) \times U(P) | x^{-1}y \in P\}$. The initial vertex of an edge (x,y) is x and the terminal vertex is y. Let E^- be a disjoint copy of E^+ and let $\varepsilon: E^+ \to E^-$ be an arbitrary bijection. The edges of Γ are $E^+ \cup E^-$. Define the opposite edge map $e \to \bar{e}$ by extending ε to an involution of $E^+ \cup E^-$.

Notice that a path p in Γ with vertices $(v_0, v_1, \cdots, v_n)$ represents a P-word $X = (v_0^{-1}v_1, v_1^{-1}v_2, \cdots, v_{n-1}^{-1}v_n)$ and that X is P-reduced iff p is a geodesic in Γ. Conversely, given a P-word $X = (x_1, \cdots, x_n)$ and a vertex $v_0 \in V$, X is represented by the path p based at v_0 with vertices $(v_0, v_0x_1, v_0x_1x_2, \cdots, v_0x_1x_2\cdots x_n)$. Given a path p in Γ with vertices $(v_0, \cdots, v_n)$, the *intermediate vertices* of p are $\{v_1, \cdots, v_{n-1}\}$. An element $x \in S$ is actually in Q iff there exists a path in Γ from 1 to x with no intermediate vertices in S. In fact, if $x \in S$ and s is any element in S, then $x \in Q$ iff there exists a path from s to sx with no intermediate vertices in S. These facts will be used implicitly throughout the proof.

We say a path p in Γ is *flat* if every vertex of p has the same P-length. Define an equivalence relation on V by $x \sim y$ if there is a flat path joining x to y. Set $\bar{V} = V/\sim$, and

$\bar{E}^+ = \{([x],[y]) \in \bar{V} \times \bar{V} | \text{ there exists } z \in [x] \text{ such that } 1_P(y) = 1_P(z)+1 \text{ and } z^{-1}y \in P\}$.

To see that $\bar{E}^+$ is well defined, suppose $y_1 \sim y_2$, $1_P(y_1) = 1_P(z)+1$ and $z^{-1}y_1 \in P$. From $y_1 \sim y_2$, $1_P(y_1) = 1_P(y_2)$, and hence $1_P(y_2) = 1_P(z)+1$. Since $z^{-1}y_1 \in P$, there exists $a \in P$ such that $y_1 = za$. Moreover, there is a flat path from y_1 to y_2 with vertices $(y_1 = b_0, b_1, \cdots, b_k = y_2)$. Let $(z_1, \cdots, z_n)$ be a P-reduced word representing z. Then $(z_1, \cdots, z_n, a)$ is P-reduced representing y_1, since $1_P(y_1) = 1_P(z)+1$. Since $y_1^{-1}b_1 \in P$, we deduce that $(z_1, \cdots, z_n, a, y_1^{-1}b_1)$ is a P-word representing b_1. Since

$1_P(b_1) = 1_P(y_1)$, it follows that $(z_1,\cdots,z_n,ay_1^{-1}b_1)$ is a P-reduced word representing b_1. Similarly, $(z_1,\cdots,z_n,ay_1^{-1}b_1(b_1^{-1}b_2)) = (z_1,\cdots,z_n,ay_1^{-1}b_2)$ is P-reduced representing b_2. Continuing in this way, we see $(z_1,\cdots,z_n,ay_1^{-1}y_2)$ is P-reduced representing y_2. Thus $z^{-1}y_2 = ay_1^{-1}y_2 \in P$, and so E^+ is well defined.

Let $E^- = \{([y],[x]) | ([x],[y]) \in E^+\}$, $E = E^+ \cup E^-$, and let $e \to \bar{e}$ be the edge reversal involution defined in the obvious way. Define T to be the graph with vertices V and edges E. Define the P-length of $[x] \in V$ to be $1_P(x)$. Given a vertex $[y] \in V$ of length $n > 1$, let $(y_1,\cdots,y_{n-1},y_n)$ represent y, and let $x = y_1 \cdots y_{n-1}$. Then $([x],[y]) \in E^+$. Since every vertex of length 1 is joined to the unique vertex $[1]$ of length 0, it follows that T is connected. (Recall that $1_P(1) = 0$ by special dispensation.)

To show that T is a tree, we must show there are no loops in T without backtracking. In other words, if p is a path in T with vertices $([v_0],\cdots,[v_n])$ such that $[v_0] = [v_n]$, then for some pair of adjacent edges e_i and e_{i+1} of p we have $e_i = \bar{e}_{i+1}$.

First consider the following situation. Let p be a path in T with edges $(e_1,\cdots,e_n)$. Suppose p is without backtracking, so that $e_i \neq \bar{e}_{i+1}$ for $i = 1,\cdots,n-1$. Suppose that $e_i \in E^+$ for some $1 \leq i < n$. We claim that $e_{i+1} \in E^+$ and hence $e_k \in E^+$ for all $i \leq k \leq n$. To see this, let $e_i = ([x],[y])$, $e_{i+1} = ([y],[z])$. If $e_{i+1} \in E^-$, then $1_P(x) = 1_P(z)$, and we may choose $x' \in [x]$, $z' \in [z]$, and $a,b \in P$ such that $x'a = z'b = y$. Let $(x_1,\cdots,x_n)$ and $(z_1,\cdots,z_n)$ be P-reduced representing x' and z' respectively. Then $(x_1,\cdots,x_n,a)$ and $(z_1,\cdots,z_n,b)$ are P-reduced words representing y. By Theorem 1, these two words differ by an interleaved product, and so $x^{-1}z \in P$. Thus $[x] = [z]$, contradicting the fact that p is without backtracking.

Now let p be any path in T without backtracking, and suppose p is a loop. Let $[v] \in V$ be the common initial and terminal vertex of p. Let $(e_1,e_2,\cdots,e_n)$ be the edges of p. It follows that for some $1 \leq l < n$, $e_1,e_2,\cdots,e_l \in E^-$ and $e_{l+1},e_{l+2},\cdots,e_n \in E^+$. Let p_1 be the path with edges $(\bar{e}_l, \bar{e}_{l-1}, \cdots, \bar{e}_1)$, and p_2 be the path with edges $(e_{l+1}, \cdots, e_n)$. Let $[x]$ be the terminal vertex of e_l. Let q be a geodesic in T from $[1]$ to $[x]$. By the method used to show T is connected, construct geodesics q_1 and q_2 in Γ from 1 to v representing qp_1 and qp_2 respectively. Then q_1 and q_2 represent P-reduced words X_1 and X_2 which in turn represent v. By Theorem 1, X_1 and X_2 differ by an interleaved product, and therefore $p_1 = p_2$. This contradicts the fact that p is without backtracking. We conclude that T is a tree.

Part 2. Exploiting the structure of T to prove a lemma

Notice that a path p in Γ represents a path $[p]$ in T. The vertices v of p correspond to the vertices $[v]$ of $[p]$, except that the vertices joined by flat subpaths of p are collapsed to a single vertex in $[p]$. Hence if $[x]$ and $[y]$ are vertices of T, and p is the (unique) geodesic from $[x]$ to $[y]$ with vertices $([x]=[v_1],[v_2],\cdots,[v_n]=[y])$, then any path p in Γ from x to y must contain vertices $v_1', \cdots, v_n'$ such that $v_i \sim v_i'$ for $i = 1,\cdots,n$. Moreover, p contains a unique vertex $[b]$ of minimal P-length among all the vertices of p. The vertex $[b]$ is called the *branch class* $B(x,y)$ of x and y.

Fact 1. Let $x,y \in U(P)$. Then any path in Γ from x to y has a vertex in $B(x,y)$.

The following refinement of these observations is necessary. Suppose that x and y are vertices of Γ and $l_P(B(x,y)) < l_P(y)$. Let $(B(x,y), [v]) \in E^+$ be the initial edge of the geodesic in T from $B(x,y)$ to $[y]$. We define the *image* of y in x to be the following subset of $B(x,y)$:

$$\mathrm{im}_x(y) = \{z \in B(x,y) | z^{-1}v \in P\} .$$

Clearly any path from x to y in Γ must have a vertex in $\mathrm{im}_x(y)$. Notice $\mathrm{im}_x(y)$ is well defined for the same reason E^+ is.

Fact 2. If $x,y \in U(P)$, $l_P(B(x,y)) < l_P(y)$, and $x \notin im_x(y)$, then any path from x to y in Γ has an *intermediate* vertex in $im_x(y)$.

Fact 2 may be elaborated on as follows.

Fact 2′. Suppose $x,y \in U(P)$ and $l_P(B(x,y)) < l_P(y)$. Suppose $[z]$ is an intermediate vertex of the geodesic from $B(x,y)$ to $[y]$ in T. Then every path from x to y has an intermediate vertex in $im_z(y)$.

If $x,y \in U(P)$, then the geodesic p from $[1]$ to $[x]$ in T may be written as the product $p_1 p_2$, where p_1 is the geodesic from $[1]$ to $B(x,y)$ and p_2 is the geodesic from $B(x,y)$ to $[x]$. Hence we have the following:

Fact 3. If $x,y \in U(P)$ then every path from 1 to x in Γ has a vertex in $B(x,y)$. Moreover, $B(x,y)$ is a vertex of the geodesic in T from $[1]$ to $[x]$.

Fact 4. Let $a,b,x,y \in U(P)$ and suppose $a,b \in im_x(y)$. Then $a^{-1}b \in P$.

Proof. Let $[v]$ be the terminal vertex of the initial edge of the geodesic from $B(x,y)$ to y. Let $(a_1,\cdots,a_n)$ and $(b_1,\cdots,b_n)$ be P-reduced representing a and b. Then $(a_1,\cdots,a_n,a^{-1}v)$ and $(b_1,\cdots,b_n,b^{-1}v)$ are P-reduced representing v, and so the result follows from Theorem 1.

LEMMA. *Recall that* $E = \{(x,y) \in Q \times Q | xy \in Q\}$. *Let* $(x,y) \in Q \times Q$ *and suppose* (x,y) *is E-reduced. Then (i)* $[x] \neq [xy]$, *(ii) the geodesic* p *from* $[x]$ *to* $[xy]$ *has all its edges in* E^+, *and (iii)* $im_x(xy) \subset S$.

Proof. Suppose (x,y) is E-reduced. Let $(x_1,\cdots,x_n)$ be a P-reduced word representing x such that $x_1 \cdots x_i \notin S$ for $i = 1,\cdots,n-1$. Thus there is a path p in Γ from 1 to $z = x_1 \cdots x_{n-1}$ with only initial vertex in S. If $[x] = [xy]$, then $([z],[x]) = ([z],[xy])$ is an edge of T. Thus $z^{-1}(xy) \in P$, contradicting the fact that $xy \notin Q$.

Now suppose (ii) is false. Fact 3 implies that every path from 1 to x in Γ must have a vertex in $[b] = B(x,xy)$. If $b \epsilon P$, set $z = 1$, otherwise for some $1 \leq i < n$ we have $z = x_1 \cdots x_i \epsilon \operatorname{im}_z(b)$. By fact 1, every path from x to xy must have a vertex in $B(x,xy)$. Let $(x = v_0, v_1, \cdots, v_k = xy)$ be such a path with no intermediate vertices in S. Then for some j, $[v_j] = B(x,xy) = [b]$. Since (ii) is false we deduce that v_j is not the initial vertex $v_0 = x$. Thus the path with vertices $(v_j, v_{j+1}, \cdots, v_k = xy)$ has only terminal vertex in S. Note $z \epsilon \operatorname{im}_z(b) = \operatorname{im}_z(v_j)$, and so $z^{-1}v_j \epsilon P$. Since there is a path from 1 to z with only initial vertex in S and a path from v_j to xy with only terminal vertex in S, it follows that $xy \epsilon Q$, a contradiction. Thus (ii) must be true.

By (ii), the geodesic in T from [1] to [xy] has the form $([1], \cdots, [v_\alpha], [x], [v_\beta], \cdots, [xy])$. Since x and y are in Q, we may choose v_α and v_β such that there is a path p_1 in Γ from 1 to v_α with only initial vertex in S, and there is a path p_2 from v_β to xy with only terminal vertex in S. Moreover, we may assume $v_\alpha \epsilon \operatorname{im}_{v_\alpha}(x)$. If (iii) is false, then for some $z \epsilon U(P)$ we have $z \epsilon \operatorname{im}_x(xy)$ and $z \notin S$. By definition $\operatorname{im}_x(v_\beta) = \operatorname{im}_x(xy)$ and thus $z \epsilon \operatorname{im}_x(v_\beta)$. Moreover, $z \epsilon [x]$ implies $v_\alpha \epsilon \operatorname{im}_{v_\alpha}(z)$. Thus $v_\alpha^{-1}z$ and $z^{-1}v_\beta$ are in P, and so there is a path p_3 from v_α to v_β with no vertices in S. Thus $p_1 p_3 p_2$ is a path from 1 to xy with no intermediate vertices in S, contradicting $xy \notin Q$.

Part 3. Verifying axiom (P4) and constructing the isomorphism.

To show (Q,E) is a pregroup, we must verify (P1) to (P4). Axioms (P1) to (P3) are easily proved. To verify (P4), suppose $(w,x,y,z)_E$, but (w,xy) is E-reduced. We must prove that $(xy,z)_E$. In the first place, we claim that $wx \epsilon \operatorname{im}_w(wxy)$. The lemma may be applied to (w,xy), and so the geodesic p from [w] to [wxy] has all its edges in E^+. Thus p is a subgeodesic of the geodesic p´ from [1] to [wxy]. By fact 3, B(wx,wxy) is some vertex of p´. If B(wx,wxy) is not a vertex of p, then the geodesic from [wx] to [wxy] has intermediate vertex [w].

Hence fact 2' implies that every path in Γ from wx to wxy has an intermediate vertex in im_w(wxy). But the lemma implies that im_w(wxy) $\subset$ S, contradicting y ϵ Q.

If B(wx,wxy) is a vertex of p but B(wx,wxy) $\neq$ [w], then we deduce that the geodesic from [1] to [wx] has [w] as an intermediate vertex. By definition, im_w(wx) = im_w(wxy), and so using fact 2' and the lemma we deduce that wx $\notin$ Q. But this contradicts the original hypothesis that $(w,x,y,z)_E$. Thus [wx] = [w]. Moreover, if wx $\notin im_w$(wxy), then fact 2 implies any path in Γ from wx to wxy must have an initial flat subpath from wx to a vertex in im_w(wxy) $\subset$ S, contradicting y ϵ Q. Thus wx ϵ im_w(wxy) as desired. Likewise, xy ϵ Q implies w ϵ im_w(wxy).

Since z ϵ Q, similar reasoning shows that B(wxy,wxyz) is a vertex of p. If [wxyz] = [w] then fact 2 implies wxyz ϵ im_w(wxy). In this event, w ϵ im_w(wxy) together with fact 4 implies xyz = w^{-1}(wxyz) ϵ P as desired. On the other hand, suppose B(wxy,wxyz) is a vertex of p other than [w]. Let ([x],[v_1],$\cdots$,[wxy]) be the vertices of p. Then [v_1] is a vertex of the geodesic q from [wx] to [wxyz]. Since yz ϵ Q, there is a path q in Γ representing q with no intermediate vertices in S. Since w ϵ im_w(v_1), we may replace the first edge of q so as to obtain a path from w to wxyz with the same intermediate vertices as q. Thus xyz ϵ Q, and so (Q,E) is a pregroup.

In fact, (Q,E) is a subpregroup of U(P), and so the inclusion map Q $\subset$ U(P) induces a natural map h: U(Q) $\to$ U(P). It is easy to see that Q generates S and therefore h(U(Q)) = S. Moreover, the lemma implies that if (x,y) ϵ Q×Q is Q-reduced, then 1_P(xy) $\geq$ 1_P(x). It follows that ker h $\subset$ Q. But h is injective on Q since the inclusion map Q $\subset$ U(P) factors through h. We conclude that h: U(Q) $\to$ S is an isomorphism. Q.E.D.

4. Examples

Example 1. Consider the following pregroup (cf. [1, 3.A.5.4]). Let G be a group, H a subgroup of G, and P a *subset* of G such that H $\subset$ P $\subset$ G,

$P = P^{-1}$, and $HP = P = PH$. Define partial multiplication in P by $D = \{(x,y) \in P \times P | xy \in H$, or $x \in H$, or $y \in H\}$. This structure yields a pregroup P, called a *quasi*-HNN pregroup. To compute the structure of U(P), we modify the structure of the tree T of section 3 so that all the elements of H have length 0. Then it is easy to see that the vertices of T are just the left cosets of H in U(P), and so U(P) acts on T. By Bass-Serre theory [1, chap. 1.5], it follows that U(P) is an HNN extension of H provided that this action is without inversion, and otherwise U(P) has a more complicated structure.

Let (P,D) be a quasi-HNN pregroup, with subgroup H and modified tree T as above. Let S be a subgroup of U(P). Then S is free if it acts freely on T without inversion. Notice that the edges of T are represented by elements of P−H. If $s \in S$ inverts an edge (aH,bH) of T, then for some $h,k \in H$ we have $s = bka^{-1}$ and $s = ahb^{-1}$. Thus $s^2 \in aHa^{-1} \cap bHb^{-1}$. Moreover, since $a^{-1}b \in P$ we see that $s \in aPa^{-1} \cap bPb^{-1}$. Similarly if $s \in U(P)$ stabilizes a vertex of T, it must lie in a conjugate of H. Thus S is free if for all $s \in S$, s is in no conjugate of P in U(P) and s^2 is in no conjugate of H in U(P).

Example 2. Let F(x,y) be the free group on two generators. Let $P = \{x, x^{-1}, y, y^{-1}, 1\}$ be the obvious pregroup such that $U(P) \approx F(x,y)$. Then the Cayley graph Γ and the tree T are both the universal cover of the figure 8. Let $S = <y^{-1}xy> \subset U(P)$ be the subgroup generated by $y^{-1}xy$. It is easy to see that every element of S is S-irreducible, according to the definition of irreducibility given at the beginning of section 3. By Theorem 3, the induced pregroup structure Q(S) is just S. However, if we conjugate S by y, then we see that the induced pregroup structure $Q(<x>) = \{x, x^{-1}, 1\}$. Thus $<x>$ is a "better" subgroup than S. I feel that pregroups may be of some use in situations where there is no other "good" way of finding a canonical representative of the conjugacy class of a subgroup.

DEPARTMENT OF MATHEMATICS
COLUMBIA UNIVERSITY
NEW YORK, NEW YORK 10027

REFERENCES

[1] Serre, J.-P.: Arbres, amalgames, SL_2. Cours au Collège de France rédigé avec Hyman Bass. Asterisque 46, chap. 1 (1977).

[2] Stallings, J.R.: Group theory and three-dimensional manifolds. Yale Monographs 4, (1971).

[3] Van der Waerden, B.L.: Free products of groups. Amer. J. Math. 70 527-528 (1948).

GROUPS WITH A RATIONAL CROSS-SECTION

Robert H. Gilman[*]

Abstract. We study groups G for which there exist a finitely generated free monoid F a projection $\pi: F \to G$, and a rational subset R of F such that π maps R bijectively to G.

§1. Introduction

Consider the group

(1) $$G = \langle a,b; aba^{-1} = b^2 \rangle.$$

Clearly G satisfies the relations

(2)
$$aa^{-1} = 1 \quad a^{-1}a = 1 \quad bb^{-1} = 1 \quad b^{-1}b = 1$$
$$ba = b^{-1}ab \quad ba^{-1} = a^{-1}b^2 \quad b^{-2}a = ab^{-1} \quad b^{-1}a^{-1} = a^{-1}b^{-2},$$

and these relations present G as a quotient of the free monoid on $\{a, a^{-1}, b, b^{-1}\}$. This presentation has some special properties. Let w be any word in a, a^{-1}, b, b^{-1}. If we substitute the right-hand side of any of the relations (2) for the left-hand side in w and repeat this procedure, then eventually we arrive at a word w^* for which no more substitutions are possible. Also w^* depends neither on the particular substitutions we made nor on w itself, but only on the element of G represented by w. We will call a presentation (of a group as a quotient of a finitely generated free monoid) with these properties a confluent presentation, and a group with a confluent presentation will be called a confluent group.

Confluent presentations are a slight generalization of the class $\mathcal{C}$ of presentations studied in [4]. Various notions of confluence have long

---

[*]I would like to thank the Rutgers University Department of Mathematics for its hospitality while this paper was being written.

been used to solve word problems in algebras ([2], [8]), and indeed the word problem for a confluent group is solvable by essentially the same algorithm used by Dehn to solve the word problem for fundamental groups of surfaces. In addition one can determine from a confluent presentation the cardinality of the group presented, and there is a procedure for attempting to turn an arbitrary finite presentation into a confluent one [4]. Other decision problems for confluent groups and monoids are discussed in [3] and [6]. In [12] it is shown how to turn any presentation of a nilpotent group into a confluent one, and the method is used to investigate free nilpotent groups. Of course the structure of example (1) is apparent without these techniques; some other examples are given in [5].

We would like to determine which groups possess confluent presentations, and whether there is a useful procedure which, given a presentation for such a group, will succeed in turning it into a confluent presentation. (The procedure in [4] may fail for groups which are in fact confluent.) Here we focus on the first problem, and it leads us to a class $\mathcal{C}$ which contains all confluent groups. In fact $\mathcal{C}$ contains other groups as well, but it seems to be easier to deal with than the class of all confluent groups. Also $\mathcal{C}$ may be interesting in itself as an example of the interplay between group theory and the theory of formal languages. Another investigation in this area is the classification of groups with context-free word problem [11].

NOTATION. All groups and monoids are taken to be finitely generated and all presentations finite. F will always denote a free monoid with free generators A, and π a surjective homomorphism. G is always a group.

§2. *Definition of* $\mathcal{C}$

DEFINITION 1. $\mathcal{C}$ is the class of all groups G for which there exist $\pi: F \to G$ with a cross-section $R \in \text{Rat}(F)$ for π.

In the definition above we mean that π maps R bijectively to G. Rat(F) denotes the rational (or regular) subsets of F.

DEFINITION 2. For any monoid M, Rat(M) is the closure of the finite subsets of M under union, product, and generation of submonoid.

Often union is denoted by $+$ and generation of submonoid by $*$. Thus $a^*(b+c)$ stands for

$$\{x | x = a^i b \text{ or } x = a^i c, \ i = 0,1,2,\cdots\}.$$

The rest of this section is devoted to a list of standard properties of rational sets; see [10, Ch. 6 §§1-3] or any text on the theory of automata and formal languages.

LEMMA 1. *If* $R \in \text{Rat}(M)$ *and* $\sigma : M \to N$ *is a homomorphism, then* $\sigma(R) \in \text{Rat}(N)$. *If* σ *is surjective, then every* $T \in \text{Rat}(N)$ *is* $\sigma(R)$ *for some* $R \in \text{Rat}(M)$.

LEMMA 2. $R \in \text{Rat}(F)$ *if and only if there is a homomorphism* $\sigma : F \to M$, *M a finite monoid, and a subset* $T \subseteq M$ *such that* $R = \sigma^{-1}(T)$.

The next lemma deals with a graph with edges labelled by elements of A. For any such graph every (finite) path γ determines a word in F by reading labels along the path. We call that word the label of γ. If γ is a path of length 0, its label is 1.

LEMMA 3. $R \in \text{Rat}(F)$ *if and only if there is a finite directed graph with a distinguished initial vertex, distinguished terminal vertices, and edges labelled by elements of* $\mathcal{A}$ *such that*

   (i) *For each vertex all outedges have distinct labels;*

   (ii) R *is the set of labels of paths from the initial vertex to any terminal vertex.*

The graph of Lemma 3 is essentially a finite deterministic automaton accepting R.

LEMMA 4. *Rat(F) is closed under intersection and complement.*

§3. *Properties of* $\mathcal{A}$

The definition of $\mathcal{A}$ (Definition 1) seems to depend on the choice of F and π (that is, on the choice of generators for G); but if $\pi_1 : F_1 \to G$ is another choice, then there is $\sigma : F \to F_1$ with $\pi_1 \sigma = \pi$. $R_1 = \sigma(R)$ is a cross-section for π_1, and by Lemma 1 $R_1 \in \text{Rat}(F_1)$. Thus the choice of generators for G does not affect the existence of a rational cross-section.

The next three results follow directly from the properties of rational sets in Section 2.

PROPOSITION 1. *$\mathcal{A}$ contains all confluent groups. In particular $\mathcal{A}$ contains all free and finite groups.*

Proof. If G is confluent, then G has a cross-section consisting of all words in F not divisible by any word in some fixed set $W = \{w_i | 1 \leq i \leq m\}$. As $F = A^*$, $R = A^* - A^* W A^* \in \text{Rat}(F)$. Thus $G \in \mathcal{A}$. The usual presentation of a free group and the multiplication table presentation of a finite group are both confluent.

PROPOSITION 2. (i) *$\mathcal{A}$ is closed under direct product, free product, and extension.*

(ii) *If $G \in \mathcal{A}$, then $G \sim Z$ (the ordinary restricted wreath product with the integers) is in $\mathcal{A}$.*

Proof. Suppose N is normal in G with $N \in \mathcal{A}$, $G/N \in \mathcal{A}$; and let R_1 and R_2 be cross-sections for N and G/N respectively with $R_i \in \text{Rat}(F_i)$, F_i free on A_i, $A_1 \cap A_2 = \emptyset$. It follows that $R = R_1 R_2 \in \text{Rat}(F)$, F free on $A = A_1 \cup A_2$; and it is easy to make R into a cross-section for G.

The other parts of the proposition are proved similarly. For (ii) use the fact that every $x \in G \sim Z$ has a unique representation

$x = i_0 g_1 i_1 \cdots i_{n-1} g_n i_n$, with $i_j \epsilon Z$, $0 \leq j \leq n$, $g_j \epsilon G - \{1\}$, $1 \leq j \leq n$, and $i_j > 0$, $1 \leq j \leq n-1$.

PROPOSITION 3. *If* $G \epsilon \mathcal{C}$ *and* G *is infinite, then* G *has an element of infinite order.*

Proof. Use Lemma 3. As R is infinite, it must contain the label of a path with a loop. Thus there exist $u, v, w \epsilon F$ with $v \neq 1$ and $uv^i w \epsilon R$ for all $i \geq 0$. Since R is a cross-section, $\pi(v)$ has infinite order.

The next proposition seems to be deeper than the preceding ones.

PROPOSITION 4. *If* $G \epsilon \mathcal{C}$ *and* H *has finite index in* G, *then* $H \epsilon \mathcal{C}$.

Proof. Choose f, π, R as in Definition 1. Let N be the largest subgroup of H a normal in G and let $\sigma : G \to G/N$ be the projection. As G/N is finite, $T = \pi^{-1}(H) = (\sigma\pi)^{-1}\sigma(H) \epsilon \mathrm{Rat}(F)$. Thus $R_1 = R \cap T \epsilon$ $\mathrm{Rat}(F)$, and by definition of T, π maps R_1 bijectively to H. In fact T is a free submonoid of F. If T were finitely generated, we would be done; but this need not be the case. As $R_1 \epsilon \mathrm{Rat}(F)$, we can use the following lemma (with R_1 substituted for R) to complete the proof.

LEMMA 5. *If* $\pi : F \to G$ *and* $R \epsilon \mathrm{Rat}(F)$, *then there exist a free monoid* F_0, $R_0 \epsilon \mathrm{Rat}(F_0)$, *a homomorphism* $\sigma : F_0 \to F$, *and a bijection* $\rho : R_0 \to R$ *such that*

(i) $\pi\sigma(F_0) = <\pi(R)>$, *the subgroup of* G *generated by* $\pi(R)$;

(ii) $\pi\sigma(r) = \pi\rho(r)$ *for* $r \epsilon R_0$.

Before giving the proof we note the following interesting corollary.

COROLLARY 1.[9, Ch. 6][1, Lemma 3.1]. *If* $X \epsilon \mathrm{Rat}(G)$, *then* $<X>$ *is finitely generated.*

Proof. Pick $\pi : F \to G$. By Lemma 1 pick $R \epsilon \mathrm{Rat}(F)$ with $\pi(R) = X$. By Lemma 5 $<X> = \pi\sigma(F_0)$.

The proof of Lemma 5 uses a technique exploited in [9]. When G is a free group and $\pi: F \to G$ corresponds to the usual free generators of G, then for the F_0 constructed in the proof the generators of F_0 project to a set of free generators of $<\pi(R)>$ with significant factors [7, Ch. 6.2]. This fact is important in [9].

Proof of Lemma 5. Let Γ be a graph which describes R as in Lemma 3. We will use v_0 to denote the initial vertex of Γ_0 and v_t to stand for an arbitrary terminal vertex. We may assume every vertex of Γ lies on a directed path from v_0 to some v_t. It follows that Γ has a directed spanning subtree, Γ_0, with root v_0. For each edge $e \in \Gamma - \Gamma_0$ let $p(e)$ be the label of the path in Γ_0 from v_0 to the source of e, and let $q(e)$ be the label of the path to the target of e. Let $a(e)$ be the label of e. For each v_t let $r(v_t)$ be the label of the path in Γ_0 from v_0 to v_t. Note that $r(v_t) \in R$.

By Lemma 3 a word w is in R if and only if it is the label of a path y in Γ from v_0 to some v_t. If such a y lies entirely in Γ_0, then $w = r(v_t)$; otherwise y decomposes into a sequence of segments in Γ_0 separated by edges $e_1, \cdots, e_n$, $n \geq 1$, in $\Gamma - \Gamma_0$. Since Γ_0 is a tree, the initial segment of y from v_0 to the source of e_1 must have label $p(e_1)$. Likewise if $s(e_i, e_{i+1})$ is the label of the segment of y from the target of e_i to the source of e_{i+1}, then

(3) $$p(e_{i+1}) = q(e_i)s(e_i, e_{i+1}).$$

Finally if the segment from the target of e_n to v_t has label $s(e_n, v_t)$, then

(4) $$r(v_t) = q(e_n)s(e_n, v_t).$$

It follows that $w \in R$ if and only if

(5) $$w = r(v_t) \text{ or}$$

(6) $\quad w = p(e_1)a(e_1)s(e_1,e_2)a(e_2)\cdots a(e_n)s(e_n,v_t), \quad n \geq 1$,

and (3) and (4) hold.

Now take all edges in $\Gamma - \Gamma_0$ and all terminal vertices as a set of symbols, and let F_0 be the free monoid on these symbols and their (formal) inverses. For each $w \in R$ let $w_0 = v_t$ if (5) holds, and let $w_0 = e_1 e_2 \cdots e_n v_t$ if (6) holds. Define $R_0 = \{w_0 | w \in R\}$. By Lemma 3(i) each $w \in R$ is the label of a unique path in Γ, whence there is just one w_0 for each $w \in R$. Likewise distinct w's yield distinct w_0's. Thus $\rho(w_0) = w$ defines a bijection from R_0 to R.

Define $\sigma: F_0 \to F$ on the generators of F_0 as follows: Let $\sigma(v_t) = r(v_t)$ and let $\sigma(e)$ be any word in F such that $\pi\sigma(e) = \pi(p(e))\pi(a(e))[\pi(q(e))]^{-1}$. Define σ on the inverses of the v_t's and e's so that $\pi\sigma(v_t^{-1}) = [\pi\sigma(v_t)]^{-1}$, $\pi\sigma(e^{-1}) = [\pi\sigma(e)]^{-1}$. $H = \pi\sigma(F_0)$ is a subgroup of G, and by (5) and (6) we see that (ii) holds and $\pi(R) \subseteq H$.

To prove (i) it suffices to show $H \subseteq \langle R \rangle$. In other words we need $\pi\sigma(v_t) \in \langle R \rangle$ for each v_t and $\pi\sigma(e) \in \langle R \rangle$ for each edge e in $\Gamma - \Gamma_0$. As $\sigma(v_t) = r(v_t) \in R$, clearly $\pi\sigma(v_t) \in \langle R \rangle$. By our initial assumption that every vertex appears on some path from v_0 to a v_t every e occurs in some decomposition (6). That is, every e occurs in some $w_0 = e_1 \cdots e_n v_t \in R_0$. But from (6) if $e_1 \cdots e_n v_t \in R_0$, then $e_2 \cdots e_n v_t \in R_0$ too. Since $\pi\sigma(e_1) = \pi\sigma(w_0)[\pi\sigma(e_2 \cdots e_n v_t)]^{-1}$, an inductive argument yields the desired conclusion.

It remains only to show that $R_0 \in \mathrm{Rat}(F_0)$. Define Γ_1 to have initial vertex v_0, terminal vertices v_t, and other vertices e for each edge e in $\Gamma - \Gamma_0$. Join v_0 to each e with an edge labelled e from v_0 to e. An edge goes from e to e' with label e' if $s(e,e')$ exists, and an edge goes from e to v_t with label v_t if $s(e,v_t)$ exists. Finally v_0 is joined to each v_t by an edge with label v_t from v_0 to v_t. Check that every path from v_0 to any v_t corresponds to a decomposition (5) or (6) and yields $w_0 \in R_0$ with $\rho(w_0) \in R$.

§4. *Conclusion*

We have not made much progress in pinning down the structure of groups in $\mathcal{C}$. One possibility for further study is to try to find cross-sections of a restricted type. For example a variation of the argument used in the proof of Lemma 5 shows

(7) If $G \in \mathcal{C}$, then G has a cross-section closed under suffixes.

Another possibility is to pick a restricted type of cross-section and study the corresponding subclass of $\mathcal{C}$. For example $T \in \text{Rat}(F)$ is bounded if there are words $w_1, \cdots, w_m$ in F such that $T \subseteq w_1^* \cdots w_m^*$. Every polycyclic by finite group has a bounded rational cross-section.

Problem. Prove that a group with a bounded rational cross-section is polycyclic by finite or find a counterexample.

DEPARTMENT OF PURE AND APPLIED MATHEMATICS
STEVENS INSTITUTE OF TECHNOLOGY
HOBOKEN, NEW JERSEY 07030

REFERENCES

[1] A. V. Anisimov and F. D. Seifert, Zur algebraischen Charakteristik der durch Kontext-freie Sprachen definierten Gruppen, Elektronische Informationsverarbeiting und Kybernetic *11* 1975, 695-702.

[2] G. Bergman, The diamond lemma for ring theory, Advances in Math. *29* 1978, 178-218.

[3] R. V. Book, The power of the Church-Rosser property for string rewriting systems, 6th Conf. on Automated Deduction, New York, 1982, Springer Lecture Notes in Computer Science *138* 1982, 360-368.

[4] R. Gilman, Presentations of groups and monoids, J. Alg. *57* 1979, 544-554.

[5] ―――, Enumerating infinitely many cosets, in "computational Group Theory," M. Atkinson, ed., Acad. Pre., London, 1984.

[6] ―――, Computations with rational subsets of confluent groups, Eurosam 84, Springer Lecture Notes in Computer Science *174* 1984, 207-212.

[7] M. Hall, "The Theory of Groups," MacMillan, New York, 1959.

[8] G. Huet and D. Oppen, Equations and rewrite rules: A survey, in: R. Book, ed., Formal Language Theory: Perspectives and Open Problems (Academic Press, New York, 1980), 349-405.

[9] P. Johansen, "An Algebraic Normal Form for Regular Events," Polyteknisk Forlag, Lyngby, Denmark 1972.

[10] G. Lallement, "Semigroups and Combinatorial Applications," John Wiley, New York, 1979.

[11] D. Muller and P. Schupp, Groups, the theory of ends, and context-free languages, J. Computer and System Sciences, 26 1983, 311-338.

[12] C. Sims, Verifying nilpotence, J. Symbolic Computation, to appear.

ON THE RATIONAL GROWTH OF VIRTUALLY NILPOTENT GROUPS

Max Benson

§1. Introduction

Suppose G is a group generated as a monoid by a finite subset S. The growth series of G, S is defined by

$$P_{G,S}(t) = (1-t)\sum_{N=0}^{\infty} \gamma_{G,S}(N)t^N$$

where $\gamma_{G,S}(N)$ is the number of elements of G which can be represented by words from S of length less than or equal to N.

For most groups very little is known about what form $P_{G,S}(t)$ takes or even how to calculate examples. In particular it is an interesting question to determine when $P_{G,S}(t)$ can be written as a rational function of t.

We will concentrate on the case when G is a finitely generated virtually nilpotent group, that is, when it contains a nilpotent subgroup of finite index. For these groups the function $\gamma_{G,S}(N)$ grows like a polynomial in N (see [5], [6], [7]). If $P_{G,S}(t)$ can be written as a rational function of t, then, according to [3] (Theorem 8.5), $\gamma_{G,S}(N)$ is periodically polynomial. This means that there is a period d and polynomials $p_1, p_2, \cdots, p_d$ such that, for all sufficiently large N, $\gamma_{G,S}(N) = p_j(N)$ for $j \equiv N \pmod{d}$. In this case the rational function representing $P_{G,S}(t)$ can be shown to have all of its poles at roots of unity.

The degree of the polynomials $p_1, p_2, \cdots, p_d$ is equal to the order of the pole at $t = 1$. H. Bass [1] (Theorem 2, p. 608) has calculated this degree. If G contains a finitely generated nilpotent subgroup H of finite index in G which has lower central series

$$H = H_1 \supset H_2 \supset \cdots \supset H_p \supset H_{p+1} = \{1\}$$

then ord $p_{G,S}(t) = \sum_{i=1}^{p} i \ \text{rank}(H_i/H_{i+1})$.

We will outline a simplified proof that $p_{G,S}(t)$ is rational whenever G is a virtual class 1 nilpotent group (see [2]) and look at a special case where $p_{G,S}(t)$ can be computed when G is a class 2 nilpotent group.

§2. *The case* $G = Z^n$

It is easy to show [2] (p. 252) that $p_{G,S}(t)$ is rational when $G = Z^n$ (more generally any finitely generated abelian group) regardless of the choice of the generating set S. In this case $p_{G,S}(t)$ turns out to be the Poincaré series of an appropriately constructed graded algebra of finite type. The growth series is seen to be rational with denominator $(1-t)^n$. Because this is based on the techniques of commutative algebra it does not seem to generalize to the nonabelian case.

We will now sketch another more geometric method of proof. It borrows many ideas from the author's earlier paper [2]. We begin by defining polyhedral families.

Any subset $P_N \subset Z^n$ which can be expressed in one of the forms

a) $\{z \in Z^n | a \cdot z = b(N)\}$
b) $\{z \in Z^n | a \cdot z > b(N)\}$
c) $\{z \in Z^n | a \cdot z \equiv b(N) \pmod{c}\}$

for some $a \in Z^n$, $c, N \in Z$, $c, N > 0$, and $b(N)$ an affine function of N with integer coefficients will be called an elementary region in Z^n.

A family of elementary regions $P_1, P_2, \cdots$ where the a, c, and $b(N)$ are fixed is called an elementary family. We define the intersection $R_N \cap S_N$ of two elementary families in Z^n to be the family given by

$R_1 \cap S_1, R_2 \cap S_2, \cdots$. A finite intersection of elementary families will be called a basic polyhedral family. Unions of basic polyhedral families can be defined in a similar manner. A finite union of basic polyhedral families will be called a polyhedral family. A polyhedral family is bounded if each member of the family is bounded. In general the members of a bounded family are not uniformly bounded.

Bounded polyhedral families can be "counted" by power series which are rational. They also behave well under Z-affine maps. The following theorems describe these properties.

(2.1) (PROPOSITION. Suppose P_N is a bounded polyhedral family in Z^n and $f(x_1, \cdots, x_n)$ is a polynomial with integer coefficients, then the generating function

$$p(t) = \sum_{N=0}^{\infty} a_N t^N$$

where

$$a_N = \sum_{x \in P_N} f(x)$$

can be written as a rational function of t.

(2.2) PROPOSITION. Suppose P_N is a polyhedral family in Z^m and $\Gamma : Z^m \to Z^n$ is a Z-affine map. Then $\Gamma(P_N)$ is a polyhedral family in Z^n.

The proofs of both of these results appear in [2] although the language is somewhat different.

(2.3) EXAMPLE. Consider the bounded polyhedral family

$$P_N = \{(x,y) \in Z^2 | x > 0\} \cap \{(x,y) \in Z^2 | y > 0\} \cap \{(x,y) \in Z^2 | x+y < N\}.$$

Each member of the family consists of the set of lattice points inside of an isosceles right triangle. The number of points in each P_N is

$\frac{(N-2)(N-1)}{2}$. Taking $f(x) = 1$, we find that

$$a_N = \frac{(N-2)(N-1)}{2}$$

$$p(t) = \frac{t^3}{(1-t)^3}.$$

If we apply the affine map $\Gamma: \mathbb{Z}^2 \to \mathbb{Z}^2$ given by $\Gamma(x,y) = (x+y+1, x-y-1)$, then P_N is sent to the polyhedral family $R_N \cap S_N$, where

$$R_N = \{(u,v) \in \mathbb{Z}^2 | u < N+1\} \cap \{(u,v) \in \mathbb{Z}^2 | u-2 > v > -u\}$$

$$S_N = \{(u,v) \in \mathbb{Z}^2 | u,v \equiv 0 \pmod{2}\} \cap \{(u,v) \in \mathbb{Z}^2 | u,v \equiv 1 \pmod{2}\}.$$

It is easy to see that this family can be written as a finite union of basic polyhedral families.

Using (2.1) and (2.2) we can now give an alternate proof that $p_{G,S}(t)$ is rational when $G = \mathbb{Z}^n$. For each $N \in \mathbb{Z}_+$, let $\Sigma(S)_N^\varepsilon$ be the set of words from $S = \{s_1, \cdots, s_m\}$ of the form $s_1^{a_1} s_2^{a_2} \cdots s_m^{a_m}$, $a_i \in \mathbb{Z}_+$ of length $\leq N$. Then $\Sigma(S)_N^\varepsilon$ can be viewed as a polyhedral family in $\mathbb{Z}^m$. The map $\Gamma: \Sigma(S)_N^\varepsilon \to \mathbb{Z}^n$ sending each word to the group element it represents is given by

$$\Gamma(a_1, a_2, \cdots, a_m) = \sum_{i=1}^m a_i s_i.$$

This Γ is a $\mathbb{Z}$-affine map and so by (2.1) $\Gamma(\Sigma(S)_N^\varepsilon)$ is a polyhedral family. Furthermore

$$\gamma_{G,S}(N) = \sum_{x \in \Gamma(\Sigma(S)_N^\varepsilon)} 1.$$

By (2.2) we conclude that $p_{G,S}(t)$ is a rational function of t.

§3. *The case when* G *is a finite extension of* Z^n

A virtually nilpotent group of class 1 is a group which contains Z^n as a subgroup of finite index. It was shown in [2] that $p_{G,S}(t)$ is rational in this case as well. The situation is much more complicated because now words in G can't be arranged into the form $s_1^{a_1} s_2^{a_2} \cdots s_m^{a_m}$ like they could be in Z^n. But there is an analogous argument that works which we will sketch here. See [2] for details.

Let $\sigma \in \Sigma(S)$, $\sigma = s_1 s_2 \cdots s_k$ be a word from S. Suppose $s_{i_1}, s_{i_2}, \cdots, s_{i_q}$, $i_1 < i_2 < \cdots < i_q$ are those s_i for which $s_i \notin Z^n$. The word

$$\pi(\sigma) = s_{i_1} s_{i_2} \cdots s_{i_q}$$

is called the pattern of σ. If $G = Z^n$, then there is only one pattern, ε, the empty word.

The set of all words patterned after π will be denoted by $\Sigma(S)^\pi$. Any element of $\Sigma(S)^\pi$ can be written in the form

$$z_1^{a_1} z_2^{a_2} \cdots z_r^{a_r} s_{i_1} z_1^{a_1} z_2^{a_2} \cdots z_r^{a_r} s_{i_2} \cdots s_{i_q} z_1^{a_1} z_2^{a_2} \cdots z_r^{a_r}$$

where the $z_1, z_2, \cdots, z_r$ are those generators of S which are in Z^n. Any such word is called arranged. The set of arranged words are in one-one correspondence with points in Z_+^m, where $m = (q+1)r$. If we restrict ourselves to arranged words of a given pattern, the situation now resembles the case for $G = Z^n$. The problem is that there are infinitely many patterns. This difficulty can be solved as follows.

We enlarge the generating set S to S^* by adding generators so that using the new generating set S^*

(1) There are a finite number of patterns $\pi_1, \pi_2, \cdots, \pi_g$, so that any element of G may be written as a minimal word in one of these patterns.

(2) $p_{G,S^*}(t) = p_{G,S}(t)$.

Proving that the generating set can be enlarged to satisfy these properties of course does take some effort.

The words $\Sigma(S^*)_N^\pi$ of a given pattern π with length $\leq N$ can be identified with a polyhedral family in some Z^m. Each word in $\Sigma(S^*)_N^\pi$ defines an element in the same coset of Z^n in G. The elements of this coset are in one-one correspondence with the points in Z^n, and the map sending a word in $\Sigma(S^*)_N^\pi$ to the corresponding point in Z^n turns out to be Z-affine.

Therefore we can conclude that the elements of any coset of Z^n in G represented by words in S^* of length $\leq N$ can be identified with a finite union of polyhedral families in Z^n (which by definition must also be polyhedral family). As a result the generating function counting the elements of this coset with length $\leq N$ will be rational. $p_{G,S}(t)$ is a finite sum of rational functions so it is also rational.

§4. *Calculation of a growth series for a class 2 nilpotent group*

In contrast to the previous cases, not much is known about how to calculate $p_{G,S}(t)$ for class 2 nilpotent groups. We will consider the case where G, S have the property that

$$[G, G] = Z$$

$$G/[G, G] \simeq Z^n$$

$$S = \{s_1, \cdots, s_n, s_1^{-1}, \cdots, s_n^{-1}\}$$

$$[s_i, s_j] \in \{-1, 0, 1\} \text{ all } i, j$$

The simplest case is when $n = 2$. The sole example for $n = 2$ is given by

$$G = \left\{ \begin{pmatrix} 1 & a & c \\ 0 & 1 & b \\ 0 & 0 & 1 \end{pmatrix} \middle| a, b, c \in Z \right\}$$

$$S = \left\{ \begin{pmatrix} 1 & 1 & 0 \\ 0 & 1 & 0 \\ 0 & 0 & 1 \end{pmatrix}, \begin{pmatrix} 1 & 0 & 0 \\ 0 & 1 & 1 \\ 0 & 0 & 1 \end{pmatrix}, \begin{pmatrix} 1 & -1 & 0 \\ 0 & 1 & 0 \\ 0 & 0 & 1 \end{pmatrix}, \begin{pmatrix} 1 & 0 & 0 \\ 0 & 1 & -1 \\ 0 & 0 & 1 \end{pmatrix} \right\}$$

We will only carry out the computation for this one case, and although the method seemed to show some promise for examples with $n > 2$, the author has not been able to do any "new" examples for $n > 2$.

We need to introduce some notation. Let $\Sigma(S)$ be the set of all words formed from S, and $\Sigma(S)_N$ be the set of all words with length $\leq N$. As before let $\Gamma : \Sigma(S) \to G$ be the map which sends a word $\sigma \in \Sigma(S)$ to the word it represents in G.

For any $g \in G$ express its class in Z^n as a linear combination of the $\bar{s}_i$, say, $\bar{g} = a_1 \bar{s}_1 + \cdots + a_n \bar{s}_n$, $a_i \in Z$. The Z-height of g is

$$z(g) = g \, s_n^{-a_n} \cdots s_2^{-a_2} s_1^{-a_1} \in Z .$$

For any $\xi \in Z^n$ we define its Z-set by

$$Z_N(\xi) = \{z(g) | g \in \Gamma(\Sigma(S)_N), \bar{g} = \xi\} .$$

Then the cumulative growth function that we want to calculate is given by

$$\gamma_{G,S}(N) = \sum_{\xi \in \Gamma(\Sigma(\bar{S})_N)} \text{card } Z_N(\xi) .$$

Thus $\gamma_{G,S}(N)$ is the sum of card $Z_N(\xi)$ over the polyhedral family $\Gamma(\Sigma(\bar{S})_N)$ in Z^n. We want to now determine the structure of $Z_N(\xi)$ so that we can find an expression for card $Z_N(\xi)$ in terms of ξ.

(4.1) PROPOSITION. $Z_N(\xi)$ *is an interval in* Z.

Proof. We shall use induction on N. When $N = 1$, the $Z_N(\xi)$ is either $\emptyset$ or else $\{0\}$, depending on ξ. When $N = 2$, then $Z_N(\xi)$ can be $\emptyset$, $\{0\}$, $\{0,1\}$, or $\{-1,0\}$, all of which are intervals.

Assume that it is true for $N \leq k$, $k \geq 2$, and we will show it must be true for $N = k+1$. Any word of length $\leq N$ starts with one of the generators s_i or s_i^{-1} on the left and ends with a word of length $\leq N-1$. For this reason

$$(4.2) \quad Z_N(\xi) = \bigcup_{i=1}^{n} \sum_{k=1}^{i-1} \xi_k [s_i, s_k] + Z_{N-1}(\xi - \bar{s}_i)$$

$$\cup \bigcup_{i=1}^{n} \sum_{k=1}^{i=1} \xi_k [s_i^{-1}, s_k] + Z_{N-1}(\xi - \bar{s}_i^{-1}).$$

Here the $\xi_k \in Z$ are determined by the relation $\bar{\xi} = \xi_1 \bar{s}_1 + \cdots + \xi_n \bar{s}_n$. Thus $Z_N(\xi)$ is a union of intervals. We only need to show that the intervals "patch" together.

Take any two nonempty intervals in this union, say,

$$I_1 = \sum_{k=1}^{i-1} \xi_k [s_i^u, s_k] + Z_{N-1}(\xi - \bar{s}_i^u)$$

$$I_2 = \sum_{k=1}^{j-1} \xi_k [s_j^v, s_k] + Z_{N-1}(\xi - \bar{s}_j^v)$$

with $u, v \in \{-1, 1\}$ and $i \geq j$. We are going to show that they overlap. Using the same reasoning as for (4.2), I_1 must contain the subinterval

$$\sum_{k=1}^{i-1} \xi_k [s_i^u, s_k] + \sum_{k=1}^{j-1} \xi_k [s_j^v, s_k] + Z_{N-1}(\xi - \bar{s}_i^u - \bar{s}_j^v)$$

and I_2 must contain

$$\sum_{k=1}^{i-1} \xi_k [s_i^u, s_k] + \sum_{k=1}^{j-1} \xi_k [s_j^v, s_k] - [s_i^u, s_j^v] + Z_{N-1}(\xi - \bar{s}_i^u - \bar{s}_j^v).$$

These two subintervals are just offset by the commutator $[s_i^u, s_j^v]$ which is either -1, 0, or 1, so the union must also be an interval. This implies that the union given in (4.2) must also be an interval.

The meaning of (4.1) is that we merely have to know the maximum and minimum values in $Z_N(\xi)$ in order to find its cardinality. It turns out that these maxima and minima occur for words of a simple form.

Any word in which each generator appears raised to some positive power at most one place in the word will be called a collected word. For example, $(s_1^{-1})^4 s_3^2 s_1^3$ is collected but $s_2 s_1^2 s_2$ is not.

(4.3) PROPOSITION. *The maximum and minimum values in $Z_N(\xi)$ are assumed as Z-heights of collected words.*

Proof. Assume not, then pick a word which has maximum (minimum) Z-height and the least number of extra terms. If powers of s_i appear twice in the word, it is easy to see that we can "move" one of the powers over the other one producing a new word with a possibly larger (smaller) Z-height and a lesser number of extra terms. This would be a contradiction.

We will now concentrate on the case when $n = 2$. We subdivide $\Gamma(\Sigma(\overline{S})_N)$ into the four coordinate quadrants. On each quadrant we can derive an expression for the form of $Z_N(\xi)$. For instance when $\xi_1 \geq 0$, $\xi_2 \geq 0$ and $Z_N(\xi) \neq \emptyset$

$$Z_N(\xi) = [\min_{t_1, t_2} z(b^{-t_2} a^{\xi_1 + t_1} b^{\xi_2 + t_2} a^{-t_1}), \max_{t_1, t_2} z(a^{-t_1} b^{\xi_2 + t_2} a^{\xi_1 + t_1} b^{-t_2})] \cap Z.$$

The maximum and minimum are taken over all integers t_1 and t_2 in the region bounded by $t_1 \geq 0$, $t_2 \geq 0$, and $|\xi_1| + |\xi_2| + 2t_1 + 2t_2 \leq N$.

Similar forms can be derived for the other three quadrants. Surprisingly when we calculate the number of heights in $Z_N(\xi)$, we get the formula

(4.4) $\quad \text{card } Z_N(\xi) = \max_{t_1, t_2} 2(|\xi_1| + t_1)(|\xi_2| + t_2) - |\xi_1 \xi_2| + 1$

valid for all $\xi \in \Gamma(\Sigma(\overline{S})_N)$.

The next step is to find this maximum. It turns out that $\Gamma(\Sigma(\overline{S})_N)$ can be subdivided by appropriate linear inequalities and congruences so that on each piece card $Z_N(\xi)$ is given by a polynomial in the coordinates of ξ and N. Since each piece is a polyhedral family, we can appeal to (2.1) to show that $p_{G,S}(t)$ must be rational.

To completely carry out the computations in order to find a formula for $p_{G,S}(t)$ is quite difficult. Less precise analysis shows that the coefficients $\gamma_{G,S}(N)$ are periodically polynomial of degree 4 and period 12 for $N \geq 4$. Using the formula (4.4) we have derived for card $Z_N(\xi)$, a simple computer program was written to calculate the values of $\gamma_{G,S}(N)$ for $0 \leq N \leq 50$. By calculating four successive differences on the coefficients, the power series was easily found. The result is given below.

(4.5) CALCULATION. The growth series for the three-dimensional upper triangular matrix group is

$$\frac{1 + t + 3t^2 + 11t^3 + 6t^4 + 14t^5 + 12t^6 + 4t^7 + 14t^8 + 14t^8 + 4t^9 + 12t^{10} + 14t^{11} + 5t^{12} + 11t^{13} + 9t^{14} - 11t^{15}}{(1-t)^4 (1-t)^{12}}.$$

The author has been unable to generalize this calculation when $n > 2$. Special cases when the group is a direct sum of abelian groups and groups isomorphic of the example given above can of course be calculated, but otherwise no further progress has been made.

The difficulty with carrying out the method employed in calculating (4.5) is when you try to subdivide $\Gamma(\Sigma(\overline{S})_N)$ into polyhedral subfamilies on which the maximum and minimum are given by the Z-heights of a pair of collected words. This subdivision could be done using quadratic inequalities, but it is not apparent that it can be done with linear ones. Linear ones are needed to guarantee that the subfamilies are polyhedral.

The results (4.1) and (4.2) also have some value for conducting numerical investigations into the forms of the growth functions. Without using these results, the calculation of $\gamma_{G,S}(N)$ for even modest values of N (e.g. $N \geq 10$) can be intractible both because of exponential time and memory space requirements. Using (4.1) and (4.2) we can compute $\gamma_{G,S}(N)$ in polynomial time and space. It should be noted that the computation still requires significant computer resources however.

It would be interesting to generalize (4.1) so that $Z_N(\xi)$ could be described for any generating set S. That may allow more efficient calculation of $\gamma_{G,S}(N)$ for many different cases.

§5. *Unsolved problems*

The growth series of the pair G, S where

$$[G,G] = Z$$

$$G/[G,G] \simeq Z^3$$

$$S = \{s_1, s_2, s_3, s_1^{-1}, s_2^{-1}, s_3^{-1}\}$$

$$[s_i, s_j] = 1 \text{ all } i, j$$

has not been calculated. This is essentially the only remaining example of this type for $n = 3$, and it would be very interesting to either be able to work out this example or else show that it is not rational.

If the growth series of this pair G, S turns out not to be rational, then it raises many questions about the example that was calculated in section 4. For instance, is the fact that its growth series is rational related to the matrix presentation? And, how significant a role is geometry playing? The example given in section 4 is the fundamental group of a three dimensional manifold. M. Grayson studied the growth functions of fundamental groups of three dimensional manifolds in [4], giving many examples of calculations.

The author wishes to thank Phil Wagreich for suggesting calculation of the matrix group example.

DEPARTMENT OF MATHEMATICAL SCIENCES
UNIVERSITY OF MINNESOTA
DULUTH, MINNESOTA 55812

REFERENCES

[1] Bass, H.: The degree of polynomial growth of finitely generated nilpotent groups. Proc. London Math. Soc. (3) 25, 603-614 (1972).

[2] Benson, M.: Growth Series of Finite Extensions of Z^n are Rational. Invent. Math. 73, 251-269 (1983).

[3] Cannon, J.W.: The growth of the closed surface groups and the compact hyperbolic Coxeter groups (unpublished).

[4] Grayson, Matthew.: Geometry and Growth in Three Dimensions, Thesis, Princeton University, 1983.

[5] Gromov, M.: Groups of polynomial growth and expanding maps. Publ. Math. I.H.E.S. 53 (1981).

[6] Milnor, J.: Growth of finitely generated solvable groups. J. Differential Geometry 2, 1-7 (1968).

[7] Wolf, J.A.: Growth of finitely generated solvable groups and curvatures of Riemannian manifolds, J. Differential Geometry 2, 421-446 (1968).

SJOGREN'S THEOREM FOR DIMENSION SUBGROUPS – THE METABELIAN CASE

Narain Gupta

1. Introduction

Let $\mathbb{Z}G$ be the integral group ring of a group G, $\Delta(G)$ its augmentation ideal and $D_k(G) = G \cap (1+\Delta^k(G))$, the k-th dimension subgroup of G. For all $k \geq 1$, $D_k(G) \geq \gamma_k(G)$, the k-th term of the lower central series of G. Rips [6] has constructed a finite 2-group G such that $D_4(G) \neq \gamma_4(G)$. A remarkable result due to Sjogren [7] states that for each $n \geq 1$, there exists an n!-number C_n (i.e. the primary decomposition uses only those primes which divide n!) such that for all G, $D_{n+2}(G)/\gamma_{n+2}(G)$ has exponent dividing C_n. An immediate consequence of this result is that if G is a p-group then $D_k(G) = \gamma_k(G)$ for all $k \leq p+1$. Another consequence is the Hall-Jennings theorem: if the lower central factors of a group G are torsion free then $D_k(G) = \gamma_k(G)$ for all $k \geq 1$. Since $\gamma_2(\gamma_2(G)) \leq \gamma_4(G)$, in view of Rips' example, the Sjogren's theorem remains equally significant when G is restricted to the class of metabelian groups. In this paper we offer, for metabelian groups G, a direct and simple proof of a result similar to Sjogren's theorem. We prove that for each $n \geq 1$, there exists an n!-number C_n^* such that for all *metabelian* groups G, $D_{n+2}(G)/\gamma_{n+2}(G)$ has exponent dividing $2C_n^*$. For a comparison of simplicity, the reader is referred to an alternative proof of Sjogren's theorem (without use of spectral sequences) given by Hartley [2]. Although our bound $2C_n^*$ is considerably smaller than Sjogren's bound C_n, our proof is valid only for metabelian groups. In addition we have a few results of independent interest concerning the dimension subgroups of metabelian groups.

2. The cyclic group ring

Let $\mathbb{Z}C$ be the group ring of an infinite cyclic group $C = <x>$ and let $e \geq 0$ be a fixed integer. For each $k \geq 1$, let $\binom{e}{k}^*$ be the additive subgroup of $\mathbb{Z}$ defined by

$$\binom{e}{k}^* = \left\langle \binom{e}{1}, \cdots, \binom{e}{k} \right\rangle ; \qquad (1)$$

and for each $n \geq 1$, let $R(n)$ be the additive subgroup of $\mathbb{Z}C$ defined by

$$R(n) = \left\langle \binom{e}{n}^*(x-1)^n, \binom{e}{n+1}^*(x-1)^{n+1}, \cdots \right\rangle , \qquad (2)$$

where $\binom{e}{i}$ is the binomial co-efficient (with $\binom{e}{i} = 0$ for $i > e$). Then clearly

$$\binom{e}{1}^* \subseteq \binom{e}{2}^* \subseteq \cdots \qquad (3)$$

and

$$(x-1)^j R(n) \subseteq R(n+j) \qquad (4)$$

for all $j \geq 1$, $n \geq 1$. The expansion

$$(x^e - 1) = e(x-1) + \sum_{k=2}^{\infty} \binom{e}{k}(x-1)^k$$

yields

$$e(x-1) \equiv (x^e - 1) \mod R(2), \qquad (5)$$

which on pre-multiplication with $(x-1)^j$, $j \geq 1$, gives using (1) and (4),

$$e(x-1)^{j+1} \equiv 0 \mod (x(e) + R(j+2)), \qquad (6)$$

where $x(e) = \text{ideal}_{\mathbb{Z}C}\{(x-1)(x^e-1)\}$. For each $k \geq 1$, define b_k to be the least common multiple of $\{1, \cdots, k\}$. Then using $\binom{e}{j} = \frac{e}{j}\binom{e-1}{j-1}$ for all j, it is readily seen that

$$b_k \binom{e}{k}^* \subseteq \binom{e}{1}^* = eZ . \tag{7}$$

For $n \geq 1$, define

$$C_n^* = b_1 \cdots b_n . \tag{8}$$

We prove by induction on $n \geq 1$, that

$$C_n^* e(x-1) \equiv C_n^*(x^e-1) \mod (x(e)+R(n+1)) . \tag{9}$$

For $n = 1$, (9) follows from (5). For the inductive step, we assume (9) for some $n \geq 1$. Then

$$b_{n+1} C_n^* e(x-1) \equiv b_{n+1} C_n^*(x^e-1)$$

modulo $(x(e) + b_{n+1} R(n+1))$. However,

$$b_{n+1} R(n+1) \subseteq b_{n+1} \binom{e}{n+1}^* (x-1)^{n+1} + R(n+2)$$

$$\subseteq \binom{e}{1}^* (x-1)^{n+1} + R(n+2), \text{ by (7)}$$

$$\subseteq x(e) + R(n+2), \text{ by (6)} .$$

Thus $C_{n+1}^* e(x-1) \equiv C_{n+1}^*(x^e-1) \mod (x(e)+R(n+2))$, as required. In particular, (9) yields for all $n \geq 1$,

$$C_n^* e(x-1) \equiv C_n^*(x^e-1) \mod (\text{ideal}_{ZC}\{(x-1)(x^e-1), (x-1)^{n+1}\}) \tag{10}$$

which upon dividing by $(x-1)$ yields the fundamental relation

$$C_n^* e \equiv C_n^* t_e(x) \mod(\text{ideal}_{ZC}\{(x^e-1), (x-1)^n\}) , \tag{11}$$

where $t_e(x) = 1 + x + \cdots + x^{e-1}$ if $e \geq 1, t_e(x) = 0$ if $e = 0$.

3. The free group rings

Let F be a free group freely generated by $x_1, \cdots, x_m$, $m \geq 2$, and let $(e_1, \cdots, e_m)$ be a fixed m-tuple of non-negative integers satisfying $e_m | e_{m-1} | \cdots | e_1$. Let $S = \langle x_1^{e_1}, \cdots, x_m^{e_m}, F' \rangle$ be the normal subgroup of F associated with the m-tuple $(e_1, \cdots, e_m)$, where $F' = \gamma_2(F)$ is the commutator subgroup of F. Let $\underline{f} = ZF(F-1)$ denote the augmentation ideal $\Delta(F)$ of the integral group ring ZF and let $\underline{s} = ZF(S-1) = ZF\Delta(S)$. [Note that $\underline{s}$ is the ideal generated by all $(x_i-1)(x_j-1) - (x_j-1)(x_i-1)$ and $x_k^{e_k}-1$.] We shall seek some detailed information about the subgroup $D_{n+2}(\underline{f}\,\underline{s}) = F \cap (1 + \underline{f}\,\underline{s} + \underline{f}^{n+2})$, $n \geq 1$, which clearly contains the subgroup $S'\gamma_{n+2}(F)$.

Let $w \in D_{n+2}(\underline{f}\,\underline{s})$. Then $w - 1 \in \underline{f}\,\underline{s} + \underline{f}^{n+2} \leq \underline{f}^2$ implies that $w \in F'$. Modulo F'', using Jacobi congruence we may write w as

$$w \equiv \prod_{1 \leq i < j \leq m} [x_i, x_j]^{p_{ij}}, \tag{12}$$

where $p_{ij} = p_{ij}(x_i, \cdots, x_m) \in ZF$ (cf. [1], Lemma 2). We rewrite w as

$$w \equiv w_1 \cdots w_{m-1} \mod F'',$$

where

$$w_i = \prod_{j=i+1}^{m} [x_i, x_j]^{p_{ij}}. \tag{13}$$

For each $i = 1, \cdots, m-1$, let $\theta_i : ZF \to ZF$ be the homomorphism given by $x_k \to 1$ for $k \leq i$, $x_k \to x_k$ for $k > i$. Since the ideals $\underline{f}$ and $\underline{s}$ are invariant under θ_i for all i, using $\theta_1, \cdots, \theta_{m-1}$, in turn, shows that if $w - 1 \in \underline{f}\,\underline{s} + \underline{f}^{n+2}$ with w as in (12) then for each $i = 1, \cdots, m-1$, $w_i - 1 \in \underline{f}\,\underline{s} + \underline{f}^{n+2}$ with w_i as in (13) [note that $ZF(F''-1) \leq \underline{f}\,\underline{s}$]. Using $[x_i, x_j] - 1 = x_i^{-1} x_j^{-1} (x_i x_j - x_j x_i) = x_i^{-1} x_j^{-1}((x_i-1)(x_j-1) - (x_j-1)(x_i-1)) \equiv$

$(x_i-1)(x_j-1) - (x_j-1)(x_i-1) \pmod{\underline{f}\,\underline{s}}$, it follows that $[x_i,x_j]^{p_{ij}} - 1 \equiv ((x_i-1)(x_j-1) - (x_j-1)(x_i-1))p_{ij} \pmod{\underline{f}\,\underline{s}}$. More generally,

$$w_i - 1 \equiv \sum_{j=i+1}^{m} ((x_i-1)(x_j-1) - (x_j-1)(x_i-1))p_{ij} \pmod{\underline{f}\,\underline{s}}.$$

Since $w_i - 1 \in \underline{f}\,\underline{s} + \underline{f}^{n+2}$, it follows that

$$\sum_{j=i+1}^{m} ((x_i-1)(x_j-1) - (x_j-1)(x_i-1))p_{ij} \in \underline{f}\,\underline{s} + \underline{f}^{n+2} \tag{14}$$

for all $i = 1, \cdots, m-1$. Since $\underline{f}$ is a free right-$\mathbf{Z}F$ module on $x_1 - 1, \cdots, x_m - 1$, it follows from (14) that

$$(x_j-1)(x_i-1)p_{ij} \in \underline{f}\,\underline{s} + \underline{f}^{n+2} \tag{15}$$

and

$$(x_i-1)\sum_{j=i+1}^{m} (x_j-1)p_{ij} \in \underline{f}\,\underline{s} + \underline{f}^{n+2} \tag{16}$$

for all $1 \leq i < j \leq m$. In particular, (15) and (16) yield

$$(x_i-1)p_{ij} \in \underline{s} + \underline{f}^{n+1} \tag{17}$$

and

$$\sum_{j=i+1}^{m} (x_j-1)p_{ij} \in \underline{s} + \underline{f}^{n+1} \tag{18}$$

for all $1 \leq i < j \leq m$. Since $\mathbf{Z}F(F'-1) \leq \underline{s}$ and $\mathbf{Z}F/\mathbf{Z}F(F'-1) \cong \mathbf{Z}F(F/F')$, we may regard (17) and (18) as results in the free abelian group ring. From (17) we conclude that for all $1 \leq i < j \leq m$,

$$p_{ij} \in t_{e_i}(x_i)\mathbf{Z}F + \underline{s} + \underline{f}^n,$$

where

$$t_{e_i}(x_i) = \begin{cases} 1 + x_i + \cdots + x_i^{e_i-1}, & e_i \geq 1 \\ 0, & e_i = 0 \end{cases} \qquad (19)$$

(cf. [1], Lemma 1). We record these observations as

LEMMA 3.1. *Let* $w \in F'$ *and* $w - 1 \in \underline{f}\underline{s} + \underline{f}^{n+2}$, $n \geq 1$. *Then* $w \equiv \pi_{1 \leq i < j \leq m} [x_i, x_j]^{p_{ij}} \pmod{F''}$, *where* $p_{ij} \in ZF$ *does not involve* $x_1, \cdots, x_{i-1}$ *for* $i \geq 2$; *and each* $p_{ij} \in t_{e_i}(x_i)ZF + \underline{s} + \underline{f}^n$, *where* $t_{e_i}(x_i)$ *is given by (19). Furthermore,* $(x_i - 1)p_{ij} \in \underline{s} + \underline{f}^{n+1}$ *and* $\sum_{j=i+1}^{m}(x_j - 1)p_{ij} \in \underline{s} + \underline{f}^{n+1}$ *for all* $1 \leq i < j \leq m$.

We can now deduce some principal results of this section.

LEMMA 3.2. *Let* w *be as in Lemma 3.1 and let* C_n^* *be as defined by (8). Then modulo* $F''\gamma_{n+2}(F)$, $w^{C_n^*} \equiv \pi_{1 \leq i < j \leq m} [x_i, x_j]^{C_{n-1}^* e_i a_{ij}} \pi_{1 \leq i \leq m} [x_i^{e_i}, \eta_i]^{C_{n-1}^*}$, *where* $a_{ij} = a_{ij}(x_1, \cdots, x_m) \in b_n Z + \underline{f}$ *and* $\eta_i \in F'$.

Proof. By Lemma 3.1, it clearly suffices to prove that $C_n^* p_{ij} \in C_n^* e_i Z + C_{n-1}^* e_i \underline{f} + C_{n-1}^* \underline{s} + \underline{f}^n$. Since $t_{e_i}(x_i) \equiv \sum_{k=1}^{n} \binom{e_i}{k}(x_i - 1)^{k-1} \pmod{\underline{f}^n}$ and $b_n \binom{e_i}{k} \in e_i Z$ for $k = 2, \cdots, n$ (by (7)), it follows that $b_n t_{e_i}(x_i) \in b_n e_i Z + e_i \underline{f} + \underline{f}^n$ and consequently $b_n p_{ij} \in e_i b_n Z + e_i \underline{f} + \underline{s} + \underline{f}^n$. Since $C_n^* = b_n C_{n-1}^*$, the result follows.

LEMMA 3.3. *Let* w *be as in Lemma 3.2.*

(i) $C_{n-1}^* e_i (x_k - 1) \equiv (x_k^{C_{n-1}^* e_i} - 1)$ modulo $\underline{f}\underline{s} + \underline{f}^n$, for $i \leq k \leq m$.

(ii) $[x_i, x_j]^{C_{n-1}^* e_i a_{ij}} - 1 \equiv ((x_i - 1)(x_j - 1) - (x_j - 1)(x_i - 1))C_{n-1}^* e_i a_{ij}$ modulo $\underline{f}^2 \underline{s} + \underline{f}^{n+2}$.

(iii) $[x_i^{e_i}, \eta_i] - 1 \equiv (x_i-1)(\eta_i^{e_i}-1)$ modulo $\underline{f}^2\underline{s}$.

Proof. Since $e_m | e_{m-1} | \cdots | e_i$, the proof of (i) follows from (10). For the proof of (ii) we first note by (i) that for $u = u(x_i, \cdots, x_m) \in \underline{f}^3$, $uC_{n-1}^* e_i \in \underline{f}^2\underline{s} + \underline{f}^{n+2}$; and secondly, modulo $\underline{f}^3 C_{n-1}^* e_i$, $[x_i, x_j]^{C_{n-1}^* e_i a_{ij}} - 1 \equiv ((x_i-1)(x_j-1) - (x_j-1)(x_i-1))C_{n-1}^* e_i a_{ij}$. Proof of (iii) follows from the fact that $(\eta_i-1)(x_i^{e_i}-1) \in \underline{f}^2\underline{s}$ and $(x_i^{e_i}-1)(\eta_i-1) \equiv (x_i-1)(\eta_i-1)e_i \equiv (x_i-1)(\eta_i^{e_i}-1)$ modulo $\underline{f}^2\underline{s}$.

LEMMA 3.4. (i) *For all* $i = 1, \cdots, m$ *and* $\eta_i \in F'$, $[x_i, \eta_i^{e_i}]^{C_{n-1}^*} \in [S, F']\gamma_{n+2}(F)$.

(ii) *For all* $1 \leq i < j \leq m$ *and* $a_{ij} \in b_n Z + \underline{f}$, $[x_i, x_j]^{C_{n-1}^* e_i a_{ij}}$
$\equiv [x_i^{e_i}, x_j]^{C_{n-1}^* a_{ij}} \equiv [x_i, x_j^{e_j}]^{C_{n-1}^*(e_i/e_j)a_{ij}}$ modulo $[s, F']\gamma_{n+2}(F)$.

Proof. For the proof of (i) we have, in turn,

$[x_i, \eta_i^{e_i}]^{C_{n-1}^*} \equiv [x_i, \eta_i]^{e_i C_{n-1}^*}$ mod F''

$\equiv [x_i, \eta_i]^{t_{e_i}(x_i)C_{n-1}^*}$ mod $[S, F']\gamma_{n+2}(F)$, by (11):

$\equiv [x_i^{e_i}, \eta_i]^{C_{n-1}^*}$ mod $[S, F']\gamma_{n+2}(F)$

$\equiv 1$ mod $[S, F']\gamma_{n+2}(F)$.

For the proof of (ii), modulo $[S, F']\gamma_{n+2}(F)$ we have $[x_i^{e_i}, x_j]^{C_{n-1}^* a_{ij}} \equiv [x_i, x_j]^{t_{e_i}(x_i)C_{n-1}^* a_{ij}} \equiv [x_i, x_j]^{e_i C_{n-1}^* a_{ij}}$, by (11). Similarly, $[x_i, x_j^{e_j}]^{C_{n-1}^*(e_i/e_j)a_{ij}} \equiv [x_i, x_j]^{t_{e_j}(x_j)C_{n-1}^*(e_i/e_j)a_{ij}} \equiv [x_i, x_j]^{e_j C_{n-1}^*(e_i/e_j)a_{ij}}$
$\equiv [x_i, x_j]^{e_i C_{n-1}^* a_{ij}}$.

Finally, we prove the following important result.

LEMMA 3.5. $[D_{n+2}(\underline{f}\underline{s}), F] \le S'\gamma_{n+3}(F)$, $n \ge 1$.

Proof. It clearly suffices to show that for each w_i of the form (13) and each x_k, $[w_i, x_k] \in S'\gamma_{n+3}(F)$. Indeed, we have mod F'',

$$[w_i, x_k] \equiv \prod_{j=i+1}^{m} [[x_i, x_j]^{p_{ij}}, x_k]$$

$$\equiv \prod_{j=i+1}^{m} [x_i, x_j, x_k]^{p_{ij}}$$

$$\equiv \prod_{j=i+1}^{m} [x_i, x_k, x_j]^{p_{ij}} [x_k, x_j, x_i]^{p_{ij}}$$

$$\equiv [x_i, x_k]^{\sum_{j=i+1}^{m} (x_j - 1) p_{ij}} [x_k, x_j]^{\sum_{j=i+1}^{m} (x_i - 1) p_{ij}}$$

$$\equiv 1 \mod [F', S]\gamma_{n+3}(F), \text{ by Lemma 3.1.}$$

4. The main result

Let G be a finitely generated metabelian group. Then G admits a pre-abelian presentation of the form

$$G = F/R = \langle x_1, \cdots, x_m; x_1^{e_1} \xi_1, \cdots, x_m^{e_m} \xi_m, \xi_{m+1}, \cdots, F'' \rangle,$$

where $e_i \ge 0$, $e_m | e_{m-1} | \cdots | e_1$, $\xi_i \in F'$ (see for instance [4], page 149). Then $R \le S = \langle x_1^{e_1}, \cdots, x_m^{e_m}, F' \rangle$ and $S' \le R$. Set

$$D_{n+2}(\underline{f}\underline{r}) = F \cap (1 + \underline{f}\underline{r} + \underline{f}^{n+2}),$$

where $\underline{r} = ZF(R-1)$. Then we prove,

THEOREM 4.1. *For each* $n \ge 1$, $D_{n+2}(\underline{f}\underline{r})/R\gamma_{n+2}(F)$ *has exponent dividing* $2C_n^*$, *where* C_n^* *is defined by* (8).

Proof. Let $w \in D_{n+2}(\underline{fr})$ i.e. $w-1 \in \underline{fr} + \underline{f}^{n+2}$. Since $\underline{r} \subseteq \underline{s}$, it follows that $w-1 \in \underline{fs} + \underline{f}^{n+2}$ and hence by Lemmas 3.1 and 3.2,

$$w^{C_n^*} \equiv \prod_{1 \leq i < j \leq m} [x_i, x_j]^{C_{n-1}^* e_i a_{ij}} \prod_{i=1}^{m} [x_i^{e_i}, \eta_i]^{C_{n-1}^*} \qquad (20)$$

modulo $F'' \gamma_{n+2}(F)$, where $a_{ij} \in b_n Z + \underline{f}$ and $\eta_i \in F'$. By Lemma 3.3 (ii), (iii), modulo $\underline{f}^2 \underline{s} + \underline{f}^{n+2}$,

$$[x_i, x_j]^{C_{n-1}^* e_i a_{ij}} - 1 \equiv ((x_i-1)(x_j-1) - (x_j-1)(x_i-1)) C_{n-1}^* e_i a_{ij}$$

and $[x_i^{e_i}, \eta_i]^{C_{n-1}^*} - 1 \equiv (x_i-1)(\eta_i^{e_i C_{n-1}^*} - 1)$. Since $w^{C_n^*} - 1 \in \underline{fr} + \underline{f}^{n+2}$ implies $w^{C_n^*} - 1 \equiv 0 \mod (\underline{fr} + \underline{f}^2 \underline{s} + \underline{f}^{n+2})$, by (20) and the above expansions, $w^{C_n^*} - 1 \equiv 0$ can be expressed as

$$\sum_{k=1}^{m} (x_k-1) \left\{ \sum_{i<k} (-1) e_i C_{n-1}^* a_{ik} (x_i-1) + \sum_{k<j} (+1) e_k C_{n-1}^* a_{kj} (x_j-1) \right.$$
$$\left. + (\eta_k^{e_k C_{n-1}^*} - 1) \right\} \equiv 0 \text{ modulo } \underline{fr} + \underline{f}^2 \underline{s} + \underline{f}^{n+2}$$

which gives, using Lemma 3.3 (i),

$$\sum_{k=1}^{m} (x_k-1)(y_k^{C_{n-1}^*} \eta_k^{e_k C_{n-1}^*} - 1) \in \underline{fr} + \underline{f}^2 \underline{s} + \underline{f}^{n+2}, \qquad (21)$$

where

$$y_k = \prod_{i<k} x_i^{-e_i a_{ik}} \prod_{k<j} x_j^{+e_j (e_k/e_j) a_{kj}}. \qquad (22)$$

Since $\underline{f}$ is a free right ZF-module on $(x_1-1), \cdots, (x_m-1)$, (21) yields

$$y_k^{C_{n-1}^*} \eta_k^{e_k C_{n-1}^*} - 1 \in \underline{r} + \underline{fs} + \underline{f}^{n+1} \qquad (23)$$

for $k = 1, \cdots, m$. Since $u \in r$ implies $u \equiv (z-1) \mod \underline{f}\underline{r}(\leq \underline{f}\underline{s})$ for some $z \in R$, it follows from (23) that for each $k = 1, \cdots, m$, there exists $z_k \in R$ such that

$$y_k^{C_{n-1}^*} e_k C_{n-1}^* z_k - 1 \in \underline{f}\underline{s} + \underline{f}^{n+1} . \tag{24}$$

Thus, by Lemma 3.5,

$$[x_k, y_k^{C_{n-1}^*} e_k C_{n-1}^* z_k] \in S'\gamma_{n+2}(F) \leq R\gamma_{n+2}(F) .$$

Since $[x_k, z_k] \in R$, $[x_k, \eta_k^{e_k C_{n-1}^*}] \in R\gamma_{n+2}(F)$, by Lemma 3.4(i), it follows that

$$[x_k, y_k^{C_{n-1}^*}] \in R\gamma_{n+2}(F)$$

for all $k = 1, \cdots, m$. Thus modulo $R\gamma_{n+2}(F)$, we have

$$1 \equiv \prod_{k=1}^m [x_k, y_k^{C_{n-1}^*}]$$

$$\equiv \prod_{k=1}^m \left[x_k, \prod_{i<k} x_i^{-e_i a_{ij}} \prod_{k<j} x_j^{e_j(e_k/e_j)a_{kj}} \right]^{C_{n-1}^*} \quad \text{by (22)}$$

$$\equiv \prod_{k=1}^m \left(\prod_{i<k} [x_i^{e_i a_{ik}}, x_k]^{C_{n-1}^*} \prod_{k<j} [x_k, x_j^{e_j(e_k/e_j)a_k}] \right)^{C_{n-1}^*}$$

$$\equiv \prod_{1 \leq i < j \leq m} [x_i^{e_i}, x_j]^{a_{ij} C_{n-1}^*} \prod_{1 \leq i < j \leq m} [x_i, x_j^{e_j}]^{(e_i/e_j)a_{ij} C_{n-1}^*}$$

$$\equiv \prod_{1 \leq i < j \leq m} [x_i, x_j]^{t_{e_i}(x_i)a_{ij} C_{n-1}^*} \prod_{1 \leq i < j \leq m} [x_i, x_j]^{t_{e_j}(x_j)(e_i/e_j)a_{ij} C_{n-1}^*}$$

$$\equiv \prod_{1 \leq i < j \leq m} [x_i, x_j]^{e_i C_{n-1}^* a_{ij}} \prod_{1 \leq i < j \leq m} [x_i, x_j]^{e_i C_{n-1}^* a_{ij}} , \quad \text{by (11)}$$

$$\equiv w^{2C_n^*}, \text{ by (20), since } [x_i^{e_i}, \eta_i] \in [S, F'] \leq R .$$

Thus $w^{2C_n^*} \in R\gamma_{n+2}(F)$ as was to be proved.

If $w - 1 \in \underline{\underline{r}} + \underline{\underline{f}}^{n+2}$ then $wz - 1 \in \underline{\underline{fr}} + \underline{\underline{f}}^{n+2}$ for some $z \in R$ and consequently, by Theorem 4.1, $(wz)^{2C_n^*} \in R\gamma_{n+2}(F)$ which, in turn, yields $w^{2C_n^*} \in R\gamma_{n+2}(F)$. Thus we have established the following main result.

THEOREM 4.2. *Let G be a finitely generated metabelian group. Then for all $n \geq 1$, $D_{n+2}(G)/\gamma_{n+2}(G)$ has exponent dividing $2C_n^*$, where $C_n^* = b_1 \cdots b_n$, $b_k = \text{lcm}\{1, \cdots, k\}$.*

5. Concluding remarks

Our bound $2C_n^*$ should be compared with Sjogren's bound $C_n = b_1^{\binom{n}{1}} \cdots b_n^{\binom{n}{n}}$ as given by Hartley [2]. Another interesting fact about the exponent of $D_{n+2}(G)/\gamma_{n+2}(G)$ is the following result.

THEOREM 5.1. *Let G be a finitely generated metabelian group with G/G' of exponent dividing e. Then $D_{n+2}(G)/\gamma_{n+2}(G)$ has exponent dividing e for all $n \geq 1$.*

Proof. It clearly suffices to show that for w_i of the form (13) with $w_i - 1 \in \underline{\underline{fs}} + \underline{\underline{f}}^{n+2}$, $w_i^{e_i} \in S'\gamma_{n+2}(F)$. Using the homomorphisms $\theta_k : \mathbb{Z}F \to \mathbb{Z}F$, as in the proof of Lemma 3.1, for $k = i, \cdots, m$, if necessary, we may assume that w_i involves each of the variables $x_i, \cdots, x_m$ in all its factors. In particular, by Lemma 3.1, modulo $[F', S]\gamma_{n+2}(F)$,

$$w_i \equiv \prod_{j=i+1}^{m} [x_i, x_j]^{(x_{i+1}-1)\cdots \widehat{(x_j-1)} \cdots (x_m-1)t_{e_i}(x_i)q_{ij}}$$

$$\equiv \prod_{j=i+1}^{m} [x_i^{e_i}, x_j, x_{i+1}, \cdots, \hat{x}_j, \cdots, x_m]^{q_{ij}}$$

$$\equiv [x_i^{e_i}, x_{i+1}, \cdots, x_m]^{r_i},$$

where $r_i = \sum_{j=i+1}^{m} q_{ij} \in ZF$. Since $w_i - 1 \in \underline{fs} + \underline{f}^{n+2}$, it follows that

$$(x_i^{e_i} - 1)(x_{i+1} - 1) \cdots (x_m - 1) r_i \in \underline{fs} + \underline{f}^{n+2}$$

and, in turn,

$$t_{e_i}(x_i)(x_{i+1} - 1) \cdots (x_m - 1) r_i \in \underline{s} + \underline{f}^{n+1} \ ;$$

$$t_{e_i}(x_i) r_i \in t_{e_{i+1}}(x_{i+1}) ZF + \cdots + t_{e_m}(x_m) ZF + \underline{s} + \underline{f}^{n+1-m+i} \ ;$$

(cf. [1], Lemma 1)

$$e_i r_i \in t_{e_{i+1}}(x_{i+1}) ZF + \cdots + t_{e_m}(x_m) ZF + \underline{s} + \underline{f}^{n+1-m+i}$$

(setting $x_i = 1$)

Now, modulo $S' \gamma_{n+2}(F)$,

$$w_i^{e_i} \equiv [x_i^{e_i}, x_{i+1}, \cdots, x_m]^{e_i r_i} \equiv 1 \ ,$$

since $[x_i^{e_i}, x_{i+1}, \cdots, x_m]^{t_{e_j}(x_j)} \equiv [x_i^{e_i}, x_{i+1}, \cdots, x_j^{e_j}, \cdots, x_m] \equiv 1 \mod S'$ for all $j = i+1, \cdots, m$. This completes the proof of Theorem 5.1.

An immediate consequence of Lemma 3.5 is the following result of independent interest due to Gupta-Passi (unpublished).

THEOREM 5.2. *If G is a metabelian group, then for all $n \geq 1$, $[D_{n+2}(G), G] = \gamma_{n+3}(G)$.*

Tahara [8] has given a characterization of $D_4(G)$. A slight modification of our argument yields an alternative description of $D_4(\underline{r}) = F \cap (1 + \underline{r} + \underline{f}^4)$ and yields $D_4^2(G) \leq \gamma_4(G)$, a result of Losey [3]. Let $w = w_1 \cdots w_{m-1} \in D_4(\underline{fs})$, where w_i are as in (13). Since modulo $[F', S] \gamma_4(F)$, for $k = i, \cdots, m$,

$$[x_i,x_j,x_k]^{t_{e_i}(x_i)^{e_i}} \equiv [x_i,x_j,x_k]^{e_i} \equiv [x_i,x_j,x_k^{e_k}]^{e_i/e_k} \equiv 1,$$

it follows from Lemma 3.1 that if $w-1 \in \underline{fs} + \underline{f}^4$ then

$$w \equiv \prod_{1 \leq i < j \leq m} [x_i,x_j]^{p_{ij}} \text{ with } p_{ij} \equiv t_{e_i}(x_i)b_{ij} \mod(\underline{s}+\underline{f}^2) \text{ where } b_{ij} \in Z.$$

Lemma 3.1 further yields $\sum_{j=i+1}^{m} (x_j-1)p_{ij} \in \underline{s} + \underline{f}^3$. Using the homomorphisms $\theta_{i+1} \cdots \theta_k \cdots \theta_m$, $k = i+1,\cdots,m$, it follows that $(x_j-1)p_{ij} \in \underline{s}+\underline{f}^3$ and, in turn $p_{ij} \in t_{e_j}(x_j)ZF + \underline{s} + \underline{f}^2$ for all $j = i+1,\cdots,m$. It follows that for $1 \leq i < j \leq m$,

$$p_{ij} \equiv t_{e_i}(x_i)b_{ij} \equiv t_{e_j}(x_j)b'_{ij} \mod(\underline{s}+\underline{f}^2) \quad (25)$$

where $b_{ij} \in Z$, $b'_{ij} = c_{ij} + (x_i-1)d_{ij}$, $c_{ij}, d_{ij} \in Z$. In particular,

$$[x_i^{e_i},x_j]^{b_{ij}} \equiv [x_i,x_j^{e_j}]^{b'_{ij}} \mod [F',S]\gamma_4(F).$$

Using $t_{e_k}(x_k) \equiv e_k + \binom{e_k}{2}(x_k-1) \mod \underline{f}^2$ for $k = i, j$, (25) yields the following arithmetic relations:

$$c_{ij} = (e_i/e_j)b_{ij}; \binom{e_j}{2}c_{ij} \equiv 0 \pmod{e_j Z}; \binom{e_i}{2}b_{ij} \equiv e_j d_{ij} \pmod{e_i Z} \quad (26)$$

Now proceeding as in the proof of Theorem 4.1, if $w \in D_4(\underline{fr})$ then $w-1 \in \underline{fr} + \underline{f}^2\underline{s} + \underline{f}^4$ implies (as in (21)) that $\sum_{k=1}^{m} (x_k-1)(y_k-1) \in \underline{fr} + \underline{f}^2\underline{s} + \underline{f}^4$, where $y_k = \prod_{i<k} x_i^{e_i b_{ik}} \prod_{k<j} x_j^{e_j b'_{kj}}$, where b_{ij}, b'_{ij} are as in (25), (26). Thus $y_k - 1 \in \underline{r} + \underline{fs} + \underline{f}^3 = \underline{r} + \underline{f}^3$ (since $\underline{s} \leq \underline{r} + \underline{f}^2$) and $y_k \in R\gamma_3(F)$ (since $D_3(G) = \gamma_3(G)$). Thus $w \in D_4(\underline{fr})$ if and only if $y_k \in R\gamma_3(F)$ for all k. Further, as in Theorem 4.1, $w^2 \equiv \prod_{k=1}^{m} [x_k,y_k] \equiv 1 \mod R\gamma_4(F)$.

Thus if F/R is a p-group, p odd, then $w^2 \equiv 1$ and $w^{p^a} \equiv 1$ (Theorem 5.1) imply $w \equiv 1 \mod R\gamma_4(F)$. If $p = 2$, then setting $e_i = 2^{a_i}$ with $a_1 \geq \cdots \geq a_m > 0$ yields the following characterization of $D_4(\underline{r})$. See also Passi [5], chapter 5.

THEOREM 5.3 (cf. Tahara [8]). *Let* $G = F/R$ *with* $F/RF' \equiv F/S$,
$S = <x_1^{e_1}, \cdots, x_m^{e_m}, F'>$, $e_m | \cdots | e_1$, $e_i = 2^{a_i}$, $a_i \geq 1$. *Then modulo* $R\gamma_4(F)$,
$D_4(\underline{r}) = F \cap (1 + \underline{r} + \underline{f}^4)$ *consists of all elements* $w \equiv \prod_{1 \leq i < j \leq m} [x_i^{e_i}, x_j]^{b_{ij}}$,
$b_{ij} \in \mathbb{Z}$, *such that*

(i) $[x_i^{e_i}, x_j]^{b_{ij}} \equiv [x_i, x_j^{e_j}]^{b'_{ij}}$ *for all* $1 \leq i < j \leq m$;

(ii) $\prod_{i < k} x_i^{-e_i b_{ik}} \prod_{k < j} x_j^{e_j b'_{kj}} \in R\gamma_3(F)$ *for all* $k = 1, \cdots, m$, *where* b'_{ij} *is as given by (25), (26).*

An immediate consequence of Theorem 5.3 is the following.

COROLLARY 5.4. *If* G *is a 2-generator group then* $D_4(G) = \gamma_4(G)$.

DEPARTMENT OF MATHEMATICS AND ASTRONOMY
UNIVERSITY OF MANITOBA
WINNIPEG, MANITOBA R3T 2N2
CANADA

REFERENCES

[1] Narain Gupta, On the dimension subgroups of metabelian groups, J. Pure Appl. Algebra 24 (1982), 1-6.

[2] B. Hartley, Dimension and lower central subgroups — Sjogren's theorem revisited, National Univ. of Singaport Lecture Notes 9 (1982), 1-19.

[3] G. Losey, N-series and filterations of the augmentation ideal, Can. J. Math. 26 (1974), 962-977.

[4] W. Magnus, A. Karrass and D. Solitar, Combinatorial group theory, Interscience, New York, 1966.

[5] I.B.S. Passi, Group rings and their augmentation ideals, Lecture Notes in Math. 715, Springer-Verlag, New York, 1979.

[6] E. Rips, On the fourth integer dimension subgroup, Israel J. Math. 12 (1972), 342-346.

[7] J. A. Sjogren, Dimension and lower central subgroups, J. Pure Appl. Algebra 14 (1979), 175-194.

[8] K. I. Tahara, On the structure of $Q_3(G)$ and the fourth dimension subgroup, Japan J. Math. 3 (1977), 381-394.

ON GROUP PRESENTATIONS, COPRODUCTS AND INVERSES

Robert Craggs[1] and James Howie[2]

There is a coproduct operation in the theory of finite group presentations, defined by taking the disjoint union of generators and corresponding disjoint union of relators. Coproduct corresponds to free product on the group level, and to one-point union of 2-complexes. A presentation $\mathcal{P}$ is said to have an *inverse* if there exists a second presentation $\mathcal{P}'$ such that the coproduct of the two satisfies the conclusion of the Andrews-Curtis Conjecture.

In this paper we characterize presentations with inverses in terms of presentations of HNN-extensions of free groups. Actually, we give a somewhat more general theorem on summands under the coproduct operation.

§1. *Preliminaries*

Throughout the paper, the notation $F(X)$ will be used for the free group with basis X, and $Cl\{Y\}$ for the normal closure of the subset Y in a free group.

Recall [2] that an *extended Nielsen transformation* on a presentation $\mathcal{P} = \{x_1, \cdots, x_n | r_1, \cdots, r_p\}$ is the composite of moves of the following types.
1. Replace some r_i by r_i^{-1}.
2. Replace some r_i by $r_k r_i$ for some $k \neq i$.
3. Replace some r_i by $x_k^{-\varepsilon} r_i x_k^{\varepsilon}$ for some k and some $\varepsilon = \pm 1$.

---

[1] Supported in part by the National Science Foundation.
[2] Supported by an SERC Advanced Fellowship.

4. Replace each r_i by $\lambda(r_i)$ for some automorphism λ of the free group on $\{x_1,\cdots,x_n\}$.

5. Add a generator x_{n+1} and a relator $r_{p+1} = x_{n+1}$.

6. The reverse of 5, if this is possible.

Two presentations which are related by a sequence of moves of types 1-3 (resp. 1-4, 1-6) are said to be Q- (resp. Q^*-, Q^{**}-) equivalent. In this paper we consider only the relation of Q^{**}-equivalence, which we abbreviate to *equivalence*, and denote $\approx$. This corresponds to equivalence of 2-complexes under 3-dimensional formal deformations [1, 5].

The *coproduct* $\mathcal{P} \cup \mathcal{P}'$ of two presentations $\mathcal{P}$, $\mathcal{P}'$ is the presentation consisting of the disjoint union of the generators of $\mathcal{P}$ and $\mathcal{P}'$, and the disjoint union of their relators. An *inverse* of a presentation $\mathcal{P}$ is a presentation $\mathcal{P}'$ such that $\mathcal{P} \cup \mathcal{P}' \approx \emptyset$, where $\emptyset$ is the empty presentation of the trivial group.

Metzler [4] noted that equivalence classes of presentations form a monoid under coproduct, and raised a number of questions about this monoid, such as the existence of nontrivial invertible elements. Of course the Andrews-Curtis Conjecture implies that no nontrivial element in this monoid is invertible, so any restrictions on the existence of inverses might be interpreted as partial evidence in support of the Andrews-Curtis Conjecture.

§2. *t-presentations*

For a fixed symbol t consider the family $\mathcal{T}$ of *t-presentations* — finite presentations of the form $\mathcal{P} = \{x_1,\cdots,x_n,t|r_1,\cdots,r_p\}$, where the exponent sum of t in each relator r_j is zero. The t-presentation $\mathcal{P}$ is *alternating* if t occurs only with exponents ± 1 in the relators r_j, and these exponents occur alternately in each r_j.

We distinguish a subclass $\mathcal{T}_{HNN}$ of $\mathcal{T}$, consisting of the natural presentations of HNN-extensions of free groups with stable letter t:

$$\mathcal{P} = \{x_1,\cdots,x_n,t|t^{-1}u_i^{-1}tv_i, 1 \leq i \leq p\},$$

where u_i and v_i freely generate rank p subgroups U and V of the free group $F = F(x_1, \cdots, x_n)$. If $p = n$ and $U = V = F$, then the correspondence $u_i \to v_i$ determines an automorphism α of F and we have a presentation

$$\mathcal{P}_\alpha = \{x_1, \cdots, x_n, t | t^{-1} x_i^{-1} t \alpha(x_i), 1 \leq i \leq p\},$$

which is easily seen to be equivalent to $\mathcal{P}$ (use Nielsen transformations to change bases).

For a presentation $\mathcal{P} \in \mathcal{T}$ we denote by $\mathcal{P}(t)$ the presentation obtained by adding t as a relator to $\mathcal{P}$. We say that $\mathcal{P}$ *represents* a presentation $\mathcal{P}_0$ (not necessarily in $\mathcal{T}$) if $\mathcal{P}(t) \approx \mathcal{P}_0$.

Two presentations $\mathcal{P}$, $\mathcal{P}' \in \mathcal{T}$ are *equivalent rel t* if they are related by a sequence of extended Nielsen transformations such that:

(i) t is fixed by the automorphism λ in any type 4 transformation;

(ii) t is not added in any type 5 transformation, nor deleted in any type 6 transformation.

REMARK 1. If $\mathcal{P} \approx \mathcal{P}'$ rel t, then $\mathcal{P}(t) \approx \mathcal{P}'(t)$.

LEMMA 2. *If $\mathcal{P}_0$ is any finite presentation, then $\mathcal{P}_0$ is represented by some $\mathcal{P} \in \mathcal{T}_{HNN}$. Furthermore if $\mathcal{P}_0$ is balanced then $\mathcal{P}$ can be chosen to be $\mathcal{P}_\alpha$ for some automorphism α.*

Proof. If $\mathcal{P}_0$ is balanced, then $\mathcal{P}_0$ is represented by $\mathcal{P}_\alpha$ for a suitable automorphism, by Craggs [3], Theorem A. The proof in the general case is modelled on the argument in [3].

By transformations of type 5 we may assume that $\mathcal{P}_0$ has at least 2 generators; and by transformations of types 1 and 2 we may assume that the set of nontrivial relators in $\mathcal{P}_0$ is Nielsen-reduced. Thus $\mathcal{P}_0$ has the form $\{x_1, \cdots, x_m | r_1, \cdots, r_p\}$ with $m \geq 2$; such that for some $q \leq p$, the set $\{r_1, \cdots, r_q\}$ is Nielsen-reduced, and $r_{q+1} = \cdots = r_p = 1$. Indeed we claim that we may further assume that $q = p$, in other words that $\mathcal{P}_0$ contains no nontrivial relators. To see this, note that

$$\mathcal{P}_0 \approx \{x_1,\cdots,x_m,y,z|r_1,\cdots,r_p,y,z\}$$

$$\approx \{x_1,\cdots,x_m,y,z|y^{-1}r_1y,\cdots,y^{-1}r_qy,y^{-q-2}zy^{q+2},\cdots,y^{-p-1}zy^{p+1},z^{-1}yz,y^{-1}zy\}$$

and that this last presentation has all the desired properties.

Define $u_i = r_i$, $u_{i+p} = x_2^{-1}x_{m+i}x_2$, $v_i = x_1^{-1}x_{m+i}x_1$, and $v_{i+p} = x_2^{-1}x_{m+i}^2 x_2$ $(1 \leq i \leq p)$.

Then, since $\{r_1,\cdots,r_p\}$ is Nielsen-reduced, the sets $\{u_1,\cdots,u_{2p}\}$ and $\{v_1,\cdots,v_{2p}\}$ freely generate free subgroups of $F(x_1,\cdots,x_{m+p})$.

Now define $\mathcal{P} = \{x_1,\cdots,x_{m+p},t|t^{-1}u_i^{-1}tv_i \ (1 \leq i \leq 2p)\}$. Then $\mathcal{P} \in \mathcal{T}_{HNN}$, and a little computation shows that $\mathcal{P}(t) \approx \mathcal{P}_0$, as required.

The following two theorems together provide a characterization of inverses of group presentations (see Corollary 5).

THEOREM 3. *Let $\mathcal{P}_1$ and $\mathcal{P}_2$ be two finite presentations of the same group, of equal deficiency. Then a necessary condition for the existence of a balanced presentation $\mathcal{P}_3$ of the trivial group, such that $\mathcal{P}_1 \approx \mathcal{P}_2 \cup \mathcal{P}_3$, is that there exist an HNN-extension G of a free group, and two t-presentations*

$$\mathcal{P}_4 = \{x_1,\cdots,x_n,t|r_1,\cdots,r_p\},$$

$$\mathcal{P}_5 = \{x_1,\cdots,x_n,t|t^{-1}u_itv_i, 1 \leq i \leq p\}$$

for G, such that:

(1) *$\mathcal{P}_4$ is alternating, and $\mathcal{P}_5 \in \mathcal{T}_{HNN}$;*
(2) *$Cl\{r_1,\cdots,r_p\} = Cl\{t^{-1}u_1^{-1}tv_1,\cdots,t^{-1}u_p^{-1}tv_p\}$;*
(3) *$\mathcal{P}_4$ represents $\mathcal{P}_1$ and $\mathcal{P}_5$ represents $\mathcal{P}_2$. Furthermore, if $\mathcal{P}_1$ and $\mathcal{P}_2$ are balanced, then $\mathcal{P}_5$ may be taken to be $\mathcal{P}_\alpha$ for some automorphism α.*

Proof. Let $\mathcal{P}_3$ be given as in the hypotheses of the theorem. By Lemma 2 we may choose $\mathcal{P}_5 \in \mathcal{T}_{HNN}$ representing $\mathcal{P}_2$, and if $\mathcal{P}_2$ is balanced we may choose $\mathcal{P}_5$ to be $\mathcal{P}_\alpha$ for some automorphism α. Furthermore the proof of Lemma 2 shows that, for any k, we may choose $\mathcal{P}_5$ to have at least k relators. We may also increase the number of generators and relators of $\mathcal{P}_3$ by moves of type 5. Hence we may assume that $\mathcal{P}_3$ and $\mathcal{P}_5$ have the same number of relators. Say

$$\mathcal{P}_3 = \{y_1,\cdots,y_p | s_1,\cdots,s_p\},$$
$$\mathcal{P}_5 = \{x_1,\cdots,x_n, t | t^{-1}u_i^{-1}tv_i, 1 \le i \le p\}.$$

Define a homomorphism

$$\phi : F(y_1,\cdots,y_p) \to F(x_1,\cdots,x_n,t)$$

by $\phi(y_i) = R_i = t^{-1}u_i^{-1}tv_i$. For each i, let w_i denote the image $\phi(s_i)$ regarded as a word in $\{x_1,\cdots,x_n,t\}$. For $\mathcal{P}_4$ take the presentation

$$\mathcal{P}_4 = \{x_1,\cdots,x_n,t | w_1,\cdots,w_p\}.$$

Now $\mathcal{P}_4$ is clearly an alternating t-presentation, and since $Cl(s_1,\cdots,s_p) = F(y_1,\cdots,y_p)$ it follows that $Cl(w_1,\cdots,w_p) = Cl(R_1,\cdots,R_p)$. It remains for us to establish that $\mathcal{P}_4$ represents $\mathcal{P}_1$. To do this we will establish that $\mathcal{P}_4 \approx \mathcal{P}_5 \cup \mathcal{P}_3$ rel t. By Remark 1 we will then have $\mathcal{P}_4(t) \approx \mathcal{P}_5(t) \cup \mathcal{P}_3 \approx \mathcal{P}_2 \cup \mathcal{P}_3 \approx \mathcal{P}_1$.

Consider the sequence of transformations:

$$\mathcal{P}_4 = \{x_1,\cdots,x_n,t | w_1,\cdots,w_p\} \xrightarrow{\quad I \quad}$$

$$\mathcal{P}_6 = \left\{ \begin{matrix} x_1,\cdots,x_n,t, \\ y_1,\cdots,y_p \end{matrix} \middle| \begin{matrix} w_1, \cdots, w_p, \\ y_1 R_1^{-1},\cdots,y_p R_p^{-1} \end{matrix} \right\} \xrightarrow{\quad II \quad}$$

$$\mathcal{P}_7 = \left\{ \begin{matrix} x_1,\cdots,x_n,t, \\ y_1,\cdots,y_p \end{matrix} \middle| \begin{matrix} s_1, \cdots, s_p, \\ y_1 R_1^{-1},\cdots,y_p R_p^{-1} \end{matrix} \right\} \xrightarrow{\quad III \quad}$$

$$\mathcal{P}_8 = \left\{ \begin{matrix} x_1,\cdots,x_n,t, \\ y_1,\cdots,y_p \end{matrix} \middle| \begin{matrix} s_1, \cdots, s_p, \\ R_1^{-1},\cdots,R_p^{-1} \end{matrix} \right\} \xrightarrow{\quad IV \quad}$$

$$\mathcal{P}_9 = \left\{ \begin{matrix} x_1,\cdots,x_n,t, \\ y_1,\cdots,y_p \end{matrix} \middle| \begin{matrix} R_1,\cdots,R_p, \\ s_1,\cdots,s_p \end{matrix} \right\}$$

$$= \mathcal{P}_5 \cup \mathcal{P}_3.$$

Step I adds new generators y_i and equates them to the old words R_i. Step II resubstitutes y_i for the corresponding syllables R_i in the relators w_i. Step III removes the letters y_i from the relators $y_i R_i^{-1}$ using the fact that each y_i belongs to $Cl(s_1, \cdots, s_p)$. Step IV inverts and permutes relators.

Each of the above steps is an extended Nielsen transformation fixing t (see for example Craggs [2], Appendix). Thus $\mathcal{P}_4 \approx \mathcal{P}_5 \cup \mathcal{P}_3$ rel t, and our theorem is proved.

THEOREM 4. *Let $\mathcal{P}_1$ and $\mathcal{P}_2$ be balanced presentations for the trivial group. If there exist t-presentations $\mathcal{P}_4 = \{x_1, \cdots, x_n, t | r_1, \cdots, r_n\}$ and $\mathcal{P}_5 = \mathcal{P}_\alpha = \{x_1, \cdots, x_n, t | t^{-1} x_i^{-1} t \alpha(x_i), 1 \leq i \leq n\}$ for a group G, satisfying conditions (1)-(3) of Theorem 3, then there exists a balanced presentation $\mathcal{P}_3$ for the trivial group such that $\mathcal{P}_1 \approx \mathcal{P}_2 \cup \mathcal{P}_3$.*

Proof. By conjugating if necessary, we may assume each relator r_i in $\mathcal{P}_4$ has the form

$$r_i = u_{i1} t^{-1} v_{i1} t u_{i2} \cdots u_{im(i)} t^{-1} v_{im(i)} t$$

where each u_{ij} and v_{ij} is a t-free word. For each i set $z_i = t^{-1} x_i t$, and replace $\mathcal{P}_4$ by the equivalent presentation

$$\mathcal{P}_6 = \{x_1, \cdots, x_n, z_1, \cdots, z_n, t | w_i, t^{-1} x_i^{-1} t z_i, 1 \leq i \leq n\}.$$

Here each w_i is obtained from r_i by replacing each segment $t^{-1} v_{ij} t$ by an appropriate word in $\{z_1, \cdots, z_n\}$. Thus in $\mathcal{P}_6$ the relators w_i are all t-free words.

We now change basis in $F = F(x_1, \cdots, x_n, z_1, \cdots, z_n)$. Set $k_i = \alpha(x_i) z_i^{-1}$. Then $\{x_1, \cdots, x_n, k_1, \cdots, k_n\}$ is a basis for F. Notice that each k_i is a consequence of the relators of $\mathcal{P}_6$, since by hypothesis the relator $t^{-1} x_i^{-1} t \alpha(x_i)$ of $\mathcal{P}_5$ is a consequence of the relators of $\mathcal{P}_4$.

Hence adding the k_i as new relators to $\mathcal{P}_6$ would give another presentation of G. We could then use the k_i to rewrite the w_i as words

w'_i in the x_i. Since G is an HNN-extension of the free group $F(x_1,\cdots,x_n)$, it follows that $w'_i = 1$ and hence $w_i \in \mathrm{Cl}\{k_1,\cdots,k_n\}$ for all i.

Now let H be the group given by the subpresentation $\{x_1,\cdots,x_n, z_1,\cdots,z_n | w_1,\cdots,w_n\}$ of $\mathcal{P}_6$, and let x'_i, z'_i denote the elements of H represented by x_i, z_i respectively. Inclusion of presentations induces a homomorphism $\theta : H \to G$ mapping x'_i to x_i and z'_i to $a(x_i)$, and hence the sets $\{x'_1,\cdots,x'_n\}$ and $\{z'_1,\cdots,z'_n\}$ are bases for free subgroups of H. But the form of $\mathcal{P}_6$ shows that G is an HNN-extension of H, and so θ is injective. Since each word k_j is t-free, it follows that $k_j \in \mathrm{Cl}\{w_1,\cdots,w_n\}$. Combining this with the above, we have $\mathrm{Cl}\{k_1,\cdots,k_n\} = \mathrm{Cl}\{w_1,\cdots,w_n\}$.

Replace $\mathcal{P}_6$ by the equivalent presentation $\mathcal{P}_7 = \{x_1,\cdots,x_n,k_1,\cdots,k_n, t|w_i, t^{-1}x_i^{-1}tz_i, 1 \leq i \leq n\}$ in which the w_i and z_i are rewritten as words in the x_i and k_i. Then remove all occurrences of the k_i from the z_i using the fact that $\mathrm{Cl}\{w_1,\cdots,w_n\} = \mathrm{Cl}\{k_1,\cdots,k_n\}$, to get the equivalent presentation

$$\mathcal{P}_8 = \{x_1,\cdots,x_n,k_1,\cdots,k_n,t|w_i, t^{-1}x_i^{-1}ta(x_i), 1 \leq i \leq n\}.$$

Now $\mathcal{P}_5(t) \approx \mathcal{P}_2$ presents the trivial group, so we may use the relators t and $t^{-1}x_i^{-1}ta(x_i)$ of $\mathcal{P}_8(t)$ to remove all occurrences of the x_i from the w_i and obtain a presentation

$$\mathcal{P}_9 \approx \mathcal{P}_8(t):$$

$$\mathcal{P}_9 = \{x_1,\cdots,x_n,k_1,\cdots,k_n,t|t,s_i,x_i^{-1}a(x_i), i \leq i \leq n\},$$

where the s_i are words in the k_i with

$$\mathrm{Cl}\{s_1,\cdots,s_n\} = \mathrm{Cl}\{k_1,\cdots,k_n\}.$$

But clearly $\mathcal{P}_1 \approx \mathcal{P}_9 \approx \mathcal{P}_2 \cup \mathcal{P}_3$, where

$$\mathcal{P}_3 = \{k_1,\cdots,k_n|s_1,\cdots,s_n\},$$

so the theorem is proved.

COROLLARY 5. *If $\mathcal{P}$ is a balanced presentation for the trivial group, then $\mathcal{P}$ has an inverse if and only if there are t-presentations $\mathcal{P}_4$ and $\mathcal{P}_5 = \mathcal{P}_\alpha$ which satisfy conditions (1)-(3) for $\mathcal{P}_1 = \emptyset$ and $\mathcal{P}_2 = \mathcal{P}$.*

ROBERT CRAGGS
DEPARTMENT OF MATHEMATICS
UNIVERSITY OF ILLINOIS
URBANA, ILLINOIS 61801
U.S.A.

JAMES HOWIE
DEPARTMENT OF MATHEMATICS
HERIOT-WATT UNIVERSITY
RICCARTON
EDINBURGH EH14 4AS
U.K.

REFERENCES

[1] J. Andrews and M. L. Curtis, Free groups and handlebodies, Proc. Amer. Math. Soc. 16 (1965), 192-195.

[2] R. Craggs, Free Heegaard diagrams and extended Nielsen transformations I, Mich. Math. J. 26 (1979), 161-186.

[3] ———, On finite presentations for groups, Proc. Amer. Math. Soc. 78 (1980), 170-174.

[4] W. Metzler, Aequivalenzklassen von Gruppenbeschreibungen, Identitaeten und einfacher Homotopietyp in niederer Dimensionen, London Math. Soc. Lecture Note Series 36 (1979), 291-326.

[5] P. Wright, Group presentations and formal deformations, Trans. Amer. Math. Soc. 208 (1975), 161-169.

ON COMPLEXES DOMINATED BY A TWO-COMPLEX

John G. Ratcliffe[*]

Let X be a connected CW-complex. In this paper we find necessary and sufficient algebraic conditions such that X is homotopy equivalent to (i) a 2-complex, (ii) a finite 2-complex, (iii) a finite 2-complex wedged with an infinite number of 2-spheres, (iv) a countable 2-complex.

In §1 (§2) we prove that if X is dominated by a (finite) 2-complex, and $f: K^1 \to X$ is a 1-connected map from a (finite) wedge of circles to X, then $\pi_2(f) = \pi_2(M_f, K^1)$ is a (finitely generated) projective crossed module, and $\pi_2(f)$ is (finitely) stably free if and only if X is homotopy equivalent to a (finite) 2-complex. In §3 we develop an obstruction theory for complexes, which are dominated by a finite 2-complex, to be homotopy equivalent to a finite 2-complex. The obstruction is the stable class of a finitely generated projective crossed module. In §4 we develop an obstruction theory for complexes, which are dominated by a 2-complex but not by a finite 2-complex, to be homotopy equivalent to a finite 2-complex wedged with an infinite number of 2-spheres. The obstruction is the stable class of a nonfinitely generated projective module. In §5 we prove countable versions of the results of §1 and §2. As an application we give an algebraic characterization of finitely generated groups which have a 2-dimensional classifying space.

Throughout the paper X is a connected CW-complex unless otherwise stated, and an n-complex is an n-dimensional CW-complex.

§1. 2-dimensionality

In [20] Wall gave obvious necessary conditions for X to be homotopy equivalent to an n-complex.

---

[*]Partially supported by the National Science Foundation.

Dn: If $\tilde{X}$ is a universal cover of X,
then $H_i(\tilde{X}) = 0$ for $i > n$; and
$H^{n+1}(X;A) = 0$ for every $\pi_1(X)$-module A.

Wall proved that X satisfies Dn, $n \geq 3$, if and only if X is homotopy equivalent to an n-complex, and X satisfies D2 if and only if X is dominated by a 2-complex. By theorems of Stallings [17] and Swan [19], X satisfies D1 if and only if X is homotopy equivalent to a 1-complex. Whether or not satisfying condition D2 is sufficient for X to be homotopy equivalent to a 2-complex is an open problem. In this section we reduce this problem to an algebraic problem. Our first result is a 2-dimensional version of Lemma 2.1 in [20]. See [14] for all crossed module definitions.

LEMMA 1.1. *If X satisfies D2, and K^1 is a wedge of circles, and $f: K^1 \to X$ maps the circles onto generators of $\pi_1(X)$, then $\pi_2(f) = \pi_2(M_f, K^1)$ is a projective crossed module; moreover, one may attach 2-cells to K^1 and extend f over these 2-cells to obtain a homotopy equivalence if and only if $\pi_2(f)$ is a free crossed module.*

Proof. To show that $\pi_2(f)$ is a projective $\pi_1(K^1)$-crossed module we only need to show that $\pi_2(f)_{ab}$ is a projective $\pi_1(X)$-module by Theorem 4.1 of [14]. Let $\tilde{K}^1$ be the covering of K^1 corresponding to the kernel of $f_*: \pi_1(K^1) \to \pi_1(X)$. Lift f to a map $\tilde{f}: \tilde{K}^1 \to \tilde{X}$ where $\tilde{X}$ is a universal covering of X. Then the mapping cylinder $M_{\tilde{f}}$ is a universal cover of M_f. The argument in the proof of Lemma 2.1 in [20] shows that $H_2(\tilde{f}) = H_2(M_{\tilde{f}}, \tilde{K}^1)$ is a projective $\pi_1(X)$-module. As $\pi_2(f)_{ab} \cong \pi_2(\tilde{f})_{ab} \cong H_2(\tilde{f})$, we have that $\pi_2(f)$ is projective.

Suppose f extends to a homotopy equivalence $h: K^2 \to X$. Let $i: K^1 \to K^2$ be the inclusion. From the exact sequence

$$0 = \pi_3(h) \to \pi_2(i) \to \pi_2(f) \to \pi_2(h) = 0$$

we see that $\pi_2(f) \cong \pi_2(i) \cong \pi_2(K^2, K^1)$ as $\pi_1(K^1)$-crossed modules. As $\pi_2(K^2, K^1)$ is free, we have that $\pi_2(f)$ is free.

Conversely, suppose $\pi_2(f)$ is free. Attach 2-cells to K^1 corresponding to a basis of $\pi_2(f)$ and extend f over these 2-cells to obtain a map $g: K^2 \to X$ which is 2-connected. Lift g to a map $\tilde{g}: \tilde{K}^2 \to \tilde{X}$ of universal coverings. Let $\tilde{K}^1$ be the 1-skeleton of $\tilde{K}^2$. Note that $\tilde{K}^1$ is the covering of K^1 corresponding to the kernel of $f_*: \pi_1(K^1) \to \pi_1(X)$. Let $\tilde{i}: \tilde{K}^1 \to \tilde{K}^2$ be the inclusion, and $\tilde{f} = \tilde{g}\tilde{i}$. As $H_2(\tilde{f}) \cong \pi_2(f)_{ab}$, we have that $H_2(\tilde{f})$ is a free $\pi_1(X)$-module. From the exact sequence

$$0 = H_3(\tilde{X}) \to H_3(\tilde{f}) \to H_2(\tilde{K}^1) = 0$$

we have that $H_3(\tilde{f}) = 0$. Consider the exact sequence

$$0 = H_3(\tilde{f}) \to H_3(\tilde{g}) \to H_2(\tilde{i}) \to H_2(\tilde{f}).$$

The last homomorphism is an isomorphism of free $\pi_1(X)$-modules; therefore, $H_3(\tilde{g}) = 0$, and g is 3-connected. From the exact sequence

$$0 = H_n(\tilde{K}^2) \to H_n(\tilde{g}) \to H_{n-1}(\tilde{X}) = 0$$

we have that g is n-connected for all $n > 3$. Thus g is a homotopy equivalence. □

Let C be an E-crossed module and $\gamma: E' \to E$ be a homomorphism. Consider the pull back diagram

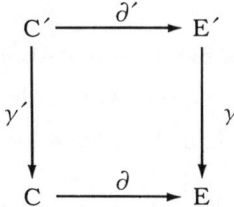

where $C' = \{(c, e') \in C \times E' | \partial c = \gamma(e')\}$. Then C' is an E'-crossed module with action of E' defined by $g' \cdot (c, e') = (\gamma(g') \cdot c, g'e'g'^{-1})$ and boundary $\partial': C' \to E'$ defined by $\partial'(c, e') = e'$.

Let F be a free group, and $\rho : E*F \to E$ be the natural retraction. The pull back $E*F$-crossed module

$$C' = \{(c,x) \in C \times (E*F) | \partial c = \rho(x)\}$$

is called the *F-stabilization* of C.

LEMMA 1.2. *Let C be a free (projective) E-crossed module, G the cokernel of $\partial : C \to E$, F a free group, and C' the F-stabilization of C. Then C' is a free (projective) $E*F$-crossed module; moreover, $C'_{ab} \cong C_{ab} \oplus B$ where B is a free G-module of the same rank as F.*

Proof. Let N be the image of $\partial : C \to E$, and N' be the image of $\partial' : C' \to E*F$. Then $N' = \rho^{-1}(N)$ and we have a commutative diagram

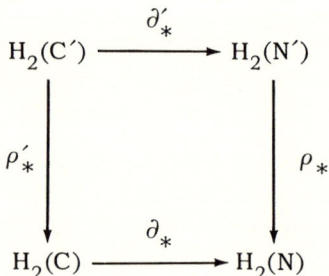

Let T be a transversal for N in E. Then $N' = N * (\underset{t \in T}{*} tFt^{-1})$. Hence ρ_* is an isomorphism. As ∂_* is trivial by Theorem 4.1 of [14], we have that ∂'_* is trivial.

For ease of notation let a bar denote abelianization. Clearly, we have $\overline{N'} \cong \overline{N} \oplus B$ where B is a free G-module of the same rank as F. Observe that $\ker \partial' = \ker \partial \times 1$ and we have a commutative diagram.

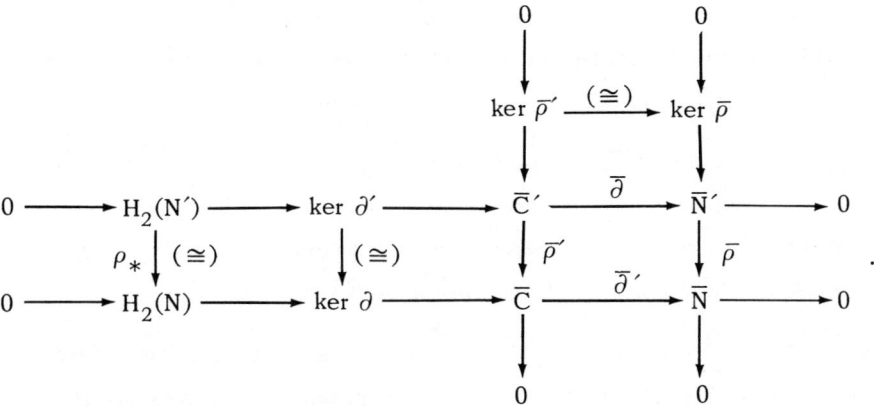

The 2nd and 3rd rows are exact since each is part of the Stallings-Stambach exact sequence [16]. The 1st row is an isomorphism by chasing the diagram. As $\bar{C}$ is a projective G-module, we have that $\bar{C}' \cong \bar{C} \oplus \ker \bar{p} \cong \bar{C} \oplus B$. Thus C' is a projective $E*F$-crossed module by Theorem 4.1 of [14].

Now assume C is free. Let $\{c_i\}$ be a basis for C, and $\{x_j\}$ be a set of free generators of F. Then C' is free with basis $\{(c_i, \partial c_i), (0, x_j)\}$ by Theorem 2.2 of [14]. □

LEMMA 1.3. *Let $f, g : (X, x_0) \to (Y, y_0)$ be maps and $H : (X, x_0) \times I \to (Y, y_0)$ a homotopy from f to g. Then H induces an isomorphism $\theta(H)$ of $\pi_1(X)$-crossed modules such that the following diagram commutes*

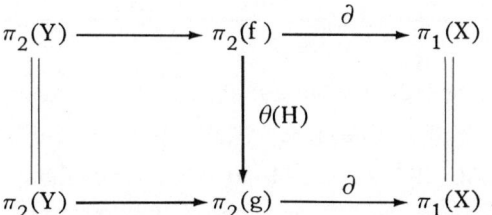

Proof. The ideal of the proof is to compare $\pi_2(f)$ and $\pi_2(g)$ with $\pi_2(H)$ and change base points from $(x_0, 0)$ to $(x_0, 1)$. □

A crossed module C is said to be *stably free* if there is a free group F such that the F-stabilization of C is free. By an argument similar to the proof of Lemma 1.2, a stably free crossed module is projective.

THEOREM 1.4. *If X satisfies D2 and K^1 is a wedge of circles and $f: K^1 \to X$ maps the circles onto generators of $\pi_1(X)$, then $\pi_2(f) = \pi_2(M_f, K^1)$ is a projective crossed module; moreover, X is homotopy equivalent to a 2-complex if and only if $\pi_2(f)$ is stably free.*

Proof. Let K^1 be a wedge of circles and $f: K^1 \to X$ map the circles onto generators of $\pi_1(X)$. Then $\pi_2(f)$ is projective crossed module by Lemma 1.1.

Suppose $\pi_2(f)$ is stably free. Then there is a free group F such that the F-stabilization of $\pi_2(f)$ is free. Let L^1 be a wedge of circles corresponding to a set of free generators of F, and $r: K^1 \vee L^1 \to K^1$ be the natural retraction. Define $f': K^1 \vee L^1 \to X$ by $f' = fr$. Then r induces a retraction $r': M_{f'} \to M_f$ defined by $r'[k,t] = [r(k),t]$ if (k,t) is in $(K^1 \vee L^1) \times I$ and $r'[x] = [x]$ if x is in X. Moreover, we have a commutative diagram

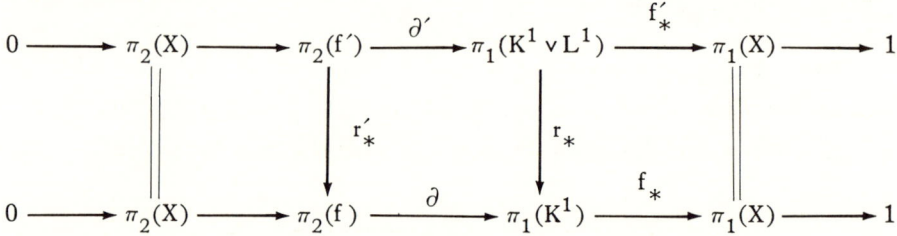

By Proposition 4.1 of [13], we have that $\pi_2(f')$ is isomorphic to the F-stabilization of $\pi_2(f)$; therefore, $\pi_2(f')$ is free. By Lemma 1.1 we may attach 2-cells to $K^1 \vee L^1$ and extend f' over these 2-cells to obtain a homotopy equivalence. Thus X is homotopy equivalent to a 2-complex.

Conversely, suppose X is homotopy equivalent to a 2-complex. Then there is a wedge of circles L^1 and a 1-connected map $g: L^1 \to X$ which can be extended over 2-cells to give a homotopy equivalence. Let $r: K^1 \vee L^1 \to K^1$ and $s: K^1 \vee L^1 \to L^1$ be the natural retractions and extend f and g to $f', g' : K^1 \vee L^1 \to X$ by letting $f' = fr$ and $g' = gs$. By the argument in the proof of Theorem 2.5 of [10], there is an automorphism α of the free group $\pi_1(K^1 \vee L^1)$ such that $f'_* = g'_* \alpha$. Clearly, α is realized by a homotopy equivalence $h: K^1 \vee L^1 \to K^1 \vee L^1$. This implies that $\pi_2(g'h) \cong \pi_2(g')$ as crossed modules. As $f'_* = (g'h)_*$, we have that f' is homotopic to $g'h$. By Lemma 1.3, $\pi_2(f') \cong \pi_2(g'h)$ as crossed modules. Thus $\pi_2(f') \cong \pi_2(g')$ as crossed modules. As above $\pi_2(f')$ and $\pi_2(g')$ are stabilizations of $\pi_2(f)$ and $\pi_2(g)$, respectively. By Lemma 1.1, we have that $\pi_2(g)$ is free, so $\pi_2(g')$ is free by Lemma 1.2. Thus $\pi_2(f)$ is stably free. □

Let $\pi_2(f)$ be as in Theorem 1.4, and assume that $\pi_2(f)$ is stably free. A shortfall of Theorem 1.4 is that it does not give a lower bound on the rank of a free group F such that the F-stabilization of $\pi_2(f)$ is free; however, it is true that if the rank of F is at least the cardinality of $\pi_1(X)$ and $\pi_2(X)$, then the F-stabilization of $\pi_2(f)$ is free. The countable case is proven in §5.

§2. *Finite 2-dimensionality*

In this section we prove a finite version of Theorem 1.4. Recall Wall's finiteness conditions for X given in [20].

F1: The group $\pi_1(X)$ is finitely generated.

F2: The group $\pi_1(X)$ is finitely presented and for any finite 2-complex K^2 and map $f: K^2 \to X$ inducing an isomorphism of fundamental groups, $\pi_2(f)$ is a finitely generated $\pi_1(X)$-module.

Fn($n \geq 3$): Condition F(n-1) holds, and for any finite (n-1)-complex K^{n-1} and (n-1)-connected map $f: K^{n-1} \to X$, $\pi_n(f)$ is a finitely generated $\pi_1(X)$-module.

Condition F2 is not in a form suitable for our purposes. We will show that F2 is equivalent to a 2-dimensional version of Fn($n \geq 3$), but first we need to prove a couple of algebraic lemmas.

LEMMA 2.1. *Let G be a finitely presented group, F a finitely generated free group, and $\eta : F \to G$ an epimorphism. Then the kernel of η is the normal closure in F of a finite set of elements.*

Proof. Let $(x_1, \cdots, x_m; r_1, \cdots, r_p)$ be a finite presentation for G under a map $\alpha : \{x_i\} \to G$ defined by $\alpha(x_i) = g_i$. Let $\{y_1, \cdots, y_n\}$ be a set of free generators of F, and let $h_j = \eta(y_j)$. Since $\{g_i\}$ generates G, there are words $\{W_j(x_i)\}$ such that $h_j = W_j(g_i)$. Introduce new generators $\{y_j\}$ and new relations $\{y_j = W_j(x_i)\}$ by a T-3 Tietze transformation [12, p. 49] to obtain a presentation for G

$$(x_i, y_j; r_k, y_j = W_j(x_i))$$

under the map $\beta : \{x_i, y_j\} \to G$ defined by $\beta(x_i) = g_i$ and $\beta(y_j) = h_j$.

Since $\{h_j\}$ generates G, there are words $\{V_i(y_j)\}$ such that $g_i = V_i(h_j)$. Next, add the relations $\{x_i = V_i(y_j)\}$ by a T-1 Tietze transformation to obtain the presentation for G

$$(x_i, y_j; r_k, y_j = W_j(x_i), x_i = V_i(y_j))$$

under the map β. By a T-4 Tietze transformation we obtain the presentation for G

$$(y_j; r_k(V_i(y_j)), y_j = W_j(V_i(y_j)))$$

under the map $\gamma : \{y_j\} \to G$ defined by $\gamma(y_j) = h_j$. Thus ker η is the normal closure in F of the finite set

$$\{r_k(V_i(y_j)), W_j(V_i(y_j)) y_j^{-1}\}. \quad \square$$

LEMMA 2.2. *Let* C *be an* E-*crossed module*, N *the image of* $\partial : C \to E$, *and* G *the cokernel of* ∂. *Then* C *is a finitely generated crossed module if and only if* C_{ab} *is a finitely generated* G-*module and* N *is the normal closure in* E *of a finite set of elements.*

Proof. If C is a finitely generated crossed module, then it is clear that C_{ab} is a finitely generated G-module and N is the normal closure in E of a finite set of elements. Conversely, suppose C_{ab} is a finitely generated G-module and N is the normal closure in E of a finite set of elements S. Lift S to a set of elements T in C, and let D be the sub-crossed module of C generated by T. Then D is a E-crossed module, and C/D is a E/∂(D)-crossed module. As ∂(D) = N, we have that C/D is actually a G-module. Thus the natural projection C $\to$ C/D induces an epimorphism $C_{ab} \to$ C/D of G-modules. Thus C/D is a finitely generated G-module. As D is a finitely generated crossed module and C/D is a finitely generated G-module, we have that C is a finitely generated crossed module. □

LEMMA 2.3. *Condition F2 holds if and only if F1 holds, and for any finite* 1-*complex* K^1 *and* 1-*connected map* $f : K^1 \to X$, $\pi_2(f)$ *is a finitely generated crossed module.*

Proof. Suppose F1 holds and for any finite 1-complex K^1 and 1-connected map $f : K^1 \to X$, $\pi_2(f)$ is a finitely generated crossed module. Now F1 says that $\pi_1(X)$ is finitely generated; hence, there is a finite wedge of circles K^1 and a 1-connected map $f : K^1 \to X$. Consider the exact sequence

$$\pi_2(f) \xrightarrow{\partial} \pi_1(K^1) \xrightarrow{f_*} \pi_1(X) \to 1 \ .$$

By assumption, $\pi_2(f)$ is a finitely generated crossed module; therefore, the kernel of f_* is the normal closure in the free group $\pi_1(K^1)$ of the image of a finite set of generators of $\pi_2(f)$. Thus $\pi_1(X)$ is finitely presented.

Let K^2 be a finite connected 2-complex and $g : K^2 \to X$ be a map inducing an isomorphism of fundamental groups. Let $i : K^1 \to K^2$ be the inclusion. Then $f = gi$ is 1-connected and we have an exact sequence

$$\pi_2(f) \to \pi_2(g) \to \pi_1(i) = 0 .$$

By assumption, $\pi_2(f)$ is a finitely generated crossed module; therefore, $\pi_2(g)$ is a finitely generated module. Thus F2 holds.

Conversely, suppose F2 holds. Let K^1 be a finite 1-complex, and $f : K^1 \to X$ be a 1-connected map. The kernel of $f_* : \pi_1(K^1) \to \pi_1(X)$ is the normal closure in $\pi_1(K^1)$ of a finite set of elements by Lemma 2.1. From the exact sequence

$$\pi_2(f) \xrightarrow{\partial} \pi_1(K^1) \xrightarrow{f_*} \pi_1(X) \longrightarrow 1$$

we see that the image of ∂ is the normal closure in $\pi_1(K^1)$ of a finite set of elements. Consequently, we may attach a finite number of 2-cells to kill these elements and extend f over these 2-cells to obtain a map $g : K^2 \to X$ inducing an isomorphism on fundamental groups. Let $i : K^1 \to K^2$ be the inclusion. Observe that the exact sequence

$$\pi_2(i) \to \pi_2(f) \to \pi_2(g) \to 0$$

induces an exact sequence of $\pi_1(K^2)$-modules

$$\pi_2(i)_{ab} \to \pi_2(f)_{ab} \to \pi_2(g) \to 0 .$$

As $\pi_2(i)_{ab} \cong \pi_2(K^2, K^1)_{ab}$, we have that $\pi_2(i)_{ab}$ is a finitely generated free module; moreover, $\pi_2(g)$ is finitely generated by F2; therefore, $\pi_2(f)_{ab}$ is finitely generated. Thus, $\pi_2(f)$ is a finitely generated crossed module by Lemma 2.2. □

A crossed module C is said to be *finitely stably free* if there is a finitely generated free group F such that the F-stabilization of C is free.

THEOREM 2.4. *If* X *is dominated by a finite* 2-*complex,* K^1 *is a finite wedge of circles, and* $f : K^1 \to X$ *maps the circles onto generators of* $\pi_1(X)$, *then* $\pi_2(f) = \pi_2(M_f, K^1)$ *is a finitely generated projective crossed module; moreover,* X *is homotopy equivalent to a finite* 2-*complex if and only if* $\pi_2(f)$ *is finitely stably free.*

Proof. As X is dominated by a finite 2-complex, X satisfies D2 and F2. Let K^1 be a finite wedge of circles, and $f : K^1 \to X$ a 1-connected map. Then $\pi_2(f)$ is finitely generated by Lemma 2.3 and projective by Lemma 1.1. The proof of Theorem 1.4 shows that X is homotopy equivalent to a finite 2-complex if and only if $\pi_2(f)$ is finitely stably free. □

Let X be the finite 2-complex which models the presentation $(a, b; a^2 b^3)$ of the trefoil knot group G. Let F be a free group of rank two generated by x and y. Define $\eta : F \to G$ by $\eta(x) = a^3$ and $\eta(y) = b^4$, and let $N = \ker \eta$. In [4] and [5] Dunwoody proved that a^3 and b^4 generate G and the relation module N_{ab} is a nonfree projective G-module with the property that

$$N_{ab} \oplus Z(G) \cong Z(G) \oplus Z(G).$$

Let K^1 be a wedge of two circles and define $f : K^1 \to X$ by mapping the circles onto loops representing a^3 and b^4. From the exact sequence

$$0 = \pi_2(X) \longrightarrow \pi_2(f) \xrightarrow{\partial} \pi_1(K^1) \xrightarrow{f_*} \pi_1(X) \longrightarrow 1$$

we have that $\pi_2(f) \cong N$ as crossed modules. Therefore, $\pi_2(f)$ is a nonfree projective crossed module. The proof of Theorem 1.4 shows that the F-stabilization of $\pi_2(f)$ is free. Thus $\pi_2(f)$ is a nonfree projective crossed module which is finitely stably free. This example shows that the stabilization in Theorem 2.4 is necessary.

3. The obstruction to finite 2-dimensionality

In [20] Wall proved that if X is dominated by a finite n-complex, $n \geq 2$, then there is an obstruction in $\widetilde{K}_0(Z\pi_1(X))$ which vanishes if and only if X is homotopy equivalent to a finite complex of dimension max $(n,3)$. It is unknown whether or not the 3 in Wall's theorem is really necessary.

In this section we find the obstruction to finite 2-dimensionality. Wall's obstruction is defined in terms of projective modules. Our obstruction is defined in terms of projective crossed modules.

Let G be a group and A a G-module. A *crossed sequence* [11] from A to G is an exact sequence

$$S : 0 \longrightarrow A \xrightarrow{\iota} C \xrightarrow{\partial} E \xrightarrow{\pi} G \longrightarrow 1 \qquad (3.1)$$

such that C is an E-crossed module with boundary ∂, the map ι is an E-homomorphism and E acts on A via π.

Clearly, S is determined up to isomorphism by its crossed module C, so crossed sequences and crossed modules are essentially the same. The advantage of crossed sequences over crossed modules is that we can fix the terminal group G in a convenient manner.

Let $\mathcal{S}_G$ be the category whose objects are crossed sequences from some G-module A to G, and whose morphisms are triples of homomorphisms (α,β,γ) such that the following diagram commutes

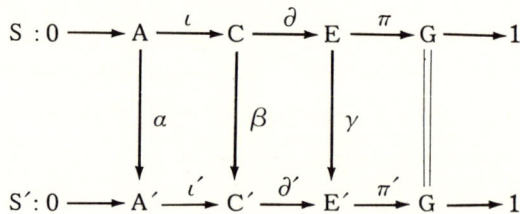

and (β,γ) is a homomorphism of crossed modules. A morphism (α,β,γ) in $\mathcal{S}_G$ is said to be an *epimorphism* if α,β,γ are all epimorphisms. A crossed sequence S in $\mathcal{S}_G$ is said to be *projective* if it is projective in the usual sense, that is, every diagram of the following form can be completed in $\mathcal{S}_G$

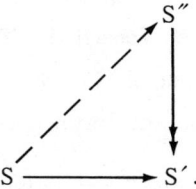

THEOREM 3.1. *A crossed sequence*

$$S: 0 \longrightarrow A \xrightarrow{\iota} C \xrightarrow{\partial} E \xrightarrow{\pi} G \longrightarrow 1$$

is projective in $\mathcal{S}_G$ *if and only if C is a projective E-crossed module and E is a free group.*

Proof. Suppose S is projective in $\mathcal{S}_G$. First we show that E is a free group. Let $\eta: F \to E$ be an epimorphism from a free group F onto E. Consider the pull back crossed sequence $S\eta$ which fits into the following pull back diagram

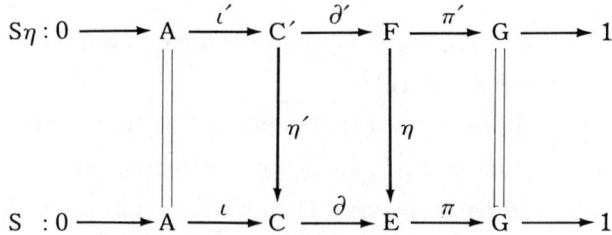

where $C' = \{(c,x) \in C \times F \mid \partial(c) = \eta(x)\}$. By the 5-lemma, η' is an epimorphism. Since S is projective, there is a morphism $(\alpha, \beta, \gamma): S \to S\eta$ such that

$$(\mathrm{id}_A, \eta', \eta)(\alpha, \beta, \gamma) = (\mathrm{id}_A, \mathrm{id}_C, \mathrm{id}_E).$$

Therefore $\eta\gamma = \mathrm{id}_E$. Thus E is a retract of a free group, and therefore is free.

Next, we show that C is a projective. Let $\{c_i\}$ be a set of generators of C, and C′ be a free crossed module [14] with basis $\{c'_i\}$ such that $\partial' c'_i = \partial c_i$ for all i. Since C′ is free, there is a E-homomorphism $\phi: C' \to C$ such that $\phi(c'_i) = c_i$ for all i. This implies ϕ is an epimorphism. Let A′ be the kernel of $\partial': C' \to E$. Then ϕ induces a homomorphism $\phi': A' \to A$ such that the following diagram commutes

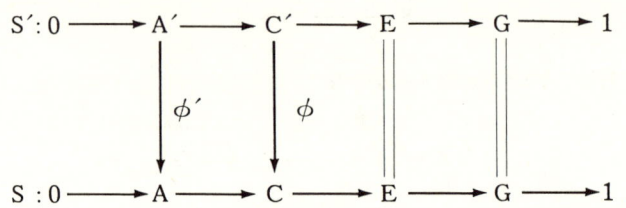

By the 5-lemma, ϕ' is an epimorphism. Since S is projective, there is a morphism $(\alpha, \beta, \gamma): S \to S'$ such that $(\phi', \phi, \mathrm{id}_E)(\alpha, \beta, \gamma) = (\mathrm{id}_A, \mathrm{id}_C, \mathrm{id}_E)$. Therefore $\phi: C' \to C$ has a right inverse β in the category of E-crossed modules. As $\ker \phi$ is contained in A′, we have a split central extension

$$0 \longrightarrow \ker \phi \longrightarrow C' \xrightarrow{\phi} C \longrightarrow 1.$$

Therefore, $C' \cong \ker \phi \times C$ as E-crossed modules. This implies C is projective by Theorem 4.1 of [14].

Conversely, suppose E is a free group and C is a projective E-crossed module. Let $(\alpha, \beta, \gamma): S \to S'$ be a morphism, and $(\alpha', \beta', \gamma'): S'' \to S'$ and epimorphism. Then we have a commutative diagram

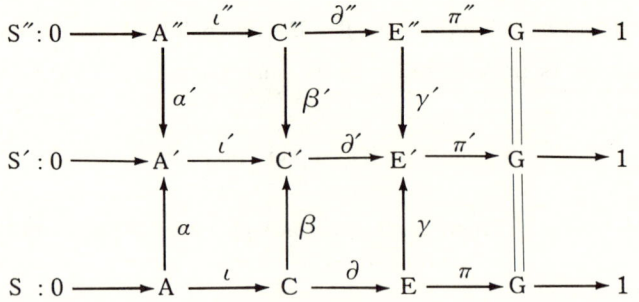

Since E is a free group, there is a homomorphism $\gamma'' : E \to E''$ such that $\gamma'\gamma'' = \gamma$. Since C is projective, there is a G-module B such that $B \times C$ is a free E-crossed module. Let $\{\hat{c}_i\}$ be a basis for $B \times C$, and $r : B \times C \to C$ be the natural projection. Let $c_i = r(\hat{c}_i)$, $e_i = \partial c_i$, $c'_i = \beta(c_i)$, $e'_i = \partial' c'_i$, and $e''_i = \gamma''(e_i)$ for all i. Since $\pi''\gamma'' = \pi$, we may choose b''_i in C'' such that $\partial'' b''_i = e''_i$. Then $c'_i \beta'(b''_i)^{-1}$ is in ker ∂'. Since α' is an epimorphism, we may choose a''_i in A'' such that $\iota'\alpha'(a''_i) = c'_i \beta'(b''_i)^{-1}$. Let $c''_i = \iota''(a''_i) b''_i$ for all i. Then $\partial'' c''_i = e''_i$ and $\beta'(c''_i) = c'_i$ for all i. Since $B \times C$ is a free crossed module, there is a homomorphism $\hat{\beta} : B \times C \to C''$ such that $\hat{\beta}(\hat{c}_i) = c''_i$ for all i and $(\hat{\beta}, \gamma'')$ is a homomorphism of crossed modules. Define $\beta'' : C \to C''$ by $\beta'' = \hat{\beta}j$ where $j : C \to B \times C$ is the natural injection. Then (β'', γ'') is a homomorphism of crossed modules.

Next, observe that $\beta'\hat{\beta}(\hat{c}_i) = \beta r(\hat{c}_i)$ for all i. Hence $\beta'\hat{\beta} = \beta r$, and $\beta'\hat{\beta}j = \beta r j$. Thus $\beta'\beta'' = \beta$. As $\partial''\beta'' = \gamma''\partial$, we have that β'' induces a homomorphism $\alpha'' : A \to A''$ such that $\beta''\iota = \iota''\alpha$. As $\iota'\alpha'\alpha'' = \iota'\alpha$, we have $\alpha'\alpha'' = \alpha$. Thus $(\alpha'', \beta'', \gamma'') : S'' \to S$ and $(\alpha', \beta', \gamma')(\alpha'', \beta'', \gamma'') = (\alpha, \beta, \gamma)$. Thus S is projective in $\mathcal{S}_G$. □

Let S be the crossed sequence (3.1), F a free group of rank n, and $\rho^n : E * F \to E$ the natural retraction. Let $S\rho^n$ be the pull pack crossed sequence formed from S by pulling back by ρ^n; and consider the pullback diagram:

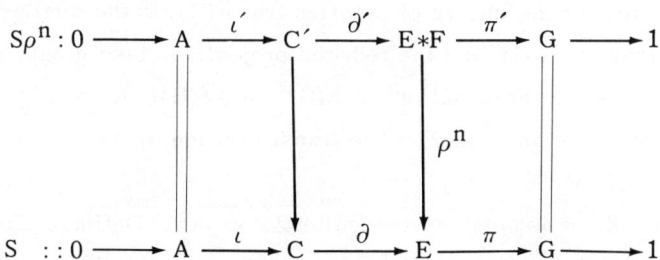

Note that C' is the F-stabilization of C. We call $S\rho^n$ the nth *stabilization* of S.

The proof of the next lemma is routine and is left to the reader.

LEMMA 3.2. *Let* S, S_1, S_2 *be crossed sequences in* $\mathcal{S}_G$, *and m and n be nonnegative integers.*
(i) *If* $S_1 \cong S_2$, *then* $S_1 \rho^n \cong S_2 \rho^n$;
(ii) $(S\rho^m)\rho^n \cong S\rho^{m+n}$. □

Suppose G is a finitely presented group. Let S be a projective crossed sequence in $\mathcal{S}_G$.

$$S: 0 \longrightarrow A \xrightarrow{\iota} C \xrightarrow{\partial} F \xrightarrow{\pi} G \longrightarrow 1. \tag{3.2}$$

Then S is said to be *finitely generated* if C is a finitely generated crossed module and F is a finitely generated free group. If S is finitely generated, then the nth stabilization $S\rho^n$ of S is also a finitely generated projective crossed sequence by Lemmas 1.2 and 2.2.

Let $K(G)$ be the set of isomorphism classes of finitely generated projective crossed sequences in $\mathcal{S}_G$. Define a relationship in $K(G)$ by $\{S_1\} \sim \{S_2\}$ if and only if there are nonnegative integers m and n such that $\{S_1 \rho^m\} = \{S_2 \rho^n\}$. This relation is well defined by Part (i) of Lemma 3.2, obviously reflexive, and transitive by Part (ii) of Lemma 3.2. Let $\widetilde{K}(G)$ be the set of equivalence classes, and let $[S]$ denote the equivalence class of $\{S\}$. If $[S_1] = [S_2]$, we say that S_1 and S_2 are *finitely stably equivalent*.

The reason for the choice of notation for $\widetilde{K}(G)$ is the similarity of the definitions of $\widetilde{K}(G)$ and the reduced projective class group $\widetilde{K}_0(Z(G))$; in fact, there is a transformation $\eta : \widetilde{K}(G) \to \widetilde{K}_0(Z(G))$ defined by $\eta[S] = [C_{ab}]$ where C is as in (3.2). The transformation η is easily shown to be surjective.

Suppose X is dominated by a finite 2-complex. Define a class $\kappa(X)$ in $\widetilde{K}(\pi_1(X))$ as follows: Let K^1 be a finite wedge of circles and $f : K^1 \to X$ map the circles onto a set of generators of $\pi_1(X)$. Then $\pi_2(f)$ is a finitely generated projective $\pi_1(K^1)$-crossed module by Theorem 2.4;

therefore, the crossed sequence

$$S_f : 0 \longrightarrow \pi_2(X) \longrightarrow \pi_2(f) \xrightarrow{\partial} \pi_1(K^1) \xrightarrow{f_*} \pi_1(X) \longrightarrow 1$$

is a finitely generated projective crossed sequence by Theorem 3.1. Let $\kappa(X) = [S_f]$.

We claim that the class $\kappa(X)$ does not depend on the choice of f. Suppose L^1 is another finite wedge of circles and $g : L^1 \to X$ maps the circles to a set of generators of $\pi_1(X)$. Extend f and g to maps $f', g' : K^1 \vee L^1 \to X$ as in the proof of Theorem 1.4. Then there is a homotopy equivalence $h : K^1 \vee L^1 \to K^1 \vee L^1$ such that $f'_* = (g'h)_*$. This implies $f' \simeq g'h$. By Lemma 1.3, there is an isomorphism $\theta : \pi_2(f') \to \pi_2(g'h)$ of $\pi_1(K^1 \vee L^1)$-crossed modules such that the following diagram commutes.

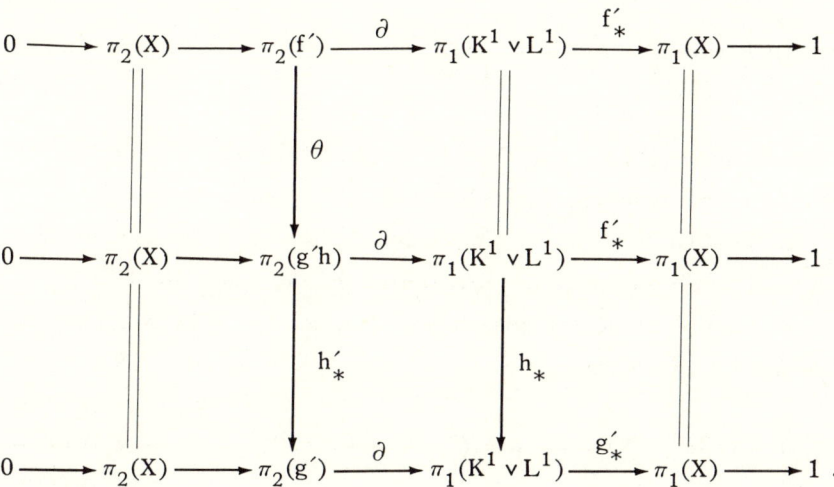

The map $h' : M_{g'h} \to M_{g'}$ is defined by $h'[k,t] = [h(k),t]$ if (k,t) is in $(K' \vee L') \times I$ and $h'[x] = [x]$ if x is in X. Thus S_f and S_g are finitely stably equivalent. This proves that $\kappa(X)$ does not depend on the choice of f.

A *G-complex* is a pair (X,ϕ) consisting of a pointed CW-complex X and an isomorphism $\phi : \pi_1(X) \to G$. Two G-complexes (X,ϕ) and (Y,ψ)

are said to be *homotopically equivalent* if there is a base point preserving homotopy equivalence $h: X \to Y$ such that $\psi h_* = \phi$. Let χ_G^2 be the set of all homotopy classes of G-complexes each of which is dominated by a finite 2-complex.

Let S be the projective crossed sequence (3.2) and $\phi: G \to G'$ an isomorphism.

Consider the crossed sequence

$$\phi S: 0 \longrightarrow A \xrightarrow{\iota} C \xrightarrow{\partial} F \xrightarrow{\phi \pi} G' \longrightarrow 1.$$

Clearly, ϕ induces a bijection

$$\phi_*: \widetilde{K}(G) \to \widetilde{K}(G')$$

defined by $\phi_*[S] = [\phi S]$.

THEOREM 3.3. *The assignment* $(X, \phi) \mapsto \phi_* \kappa(X)$ *induces a bijection from* χ_G^2 *to* $\widetilde{K}(G)$.

Proof. Suppose (X, ϕ) and (Y, ψ) represent the same class in χ_G^2. Then there is a homotopy equivalence $h: X \to Y$ such that $\psi h_* = \phi$. Let K^1 be a finite wedge of circles and $f: K^1 \to X$ a 1-connected map. Then $\kappa(X)$ is represented by S_f. Define $h': M_f \to M_{hf}$ by $h'[k,t] = [k,t]$ if (k,t) is in $K^1 \times I$ and $h'[x] = [h(x)]$ if x is in X. Then the following diagram commutes

$$
\begin{array}{ccccccccccc}
h_* S_f: 0 & \longrightarrow & \pi_2(X) & \longrightarrow & \pi_2(f) & \xrightarrow{\partial} & \pi_1(K^1) & \xrightarrow{h_* f_*} & \pi_1(Y) & \longrightarrow & 1 \\
& & \downarrow h_* & & \downarrow h'_* & & \| & & \| & & \\
S_{hf}: 0 & \longrightarrow & \pi_2(Y) & \longrightarrow & \pi_2(hf) & \xrightarrow{\partial} & \pi_1(K^1) & \xrightarrow{(hf)_*} & \pi_1(Y) & \longrightarrow & 1
\end{array}
$$

Thus $h_* S_f$ and S_{hf} are equivalent. Observe that

$$\phi_* \kappa(X) = [\phi S_f]$$
$$= [\psi h_* S_f]$$
$$= \psi_* [h_* S_f]$$
$$= \psi_* [S_{hf}] = \psi_* \kappa(Y).$$

Thus the assignment $(X,\phi) \mapsto \phi_* \kappa(X)$ induces a function $\kappa : \chi_G^2 \to \widetilde{K}(G)$.

To see that κ is injective, suppose (X,ϕ) and (Y,ψ) represent classes in χ_G^2 and $\phi_* \kappa(X) = \psi_* \kappa(Y)$. This implies that there are finite wedges of circles K^1 and L^1, and 1-connected maps $f : K^1 \to X$ and $g : L^1 \to Y$, and an equivalence $(\alpha,\beta,\gamma) : \phi S_f \to \psi S_g$. Thus we have a commutative diagram

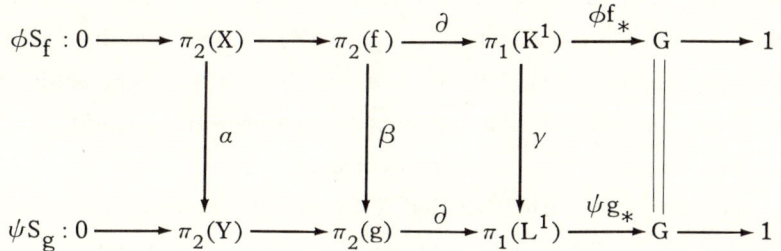

We may assume that f and g are cellular. Let $f^1 : K^1 \to X^1$ be the restriction of f, and $\rho : (M_f, K^1) \to (X, X^1)$ be the natural retraction. Then we have a commutative diagram

$$\begin{array}{ccccccccccc}
S_f : & 0 & \to & \pi_2(X) & \to & \pi_2(f) & \xrightarrow{\partial} & \pi_1(K^1) & \xrightarrow{f_*} & \pi_1(X) & \to 1 \\
& & & \| & & \downarrow \rho_* & & \downarrow f^1_* & & \| & \\
S_X : & 0 & \to & \pi_2(X) & \to & \pi_2(X, X^1) & \xrightarrow{\partial} & \pi_1(X^1) & \to & \pi_1(X) & \to 1.
\end{array}$$

By Theorem 9.4 of [13] the crossed sequences S_f and S_X represent the same element of $H^3(\pi_1(X), \pi_2(X))$. As S_X represents the k-invariant k(X) of X in $H^3(\pi_1(X), \pi_2(X))$, we have that S_f does also. Similarly, S_g represents k(Y) in $H^3(\pi_1(Y), \pi_2(Y))$.

Observe that the following diagram commutes:

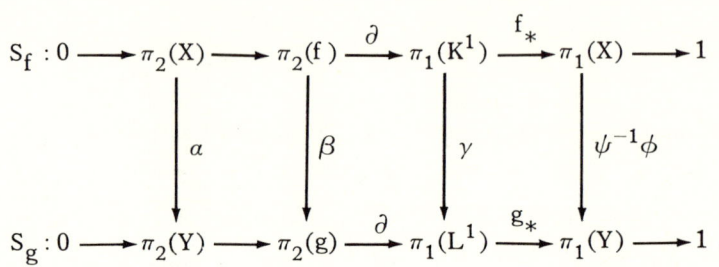

By Proposition 9.3 of [13], the pair $(\psi^{-1}\phi, \alpha)$ is a homomorphism [11] of algebraic 3-types. By Theorem 3 of [11], the pair $(\psi^{-1}\phi, \alpha)$ is realized by a map $h: X \to Y$, whence $h_* = \psi^{-1}\phi: \pi_1(X) \to \pi_1(Y)$ and $h_* = \alpha: \pi_2(X) \to \pi_2(Y)$. By Theorem 1 of [22] we have that h is a homotopy equivalence. As $\psi h_* = \phi$, the map h is a homotopy equivalence from (X,ϕ) to (Y,ψ). This proves that $\kappa: \chi_G^2 \to \tilde{K}(G)$ is injection.

To see that κ is surjective, let S be the projective crossed sequence (3.2), and suppose S is finitely generated. Then F is a finitely generated free group, and C is a finitely generated projective F-crossed module. By Theorem 4.1 of [14], there is a projective G-module P such that $P \times C$ is a free F-crossed module. Let K^1 be a finite wedge of circles corresponding to a basis of F. Attach 2-cells to K^1 according to the boundaries of a basis of $P \times C$ to obtain K^2. Then $\pi_2(K^2, K^1) \cong P \times C$. By Eilenberg's Lemma [1] there is a free G-module B such that $B \oplus P \cong B$. Now wedge on 2-spheres onto K^2 corresponding to a basis of B to obtain X^2. Then $\pi_2(X^2, X^1) \cong (B \oplus P) \times C \cong B \times C$. Next, attach 3-cells onto X^2 to kill the elements of $\pi_2(X^2)$ which correspond to a basis of $B \oplus P$ to obtain X. Then $\pi_2(X, X^1) \cong C$.

Observe that X satisfies D3, since X is 3-dimensional. Let $\tilde{X}$ be the universal cover of X, and consider the boundary

$$\partial_3 : H_3(\tilde{X}^3, \tilde{X}^2) \to H_2(\tilde{X}^2, \tilde{X}^1).$$

Note that $H_2(\tilde{X}^2, \tilde{X}^1) \cong B \oplus C_{ab}$ and ∂_3 maps the cellular basis of $H_3(\tilde{X}^3, \tilde{X}^2)$ onto elements corresponding to a basis of B, so ∂_3 is a monomorphism; therefore, $H_3(\tilde{X}) = 0$. By Theorem 6 of [21] we have that X satisfies D2. Observe that X satisfies F1, since $\pi_1(X) \cong G$ and G is finitely generated; and X satisfies F2, since $\pi_2(X, X^1) \cong C$ and C is finitely generated. Thus X satisfies D2 and F2. By Theorem F of [20], we have that X is dominated by a finite 2-complex. Finally, there are isomorphisms $\alpha, \beta, \gamma, \phi$ such that the following diagram commutes

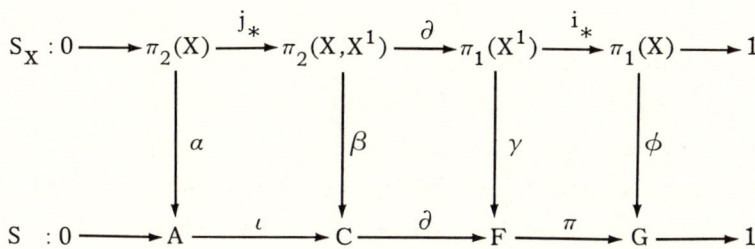

and (β, γ) is a homomorphism of crossed modules. This implies that ϕS_X is equivalent to S. As $S_i \cong S_X$, we have that $\phi_* \kappa(X) = [S]$. This proves that $\kappa : \chi_G^2 \to \tilde{K}(G)$ is surjective. □

Let S be the crossed sequence (3.1). Then S is said to be *free* if C is a free E-crossed module and E is a free group. By Theorem 3.1 every free crossed sequence is projective. A class in $\tilde{K}(G)$ is said to be *free* if it is represented by a free crossed sequence. The next theorem follows immediately from Theorems 2.4 and 3.3. Compare with Theorem F of [20].

THEOREM 3.4. *If X is dominated by a finite 2-complex, then X determines an obstruction $\kappa(X)$ in $\tilde{K}(\pi_1(X))$, depending only on the homotopy type of X, such that $\kappa(X)$ is free if and only if X is homotopy equivalent to a finite 2-complex. Every element of $\tilde{K}(\pi_1(X))$ is the obstruction of a complex dominated by a finite 2-complex.* □

Suppose X is dominated by a finite 2-complex. Then the transformation

$$\eta : \tilde{K}(\pi_1(X)) \to \tilde{K}_0(Z\pi_1(X))$$

maps $\kappa(X)$ onto the Wall obstruction of X. In [6] Dyer has shown that in all likelihood there is a finite superperfect group G which admits a finite 3-dimensional G-complex X which is dominated by a finite 2-complex but is not homotopy equivalent to a finite 2-complex. For such an X, the obstruction $\kappa(X)$ is nonfree even though its Wall obstruction is zero. In contrast, Browning [3] proved that if G is finite abelian, then no such example can arise.

§4. *Almost finite 2-dimensionality*

A 2-complex K^2 is said to be *almost finite* if there is a finite 2-complex L^2 and an infinite wedge of 2-spheres W^2 such that $K^2 = L^2 \vee W^2$. Let L_1^2 and L_2^2 be finite 2-complexes with isomorphic fundamental groups. Then by Theorem 6 of J.H.C. Whitehead [23] there are finite wedges of 2-spheres W_1^2 and W_2^2 such that $L_1^2 \vee W_1^2 \simeq L_2^2 \vee W_2^2$. Thus if W^2 is an infinite wedge of 2-spheres, then $L_1^2 \vee W^2 \simeq L_2^2 \vee W^2$. This implies that the homotopy type of an almost finite 2-complex K^2 depends only on the isomorphism type of $\pi_1(K^2)$ and the cardinality of $\pi_2(K^2)$.

In this section we give four necessary and sufficient conditions for X to be homotopy equivalent to an almost finite 2-complex. The first three are the obvious necessary conditions that $\pi_1(X)$ is finitely presented and X satisfies D2 but not F2. The fourth condition is that a certain big projective $\pi_1(X)$-module is free. A module is said to be *big* if it is nonfinitely generated.

LEMMA 4.1. *Let A be a module with a finite set of generators $\{a_1, \cdots, a_n\}$, and B be a big free module. If $\eta : B \to A$ is an epimorphism, then B has a basis $\{b_i\}_{i \in I}$ such that $\eta(b_{i_j}) = a_j$ for $j = 1, \cdots, n$, and $\eta(b_i) = 0$ for $i \neq i_j$, $j = 1, \cdots, n$.*

Proof. Let $\{b_i\}_{i \in I}$ be a basis for B. Choose $a'_1, \cdots, a'_n$ in B such that $\eta(a'_j) = a_j$. Then there is a finite subset I_j of I such that a'_j is spanned by $\{b_i | i \in I_j\}$. Then $I_0 = \bigcup_{i=1}^{n} I_i$ is a finite set of indices. Let B_0 be the submodule of B generated by $\{b_i | i \in I_0\}$. Then $\eta(B_0) = A$. Let i be an index not in I_0. Then there is a b'_i in B_0 such that $\eta(b'_i) = \eta(b_i)$. Replace b_i by $b_i - b'_i$. Then we still have a basis, and $\eta(b_i) = 0$ for all i in $I - I_0$. As I is infinite, we may choose $i_1, \cdots, i_n$ in $I - I_0$. Next, replace b_{i_j} with $b_{i_j} + a'_j$. Then we still have a basis and $\eta(b_{i_j}) = a_j$ for $j = 1, \cdots, n$. Let B_1 be the submodule of B generated by $\{b_{i_1}, \cdots, b_{i_n}\}$. Then $\eta(B_1) = A$. Let $i \in I_0$. Since $\eta(B_1) = A$, there is a $b'_i \in B_1$ such that $\eta(b'_i) = \eta(b_i)$. Replace b_i by $b_i - b'_i$. Then we still have a basis, and $\eta(b_{i_j}) = a_j$ for $j = 1, \cdots, n$, and $\eta(b_i) = 0$ for $i \neq i_j$. □

LEMMA 4.2. *Let B be a big projective module. Then B is free if and only if there is a finitely generated free module A such that $A \oplus B$ is free.*

Proof. Let A be a finitely generated free module such that $F = A \oplus B$ is free. Then we have an exact sequence

$$0 \longrightarrow B \longrightarrow F \xrightarrow{\eta} A \longrightarrow 0.$$

Let $\{a_1, \cdots, a_n\}$ be a basis for A. By Lemma 4.1, there is a basis $\{b_i\}$ for F such that $\eta(b_{i_j}) = a_j$ for $j = 1, \cdots, n$, and $\eta(b_i) = 0$ for $i \neq i_j$, $j = 1, \cdots, n$. Let C be the submodule of F generated by $\{b_i | i \neq i_j, j = 1, \cdots, n\}$ then we have an exact sequence

$$0 \to B/C \to F/C \to A \to 0.$$

The sequence splits, since A is free. As F/C is free, we have that B/C is projective. Hence the following exact sequence splits.

$$0 \to C \to B \to B/C .$$

Thus B has a direct summand C isomorphic to F. By a lemma of Beck [2], $B \cong F$. □

THEOREM 4.3. *If X satisfies D2 but not F2, $\pi_1(X)$ is finitely presented, K^1 is a finite wedge of circles, and $f: K^1 \to X$ maps the circles onto generators of $\pi_1(X)$, then $\pi_2(f)_{ab}$ is a big projective $\pi_1(X)$-module; moreover, X is homotopy equivalent to an almost finite 2-complex if and only if $\pi_2(f)_{ab}$ is free.*

Proof. Let $\{x_1, \cdots, x_n\}$ be a finite set of generators of $\pi_1(X)$, and K^1 be a wedge of n circles. Define $f: K^1 \to X$ by mapping the ith circle to a loop representing x_i. By Lemmas 1.1 and 2.3, we have that $\pi_2(f)$ is a big projective $\pi_1(K^1)$-crossed module. Let N be the kernel of $f_*: \pi_1(K^1) \to \pi_1(X)$. By Lemma 2.1, N is the normal closure in $\pi_1(K^1)$ of a finite set $\{r_1, \cdots, r_n\}$ of elements. By Lemma 2.2, $\pi_2(f)_{ab}$ is a big projective $\pi_1(X)$-module.

Suppose $\pi_2(f)_{ab}$ is free. Let $C = \pi_2(f)$, and let a bar denote abelianization. Observe that we have a commutative diagram

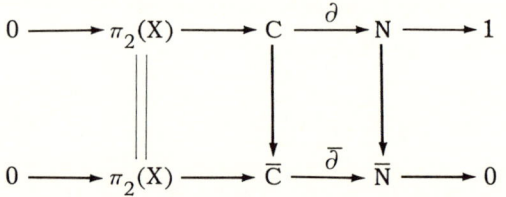

The bottom row is exact, since $H_2(N) = 0$. By Proposition 4.1 of [13], we may assume

$$C = \{(\bar{c}, r) \in \bar{C} \times \bar{N} \mid \bar{\partial}(c) = \bar{r}\}$$

and the above diagram is the pullback diagram. By Lemma 4.1, there is a basis $\{b_i\}$ for $\bar{C}$ such that $\bar{\partial}(b_{i_j}) = \bar{r}_j$ for $j = 1, \cdots, n$, and $\bar{\partial}(b_i) = 0$

for $i \neq i_j$. Let $c_{i_j} = (b_{i_j}, r_j)$ for $j = 1, \cdots, n$ and $c_i = (b_i, 1)$ for $i \neq i_j$. Then $\{\bar{c}_i\} = \{b_i\}$ is a basis for $\bar{C}$, and N is the normal closure of $\{\partial c_i\} = \{r_j\}$ in the free group $\pi_1(K^1)$. By Theorem 2.2 of [14], C is a free $\pi_1(K^1)$-crossed module with basis $\{c_i\}$. By Lemma 1.1 we may attach 2-cells to K^1 corresponding to $\{c_i\}$ and extend f over these 2-cells to obtain a homotopy equivalence; moreover, the 2-cells corresponding to the c_i such that $\partial c_i = 1$ may be attached at the base point to give 2-spheres. Thus X is homotopy equivalent to an almost finite 2-complex.

Conversely, suppose X is homotopy equivalent to an almost finite 2-complex. Then there is a finite wedge of circles L^1 and a 1-connected map $g: L^1 \to X$ which can be extended over 2-cells to give a homotopy equivalence. By the argument in the proof of Theorem 1.4, $\pi_2(f)$ and $\pi_2(g)$ are finitely stably equivalent. As $\pi_2(g)$ is free, we have that $\pi_2(f)_{ab}$ is finitely stably free. By Lemma 4.2, $\pi_2(f)_{ab}$ is free. □

Let G be a finitely presented group, and $\mathcal{P}(G)$ be the set of isomorphism classes of big projective G-modules. By a theorem of Kaplansky [8], any projective module is a direct sum of countably generated modules. Thus $\mathcal{P}(G)$ is completely determined by the set of isomorphism classes of countable big projective G-modules.

Define a relationship in $\mathcal{P}(G)$ by $\{B_1\} \sim \{B_2\}$ if and only if there are finitely generated free G-modules A_1 and A_2 such that $\{A_1 \oplus B_1\} = \{A_2 \oplus B_2\}$. This is obviously an equivalence relationship. Let $\widetilde{\mathcal{P}}(G)$ be the set of equivalence classes, and let $[B]$ denote the equivalence class of $\{B\}$. A class in $\widetilde{\mathcal{P}}(G)$ is said to be *free* if it is represented by a free module.

Suppose X satisfies D2 but not F2 and $\pi_1(X)$ is finitely presented. Define a class $\kappa(X)$ in $\widetilde{\mathcal{P}}(\pi_1(X))$ as follows: Let K^1 be a finite wedge of circles and $f: K^1 \to X$ map the circles onto generators of $\pi_1(X)$. Then $\pi_2(f)_{ab}$ is a big projective $\pi_1(X)$-module by Theorem 4.3. Let $\kappa(X) = [\pi_2(f)_{ab}]$. The same argument as in §3 shows that $\kappa(X)$ does not depend on the choice of f.

THEOREM 4.4. *If X satisfies D2 but not F2 and $\pi_1(X)$ is finitely presented, then X determines an obstruction $\kappa(X)$ in $\widetilde{\mathcal{P}}(\pi_1(X))$, depending only on the homotopy type of X, such that $\kappa(X)$ is free if and only if X is homotopy equivalent to an almost finite 2-complex. Every element of $\widetilde{\mathcal{P}}(\pi_1(X))$ is the obstruction of a complex satisfying D2 but not F2.*

Proof. The argument in the proof of Theorem 3.3 shows that $\kappa(X)$ depends only on the homotopy type of X. By Lemma 4.2 and Theorem 4.3, X is homotopy equivalent to an almost finite two complex if and only if $\kappa(X)$ is free.

Let B be a big projective $\pi_1(X)$-module, and K^2 be a finite 2-complex which models a finite presentation of $\pi_1(X)$. Consider the projective crossed sequence

$$0 \longrightarrow B \times \pi_2(K^2) \longrightarrow B \times \pi_2(K^2, K^1) \xrightarrow{\partial} \pi_2(K^1) \longrightarrow \pi_1(K^2) \longrightarrow 1 \ .$$

Observe that $[B] = [B \times \pi_2(K^2, K^1)_{ab}]$, since $\pi_2(K^2, K^1)_{ab}$ is a finitely generated free module. The argument in the proof of Theorem 3.3 shows that [B] is realized as the obstruction of a complex satisfying D2 but not F2. □

Let G be a polycyclic group. By Theorem 2.3 of P. E. Smith [15], every big projective G-module is free. In contrast, Swan [18] and Linnell [9] proved that a finite group G has the property that all big projective G-modules are free if and only if G is solvable.

§5. *Countable 2-dimensionality*

Consider Wall's countability conditions for X given in [20].

C1: The group $\pi_1(X)$ is countable.

Cn($n \geq 2$): C(n-1) holds and $H_n(\widetilde{X})$ is countable where $\widetilde{X}$ is a universal cover of X.

Note that condition C2 is equivalent to the condition that the groups $\pi_1(X)$ and $\pi_2(X)$ are countable.

THEOREM 5.1. *Let K^2 be a 2-complex. Then K^2 is homotopy equivalent to a countable 2-complex if and only if K^2 satisfies C2.*

Proof. If K^2 is homotopy equivalent to a countable 2-complex, then clearly K^2 satisfies C2. Conversely, suppose K^2 satisfies C2. Since collapsing a maximal tree does not change the homotopy type, we may assume without loss of generality that K^2 has only one 0-cell. Then the 1-cells and 2-cells of K^2 determines a group presentation $(X;R)$ for $\pi_1(K^2)$. The elements of R are cyclically reduced words in the free group F generated by X. The elements of X correspond to the homotopy classes of the 1-cells in $\pi_1(K^1)$, and the elements of R correspond to the homotopy classes in $\pi_1(K^1)$ of the boundaries of the 2-cells of K^2.

Let $\tilde{K}^2$ be a universal cover of K^2. Consider the second boundary of the cellular chain complex of $\tilde{K}^2$

$$\partial_2 : H_2(\tilde{K}^2, \tilde{K}^1) \to H_1(\tilde{K}^1, \tilde{K}^0) .$$

The domain of ∂_2 is a free $\pi_1(K^2)$-module with basis $\mathcal{R}$ corresponding to the 2-cells of K^2. By assumption $H_2(\tilde{K}^2) \cong \ker \partial_2$ is countable; hence, there is a countable subset $\mathcal{R}_0$ of $\mathcal{R}$ such that $\ker \partial_2$ is in the submodule generated by $\mathcal{R}_0$. The set $\mathcal{R}_0$ corresponds to a countable subset R_0 of R. Each element of R is a word in only finitely many elements of X. Therefore, there is a countable subset X_0 of X such that R_0 is contained in the subgroup generated by X_0.

By assumption $\pi_1(K^2)$ is countable, so there is a countable subset X_1 of X such that X_1 contains X_0, and X_1 represents a set of generators of $\pi_1(K^2)$.

Let F_1 be the subgroup of F generated by X_1, and N_1 be the kernel of the presentation epimorphism $F_1 \to \pi_1(K^2)$. Then N_1 is countable, so there is a countable subset R_1 of R such that R_1 contains R_0 and N_1 is the normal closure of R_1 in F. Each element of N_1 is a finite product of conjugates of elements of R_1 and their inverses, so there is a countable subset X_2 of X such that X_2 contains X_1, R_1 is contained in the subgroup F_2 generated by X_2, and N_1 is contained in the normal closure of R_1 in F_2. By induction we construct infinite sequences

$$X_1 \subset X_2 \subset \cdots$$
$$R_1 \subset R_2 \subset \cdots$$
$$F_1 \subset F_2 \subset \cdots$$

of countable subsets of X, R, F respectively such that F_i is the subgroup of F generated by X_i, $R_i \subset F_{i+1}$, and the kernel N_i of the presentation epimorphism $F_i \to \pi_1(K^2)$ is in the normal closure of R_i in F_{i+1}. Let $X_\omega = \bigcup_{i=1}^\infty X_i$, $R_\omega = \bigcup_{i=1}^\infty R_i$, and $F_\omega = \bigcup_{i=1}^\infty F_i$. Then X_ω and R_ω are countable subsets of X and R, respectively, and F_ω is the subgroup of F generated by X_ω.

Let u be in the kernel of the presentation epimorphism $F_\omega \to \pi_1(K^2)$. Then u is in F_i for some i, and therefore is in N_i. By construction u is in the normal closure of R_i in F_{i+1}. Thus u is in the normal closure of R_ω in F_ω. This shows that $(X_\omega; R_\omega)$ is a presentation for $\pi_1(K^2)$.

Let L^2 be the subcomplex of K^2 corresponding to $(X_\omega; R_\omega)$. Then L^2 is countable. Let $i: L^2 \to K^2$ be the inclusion. It follows from what we have just proven that $i_*: \pi_1(L^2) \to \pi_1(K^2)$ is an isomorphism. Let $\tilde{L}^2$ be the subcomplex of $\tilde{K}^2$ over L^2. Then $\tilde{L}^2$ is a universal covering of L^2, and we have a commutative diagram induced by inclusion

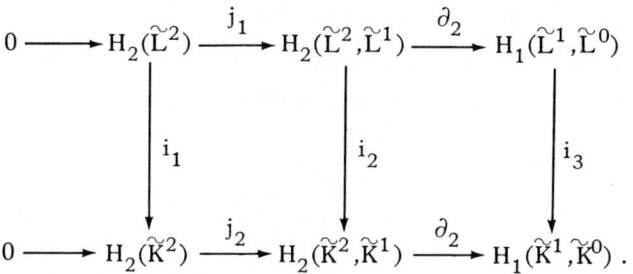

The homomorphism i_2 maps $H_2(\tilde{L}^2,\tilde{L}^1)$ isomorphically onto the free submodule generated by the subset $\mathcal{R}_\omega$ of $\mathcal{R}$ corresponding to R_ω. As $\mathcal{R}_0 \subset \mathcal{R}_\omega$, we have that the image of j_2 is contained in the image of i_2. The group $H_1(\tilde{K}^1,\tilde{K}^0)$ is a free $\pi_1(K^2)$-module generated by a basis $\mathcal{X}$ corresponding to the 1-cells of K^2, and i_3 maps $H_1(\tilde{L}^1,\tilde{L}^0)$ isomorphically onto the free submodule generated by the subset $\mathcal{X}_\omega$ of $\mathcal{X}$ corresponding to X_ω. By chasing the diagram we have that i_1 is an isomorphism. Thus $i_*: \pi_2(L^2) \to \pi_2(K^2)$ is an isomorphism. Therefore, $i: L^2 \to K^2$ is a homotopy equivalence by Theorem 1 of [22]. □

A crossed module C is said to be *countably stably free* if there is a countable free group F such that the F-stabilization of C is free.

THEOREM 5.2. *If X is dominated by a countable 2-complex, K^1 is a countable wedge of circles, and $f: K^1 \to X$ maps the circles onto generators of $\pi_1(X)$, then $\pi_2(f)$ is a countable projective crossed module, and the following are equivalent:*

(1) *X is homotopy equivalent to a 2-complex;*
(2) *X is homotopy equivalent to a countable 2-complex;*
(3) *$\pi_2(f)$ is countably stably free.*

Proof. As X is dominated by a countable 2-complex, X satisfies C2 and D2. Let K^1 be a countable wedge of circles and $f: K^1 \to X$ a 1-connected map. Then $\pi_2(f)$ is projective by Lemma 1.1. From the exact sequence

$$0 \longrightarrow \pi_2(X) \longrightarrow \pi_2(f) \overset{\partial}{\longrightarrow} \pi_1(K^1) \overset{f_*}{\longrightarrow} \pi_1(X) \longrightarrow 1$$

we see that $\pi_2(f)$ is countable. If $\pi_2(f)$ is countably stably free, then X is homotopy equivalent to a countable 2-complex by the proof of Theorem 1.4.

Conversely, suppose X is homotopy equivalent to a 2-complex. Then X is homotopy equivalent to a countable 2-complex by Theorem 5.1. The proof of Theorem 1.4 shows that $\pi_2(f)$ is countably stably free. □

THEOREM 5.3. *If X is dominated by a countable 2-complex, $\pi_1(X)$ is finitely generated, K^1 is a countably infinite wedge of circles, and $f: K^1 \to X$ is a 1-connected map, then $\pi_2(f)_{ab}$ is a free $\pi_1(X)$-module; moreover, the following are equivalent:*

(1) *X is homotopy equivalent to a 2-complex;*
(2) *X is homotopy equivalent to a countable 2-complex;*
(3) *$\pi_2(f)$ is a free crossed module;*
(4) *$\pi_2(f)_{ab}$ has a basis which lifts to a set of generators of $\pi_2(f)$.*

Proof. Let $K^1 = \bigvee_{i=1}^{\infty} S_i^1$ and $f: K^1 \to X$ be a 1-connected map. Let a_i be the homotopy class of the circle S_i^1 in $\pi_1(K^1)$. Then $\pi_1(K^1)$ is a free group with basis $\{a_i\}$. Since $\pi_1(X)$ is finitely generated, there is an n such that $\pi_1(X)$ is generated by $f_*(a_1), \cdots, f_*(a_n)$. Hence for each $i > n$ there is an element b_i in $\langle a_1, \cdots, a_n \rangle$ such that $f_*(a_i) = f_*(b_i)$. Define an automorphism α of $\pi_1(K^1)$ by

$$\alpha(a_i) = \begin{cases} a_i & \text{if } i = 1, \cdots, n \\ a_i b_i^{-1} & \text{if } i > n. \end{cases}$$

Clearly, α is realized by a homotopy equivalence $h: K^1 \to K^1$. Let $f_n: \bigvee_{i=1}^{n} S^1 \to X$ be the restriction of f, and $r_n: K^1 \to \bigvee_{i=1}^{n} S_i^1$ be the natural retraction. Define $f'_n: K^1 \to X$ by $f'_n = f_n r_n$. Then $(f'_n)_* = (fh)_*$. This implies $f'_n \simeq fh$; therefore $\pi_2(f) \cong \pi_2(fh) \cong \pi_2(f'_n)$ as crossed modules. By the argument in the proof of Theorem 1.4, we have that $\pi_2(f'_n)$ is isomorphic to the $\pi_1(\bigvee_{i=n+1}^{\infty} S^1)$-stabilization of $\pi_2(f_n)$. This implies that there is a free $\pi_1(X)$-module B of countably infinite rank such that $\pi_2(f)_{ab} \cong \pi_2(f_n)_{ab} \oplus B$. By Theorem 5.2, $\pi_2(f_n)_{ab}$ is a countable projective $\pi_1(X)$-module. By Eilenberg's Lemma [1], $\pi_2(f)_{ab}$ is free.

Conditions (1) and (2) are equivalent by Theorem 5.1. Suppose (2) holds. Then there is a 1-connected map $g: K^1 \to X$ which can be extended over 2-cells to a homotopy equivalence. Since $\pi_1(X)$ is finitely generated, there is an automorphism β of the free group $\pi_1(K^1)$ such that $f_* = g_* \beta$. By the argument in the proof of Theorem 1.4, $\pi_2(f) \cong \pi_2(g)$ as crossed modules. As $\pi_2(g)$ is free, we have that $\pi_2(f)$ is free. Thus (2) implies (3). By Theorem 5.2 we have that (3) implies (2). Conditions (3) and (4) are equivalent by Theorem 2.2 of [14]. □

In [7] Eilenberg and Ganea conjectured that the cohomological dimension (cd) of a group is equal to its geometric dimension (gd). This conjecture has been proven for all groups G such that $cd G \neq 2$. As an application we give a necessary and sufficient condition such that $cd G = 2 = gd G$ when G is finitely generated.

COROLLARY 5.4. *Let G be a finitely generated group, F a free group of countably infinite rank, $\eta: F \to G$ an epimorphism, and N the kernel of η. If $cd G = 2$, then N_{ab} is a free G-module; moreover, $gd G = 2$ if and only if N_{ab} has a basis which lifts to a set of normal generators of N in F.*

Proof. Let X be a $K(G,1)$. Then $cdG = 2$ implies X satisfies D2. Realize the epimorphism $\eta : F \to G$ by a map $f : K^1 \to X$ from a countably infinite wedge of circles to X. From the exact sequence

$$0 = \pi_2(X) \longrightarrow \pi_2(f) \overset{\partial}{\longrightarrow} \pi_1(K^1) \overset{f_*}{\longrightarrow} \pi_1(X) \longrightarrow 1$$

we see that $\pi_2(f) \cong N$ as crossed modules. By Theorem 5.3, N_{ab} is a free G-module, and X is homotopy equivalent to a 2-complex if and only if N_{ab} has a basis which lifts to a set of elements of N whose normal closure in F is N. □

The following question is still unanswered. Is there a complex X which is dominated by a 2-complex but is not homotopy equivalent to a 2-complex?

Consider the following two examples:

(1) Let G be a finite cyclic group such that $\widetilde{K}_0(\mathbf{Z}(G))$ is nonzero, and X be a G-complex which is dominated by a finite 2-complex, and such that its Wall obstruction is nonzero.

(2) Let G be a finite nonsolvable group, and X be a G-complex which satisfies D2 but not F2, and such that its obstruction $\kappa(X)$ in $\widetilde{\mathcal{P}}(G)$ is represented by a nonfree, countable G-module.

Let X be example (1) or (2). Then $\pi_1(X)$ is a finite group, X is dominated by a countable 2-complex, but X is not homotopy equivalent to a 2-complex with a finite 1-skeleton. Thus either X is not homotopy equivalent to a 2-complex or X is homotopy equivalent to a countable 2-complex with an infinite number of 1-cells and 2-cells, and for each finite wedge of circles K^1 and 1-connected map $f : K^1 \to X$, the crossed module $\pi_2(f)$ is countably stably free but not finitely stably free. In order to determine which is the case, one needs to understand the countably infinite presentations of G.

DEPARTMENT OF MATHEMATICS
VANDERBILT UNIVERSITY
NASHVILLE, TENNESSEE 37235

REFERENCES

[1] H. Bass, Big projective modules are free, Illinois J. Math. 7 (1963), 24-31.

[2] I. Beck, Projective and free modules, Math. Z. (1972), 231-234.

[3] W. Browning, Finite CW-complexes of cohomological dimension two with finite abelian π_1, Preprint (1979), ETH, Zürich.

[4] M. J. Dunwoody, Relation modules, Bull. London Math. Soc. 4 (1972), 151-155.

[5] ———, The homotopy type of a two-dimensional complex, Bull. London Math. Soc. 8 (1976), 282-285.

[6] M. N. Dyer, On constructing complexes dominated by two-complexes. Preprint (1984), University of Oregon.

[7] S. Eilenberg and T. Ganea, On the Lusternik-Schnirelmann category of abstract groups, Ann. Math. 65 (1957), 517-518.

[8] I. Kaplansky, Projective modules, Ann. Math. 68 (1958), 372-377.

[9] P. A. Linnell, Nonfree projective modules for integral group rings, Bull. London Math. Soc. 14 (1982), 124-126.

[10] R. C. Lyndon, Dependence and independence in free groups, J. Reine Angew. Math. 210 (1962), 148-174.

[11] S. Mac Lane and J. H. C. Whitehead, On the 3-type of a complex, Proc. Nat. Acad. Sci. U.S.A. 36 (1950), 41-48.

[12] W. Magnus, A. Karrass, and D. Solitar, Combinatorial Group Theory, Interscience, New York (1966).

[13] J. G. Ratcliffe, Crossed extensions, Trans. Amer. Math. Soc. 257 (1980), 73-89.

[14] ———, Free and projective crossed modules, J. London Math. Soc. 22 (1980), 66-74.

[15] P. F. Smith, A note on idempotent ideals in group rings, Arch. Math. 27 (1976), 22-27.

[16] J. R. Stallings, Homology and central series of groups, J. Algebra 2 (1965), 170-181.

[17] ———, On torsion-free groups with infinitely many ends, Ann. Math. 88 (1968), 312-334.

[18] R. G. Swan, The Grothendieck ring of a finite group, Topology 2 (1963), 85-110.

[19] ———, Groups of cohomological dimension one, J. Algebra, 12 (1969), 585-610.

[20] C. T. C. Wall, Finiteness conditions for CW-complexes, Ann. Math. 81 (1965), 56-69.

[21] ———, Finiteness conditions for CW-complexes II, Proc. Royal Soc. 295 (1966), 129-139.

[22] J. H. C. Whitehead, On the homotopy type of ANR's, Bull. Amer. Math. Soc. 54 (1948), 1133-1144.

[23] ———, Combinatorial Homotopy I, Bull. Amer. Math. Soc. 55 (1949), 213-245.

SUBCOMPLEXES OF TWO-COMPLEXES AND PROJECTIVE CROSSED MODULES

Micheal Dyer

The purpose of this note is to demonstrate that projective crossed modules, first studied by J. Ratcliffe in his thesis [4], occur in a very natural setting, namely, if X is a connected subcomplex of a 2-dimensional CW-complex Y, then the boundary map $\partial : \pi_2(Y,X) = C \to \pi_1(X) = G$ has the structure of a *projective* G-crossed module. We apply this to special cases, giving some (perhaps interesting) constructions of projective Q-modules, where Q is the group $G/\operatorname{im} \partial$.

This theorem generalizes a special case of a theorem of J. H. C. Whitehead [W]: If X is a connected CW-complex and Y is obtained from X by attaching only 2-cells to X, then $\partial : C \to G$ is a *free* G-crossed module, with basis corresponding to the characteristic maps of the attached two-cells. Proofs of this theorem are given in [W], [R], and [BH].

We show by a counterexample that if X is *not* a 2-complex, and Y is obtained from X by attaching cells of dimension one *and* two, then $\partial : C \to G$ is not necessarily projective.

In what follows a $[G,m]$-complex is a connected CW-complex of dimension $\leq m$ with $\pi_1 X \approx G$ and $\pi_i X = 0$ for $1 < i < m$ ($m \geq 2$).

1. *Basic notions*

By a *crossed module* we mean a pair of groups C, G and a homomorphism $\partial : C \to G$, together with an action of G on C which satisfies the conditions

(a) $\partial(g \cdot c) = g \partial(c) g^{-1}$ $(g \in G, c \in C)$
(b) $(\partial c_2) \cdot c_1 = c_2 c_1 c_2^{-1}$ $(c_1, c_2 \in C)$.

For example, if $X \subset Y$ are topological spaces, then $\partial : \pi_2(Y,X) \to \pi_1(X)$ is well known to be a $\pi_1(X)$-crossed module.

It follows from the definition that $K = \ker \partial$ is contained in the center of C and that $N = \operatorname{im} \partial$ is a normal subgroup of G. Let $Q = G/N$. Furthermore, the abelianization C^{ab} of C has the structure of a Q-module with action induced from the action of G on C.

A morphism of crossed-modules is a commutative square of group homomorphisms

(1.1)
$$\begin{array}{ccc} C & \xrightarrow{a} & C' \\ \partial \downarrow & & \downarrow \partial' \\ G & \xrightarrow{\beta} & H \end{array}$$

for which $a(g \cdot c) = \beta(g) \cdot a(c)$ ($g \in G$, $c \in C$). Let $\mathcal{CM}$ denote this category.

If, in 1.1, β is the identity on $G = H$, we say that a is a G-crossed module homomorphism and denote this category by $\mathcal{CM}_G$.

A *projective* G-crossed module $\partial : C \to G$ is a projective in the category $\mathcal{CM}_G$; that is, given a G-crossed module epimorphism (= surjection) $f : A \twoheadrightarrow B$ and any $g : C \to B$ in $\mathcal{CM}_G$, there is an $h : C \to A$ in $\mathcal{CM}_G$ such that $fh = g$. This has been characterized by J. Ratcliffe [R] as follows. Let $N = \operatorname{im} \partial$ and $Q = G/N$.

THEOREM 1.1. *A G-crossed module* $\partial : C \to G$ *is projective iff (1) the abelianization* C^{ab} *is a projective Q-module and (2) the map* $H_2 d : H_2 C \to H_2 N$ *is trivial, where* $d : C \twoheadrightarrow N$ *is* ∂.

We note that free G-crossed modules are projective.

2. *Subcomplexes of 2-complexes yield projective crossed modules*

Let X be a connected subcomplex of a connected 2-complex Y, where $\pi_1 X = G$, $\pi_1 Y = H$ and $i : X < Y$ be the inclusion map. Pick any base point $*$ in X^0 and use it for the homotopy groups.

THEOREM 2.1. *The triple* $(\partial, \pi_2(Y,X), \pi_1 X = G)$ *is a projective G-crossed module.*

Proof. Let $C = \pi_2(Y,X)$, $Q = \operatorname{im} \pi_1(i)$, and $N = \ker \pi_1(i)$. We will show that (1) $H_2 \partial : H_2 C \to H_2 N$ is trivial and that (2) the abelianization C^{ab} of C is a projective Q-module.

(1) Let Y^1 be the 1-skeleton of Y and consider the commutative diagram:

(2.2)
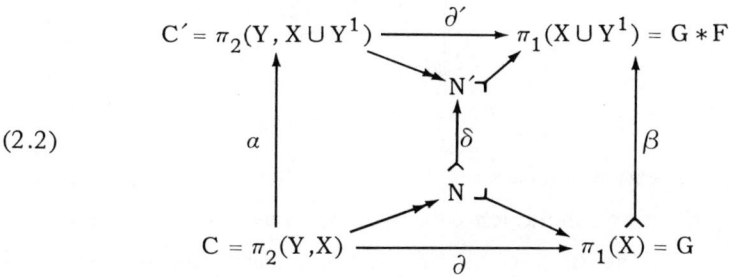

where F is a free group and β is the inclusion $G \to G * F$. We claim that $N = N' \cap G$, if we identify G with βG. Clearly $N \subset N' \cap G$. To see the other inclusion, observe that N' is the normal closure of words $[r_i] = \partial e_i$ corresponding to the boundaries ∂e_i of the 2-cells e_i in Y outside those of X. Thus an element $[\omega] \in N' \cap G$ can be written as

$$[\omega] = \prod_{j=1}^{k} [a_j][r_{i_j}]^{\epsilon_j}[a_j]^{-1} \qquad ([a_j] \in G * F, \ \epsilon_j = \pm 1)$$

where the loop ω is homotopic (rel $*$) to a loop ω' in X^1 inside $X \cup Y^1$. We may build a map $\eta : (B^2, S^1) \to (Y, X)$, described pictorially by figure 2.1, where the disc r_{i_j} maps over the 2-cell e_{i_j} and the residual space $B^2 - \{\text{lollypops}\}$ is mapped via the homotopy between ω and ω' inside $X \cup Y^1$. Hence $[\eta]$ is a member of $\pi_2(Y,X)$ whose boundary $\partial[\eta] = [\omega]$ and the claim is proved.

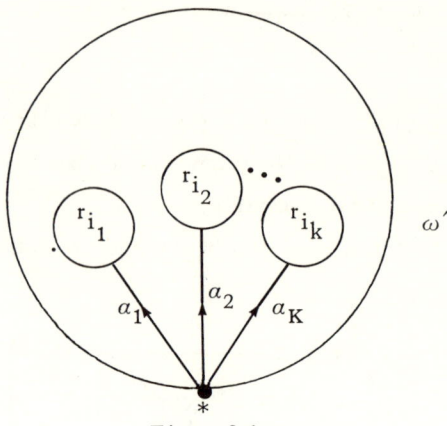

Figure 2.1

By Kurosh's theorem [S, Theorem 14], N is then a free summand of N' and $\delta: N \to N'$ is the inclusion onto that free summand. Applying the functor $H_2(-, \mathbb{Z})$ to the diagram 2.2 we see the commutative square:

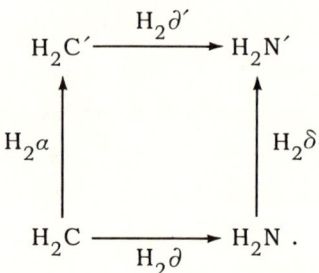

By Theorem 14.2 of [HS], $H_2\delta$ is a monomorphism. By the theorem of Whitehead [W], $\partial': \pi_2(Y, X \cup Y^1) \to \pi_1(X \cup Y^1)$ is a free crossed module, so $H_2\partial' = 0$ by 1.1. Hence $H_2\delta H_2\partial = 0$, so $H_2\partial = 0$.

(2) Let X_N be the covering of X corresponding to the subgroup N. Consider the pair $(\tilde{Y}, X_N)$, where $\tilde{Y}$ is the universal cover of Y and X_N may be identified as a subcomplex of $\tilde{Y}$. Note that both $\tilde{Y}$ and X_N are stable under the action of $Q < H = \pi_1 Y$.

We may identify $\pi_2(Y, X)$ with $\pi_2(\tilde{Y}, p^{-1}X)$ using the covering map $p: \tilde{Y} \to Y$. This in turn may be identified with $\pi_2(\tilde{Y}, X_N)$, where X_N is the component of $p^{-1}X$ containing the base point. We will show that $H_2(\tilde{Y},X_N)$ is the abelianization of $\pi_2(\tilde{Y},X_N)$ (under the Hurewicz map) by the following

PROPOSITION 2.3. *Let (W,A) be a topological pair with W simply connected and A path connected. Then the Hurewicz map $h : \pi_2(W,A) \to H_2(W,A)$ is surjective with the kernel of h equal to the commutator subgroup π_2' of $\pi_2 = \pi_2(W,A)$.*

Proof. The map h is surjective because (W,A) is 1-connected (or by using a simple diagram chase). In order to see that $\ker h = \pi_2'$, we use the relative Hurewicz theorem, which says that $\ker h$ is the normal closure in π_2 of $\{(g \cdot c)c^{-1} | c \in \pi_2, g \in \pi_1 A\}$. Consider the surjection $\partial_2 : \pi_2(W,A) \to \pi_1 A$. For any $g \in \pi_1 A$ there is a $d \in \pi_2$ with $\partial_2 d = g$. Thus $(g \cdot c)c^{-1} = (\partial d)c^{-1} = dcd^{-1}c^{-1} \in \pi_2'$. Thus $\ker h \subseteq \pi_2'$; the reverse inclusion is obvious because $H_2(W,A)$ is abelian. □

Clearly, $H_2(\tilde{Y},X_N)$ has a Q-module structure. We will show that this module is in fact a projective Q-module. The proof has already appeared in [BDS]; we give it here for the reader's convenience.

Consider the chain complex $C_*(\tilde{Y},X_N)$ of free Q-modules given by

$$C_2(\tilde{Y})/C_2(X_N) \xrightarrow{\bar{\partial}_2} C_1(\tilde{Y})/C_1(X_N) \xrightarrow{\bar{\partial}_1} C_0(\tilde{Y})/C_0(X_N)$$

Now, X_N is *connected* implies that $H_0(\tilde{Y},X_N) = 0$, hence $\bar{\partial}_1$ is surjective. Also, $H_1 \tilde{Y} = 0$ implies $H_1(\tilde{Y},X_N) = 0$, so $\operatorname{im} \bar{\partial}_2 = \ker \bar{\partial}_1$. Thus $H_2(\tilde{Y},X_N) = \ker \bar{\partial}_2$ is a projective Q-module, being a direct summand of the free Q-module $C_2(\tilde{Y})/C_2(X_N)$. This completes the proof that $\partial : C \to G$ is a projective G-crossed module. □

We note that a normal subgroup $L \triangleleft G$ is a projective G-crossed module iff $H_1 L$ is a projective G/L-module and $H_2 L = 0$.

COROLLARY 2.4. *If X is a $[G,2]$-subcomplex of the aspherical $[H,2]$-complex Y, and $i: X \to Y$ is the inclusion map, then the normal subgroup $L = \ker\{i_\# : G \to H\}$ is a projective G-crossed module.*

Proof. We show that if $i_{2\#} : \pi_2 X \to \pi_2 Y$ is surjective, then L is a projective G-crossed module. This follows from the homotopy exact sequence of the pair (Y,X) and Theorem 2.1. □

EXAMPLE 2.5. As an example, let $X = S^1$ be the 1-skeleton of $RP^2 = S^1 \cup e^2$ and let Y be $RP^2 \vee S^1$. Then $\pi_1 X \to \pi_1 Y$ is just the map $Z \to Z_2 * Z$ having image $Q = Z_2(x)$. Because $\pi_1 X$ is infinite cyclic, then $\pi_2(Y,X)$ is abelian; we will show directly that $\pi_2(Y,X)$ is a *free* Q-module. Let $W = Y \cup e^2 = RP^2 \vee B^2 \simeq RP^2$. Then the following diagram commutes

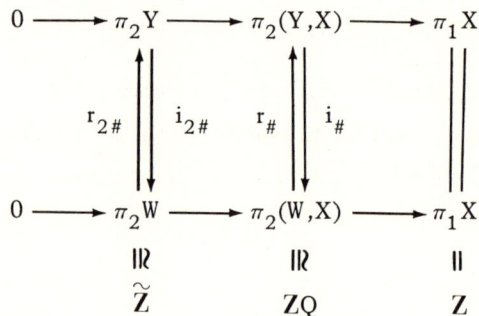

where $i: Y \hookrightarrow W$ is the inclusion and $r: W \to RP^2 \hookrightarrow Y$ is a retraction followed by an inclusion. Notice that $\pi_1 X \cong Z$, with trivial Q-action, $\pi_2(W,X)$ (by Whitehead's theorem) is isomorphic to ZQ and $\pi_2 W \cong \tilde{Z}$ ($= Z$ with non-trivial Q-action). Because $ir \simeq id_W$, we see that $\pi_2(Y,X) \cong ZQ \oplus A$ where $A = \ker i_\# = \ker i_{2\#}$.

Let $\mathcal{Q} = (Z_2 * Z)/Q$ be the set of left cosets ωQ with $\omega \in Z_2 * Z$. By looking carefully at the universal covering for $Y = RP^2 \vee S^1$, we may identify as $\pi_2 Y \approx Z\pi_1 Y \otimes_{ZQ} \tilde{Z}$ (as a left $Z\pi_1 Y$-module). This is also isomorphic, as ZQ-modules, to $\pi_2 Y \cong \bigoplus_{a \in \mathcal{Q}} (\tilde{Z})_a$. Hence, we may identify

A with the submodule of $\pi_2 Y \cong \oplus_{a \in \mathcal{Q}} (\widetilde{Z})_a$ consisting of all elements $\Sigma n_a \cdot a$ such that $\Sigma n_a = 0$. By sending $\sum_{a \in \mathcal{Q}} n_a \cdot a \to \sum_{a \neq Q} n_a \cdot a$, one may show that A is ZQ-isomorphic to the submodule $\widetilde{A} = \oplus_{a \in \mathcal{Q} - \{Q\}} (\widetilde{Z})_a$ of $\pi_2 Y$.

In order to show that $\widetilde{A}$ is free, we study the universal covering $p: \widetilde{Y} \to Y$. Let $\widetilde{RP^2}$ be the component of $p^{-1}(RP^2)$ containing the base point x_0. Then $\widetilde{A} = H_2(\widetilde{Y}, \widetilde{RP^2})$ as a ZQ-module.

Now consider the action of Q on $\widetilde{Y}$ (see Figure 2.2).

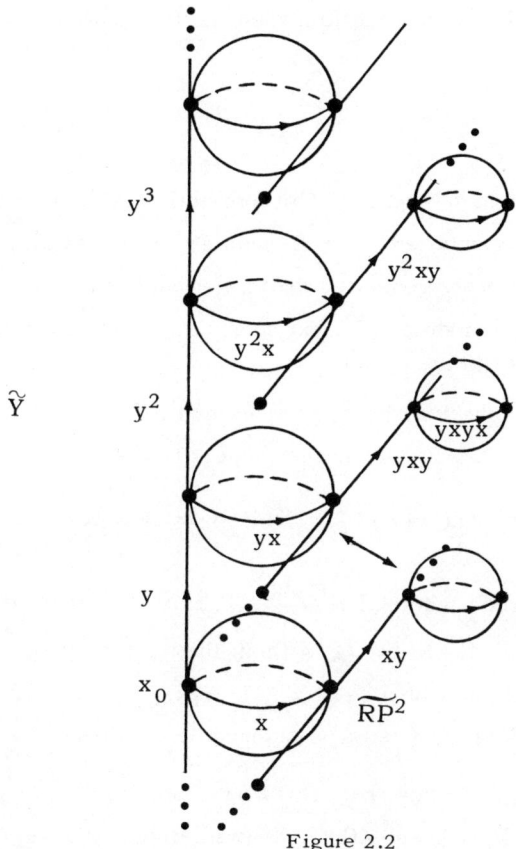

Figure 2.2

Let each sphere be indexed by the address of its left vertex which is a coset $\omega(x,y) \cdot Q$ in $\mathcal{Q}$. Thus, under the action of $x \in Q$, the sphere

$\omega(x,y) \cdot Q$ is carried (after interchanging hemispheres) to the sphere $xw(x,y)Q$. Because each sphere (with address $\neq Q$) represents a copy of $\tilde{Z}$ in $\tilde{A}$, a simple argument shows that, for any $a \neq Q$ in $\tilde{\mathcal{C}}$, the direct summand $(\tilde{Z})_a \oplus (\tilde{Z})_{xa}$ is isomorphic to ZQ (identify $(1,0)$ with $1 \in ZQ$ and $(0,1)$ with $x \in ZQ$.) Hence, $\pi_2(Y,X)$ is a free Q-module.

I would like to thank James Howie for bringing this example to my attention.

EXAMPLE 2.6. Suppose $i: X < Y$ is an inclusion of two-complexes. If $\pi_1(i): \pi_1 X = G \to \pi_1 Y = H$ is an injection, then the following is an exact sequence of G-modules

(2.7) $$\pi_2 X \rightarrowtail \pi_2 Y \twoheadrightarrow \pi_2(Y,X)$$

with $\pi_2(Y,X)$ a projective G-module. This projective is finitely generated if $\pi_2 Y$ is finitely generated (as a G-module). For example, if $[H:G] < \infty$ and $\pi_2 Y$ finitely generated as an H-module then $\pi_2 Y$ is finitely generated as a G-module. Furthermore, as G-modules, the sequence 2.7 splits.

Now let Z be an aspherical $[K,2]$-complex and let X be any $[G,2]$-complex. Let $Y = X \vee Z$. Then the (huge!) G-module

$$\pi_2(Y,X) = P = (Z(G*K) \otimes_G \pi_2 X)/\pi_2 X \approx \bigoplus_{i \in I} \pi_2 X_i$$

is projective (in fact, free), where $I = \dfrac{G*K}{G} - \{G\}$ is the set of left cosets $\{\omega G | \omega \in G*K\}$ of $G*K$ by G without the trivial coset. The G-module structure on P is given by $g(x_{\omega G}) = (gx)_{g\omega G}$ ($x \in \pi_2 X$, $g \in G$, $\omega \in G*K - G$). We think of $\pi_2 X$ as a G-submodule of $Z(G*K) \underset{G}{\otimes} \pi_2 X$ via $x \mapsto 1 \oplus x$ ($x \in \pi_2 X$). To see that P is *free* observe that $\bigoplus_{i \in I} \pi_2 X_i \cong \bigoplus_{j \in GI} ZG \otimes \pi_2 X_j$. Then $ZG \otimes \pi_2 X$ (with diagonal action) is isomorphic to $ZG \otimes (\pi_2 X)_0$ as a G-module, where $(\pi_2 X)_0$ is the underlying free abelian group of $\pi_2 X$, which is clearly free [HS, p. 212].

3. A counterexample

LEMMA 3.1. *Suppose X is a connected subcomplex of the 2-complex Y, $i: X < Y$ is the inclusion and that $\pi_1(i): \pi_1 X \to \pi_1 Y$ is an injection. Then $\pi_2(Y,X)$ is a projective $\pi_1 X$-module.*

Proof. The fact that $\pi_1(i)$ is injective and X,Y are 2-complexes implies that the sequence of $\pi_1 X$-modules $\pi_2 X \rightarrowtail \pi_2 Y \twoheadrightarrow \pi_2(Y,X) = C$ is exact; hence, $\pi_2(Y,X)$ is abelian. That $C = C^{ab}$ is a projective $\pi_1 X$-module now follows from Theorem 2.1. □

COUNTEREXAMPLE 3.2. Let $X = RP^3$ be real projective three-space, with a CW-structure having one cell in each dimension. We may choose its two-skeleton X^2 as the realization of the presentation $(x : x^2)$. We form $Y = (X \vee S^1) \cup e^2$ by adjoining cells to give $Y^2 = |(x,y : x^2, [x,y])|$.

If $i: X < Y$ is the inclusion, then $\pi_1(i): \pi_1 X = \mathfrak{N}_2 \to \pi_1 Y = \mathfrak{N} \oplus \mathfrak{N}_2 = H$ is injective, where $\mathfrak{N}$ (respectively $\mathfrak{N}_2$) is the infinite (finite) cyclic group of infinite order (order 2) generated by $y(x)$. The chain map $i_\# : C_*(\tilde{X}) \to C_*(\tilde{Y})$ is given as follows:

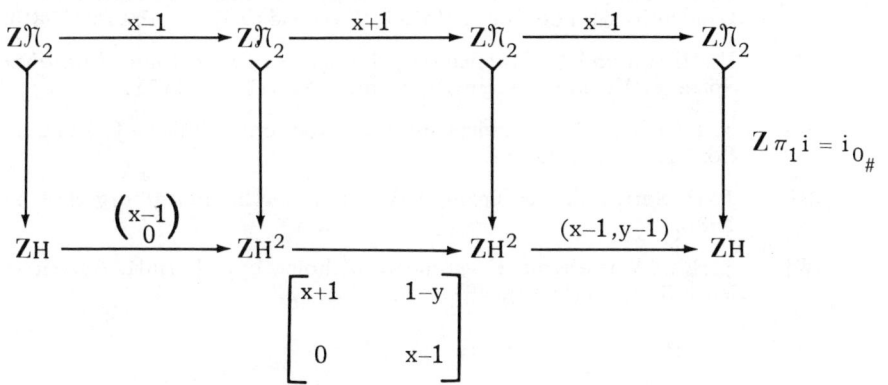

Because $\pi_1 X \to \pi_1 Y$ is injective, we see that the sequence

$$\pi_2 X \longrightarrow \pi_2 Y \twoheadrightarrow \pi_2(Y,X)$$

is exact; but $\pi_2 X = 0$ implies that $\pi_2 Y \cong \pi_2(Y,X) \cong H_2(\widetilde{Y},\widetilde{X})$. We will compute $\pi_2 Y$ and show it is *not* a projective $Z\mathcal{H}_2$-module.

A direct calculation shows that $\pi_2 Y^2$ is isomorphic (over ZH) to the submodule of ZH^2 generated by $\{(1-x,0),(y-1,1+x)\}$ and the image of $C_3 \widetilde{Y}$ in $C_2 \widetilde{Y}$ is generated by $(x-1,0)$. Thus

$$\pi_2 Y \cong H_2 \widetilde{Y} \cong \frac{ZH(1-x,0) + ZH(y-1,1+x)}{ZH(1-x,0)} \cong ZH/ZH(x-1)$$

as H-modules.

But $ZH/ZH(x-1) = H_0(\mathcal{H}_2, ZH)$ is a *trivial* $Z\mathcal{H}_2$-module; hence cannot be projective.

DEPARTMENT OF MATHEMATICS
UNIVERSITY OF OREGON
EUGENE, OREGON 97403-1222

BIBLIOGRAPHY

[BDS] J. Brandenburg, M. Dyer, and R. Strebel, On J. H. C. Whitehead's aspherical question II, Contemp. Math. 20 (1983), 65-78.

[BH] R. Brown, J. Hubschmann, Identities among relations, Low Dimensional Topology, Vol. 1, Cambridge Univ. Press (1980).

[HS] P. Hilton and U. Stammbach, *A course in homological algebra*, Springer-Verlag, Berlin Heidelberg New York, 1971.

[R] J. Ratcliffe, Free and projective crossed modules, J. London Math Soc. 22 (1980), 66-74.

[S] J.-P. Serre, *Trees*, Springer-Verlag, Berlin-Heidelberg-New York, 1980.

[W] J. H. C. Whitehead, Combinatorial homotopy II, Bull. American Math Soc. 55 (1949), 453-496.

LENGTH FUNCTIONS OF GROUP ACTIONS ON Λ-TREES

Roger Alperin and Hyman Bass

CONTENTS

0. Introduction

Chapter I. Λ-trees

1. Ordered abelian groups
2. Λ-trees
3. A Chiswell style construction of rooted Λ-trees
A. Appendix: The tree X_v of a valuation v.
4. Base change

Chapter II. Tree actions and length functions

5. Lyndon length functions
6. Automorphisms of Λ-trees; hyperbolic length
B. Appendix: Hyperbolic length of $s \in GL_2(F)$
7. The main theorems
8. The hyperbolic length of a product
9. In search of hyperbolic length axioms
   References

0. *Introduction*

This is a long and technical paper. Hopefully this introduction will provide some excuse for it.

Combinatorial presentations of a group Γ, as a free group, amalgamated free product, HNN-extension, or various generalizations, are all subsumed in Serre's book, *Trees*, under the general notion of "fundamental group of a graph of groups." In *Trees* it is shown that such presentations of Γ are essentially equivalent to actions (without "inversions") of Γ on (simplicial) trees, which we here call "Z-trees," for reasons explained below.

Trees further describes the following natural source of such tree actions. Let F be a field with a discrete valuation v. Then there is a Z-tree X_v on which $SL_2(F)$ naturally acts without inversions; X_v is actually the "Bruhat-Tits building" of $SL_2(F)$ at v.

Let M be a compact smooth manifold of dimension n = 2 or 3. Then $\Gamma = \pi_1(M)$ acts on the universal cover $\tilde{M}$. A hyperbolic structure on M leads, via this action, to a representation $\bar{\rho}: \Gamma \to PSL_2(K)$; $K = \mathbf{R}$ for n = 2, and $K = \mathbf{C}$ for n = 3. This can be lifted to $\rho: \Gamma \to SL_2(K)$. There exist (many) pairs (F,v) such that $F \subset K$ is a finitely generated extension of $\mathbf{Q}$; $\rho(\Gamma) \subset SL_2(F)$, and v is a discrete valuation of F. Then Γ acts on the tree X_v, whence the possibility of deriving combinatorial information about Γ, by the methods discussed above. This, in turn, can often yield geometric information, by methods due to Stallings and others.

This procedure was first exploited in the proof of the Smith Conjecture [MB]. Subsequently it was developed more systematically by Culler and Shalen [CS], using the algebraic variety of classes of semi-simple representations $\Gamma \to SL_2$. Their methods strongly suggested introducing a Bruhat-Tits "tree" X_v even when the valuation v is no longer discrete, i.e. the ordered abelian value group Λ of v is not necessarily $\mathbf{Z}$, nor even archimedean (i.e. embeddable in $\mathbf{R}$). The appropriate notion of "Λ-tree" for this purpose was introduced by Morgan-Shalen [MS]; simplicial trees correspond to the case $\Lambda = \mathbf{Z}$. A Λ-tree is a metric space X whose metric takes values d(x,y) in Λ, and is subject to certain tree axioms. From this point of view we have to replace a simplicial tree ($\Lambda = \mathbf{Z}$) by its set of vertices, with integer distances.

Suppose that a group Γ acts on a Λ-tree X (always by isometries). Choose $x \in X$ and define $L = L_x : \Gamma \to \Lambda$ by $L(s) = d(x, sx)$. This satisfies:

(L0) $L(1) = 0$,

(L1) $L(s) = L(s^{-1})$,

and, putting $Y(s,t) = \frac{1}{2}(L(s) + L(t) - L(s^{-1}t))$,

(L2) $\qquad Y(r,t) \geq \min(Y(r,s), Y(s,t))$,

and

(L3) $\qquad Y(s,t) \in \Lambda$.

Functions satisfying these conditions (with $\Lambda = \mathbf{Z}$) were introduced by Lyndon [L] in order to axiomatize certain cancellation arguments in combinatorial group theory. Chiswell showed in [C1] that in fact every such abstract integer valued "*Lyndon length function*" on Γ arises, essentially uniquely, as above from an action of Γ on a $\mathbf{Z}$-tree with base point. One of the results presented here (Theorem (5.4)) extends Chiswell's theorem to arbitrary Λ. (The generalization is not altogether routine.) Thus there is an equivalence between Λ-valued Lyndon length functions on Γ and Γ-actions on Λ-trees with base point ("rooted Λ-trees").

One of our main aims is to establish an analogous result for "*hyperbolic length functions.*" The hyperbolic length $\ell(s)$ of an automorphism s of a Λ-tree X is defined as follows. If s has no fixed points but s^2 has one then we call s an *inversion* and put $\ell(s) = 0$ and $A_s = \emptyset$. If s is not an inversion then $\ell(s) = \min_{x \in X} d(x, sx)$ exists, and we put $A_s = \{x \in X | d(x, sx) = \ell(s)\}$. When $\ell(s) = 0$, A_s is its tree of fixed points. If $\ell(s) > 0$ then A_s is a linear tree (isomorphic to a subtree of Λ), called the s-axis, on which s induces a translation of amplitude $\ell(s)$. This classification of Λ-tree automorphisms is due to Tits for $\Lambda = \mathbf{Z}$, and to Morgan-Shalen [MS] in general.

If a group Γ acts on a Λ-tree X we thus obtain a hyperbolic length function $\ell = \ell_X : \Gamma \to \Lambda$, which is a class function on Γ. We like to think of ℓ_X as the tree theoretic analogue of the character of a linear representation. Indeed, if v is a valuation on a field F and

$s = \begin{bmatrix} a & b \\ c & d \end{bmatrix} \epsilon \, SL_2(F)$ has trace $Tr(s) = a + d$ then (see [MS], II. 3.15; or Proposition (B.7) below),

$$\ell_{X_v}(s) = -2 \min(0, v(Tr(s))) .$$

Hyperbolic length functions are central to the work of Morgan and Shalen. For example Morgan and Shalen show that, in the Thurston compactification of the space of homotopy hyperbolic structures on an n-manifold M (n = 2 or 3), the ideal points can be parametrized by certain "projectivized" real valued hyperbolic length functions on $\Gamma = \pi_1(M)$.

We address here the following mutually related questions. (I) Given an action of a group Γ on a Λ-tree X, to what extent does $\ell_X : \Gamma \to \Lambda$ determine X and the action? (II) Which functions $\ell : \Gamma \to \Lambda$ are hyperbolic length functions?

To describe the answers we first point out a degenerate case. Let $h : \Gamma \to \Lambda$ be a group homomorphism, and put $\ell_h(s) = |h(s)|$. Clearly this is the hyperbolic length function of Γ acting on $X = \Lambda$ by translations, via h. Such hyperbolic length functions are called *abelian*; they are characterized by the condition that $\ell(st) \leq \ell(s) + \ell(t)$ for all $s, t \, \epsilon \, \Gamma$. Abelian hyperbolic length functions can arise from tree actions more complicated than the example above, but a rough classification of the possibilities can be given (Theorems (7.5) and (7.6)).

The main uniqueness result (Theorem (7.13)) concerns actions without inversions of Γ on a Λ-tree X with non abelian hyperbolic length function ℓ_X. In this case there is a unique minimal Γ-invariant subtree X_{min} of X. If Y is another Λ-tree with Γ-action and if $\ell_Y = \ell_X$ then there is a unique Γ-equivariant isomorphism $Y_{min} \to X_{min}$, provided that Γ also acts without inversions on Y. Moreover the last condition is automatic except in the "dihedral case," i.e. when X_{min} is a linear tree.

Theorem (7.13) accomplishes for hyperbolic length functions what Chiswell's Theorem, generalized to Λ-trees, does for Lyndon length functions. Indeed we use the Chiswell Theorem to prove Theorem (7.13). Rather than reconstruct X directly from ℓ_X we first show how, for a carefully selected base point $x_u \in X_{min}$, to express the Lyndon length function L_{x_u} in terms of ℓ_X (see (7.9)). Then the proof is completed by invoking the Chiswell Theorem.

This procedure simultaneously gives a response, in principle, to question (II) above. On the one hand we have axioms (L0),$\cdots$,(L3) above for Lyndon length functions. On the other hand we have, at least in the non abelian case, formula (7.9) for L_{x_u} in terms of ℓ_X. Now given an abstract function $\ell : \Gamma \to \Lambda$ which is "non-abelian" ($\ell(uv) > \ell(u) + \ell(v)$ for some $u, v \in \Gamma$) we can define a function $L : \Gamma \to \Lambda$ by the analogue of (7.9), and conclude that ℓ is a hyperbolic length function iff L satisfies (L0),$\cdots$,(L3). Unfortunately (L2) leads to an excessively complicated axiom. A more economical approach to this problem is explored in Section 9.*

The authors learned late in the course of this work that Culler and Morgan [CM] had been working independently, and simultaneously, on the same problems, except that they consider only the case $\Lambda = R$. They independently obtained the uniqueness theorem ((7.13)) in the case $\Lambda = R$. Their approach differs in some details from ours, but they rely, as we do, on a version of Chiswell's Theorem, already proved by Chiswell in the case $\Lambda = R$.

The uniqueness theorem permits one to parametrize classes of tree actions by hyperbolic length functions, thus furnishing the means to speak of "nearby actions," and to discuss deformation theoretic questions. For example:

1. When can an action be well approximated by a simplicial one?

---

*Added in proof: Walter Parry has recently confirmed a simple set of hyperbolic length axioms, conjectured by Culler-Morgan [CM].

2. What properties of an action (e.g. "nice" stabilizers of points or edges) are open (preserved by nearby actions)? Or closed (preserved by limit actions)?

3. What can one say about groups that act freely on Λ-trees?

Questions 1 and 2 were suggested by Morgan. Aspects of question 2 are treated in his work with Culler cited above. Question 3, in a different guise, goes back to Lyndon (cf. [AM]). Indeed, question 3 is a very special case of the following:

FUNDAMENTAL PROBLEM. Find the group theoretic information carried by a Λ-tree action, analogous to that presented in *Trees* for the case $\Lambda = Z$.

Morgan's questions above provide one possible program for exploiting *Trees* in the general case. First show that a Λ-tree action is well approximated by a simplicial one, to which *Trees* applies. Then show that certain properties can be recovered from those of such nearby actions.

We introduce here a technical tool, *base change*, that could be helpful in treating such questions. Specifically, if $h : \Lambda \to \Lambda'$ is a homomorphism of ordered abelian groups ($a \geq 0 \Rightarrow h(a) \geq 0$) then to each Λ-tree X we can functorially attach a Λ'-tree $X \otimes_\Lambda \Lambda'$. When $\Lambda = Z$, $X \otimes_Z R$ is like the geometric realization of X, while $X \otimes_Z \left(\frac{1}{2} Z\right)$ is like the barycentric subdivision of X. With this notion we can formulate and prove "rationality properties" of tree actions. For example suppose that Γ acts without inversions on a Λ-tree X and that ℓ_X is non-abelian. Let Λ_0 be a subgroup of Λ such that $\ell_X(\Gamma) \cap 2\Lambda \subset 2\Lambda_0$. Then there is a Λ_0-tree X_0 on which Γ acts and a Γ-equivariant isomorphism $X_0 \otimes_{\Lambda_0} \Lambda \to X_{min}$. When Λ_0 is cyclic this makes precise the sense in which the action arises from a simplicial one.

With the foregoing discussion, the layout of the paper can be discerned now from the table of contents. The exposition is essentially self-contained, which somewhat accounts for its length.

We express here our deep gratitude to John Morgan, not only for his work with Shalen, which inspired our own but also for his interest, encouragement, and insightful comments on several aspects of this work. In addition we wish to thank Steve Gersten and John Stallings, whose conference on combinatorial group theory, and whose editorial prodding, precipitated our efforts. Finally, we acknowledge the scholarly referee who provided numerous useful suggestions, and an important correction to the original manuscript which is well appreciated by the authors.

CHAPTER I. Λ-TREES

1. Ordered abelian groups

Let Λ be a (totally) ordered abelian group, written additively. For $a \in \Lambda$ we put $|a| = \max(a, -a)$. Then $|a| \geq 0$, with equality iff $a = 0$; $|-a| = |a|$; and $|a \pm b| \leq |a| + |b|$.

If $a, b \in \Lambda$, $a \leq b$, we put

$$[a,b] = [b,a] = \{x \in \Lambda | a \leq x \leq b\},$$

and call this the *closed segment* between a and b. Note that $[a,b] = \{x \in \Lambda | |a-x| + |x-b| = |a-b|\}$. A subset $A \subset \Lambda$ will be called *convex* if $a, b \in A \Rightarrow [a,b] \subset A$. If A is a convex subgroup of Λ (these are sometimes called "isolated subgroups" in the literature) then the order on Λ naturally defines one also on Λ/A so that $\Lambda \to \Lambda/A$ preserves "$\leq$."

The convex subgroups of Λ are easily seen to be totally ordered by inclusion. If there are finitely many, say $r + 1$, of them, $0 = \Lambda_0 \subset \Lambda_1 \subset \cdots \subset \Lambda_r = \Lambda$, then we say that Λ has finite *rank* r. The quotients Λ_i/Λ_{i-1} have rank 1. Groups Λ of rank 1 admit an order preserving embedding $\Lambda \to \mathbf{R}$ which is unique up to a real factor > 0.

For later application, in Sections 7 and 9, we shall need the following characterization of absolute values of homomorphisms. The remainder of this section may be skipped until Theorem (7.6).

Let Γ be a possibly noncommutative group, written multiplicatively, and let $h: \Gamma \to \Lambda$ be a homomorphism, where Λ is an ordered abelian group, as above. Put $\ell(s) = |h(s)|$ for $s \in \Gamma$. Then clearly

(1.1) $$0 = \ell(1) \leq \ell(s) = \ell(s^{-1}), \qquad \forall\, s \in \Gamma$$

and

(1.2) $$\ell(sts^{-1}) = \ell(t), \qquad \forall\, s, t \in \Gamma$$

i.e.

(1.2)' $$\ell(st) = \ell(ts) \qquad \forall\, s, t \in \Gamma.$$

We have $\ell(st) = \ell(s) + \ell(t)$ if $h(s)$ and $h(t)$ have the same sign, and $\ell(st) = |\ell(s) - \ell(t)|$ otherwise. Note further that for $a \geq b \geq 0$ in Λ we have $a + b - |a-b| = 2b$, and so $\ell(s) + \ell(t) - |\ell(s) - \ell(t)| = 2\min(\ell(s), \ell(t))$. Putting

$$\beta(s,t) = \ell(st) - \ell(s) - \ell(t)$$

we thus have

$$\beta(s,t) = \begin{cases} 0 & h(s), h(t) \text{ of same sign} \\ -2\min(\ell(s), \ell(t)) & \text{otherwise.} \end{cases}$$

Hence

(1.3) $$\{\beta(s,t), \beta(s, t^{-1})\} = \{0, -2\min(\ell(s), \ell(t))\} \qquad \forall\, s, t \in \Gamma.$$

(1.4) ABSOLUTE HOMOMORPHISM LEMMA. *Let Λ be an ordered abelian group, let Γ be a group, and let $\ell: \Gamma \to \Lambda$ be a function satisfying (1.1), (1.2), and (1.3) above. Then there is a homomorphism $h: \Gamma \to \Lambda$ such that $\ell(s) = |h(s)|$ As $\in \Gamma$. Moreover $\pm h$ are the only such homomorphisms.*

Proof. Condition (1.3) is equivalent to

(1.3) (a) $\ell(st) = \begin{cases} \ell(s) + \ell(t) & (\beta(s,t) = 0) \\ \text{or} \\ |\ell(s) - \ell(t)| & (\beta(s,t) = -2 \min(\ell(s), \ell(t))) \end{cases}$

and

(1.3) (b) $\ell(st) = \ell(s) + \ell(t) \Longleftrightarrow \ell(st^{-1}) = |\ell(s) - \ell(t^{-1})|$.

In particular $\ell(st) \leq \ell(s) + \ell(t)$, so

$$N = \{s \in \Gamma | \ell(s) = 0\}$$

is a subgroup of Γ which, by (1.2), is normal. If $s \in N$ and $t \in \Gamma$ then $\ell(st) \leq \ell(s) + \ell(t) = \ell(t) = \ell(s^{-1}st) \leq \ell(s^{-1}) + \ell(st) = \ell(st)$, so $\ell(st) = \ell(t)$. Thus ℓ factors through a map $\ell' : \Gamma/N \to \Lambda$, which clearly still satisfies (1.1), (1.2), and (1.3), as well as: $\ell'(s) = 0 \Longrightarrow s = 1$. Replacing (Γ, ℓ) by $(\Gamma/N, \ell')$, therefore, we reduce to the case where we further have,

(1.5) $\ell(s) > 0 \quad \forall s \neq 1$.

Suppose that $\ell(s) = \ell(t)$. From (1.3) (a) and (b) and (1.5) we see that $t = s^{\pm 1}$; thus

(1.6) $\ell(s) = \ell(t) \Longrightarrow t = s^{\pm 1}$.

Further, $\beta(s,t) = \beta(s,t^{-1})$ iff $\min(\ell(s), \ell(t)) = 0$, by (1.3), whence, by (1.5),

(1.7) $\beta(s,t) = \beta(s,t^{-1})$ iff $s = 1$ or $t = 1$.

In particular $t^2 = 1 \Longrightarrow t = 1$ since $\beta(t,t) = \beta(t,t^{-1})$ in this case.

Assume that $\Gamma \neq \{1\}$; otherwise all assertions of the lemma are trivial. Choose $s \neq 1$ in Γ, and put

$$\Gamma_s = \{t \in \Gamma | \beta(s,t) = 0\}$$
$$= \{t \in \Gamma | \ell(st) = \ell(s) + \ell(t)\} .$$

We claim:

(1.8) $\quad\quad\quad\quad \Gamma_s \cup \Gamma_s^{-1} = \Gamma \quad \text{and} \quad \Gamma_s \cap \Gamma_s^{-1} = \{1\}.$

(1.9) $\quad\quad\quad\quad\quad\quad \Gamma_s \cdot \Gamma_s \subset \Gamma_s$

(1.10) $\quad\quad\quad\quad \ell(tu) = \ell(t) + \ell(u) \quad\quad \forall t, u \in \Gamma_s.$

Clearly (1.8) follows from (1.3) and (1.7). To establish (1.9) and (1.10) we must show, given $t, u \in \Gamma_s$, that $tu \in \Gamma_s$, i.e. $\beta(s, tu) = 0$, and that $\beta(t, u) = 0$. Put

$$a = \ell(s) + \ell(t) + \ell(u)$$
$$b = |\ell(s) + \ell(t) - \ell(u)|$$
$$c = |\ell(s) + \ell(u) - \ell(t)|.$$

We have the following alternatives (using (1.2)′).

$$\ell(stu) = \begin{cases} \ell(s) + \ell(tu) \\ \text{or} \\ |\ell(s) - \ell(tu)| \end{cases}$$

$$= \begin{cases} \ell(st) + \ell(u) = \ell(s) + \ell(t) + \ell(u) = a \\ \text{or} \\ |\ell(st) - \ell(u)| = |\ell(s) + \ell(t) - \ell(u)| = b \end{cases}$$

$$= \ell(ust) = \begin{cases} \ell(us) + \ell(t) = \ell(u) + \ell(s) + \ell(t) = a \\ \text{or} \\ |\ell(us) - \ell(t)| = |\ell(u) + \ell(s) - \ell(t)| = c. \end{cases}$$

Suppose that $\ell(stu) = a$. Since $|\ell(s) - \ell(tu)| \leq \ell(s) + \ell(tu) \leq \ell(s) + \ell(t) + \ell(u) = a$ we must then have $\ell(stu) = \ell(s) + \ell(tu)$ and $\ell(tu) = \ell(t) + \ell(u)$, whence our claim.

Suppose that $\ell(stu) \neq a$. Then we must have $\ell(stu) = b = c$. From $b = c$ it follows that $\ell(t) = \ell(u)$, and further that $\ell(stu) = \ell(s)$ and so $\ell(tu) = 0$, whence $tu = 1$, by (1.5), and $u = t^{-1} \in \Gamma_s \cap \Gamma_s^{-1} = \{1\}$, by (1.8). In this case we have $\ell(t) = \ell(u) = 0$ so $\ell(stu) = \ell(s) = a$, contrary to assumption.

Now define $h : \Gamma \to \Lambda$ by

$$h(t) = \begin{cases} \ell(t) & \text{if } t \in \Gamma_s \\ -\ell(t) & \text{if } t^{-1} \in \Gamma_s \end{cases}$$

It follows from (1.8) that h is well defined. Clearly $h(t^{-1}) = -h(t)$, and $\ell(t) = |h(t)|$. It remains to show that $h(tu) = h(t) + h(u)$ for $t, u \in \Gamma$. Let $t, u \in \Gamma_s$. Then $h(tu) = h(t) + h(u)$ by (1.9) and (1.10). We further have $h(u^{-1}t^{-1}) = h((tu)^{-1}) = -\ell((tu)^{-1}) = -\ell(tu) = -(\ell(t)+\ell(u))$ (by (1.10)) $= -\ell(t) - \ell(u) = h(t^{-1}) + h(u^{-1})$.

If $tu^{-1} \in \Gamma_s$ then $h(t) = h(tu^{-1}u) = h(tu^{-1}) + h(u)$ so $h(tu^{-1}) = h(t) - h(u) = h(t) + h(u^{-1})$. Consequently $h(ut^{-1}) = -h(tu^{-1}) = -(h(t)-h(u)) = h(u) - h(t) = h(u) + h(t^{-1})$.

Similarly, if $ut^{-1} \in \Gamma_s$ then $h(ut^{-1}) = h(u) + h(t^{-1}) = h(t^{-1}u)$. Thus h is a homomorphism.

Finally, to show the uniqueness of h up to sign it suffices to show that if $h' : \Gamma \to \Lambda$ is a homomorphism such that $\ell(t) = |h'(t)|$ $t \in \Gamma$, and if $h'(s) > 0$, then $\{t \in \Gamma | h'(t) \geq 0\}$ coincides with $\Gamma_s = \{t | h(t) \geq 0\} = \{t | \ell(st) = \ell(s) + \ell(t)\}$. Indeed we have $\ell(st) = |h'(st)| = |h'(s) + h'(t)| \leq |h'(s)| + |h'(t)| = \ell(s) + \ell(t)$, with equality iff $h'(t) \geq 0$, since $h'(s) > 0$.

This completes the proof of the absolute homomorphism lemma.

2. Λ-*trees*

Let Λ be an ordered abelian group.

By a Λ-*metric space* we understand a set X with a "distance" function $d : X \times X \to \Lambda$, satisfying:

(2.1) $\quad d(x,y) \geq 0$, with equality iff $x = y$

(2.2) $\quad d(x,y) = d(y,x)$

(2.3) $\quad d(x,z) \leq d(x,y) + d(y,z)$.

A (Λ-) *metric morphism* is a distance preserving map of such spaces; note that such a map must be injective.

(2.4) EXAMPLE. Take $X = \Lambda$ with $d(x,y) = |x-y|$. Among the isometries of Λ are the *translations*, $t_a(x) = a+x$, and *reflections*, $r_a(x) = a-x$ ($a \in \Lambda$). The multiplication table is as follows:

$$t_a t_b = t_{a+b}$$

$$t_a r_b = r_{a+b}$$

$$r_a t_b = r_{a-b}$$

$$r_a r_b = t_{a-b}.$$

In particular $r_a^2 = t_0 = 1$, and $r_a t_b r_a = t_{-b} = t_b^{-1}$. A translation t_a is order preserving, and has no fixed points if $a \neq 0$. A reflection r_a is order reversing; if $a \in 2\Lambda$ then r_a has the unique fixed point, $a/2$; if $a \notin 2\Lambda$ then r_a has no fixed points, and we call r_a an *inversion*.

(2.5) PROPOSITION. (a) *The isometry group* $G = \text{Aut}_{\text{metric}}(\Lambda)$ *consists entirely of translations and reflections. It is the semi-direct product of the normal subgroup,* t_Λ, *of translations, with the group* $\{1, r_0\}$, *where* $r_0(x) = -x$.

(b) *Let* A *be a subset of* Λ *and* $f : A \to \Lambda$ *a metric morphism. Then there is a* $g \in G$ *such that* $f(x) = g(x)$ *for all* $x \in A$. *If* $\text{Card}(A) \geq 2$ *then* g *is unique.*

It suffices to prove that if $A \subset \Lambda$ and if $f : A \to \Lambda$ is a metric morphism, then there is a $g \in G$ which is either a translation or a reflection such that $f = g|A$. Since such a $g \neq 1$ can never have two fixed points (see (2.4)) the uniqueness of g when Card $A \geq 2$ also follows.

The case $A = \emptyset$ is vacuous. Otherwise choose $a \in A$, put $b = f(a) - a$, and $f' = t_{-b} \circ f$. Then $f'(a) = a$. If $A = \{a\}$ then $f = t_b|A$, and we're done. Otherwise choose $c \neq a$ in A. Since $|c-a| = |f'(c) - f'(a)| = |f'(c) - a|$ we must have $f'(c) = c$ or $f'(c) = 2a - c = r_{2a}(c)$. Put $g = t_b = t_{f(a)-a}$ in the first case, and $g = t_b r_{2a} = r_{f(a)+a}$ in the second. Then $f'' = g^{-1} \circ f$ fixes a and c. If $d \in A$ then $|f''(d) - a| = |d-a|$ so $f''(d) = d$ or $2a - d$, and $|f''(d) - c| = |d-c|$ so $f''(d) = d$ or $2c - d$. Since $2a - d$ and $2c - d$ are distinct we must have $f''(d) = d$. Thus $f = g|A$, as claimed.

(2.6) COROLLARY. *Let A be a subset of Λ with at least two elements. The stabilizer Γ_A of A in $\mathrm{Aut}_{\mathrm{metric}}(\Lambda)$ is isomorphic, by restriction to A, to the full group $\mathrm{Aut}_{\mathrm{metric}}(A)$. If $\Lambda_0 = \{a \in \Lambda | a + A = A\}$ then t_{Λ_0} is the translation group of A. If Γ_A contains a reflection r_c (i.e. $c - A = A$) then the full set of reflections in Γ_A is $r_{c+\Lambda_0}$, and $\Gamma_A = t_{\Lambda_0} \rtimes \{1, r_c\}$.*

EXAMPLES. If A has a least element, a, then $\Lambda_0 = 0$, so either $\Gamma_A = \{1\}$, or $\Gamma_A = \{1, r_c\}$, and A is a symmetric subset of the interval $[a, c-a]$ in Λ.

(2.7) DEFINITION. Call a Λ-metric space (X,d) *geodesically linear* if, given $x, y \in X$, there is a unique metric morphism $\alpha : [0, d(x,y)] \to X$ such that $\alpha(0) = x$ and $\alpha(d(x,y)) = y$. We then denote by $[x,y]$ the image of α, and call it the *closed segment* in X between x and y. We call a subset $A \subset X$ *convex* if $x, y \in A \Rightarrow [x,y] \subset A$. Clearly closed segments are convex, as is any intersection of convex subsets. We call $A \subset X$ *convex closed* if the intersection of A with any closed segment is either empty or a closed segment.

(2.8) EXAMPLE. Take $\Lambda = \mathbf{R}$ and $X = \mathbf{R}^n$ with its euclidean metric. Convexity has here its usual meaning.

(2.9) DEFINITION. By a Λ-*tree* we understand a Λ-metric space (X,d) with $X \neq \emptyset$ such that

(a) (X,d) is geodesically linear (cf. (2.7)).

(b) If $x,y,z \in X$ then $[x,y] \cap [x,z] = [x,w]$ for some w. This w is clearly unique, and we write

$$w = Y(y,x,z)$$

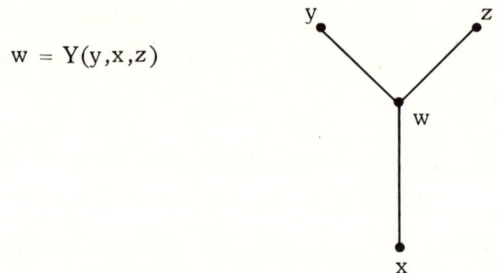

(c) If $x,y,z \in X$ and $[x,y] \cap [y,z] = \{y\}$ then $[x,y] \cup [y,z] = [x,z]$.

(2.10) REMARKS. 1. Axiom (b) is a kind of closure property. Indeed, $[x,y] \cap [x,z]$, being a convex subset containing x, of the interval $[x,y]$, must be a nested union of intervals $[x,u]$; axiom (b) requires the existence of a maximum such u. This axiom holds in examples (2.8), so (a) + (b) is far from making (X,d) "tree-like." This is accomplished by axiom (c).

2. If $x_0, x_1, \cdots, x_n \in X$ we shall write $[x_0, x_n] = [x_0, x_1, \cdots, x_n]$ if, for the metric morphism $\alpha : [0, d(x_0, x_n)] \to X$ such that $\alpha(0) = x_0$, $\alpha(d(x_0, x_n)) = x_n$, we have $x_i = \alpha(a_i)$, where $0 = a_0 \leq a_1 \leq \cdots \leq a_n = d(x_0, x_n)$ in Λ; i.e. if $x_0, x_1, \cdots, x_n$ lie in linear order along the segment $[x_0, x_n]$. With this notation, axiom (c) may be restated as follows:

$$[x,y] \cap [y,z] = \{y\} \implies [x,z] = [x,y,z].$$

3. A non-empty subset of X is a *subtree* iff it is convex. Thus an intersection of subtrees is again a subtree, or else empty. The subtree

spanned by a non-empty subset Y of X is the intersection of all subtrees containing Y. By a *closed subtree* we mean a nonempty convex closed subset of X (cf. (2.7)). For example closed segments are convex closed; see Corollary (2.16) below.

(2.11) THE Y-PROPOSITION. *Let* X *be a* Λ*-tree, let* x,y,z $\in$ X *, and put* w = Y(y,x,z).

(a) *We have* (i) $[y,w] \cap [w,z] = \{w\}$,

(ii) $[y,z] = [y,w,z]$,

and

(iii) $d(y,z) = d(y,x) + d(z,x) - 2d(w,x)$.

(b) w = Y(y,x,z) *depends only on* $\{x,y,z\}$*, not on the order in which the elements are listed.*

Proof. (a) $[y,w] \cap [w,z] \subset [y,x] \cap [z,x] = [w,x]$, so $[y,w] \cap [w,z] \subset [y,w] \cap [w,x] = \{w\}$, whence assertion (i) of (a). Assertion (ii) follows then from axiom (2.9)(c) (cf. Remark (2.10)2). The distance formula (iii) then follows since $d(y,x) + d(z,x) - 2d(w,x) = (d(y,x) - d(w,x)) + (d(z,x) - d(w,x)) = d(y,w) + d(w,z) = d(y,z)$.

(b) Clearly Y(y,x,z) = Y(z,x,y). Since a transposition and a 3-cycle generate the permutation group of $\{x,y,z\}$, it suffices to show that Y(y,x,z) = Y(x,z,y). We have

$$[z,y] \cap [z,x] = ([z,w] \cup [w,y]) \cap [z,x]$$
$$= [z,w] \cup ([w,y] \cap [z,x])$$
$$= [z,w] \cup ([w,y] \cap [z,w]) \cup ([w,y] \cap [w,x]).$$

The middle term is $\{w\}$ by part (a), and the third term is $\{w\}$ since w $\in$ [x,y]. Hence $[z,y] \cap [z,x] = [z,w]$, so, by definition, w = Y(x,z,y), as claimed.

(2.12) COROLLARY. *With notation as in the Y-Proposition, the following conditions are equivalent*

(a) $[y,x] \cap [x,z] = \{x\}$

(b) $x \in [y,z]$, i.e. $[y,z] = [y,x,z] = [y,x] \cup [x,z]$

(c) $Y(y,x,z) = x$

(d) $d(y,z) = d(y,x) + d(x,z)$.

Proof. (a) $\Rightarrow$ (b) by axiom (2.9) (c), (b) $\Rightarrow$ (c) is clear, (c) $\Leftrightarrow$ (d) by (a) (iii) above, and (c) $\Rightarrow$ (a) is clear.

(2.13) LEMMA. *Let $S_1, \cdots, S_n$ be subtrees of the Λ-tree X. (a) If $S_i \cap S_{i+1} \neq \emptyset$ ($i = 1, \cdots, n-1$) then $S_1 \cup \cdots \cup S_n$ is a subtree. (b) If $S_i \cap S_j \neq \emptyset$ ($1 \leq i < j \leq n$) then $S_1 \cap \cdots \cap S_n \neq \emptyset$.*

(a) Induction on n, the case $n = 1$ being trivial. To show $S_1 \cup S_2$ is a subtree it suffices to show that if $x_i \in S_i$ ($i = 1, 2$) then $[x_1, x_2] \subset S_1 \cup S_2$. Choose $y \in S_1 \cap S_2$. Then $[x_i, y] \subset S_i$ ($i = 1, 2$). Put $w = Y(x_1, y, x_2) \in [x_1, y] \cap [x_2, y] \subset S_1 \cap S_2$. Then, by the Y-Proposition (2.11), $[x_1, x_2] = [x_1, w] \cup [w, x_2] \subset S_1 \cup S_2$.

The general case follows by applying induction and the case $n = 2$ to $S_1 \cup \cdots \cup S_{n-1}$ and S_n.

(b) Induction on n, the cases $n \leq 2$ being trivial. If $n = 3$ choose $x_{ij} \in S_i \cap S_j$ ($1 \leq i < j \leq 3$) and put $y = Y(x_{12}, x_{13}, x_{23})$. Then $y \in [x_{12}, x_{13}] \subset S_1$, $y \in [x_{12}, x_{23}] \subset S_2$, and $y \in [x_{13}, x_{23}] \subset S_3$. If $n > 3$ put $S_3' = S_3 \cap \cdots \cap S_n$. Then $S_1 \cap S_2 \neq \emptyset$ by hypothesis and $S_1 \cap S_3' \neq \emptyset$ and $S_2 \cap S_3' \neq \emptyset$ by induction, whence $S_1 \cap S_2 \cap S_3' \neq \emptyset$.

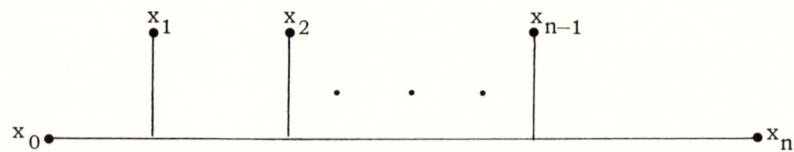

(2.14) PIECEWISE GEODESIC PROPOSITION. *Let X be a Λ-tree and $x_0, x_1, \cdots, x_n$ points of X.*

(a) (i) $[x_0, x_n] \subset \bigcup_{i=1}^{n} [x_{i-1}, x_i]$

and

(ii) $d(x_0, x_n) \leq \sum_{i=1}^{n} d(x_{i-1}, x_i)$.

(b) *Suppose that* $d(x_0, x_n) = \sum_{i=1}^{n} d(x_{i-1}, x_i)$. *Then*

(i) $[x_{i-1}, x_i] \cap [x_i, x_{i+1}] = \{x_i\}$ $(i = 1, \cdots, n-1)$

and

(ii) $[x_0, x_n] = \bigcup_{i=1}^{n} [x_{i-1}, x_i] = [x_0, x_1, \cdots, x_n]$.

(c) *Suppose that* $[x_{i-1}, x_i] \cap [x_i, x_{i+1}] = \{x_i\}$ $(i = 1, \cdots, n-1)$ *and* $x_i \neq x_{i+1}$ $(i = 1, \cdots, n-2)$. *Then*

$$d(x_0, x_n) = \sum_{i=1}^{n} d(x_{i-1}, x_i)$$

and (by (b)) $[x_0, x_n] = [x_0, x_1, \cdots, x_n]$.

REMARK. The necessity of the hypothesis "$x_i \neq x_{i+1}$" in (c) can be seen from the example x_0, x_1, x_2, x_3 with $x_0 = x_3 \neq x_1 = x_2$.

Proof. (a) (i) follows from Lemma (2.13) and (2.9) (a), while (a) (ii) is just the triangle inequality.

We prove (b) by induction on n, the case $n = 1$ being trivial. Suppose that $n \geq 2$. The case $n = 2$ follows from Corollary (2.12). In general we have

$d(x_0, x_n) \leq d(x_0, x_1) + d(x_1, x_n) \leq d(x_0, x_1) + \sum_{i=1}^{n} d(x_{i-1}, x_i) = d(x_0, x_n)$.

Hence $d(x_0, x_n) = d(x_0, x_1) + d(x_1, x_n)$ and $d(x_1, x_n) = \sum_{i=2}^{n} d(x_{i-1}, x_i)$. By induction we conclude that $[x_0, x_1] \cap [x_1, x_n] = \{x_1\}$, $[x_0, x_n] = [x_0, x_1, x_n]$, and $[x_{i-1}, x_i] \cap [x_i, x_{i+1}] = \{x_i\}$ $(i = 2, \cdots, n-1)$ and $[x_1, x_n] = [x_1, x_2, \cdots, x_n]$. This clearly implies (b).

We similarly prove (c) by induction on n, the case $n = 1$ being trivial, and the case $n = 2$ following from Corollary (2.12). Suppose that $n \geq 2$. By induction we have $d(x_1, x_n) = \sum_{i=2}^{n} d(x_{i-1}, x_i)$ and $[x_1, x_n] = [x_1, x_2, \cdots, x_n]$. Let $w = Y(x_0, x_1, x_n)$, so that $[x_1, w] = [x_0, x_1] \cap [x_1, x_n]$. Since, by hypothesis, $[x_0, x_1] \cap [x_1, x_2] = \{x_1\}$ we have $[x_1, w] \cap [x_1, x_2] = \{x_1\}$ inside the interval $[x_1, x_n] = [x_1, x_2, \cdots, x_n]$. By assumption, $x_1 \neq x_2$, so this is possible only if $w = x_1$. Thus $[x_0, x_1] \cap [x_1, x_n] = \{x_1\}$. The case $n = 2$ now implies that $d(x_0, x_n) = d(x_0, x_1) + d(x_1, x_n) = \sum_{i=1}^{n} d(x_{i-1}, x_i)$, as claimed.

The next proposition concerns four points x_0, x_1, x_2, x_3 in a Λ-tree X. (Cf. [12], Theorem 1.) For points $u, v \in X$ we shall (provisionally) write $u \leq v$ (resp., $u < v$) to mean that $u \in [x_0, v]$ (resp., that $u \in [x_0, v]$ and $u \neq v$).

(2.15) THE H-PROPOSITION. Let x_0, x_1, x_2, x_3 be points of a Λ-tree X, and put $y_{ij} = Y(x_i, x_0, x_j)$ ($= y_{ji}$). On the segment $[x_0, x_2]$ containing y_{12} and y_{23} we have one of the following cases:

(i) $y_{12} < y_{23}$;   (ii) $y_{23} < y_{12}$;   (iii) $y_{12} = y_{23}$

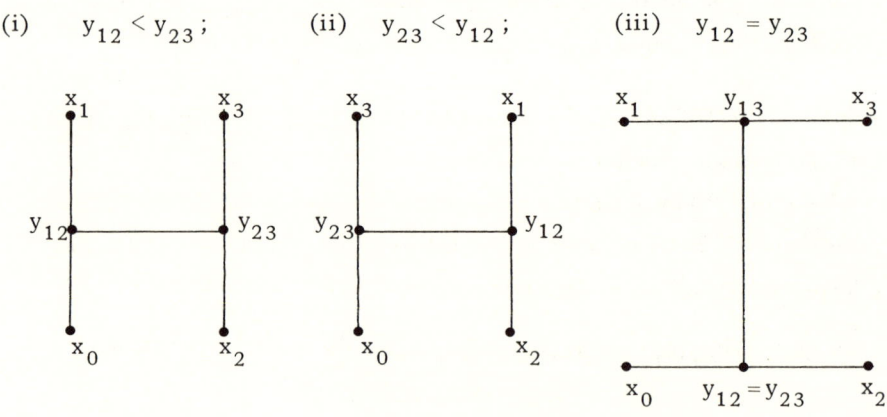

In case (i), $y_{13} = y_{12} < y_{23}$ and $[x_0, x_1] \cap [x_2, x_3] = \emptyset$
In case (ii), $y_{13} = y_{23} < y_{12}$ and $[x_0, x_1] \cap [x_2, x_3] = [y_{23}, y_{12}]$

In case (iii), $y_{12} = y_{23} \leq y_{13}$ *and* $[x_0,x_1] \cap [x_2,x_3] = [y_{12},y_{13}]$. *In all cases we have*

$$d(y_{13},x_0) \geq \min(d(y_{12},x_0), d(y_{23},x_0)).$$

Proof. Consider case (i): $y_{12} < y_{23}$. Then $[x_0,x_1] = [x_0,y_{12},x_1]$,

$[x_1,x_2] = [x_0,y_{12},y_{23},x_2]$

$[x_0,x_3] = [x_0,y_{12},y_{23},x_3]$

and

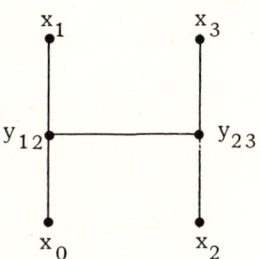

$$[y_{12},x_1] \cap [y_{12},y_{23}] \subset [y_{12},x_1] \cap [y_{12},x_2] = \{y_{12}\},$$

so, by the Piecewise Geodesic Proposition (2.14),

$$[x_1,x_3] = [x_1,y_{12},y_{23},x_3].$$

We have $y_{12} \in [x_0,x_1] \cap [x_0,x_3]$ so $y_{12} \leq y_{13}$. On the other hand $y_{12} < y_{23}$ so $y_{23} \notin [x_0,x_1]$, hence $y_{13} < y_{23}$. Hence $[x_0,y_{13}] \subset [x_0,x_1] \cap [x_0,y_{23}] \subset [x_0,x_1] \cap [x_0,x_2] = [x_0,y_{12}]$, so $y_{13} \leq y_{12}$, whence $y_{13} = y_{12}$. Further $[x_0,x_1[\cap [x_2,x_3] = [x_0,y_{12},x_1] \cap [x_2,y_{23},x_3]$ (Y-Proposition). Since $y_{12} < y_{23}$, $[x_0,y_{12}]$ is disjoint from $[y_{23},x_2]$ and $[y_{23},x_3]$, hence from $[x_2,x_3]$. On the other hand $[x_0,x_1]$ meets $[x_0,x_2] \cup [x_0,x_3]$, which contains $[x_2,x_3]$, in $[x_0,y_{12}] = [x_0,y_{13}]$, which, as we just showed, is disjoint from $[x_2,x_3]$. Hence $[x_0,x_1] \cap [x_2,x_3] = \emptyset$, as claimed.

Consider case (ii): $y_{23} < y_{12}$. As in case (i) we have $y_{13} = y_{23}$, $[x_0,x_1] = [x_0,y_{23},y_{12},x_1]$ and $[x_2,x_3] = [x_2,y_{12},x_{23},x_3]$. We have $[y_2,y_{12}] \subset [x_0,x_1] \cap [x_2,x_3] \subset [x_0,x_1] \cap ([x_0,x_2] \cup [x_0,x_3]) = [x_0,y_{12}] \cup [x_0,y_{13}] = [x_0,y_{23},y_{12}]$. Since $[x_0,y_{23}] \cap [x_2,x_3] =$

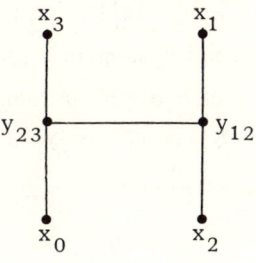

$[x_0,y_{23}] \cap ([y_{23},x_2] \cup [y_{23},x_3]) = \{y_{23}\}$, it follows, as claimed, that $[x_0,x_1] \cap [x_2,x_3] = [y_{23},y_{12}]$.

Consider case (iii): $y_{12} = y_{23}$.
We have $y_{12} = y_{23} \in [x_0,x_1] \cap [x_0,x_3]$ so $y_{12} = y_{23} \leq y_{13}$. We have

$$[x_0,x_1] = [x_0,y_{23},y_{13},x_1]$$

and

$$[x_2,x_3] = [x_2,y_{23},y_{13},x_3]$$

so

$$[y_{23},y_{13}] \subset [x_0,x_1] \cap [x_2,x_3]$$
$$\subset [x_0,x_1] \cap ([x_0,x_2] \cup [x_0,x_3])$$
$$= [x_0,y_{12}] \cup [x_0,y_{13}] = [x_0,y_{12},y_{13}] = [x_0,y_{23},y_{13}].$$

Further

$$[x_0,y_{23}] \cap [x_2,x_3] = [x_0,y_{23}] \cap ([y_{23},x_2] \cup [y_{23},x_3])$$
$$= \{y_{23}\}.$$

Hence $[x_0,x_1] \cap [x_2,x_3] = [y_{23},y_{13}]$ as claimed.

The final assertion of the proposition follows from the description of y_{13} in the three cases.

(2.16) COROLLARY. *Let* x_0,x_1,x_2,x_3 *be points of a Λ-tree* X.

(a) $[x_0,x_1] \cap [x_2,x_3]$ *is either empty or a closed segment. I.e. a closed segment is a closed subtree of* X. *Hence a non-empty intersection of two closed subtrees is a closed subtree.*

(b) *If* $d(x_0,x_2) + d(x_1,x_3) < d(x_0,x_1)$ *then* $[x_0,x_1] \cap [x_2,x_3] = [y_{12},y_{13}]$, *and*

$$d(y_{12},y_{13}) \geq d(x_0,x_1) - (d(x_0,x_2)+d(x_1,x_3)) > 0.$$

Assertion (a) is immediate from (2.15). In case (b) the assumption implies that $[x_0,x_1]$ cannot be contained in $[x_0,x_2] \cup [x_1,x_3]$. Inspection shows that we must be in case (iii) of (2.15), whence $y_{12} = y_{23} \leq y_{13}$, $[x_0,x_1] \cap [x_2,x_3] = [y_{12},y_{13}]$ and $[x_0,x_1] = [x_0,y_{12},y_{13},x_1]$, so $d(y_{12},y_{13}) = d(x_0,x_1) - (d(x_0,y_{12})+d(y_{13},x_1))$. We have $d(y_{13},x_1) \leq d(x_3,x_1)$ and $d(x_0,y_{12}) \leq d(x_0,x_2)$, whence the final inequalities.

(2.17) THE BRIDGE PROPOSITION. *Let* X_0, X_1 *be disjoint subtrees of a* Λ-*tree* X.

(a) *If* $x_0,x_2 \in X_0$ *and* $x_1,x_3 \in X_1$ *then* $[x_0,x_1] \cap [x_2,x_3] = [y_0,y_1]$, *a closed segment with* $y_i \in X_i$ $(i = 0,1)$.

(b) *Suppose that* X_0 *is a closed subtree of* X (*cf.* (2.7)). *Then there is a unique* $z_0 \in X_0$, *denoted* $\mathrm{pr}_{X_0}(X_1)$, *such that* $\forall (x_0,x_1) \in X_0 \times X_1$ *we have* $z_0 \in [x_0,x_1]$; *furthermore* $[x_0,z_0] = [x_0,x_1] \cap X_0$.

(c) *Suppose further that* X_1 *is a closed subtree of* X, *and put* $z_1 = \mathrm{pr}_{X_1}(X_0)$. *Then* $[z_0,z_1] \subset [x_0,x_1]$ *for all* $(x_0,x_1) \in X_0 \times X_1$. *We denote* $[z_0,z_1]$ *by* $[X_0,X_1]$, *and call it the "bridge" between* X_0 *and* X_1, *and put* $d(X_0,X_1) = d(z_0,z_1)$.
We have $[z_0,z_1] \cap X_i = \{z_i\}$ $(i = 0,1)$.

REMARK. If $X_0 \cap X_1 \neq \emptyset$ then $X_0 \cap X_1$ is a subtree of X, and we put $d(X_0,X_1) = 0$. If $X_0 = \{x_0\}$ and $x_0 \in X_1$ we define the bridge from x_0 to X_1 to be $[x_0,x_0]$.

Proof. (a) With the notation $y_{ij} = Y(x_i,x_0,x_j)$ of (2.15) we have $y_{12}, y_{23} \in [x_0,x_2] \subset X_0$, while $y_{13} \in [x_1,x_3] \subset X_1$. Since, by (2.15), two of y_{12}, y_{23}, y_{13} are equal, we must have $y_{12} = y_{23}$, case (iii) of (2.15), and so $[x_0,x_1] \cap [x_2,x_3] = [y_0,y_1]$ where $y_0 = y_{12} = y_{23}$ and $y_1 = y_{13}$.

(b) Assume that X_0 is a closed subtree. Let $x_i \in x_i$ $(i = 0,1)$. By definition, $[x_0,x_1] \cap X_0$ is a closed segment, say $[x_0,z_0]$. Suppose that $x_2 \in X_0$ and $x_3 \in X_1$, and put $[x_0,x_1] \cap [x_2,x_3] = [y_0,y_1]$, as in (a). We claim that $z_0 \in [x_2,x_3]$. We have $[x_0,y_0] \subset [x_0,x_1] \cap X_0 = [x_0,z_0]$. On the other hand $[y_1,x_1] \subset X_1$ is disjoint from X_0, so $[x_0,z_0] = [x_0,x_1] \cap X_0 = [x_0,y_1] \cap X_0$, so we have $[x_0,y_0] \subset [x_0,z_0] \subset [x_0,y_1]$, whence $z_0 \in [y_0,y_1] \subset [x_2,x_3]$. The uniqueness of z_0 follows since $[z_0,x_1] \cap X_0 = \{z_0\}$. This proves (b), and (c) is an immediate consequence of (b).

(2.18) DEFINITION. Call a Λ-tree X *linear* if X is isomorphic to a subtree of Λ. Call a collection of points of a Λ-tree X *collinear* if they all belong to a linear subtree.

(2.19) REMARK. In view of Corollary (2.6), every automorphism of a linear tree is equivalent to either a translation $x \mapsto c+x$, or one of the form $x \mapsto c-x$, which is a reflection in the fixed point $c/2$ if $c/2 \in \Lambda$, or an inversion if not.

(2.20) PROPOSITION. *A tree X is linear iff every three of its points are collinear. More generally, if X is spanned by a set S and any three points of S are collinear, then X is linear.*

Proof. Choose $x_0 \in S$. For $x,y \in S - \{x_0\}$ write $x \sim y$ if $x_0 \notin [x,y]$. This is clearly an equivalence relation. Let $x,y,z \in S - \{x_0\}$ be distinct. They span a segment, being collinear say with endpoints x and z. Then $[x,y] \cap [y,z] = \{y\}$ and $x_0 \ne y$ so $x_0 \notin [x,y]$ or $x_0 \notin [y,z]$, i.e. either $x \sim y$ or $y \sim z$. Thus there are at most two equivalence classes. If $S = \{x_0\}$ there is nothing to prove. Otherwise let S_+ be one equivalence class of $S - \{x_0\}$. If there is another, denote it by S_-; otherwise put $S_- = \emptyset$. Then S is the disjoint union of S_+, $\{x_0\}$, and S_-. For $x \in S$ define $\bar{x} \in \Lambda$ by

$$\bar{x} = \begin{cases} d(x,x_0) & \text{if } x \in S_+ \cup \{x_0\} \\ -d(x,x_0) & \text{if } x \in S_- . \end{cases}$$

Let $a_x : [0,\bar{x}] \to X$ be the metric morphism such that $a_x(0) = x_0$ and $a_x(\bar{x}) = x$. If $x,y \in S$ then a_x and a_y agree on $[0,\bar{x}] \cap [0,\bar{y}]$. For example, if $0 \leq \bar{x} \leq \bar{y}$ then $x,y \in S_+ \cup [x_0]$ and $x \in [x_0,y]$, whence the claim. Similarly if $\bar{x} \leq \bar{y} \leq 0$. Suppose that $\bar{x} < 0 \leq \bar{y}$. Then $[x,y] = [x,x_0,y]$ and a_x and a_y clearly combine to define a metric morphism from $[\bar{x},\bar{y}] = [\bar{x},0] \cup [0,\bar{y}]$ to $[x,y]$. Thus the a_x ($x \in S$) combine to define a metric morphism from $A = \cup_{x \in S} [0,\bar{x}] \subset \Lambda$ to X with image the subtree spanned by S. This proves the proposition.

(2.21) DEFINITIONS. Let X be a Λ-tree. We call $e \in X$ an *end point* of X if, whenever $e \in [x,y] \subset X$ either $e = x$ or $e = y$. Let $x \in X$. A linear subtree L of X having x as an end point will be called a *linear subtree from* x. Such an L carries a natural linear ordering with x as least element (induced by a suitable embedding of L in Λ). If $y \in L$ then $L_y = \{z \in L | y \leq z\}$ is a linear subtree from y, $L = [x,y] \cup L_y$, and $[x,y] \cap L_y = \{y\}$. Conversely, if L' is a linear subtree from y and if $[x,y] \cap L' = \{y\}$ then $[x,y] \cup L'$ is a linear subtree from x. A maximal linear subtree L from x in X is called an *X-ray from* x. If $y \in L$ then L_y is also an X-ray (from y), unless $L = [x,y]$.

(2.22) PROPOSITION. (a) *Let L be a linear subtree from x in X and $y,z \in X$ such that $L \cap [y,z] \neq \emptyset$. If $L \cap [y,z]$ is bounded above in L then $L \cap [y,z]$ is a closed segment. Otherwise $L \cap [y,z] = L_v$ for some $v \in L$. The latter case cannot occur if L is an X-ray; thus X-rays are closed subtrees of X.*

(b) *Let L, L' be linear subtrees from x, x', respectively, such that $L \cap L' \neq \emptyset$. Suppose that L and L' are closed subtrees of X, e.g. X-rays. Then $L \cap L'$ is either a closed segment or $L \cap L' = L_v = L'_v$ for some v.*

(c) *The relation, "$L \cap L' = L_v = L'_v$ for some v," is an equivalence relation on the set of X-rays.*

(a) Choose $u \in L \cap [y,z]$. If $L \cap [y,z]$ is bounded above in L by $w \in L$ then $L \cap [y,z] = [x,w] \cap [y,z]$ is a closed segment by (2.16) (a). Otherwise $L \cap [y,z]$ contains L_u so $L_u \subset [u,y]$ or $L_u \subset [u,z]$, say the latter. We have $L \cap [y,z] = ([x,u] \cap [y,z]) \cup L_u = [v,u] \cup L_u = L_v$ for some $v \in [x,u]$, by (2.16) (a) again. Since $L_u \subset [u,z]$ and L_u is not bounded above in L in the case at hand, $L_u \neq [u,z]$ so L_u is not an X-ray, hence L is not an X-ray; whence (a).

(b) If $L \cap L'$ is bounded above in L' by $y' \in L'$ then $L \cap L' = L \cap [x',y']$ is a closed segment since L is a closed subtree. Suppose now that $L \cap L'$ is unbounded above in both L and L'. Choose $u \in L \cap L'$. Then $L_u \subset L \cap L'$ so $L \cap L' = ([x,u] \cap L') \cup L_u$. Since L' is a closed subtree, $[x,u] \cap L' = [v,u]$ for some $v \in [x,u] \cap L'$. Thus $L \cap L' = [v,u] \cup L_u = L_v$. Similarly $L \cap L' = L'_{v'}$ for some $v' \in L'$. Since $L \cap L'$ is unbounded above in L (and L') in the case at hand it cannot have two end points, so $v = v'$.

(c) Only transitivity is not immediate. Let L, L', L'' be X-rays from x, x', x'', respectively such that $L \cap L' = L_v = L'_v$ and $L' \cap L'' = L'_{v'} = L''_{v'}$ for some v, v'.

Case 1. $L = [x,e]$, where e is an end point of X and $x \neq e$. Then $L_v = [v,e] = L'_v$ so clearly $L' = [x',e]$, $x' \neq e$. Further $v \neq e$; otherwise $e \in [x,x']$, contradicting the fact that e is an end point. Similarly $L'' = [x'',e]$ and $L'_{v'} = L''_{v'} = [v',e]$ with $v' \neq e$. Put $w = \max(v,v')$ in L'. Then $L \cap L'' \supset L \cap L' \cap L'' = [v,e] \cap [v',e] = [w,e]$. It follows easily that $L \cap L'' = [u,e] = L_u = L''_u$ for $u = Y(x,x'',w)$.

Case 2. L is not a closed segment; hence, by case 1, neither is L' or L''. Put $w = \max(v,v')$ in L'. Then $L \cap L'' \supset L \cap L' \cap L'' = L'_v \cap L'_{v'} = L'_w = L_w = L''_w$. It follows now from (b) that $L \cap L'' = L_u = L''_u$ for some u.

(2.23) DEFINITIONS. The equivalence classes of X-rays for the relation, "$L \cap L' = L_v = L'_v$ for some v," will be called *ends* of X; we write Ends(X) for the set of them. If an X-ray L represents an end ε we say that "L ends at ε." We say that ε *belongs to or is contained in a subtree* Y of X if ε is represented by an X-ray $L \subset Y$. (Note that Y-rays need not be X-rays.)

Each end point e of X (cf. (2.21)) defines and end ε_e whose X-rays are the closed segments [x,e], $x \in X$, $x \neq e$. We call ε_e a *closed end* of X. For all $x \in X$, including x = e, we write $[x, \varepsilon_e)$ for [x,e]. An *open end* ε is one which is not closed. The notation $[x, \varepsilon)$ in this case is established in the next proposition.

(2.24) PROPOSITION. (a) *If $x \in X$ and ε is an open end of X there is a unique X-ray from x ending at ε; we denote it $[x, \varepsilon)$.*

(b) *Let ε, ε' be distinct ends of X. There is an $x \in X$ such that $[x, \varepsilon) \cap [x, \varepsilon') = \{x\}$. The tree $[x, \varepsilon) \cup [x, \varepsilon')$ is linear and independent of x. We denote it $(\varepsilon, \varepsilon')$, and call it the ε-ε'-axis.*

(c) *Let Y be a subtree of X. If an end ε of X belongs to Y then $[y, \varepsilon) \subset Y$ for all $y \in Y$. If $\varepsilon, \varepsilon'$ are distinct ends of X belonging to Y then $(\varepsilon, \varepsilon') \subset Y$. Let Y be a closed subtree of X.*

Let Y be a closed subtree of X.

(d) *An open end of Y is also an open end of X.*

(e) *Suppose, on the other hand, that ε is an end of X, and let $y \in Y$. Either ε belongs to Y, and $[y, \varepsilon) \subset Y$, or else $[y, \varepsilon) \cap Y = [y, y_\varepsilon]$ for a $y_\varepsilon \in Y$ which is independent of y. (We call $[y_\varepsilon, \varepsilon)$ the "bridge from Y to ε.")*

Proof. (a) Let ε be represented by a ray L. Let [x,y] be the bridge from x to L (which exists because L is closed, by (2.22) (a) above). Then clearly $L' = [x,y] \cup L_y$ is an X-ray from x ending at ε. If L" is another then $L' \cap L''$ is an X-ray contained in L' and containing x so $L' \cap L'' = L'$. Similarly $L' \cap L'' = L''$, whence (a).

(b) Choose $y \in X$. Then $[y, \varepsilon) \cap [y, \varepsilon') = [y,x]$ for some x, by (2.22) (b). Clearly then $[x, \varepsilon) \cap [x, \varepsilon') = x$. It follows easily from (2.20) that $[x, \varepsilon) \cup [x, \varepsilon')$ is linear; clearly it contains the ends ε and ε'. To prove uniqueness and the last part of (c) at the same time we shall show that if Y is a subtree containing ε and ε' then Y contains $[x, \varepsilon) \cup [x, \varepsilon')$. It follows from the definition of ends that Y contains $[y, \varepsilon)$ and $[y', \varepsilon')$ for some $y \in [x, \varepsilon)$ and $y' \in [x, \varepsilon')$. But then Y contains $[y, y'] = [y, x, y']$ and hence also $[y, \varepsilon) \cup [y, y'] \cup [y', \varepsilon') = [x, \varepsilon) \cup [x, \varepsilon')$. To prove the first part of (c), suppose that $y \in Y$ and $[y',\varepsilon) \subset Y$ for some $y' \in Y$. We have $[y,\varepsilon) \cap [y',\varepsilon) = [u,\varepsilon)$ for some u. Clearly $u \in Y$ so $[y,\varepsilon) = [y,u] \cup [u,\varepsilon) \subset Y$.

(d) Let ε be an open end of Y and $y \in Y$. We must show that $[y,\varepsilon)$ (in Y) is an X-ray. If not then $[y,\varepsilon) \subset [y,z]$ for some $z \in X$. But Y is closed so $[y,z] \cap Y = [y,w]$ for some $w \in Y$, and the inclusion $[y,\varepsilon) \subset [y,w] \subset Y$ contradicts the fact that $[y,\varepsilon)$ is a Y-ray.

(e) Let ε be an end of X not belonging to Y, and let $y \in Y$. Then there is a $z \in [y,\varepsilon)$, $z \notin Y$. Clearly $[y,\varepsilon) \cap Y = [y,z] \cap Y = [y,w]$ for some w (Y is closed). Similarly, if $y' \in Y$ then $[y',\varepsilon) \cap Y = [y',w']$ for some $w' \in Y$. It remains to show that $w = w'$. Choose $z \in [y,\varepsilon) \cap [y',\varepsilon)$, and put $u = Y(y,z,y')$. Then $[y,y'] = [y,u,y']$ so $u \in Y$, and $[y,z] \cap [y',z] = [u,z]$ so $u \in [y,\varepsilon) \cap [y',\varepsilon)$. Now clearly $[u,\varepsilon) \cap Y = [u,w] = [u,w']$ so $w = w'$.

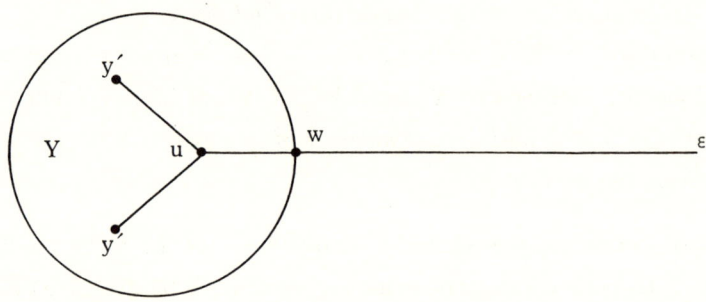

(2.25) DEFINITION. Let X be a Λ-tree. By a *cut* of X we mean a pair (X_0,X_1) of subtrees of X such that $X = X_0 \amalg X_1$ (disjoint union). (This is related to the notion of an "edge" of X, introduced by Morgan and Otal in [MO], Ch. II.)

(2.26) PROPOSITION. *Let* $X = X_0 \amalg X_1$ *be a cut of* X. (a) *If* X_0 *and* X_1 *are closed subtrees let* $[e_0,e_1]$ *be the bridge from* X_0 *to* X_1 *(cf. (2.17)). Then* $d(e_0,e_1)$ *is the least element* > 0 *in* Λ.

(b) *Suppose that X_0 is not a closed subtree of X. Then there is a unique (open) end ε_0 of X_0 such that for all $x_i \in X_i$ $(i = 0,1)$, $[x_0, x_1] \cap X_0 = [x_0, \varepsilon_0)$.*

(a) Since $[e_0, e_1] \cap X_i = \{e_i\}$ we have $[e_0, e_1] = \{e_0, e_1\}$ so $[0, d(e_0, e_1)]$ in Λ has only two points.

(b) Since X_0 is not closed we can find a closed segment $[x_0, x_1]$ $(x_i \in X_i)$ such that $L = [x_0, x_1] \cap X_0$ is not a closed segment. Clearly L is a linear subtree from x_0 in X_0; we claim that it is a maximal one. Otherwise $L \subset [x_0, y]$ for some $y \in X_0$. Put $z = Y(x_0, y, x_1)$. Then $[x_0, z] = [x_0, x_1] \cap [x_0, y] = ([x_0, x_1] \cap X_0) \cap [x_0, y] = L \cap [x_0, y] = L$; contradiction. Thus L is an X_0-ray ending at an open end ε_0 of X_0. Let $x'_i \in X_i$ $(i = 0,1)$ and $L' = [x'_0, x'_1] \cap X_0$. By (2.17) (a) $[x_0, x_1] \cap [x'_0, x'_1] = [y_0, y_1]$ for some $y_i \in X_i$ $(i = 0,1)$. Then $L \cap L' = ([x_0, x_1] \cap X_0) \cap ([x'_0, x'_1] \cap X_0) = [y_0, y_1] \cap X_0 = L_{y_0} = L'_{y_0}$. Thus L' is an X_0-ray ending at ε_0 for all $x'_i \in X_i$ $(i = 0,1): L' = [x'_0, \varepsilon_0)$.

(2.27) TERMINOLOGY. If neither X_0 nor X_1 is a closed subtree we call (X_0, X_1) an *open cut* of X. In this case we have open ends ε_i of X_i $(i = 0,1)$ and we say that $X = X_0 \amalg X_1$ "makes the ends ε_0 and ε_1 meet." If $x_0 \in X_0$ and $x_1 \in X_1$ then we have $[x_0, x_1] = [x_0, \varepsilon_0) \amalg [x_1, \varepsilon_1)$.

(2.28) THEOREM. *Let $(A_s)_{s \in \Gamma}$ be a family of subtrees of X such that $A_s \cap A_t \neq \emptyset$ for all $s, t \in \Gamma$*

(a) *Any finite number of the $A_s (s \in \Gamma)$ have nonempty intersection.*

Suppose now that $\cap_{s \in \Gamma} A_s = \emptyset$.

(b) *There is at most one end of X belonging to all of the $A_s (s \in \Gamma)$; any such end ε must be open, and $[x, \varepsilon) \subset A_s$ whenever $x \in A_s$.*

(c) *Suppose that there is no end of X common to all of the A_s, and that each A_s is a closed subtree of X. There is a unique cut $X = X_0 \amalg X_1$ of X (cf. (2.25)) such that $A_s \cap X_i \neq \emptyset$ $(i = 0,1)$ for all $s \in \Gamma$. The cut is open (cf. (2.27)) and makes open ends ε_i of X_i $(i = 0,1)$ meet. For all $s \in \Gamma$, and all $x \in A_s \cap X_i$, $A_s \cap X_i$ contains $[x, \varepsilon_i)$.*

If $\Lambda = Z$ or R then case (c) cannot occur.

(a) follows from (2.13) (b).

(b) If $\varepsilon, \varepsilon'$ are distinct ends of X belonging to every A_s then, by (2.24) (c), $(\varepsilon, \varepsilon') \subset A_s$ for all s, contradicting the assumption that $\cap_s A_s = \emptyset$. Similarly if each A_s contains a closed end ε_e (cf. (2.23)) then we obtain the contradiction $e \in \cap_s A_s$.

(c) Given $x \in X$, define $x(s) \in A_s$ so that $[x, x(s)]$ is the bridge from x to A_s. If Y is a closed subtree of A_s and if $[x,y]$ is the bridge from x to Y then one sees easily that $[x,y] = [x, x(s), y]$. Let F be a finite subset of Γ and put $Y = \cap_{s \in F} A_s$ ($\neq \emptyset$, by (a)). Then, as above, the bridge $[x,y]$ contains all $x(s)$ ($s \in F$). It follows that any three points of $\{x\} \cup \{x(s) | s \in \Gamma\}$ are collinear so, by (2.20), the tree they span,

$$L_x = \cup_{s \in \Gamma} [x, x(s)]$$

is a linear tree, evidently with x as an end point. If $z \in L_x$ and $x(s) \leq z$ then $z \in [x(s), x(t)]$ for some $t \in \Gamma$. Take $Y = A_s \cap A_t$ and $[x,y]$ the bridge from x to Y. Then, as we saw above, $[x(s), x(t)] \subset [x,y]$ and so $[x(s), x(t)] \subset [x(s), y] \subset A_s$. This shows that

(1) $$L_x \cap A_s = L_{x(s)} = \{z \in L_x | x(s) \leq z\}.$$

Let $w \in L_x$. If $w \geq x(s)$ then, by (1), $w \in A_s$ so $w = w(s)$. If $w \leq x(s)$, i.e. $w \in [x, x(s)]$ then $[w, x(s)]$ is the bridge from w to A_s, i.e. $w(s) = x(s)$. Thus $L_w = \cup_s [w, w(s)] = \cup_{w \notin A_s} [w, w(s)] = \cup_{w \notin A_s} [w, x(s)] = \{z \in L_x | w \leq z\}$, i.e.

(2) $$\text{if } w \in L_x \text{ then } L_w = \{z \in L_x | w \leq z\}.$$

Let $x_0, x_1 \in X$. Suppose first that $L_{x_0} \cap L_{x_1} \neq \emptyset$. Choose $y \in L_{x_0} \cap L_{x_1}$. Then $L_y \subset L_{x_0} \cap L_{x_1}$, by (2). Moreover $L_{x_0} \cap L_{x_1} = ([x_0, y] \cap [x_1, y]) \cup L_y = [z, y] \cup L_y = L_z$ for some $z (= Y(x_0, y, x_1))$.

(3) If $L_{x_0} \cap L_{x_1} \neq \emptyset$ then $L_{x_0} \cap L_{x_1} = L_y$ for some $y \in [x_0, x_1]$.

Next suppose that $L_{x_0} \cap L_{x_1} = \emptyset$. If $s \in \Gamma$ then $[x_0(s), x_1(s)] \subset A_s$ with $x_0(s) \neq x_1(s)$, while $[x_i, x_i(s)] \cap A_s = \{x_i(s)\}$ $(i = 0,1)$, whence $[x_0, x_1] = [x_0, x_0(s), x_1(s), x_1]$ by (2.14) (c). Thus $[x_0, x_1]$ contains $[x_i, x_i(s)]$ $(i = 0,1)$ for all $s \in \Gamma$, whence $[x_0, x_1] \supset L_{x_0} \cup L_{x_1}$. If $y \in [x_0, x_1]$ and $y \notin L_{x_0} \cup L_{x_1}$, then we must have $y \in [x_0(s), x_1(s)] \subset A_s$ for all $s \in \Gamma$, contradicting $\cap_s A_s = \emptyset$. Thus

(4) If $L_{x_0} \cap L_{x_1} = \emptyset$ then $[x_0, x_1] = L_{x_0} \cup L_{x_1}$.

Suppose that $x_0, x_1, x_2 \in X$ and $L_{x_i} \cap L_{x_j} = \emptyset$ for $i \neq j$. Put $y = Y(x_0, x_1, x_2)$. Then $y \in [x_i, x_j] = L_{x_i} \cup L_{x_j}$ for $i \neq j$. But $(L_{x_0} \cup L_{x_1}) \cap (L_{x_0} \cup L_{x_2}) \cap (L_{x_1} \cup L_{x_2}) = \emptyset$; contradiction. Thus:

(5) If $x_0, x_1, x_2 \in X$ then $L_{x_i} \cap L_{x_j} \neq \emptyset$ for some $i \neq j$.

Now fix an $x_0 \in X$ and put

$$X_0 = \{x \in X | L_x \cap L_{x_0} \neq \emptyset\}.$$

Let $x, y \in X_0$. Clearly $L_x \cup L_y \subset X_0$. By (3) we have $L_x \cap L_{x_0} = L_{x'}$ and $L_y \cap L_{x_0} = L_{y'}$ for some x', y', hence $L_x \cap L_y \neq \emptyset$, so X_0 is a subtree of X, and further $L_x \cap L_y = L_z$ for some z.

We claim that L_x is a maximal linear subtree of X_0 from x, hence an X_0-ray. Otherwise $L_x \subset [x, w]$ for some $w \in X_0$. We saw above that $L_x \cap L_w = L_u$ for some u. Choose $s \in \Gamma$ so that $u \notin A_s$. Then $x(s) = w(s) = u(s)$ does not belong to $[x, u] \cup [u, w] = [x, w]$, contradicting our assumption that $L_x \subset [x, w]$. Thus L_x is, as claimed, an X_0-ray, and the same argument shows that it is not a closed segment. Hence L_x

ends at an open end ε_0 of X_0 which is clearly independent of $x \in X_0$. Evidently ε_0 is contained in $A_s \cap X_0$ for all $s \in \Gamma$. In fact $L_x = [x, \varepsilon_0) \subset A_s \cap X_0$ for all $x \in A_s \cap X_0$.

If there is no end of X common to all A_s then $X_0 \neq X$. Choose $x_1 \in X$, $x_1 \notin X_0$. Put

$$X_1 = \{x \in X | L_x \cap L_{x_1} \neq \emptyset\}.$$

Then, as above, X_1 is a subtree of X, and the L_x ($x \in X_1$) are X_1-rays ending at the same open end ε_1 of X_1, which belongs to $A_s \cap X_1$ for all $s \in \Gamma$. Since $L_{x_0} \cap L_{x_1} = \emptyset$ it follows that $X_0 \cap X_1 = \emptyset$. Finally it follows from (5) that $X_0 \cup X_1 = X$. Further, $[x_0, x_1] \cap X_i = [x_i, \varepsilon_i)$ ($i = 0,1$). Thus $X = X_0 \amalg X_1$ is an open cut of X, making the ends ε_0 and ε_1 meet.

Suppose that $X = X_0' \amalg X_1'$ is another cut of X such that $A_s \cap X_i' \neq \emptyset$ ($i = 0,1$) for all $s \in \Gamma$. Let $x \in X_i'$. Since $A_s \cap X_i' \neq \emptyset$ it must contain the bridge $[x, x(s)]$ from x to A_s, so $L_x \subset X_i'$. If therefore $x_0 \in X_0'$ then it follows that $X_0 \subset X_0'$. If $x_1 \in X_0'$ also then $X = X_0 \cup X_1 \subset X_0'$; contradiction. Thus $x_1 \in X_1'$, hence $X_1 \subset X_1'$, and so $X_i = X_i'$ ($i = 0,1$). This proves (c).

Suppose we have case (c). Choose $x_i \in X_i$ ($i = 0,1$). Then we see that $[x_0, x_1] = L_{x_0} \amalg L_{x_1}$, which is an open cut of $[x_0, x_1]$. That this cannot occur when $\Lambda = Z$ or R follows from the next lemma.

(2.29) LEMMA. *The following conditions on the ordered group Λ are equivalent.*

(a) *No closed segment $[x_0, x_1] \subset \Lambda$ admits an open cut.*

(b) *Λ is isomorphic, as ordered group, to $\{0\}$, Z, or R.*

(b) $\Longrightarrow$ (a) is clear.

(a) $\Longrightarrow$ (b). If $0 \neq \Lambda_0 \subsetneq \Lambda$ is a convex subgroup choose $x_0 \notin \Lambda_0$, $x_0 > 0$. In $[-x_0, x_0]$ let X_1 denote the set of upper bounds for Λ_0, and X_0 the complement. Then $[-x_0, x_0] = X_0 \amalg X_1$ is an open cut. Thus, by (a), no such Λ_0 exists, i.e. Λ is archimedean, hence embeddable in $\mathbf{R}$. If Λ is not cyclic then Λ is dense in $\mathbf{R}$. If $y \in \mathbf{R}$, $y \notin \Lambda$ choose $x_0, x_1 \in \Lambda$ so that $x_0 < y < x_1$. Then $[x_0, x_1]_\Lambda = [x_0, y]_\Lambda \amalg [y, x_1]_\Lambda$ is an open cut of $[x_0, x_1]_\Lambda$; contradiction. Thus Λ, if not cyclic, equals $\mathbf{R}$.

(2.30) LEMMA. *Let* L *be a linear* Λ-*tree.*

(a) *Given a subset* A *of* L *with at least two elements, any* Λ-*metric morphism* $\alpha_0 : A \to \Lambda$ *extends uniquely to a* Λ-*metric morphism* $\alpha : L \to \Lambda$.

(b) *Suppose that* $L = L_0 \cup L_1$ *where* L_0, L_1 *are subtrees such that* Card $(L_0 \cap L_1) \geq 2$. *If* $\alpha : L \to \Lambda$ *and* $\alpha | L_i$ *is* Λ-*metric for* $i = 0, 1$ *then* α *is* Λ-*metric.*

We can assume that $L \subset \Lambda$.

(a) By (2.5) (b) $\alpha_0 = \beta | A$ for a unique $\beta \in \text{Aut}(\Lambda)$ ($= \text{Aut}_{\text{metric}}(\Lambda)$). Then $\alpha = \beta | L : L \to \Lambda$ extends α_0, and uniqueness follows likewise from (2.5) (b).

(b) Modifying α by an isometry of Λ we can assume that α fixes two points of $L_0 \cap L_1$. Then $\alpha(x) = x$ for all $x \in L_i$, by part (a), for $i = 0, 1$.

(2.31) PROPOSITION. *Let* X *be a* Λ-*tree and* B *a subset with at least two elements. There is a map* $\alpha : X \to \Lambda$ *that is* Λ-*metric on every linear subtree containing* B. *On the union of such linear subtrees,* α *is uniquely determined modulo* $\text{Aut}(\Lambda)$ ($= \text{Aut}_{\text{metric}}(\Lambda)$), *and hence is uniquely determined by its values at any pair of distinct points of* B.

If B is contained in no linear subtree of X, any map α will do (e.g. a constant map). Suppose the contrary. Then there is a Λ-metric embedding $\alpha_B : B \to \Lambda$. If L is a linear subtree containing B there is a unique Λ-metric morphism $\alpha_L : L \to \Lambda$ extending α_B, by (2.30) (a). Let L' be another linear subtree containing B. Then $\alpha_{L'} | L \cap L'$ and $\alpha_L | L \cap L'$ agree on B, hence coincide by (2.30) (a). Thus the α_L's assemble to define a map α' from the union X' of the linear subtrees containing B into Λ; by (2.5) (b) α' is unique modulo Aut(Λ). Any extension of α' to all of X will do now.

One reasonable choice is to first project X onto X', using (2.17) (b) after verifying that X' is closed, and then applying α'.

(2.32) COROLLARY. *Let s be an automorphism of X such that $sB = B$. Let α be as in (2.31). Then there is a unique $\phi \in \text{Aut}(\Lambda)$ such that $\alpha(sx) = \phi(\alpha(x))$ for all x in the union of the linear subtrees containing B.*

Since $\beta(x) = \alpha(sx)$ has the same properties as α, $\beta = \phi \circ \alpha$ on the union of the linear subtrees containing B for a unique $\phi \in \text{Aut}(\Lambda)$, by the uniqueness part of (2.31).

(2.33) PROPOSITION. *Let X be a Λ-tree and ε an open end of X. There is a map $\alpha : X \to \Lambda$ whose restriction to each X-ray $[x, \varepsilon)$ is Λ-metric. Moreover α is unique modulo Aut(Λ), hence determined by its values at any pair of distinct points of any X-ray $[x, \varepsilon)$.*

Fix $x_0 \in X$ and a Λ-metric morphism $\alpha_0 : [x_0, \varepsilon) \to \Lambda$. If $x \in X$ then $[x, \varepsilon) \cap [x_0, \varepsilon) = [x', \varepsilon)$ for some x' ((2.22) (b)), and, by (2.30), there is a unique Λ-metric $\alpha_x : [x, \varepsilon) \to \Lambda$ agreeing with α_0 on $[x', \varepsilon)$. If $x, y \in X$ then α_x and α_y agree on $[x, \varepsilon) \cap [y, \varepsilon) \cap [x_0, \varepsilon)$, and hence agree on $[x, \varepsilon) \cap [y, \varepsilon)$, by (2.30) again. Thus the α_x assemble to define $\alpha : X \to \Lambda$ with the required property. Since α was uniquely determined by α_0, and α_0 is unique modulo Aut(Λ), the same is true of α.

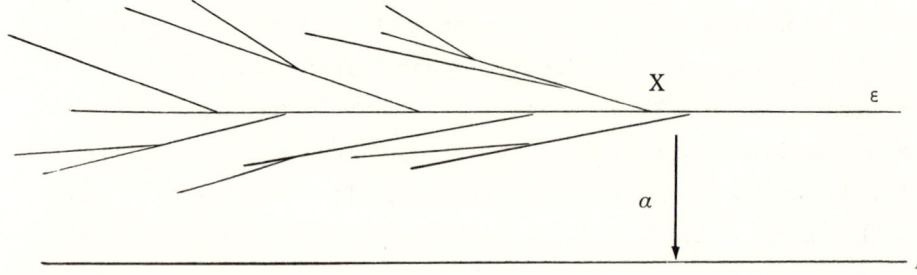

(2.34) COROLLARY. *Let s be an automorphism of* X *that fixes* ε, *and let* $a: X \to \Lambda$ *be as in (2.33). There is a unique* $\ell \in \Lambda$ *such that* $a(sx) = a(x) + \ell$ *for all* $x \in X$.

Since $x \mapsto a(sx)$ has the same properties as a it follows from (2.33) that $a(sx) = \phi(a(x))$ for a unique $\phi \in \mathrm{Aut}(\Lambda)$. We need only show that ϕ is a translation, $\phi(a) = \ell + a$, and not a reflection, $\phi(a) = \ell - a$. Choose $x_0 \in X$. On $[x_0, \varepsilon)$, a either preserves or reverses the natural order on $[x_0, \varepsilon)$ (with x_0 as least element); say a preserves order. For any $x \in X$, $[x, \varepsilon) \cap [x_0, \varepsilon) = [y, \varepsilon)$ for some y, so a must likewise preserve order on $[x, \varepsilon)$; in particular a preserves order on $[sx_0, \varepsilon) = s[x_0, \varepsilon)$. But if ϕ were a reflection, $\phi \circ a$ would reverse order on $[x_0, \varepsilon)$, so a would reverse order on $[sx_0, \varepsilon)$; contradiction.

(2.35) PROPOSITION. *Let* X *be a* Λ*-tree and* $X = X_0 \amalg X_1$ *an open cut of* X *making the open ends* ε_i *of* X_i ($i = 0,1$) *meet. There is a map* $a: X \to \Lambda$ *such that, for all* $(x_0, x_1) \in X_0 \times X_1$, $a|[x_0, x_1]$ *is* Λ*-metric. Moreover* a *is unique modulo* $\mathrm{Aut}(\Lambda)$, *hence determined by it values at any pair of distinct points of any closed segment* $[x_0, x_1]$ *as above.*

Fix $(x_0, x_1) \in X_0 \times X_1$ and a Λ-metric morphism $a_{01}: [x_0, x_1] \to \Lambda$. We have $[x_0, x_1] = [x_0, \varepsilon_0) \amalg [x_1, \varepsilon_1)$ (cf. (2.27)), and, by (2.33), $a_{01}|[x_i, \varepsilon_i)$ extends uniquely to a map $a_i: X_i \to \Lambda$ which is Λ-metric on each X_i-ray $[y_i, \varepsilon_i)$. This gives a map $a = a_0 \amalg a_1: X = X_0 \amalg X_1 \to \Lambda$. Let $y_i \in X_i$ ($i = 0,1$). Then $[y_0, y_1] \cap [x_0, x_1] = [z_0, z_1]$ for some $z_i \in X_i$ ($i = 0,1$), by (2.17) (a). We can cover $[y_0, y_1]$ by $[y_0, \varepsilon_0)$, $[z_0, z_1]$ and $[y_1, \varepsilon_1]$, and a is Λ-metric on each piece. Applying (2.30) (b) twice we see that a is Λ-metric on $[y_0, y_1]$. Finally, a is uniquely determined by a_{01}; a_{01} is unique modulo $\mathrm{Aut}(\Lambda)$; hence a is unique modulo $\mathrm{Aut}(\Lambda)$.

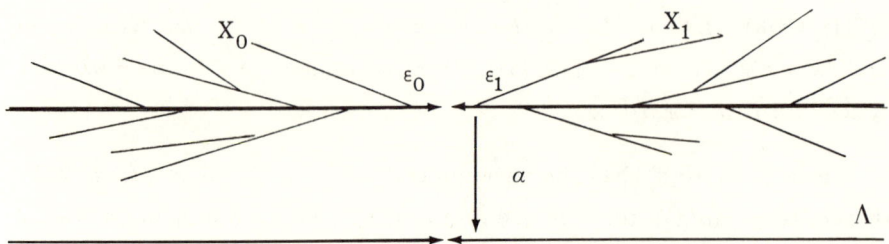

(2.36) COROLLARY. *Let s be an automorphism of* $X = X_0 \amalg X_1$ *that stabilizes* X_0 *and* X_1. *Then s fixes* $\varepsilon_i \in \text{Ends}(X_i)$ $(i = 0,1)$. *Given a as in (2.35), there is a unique* $\ell \in \Lambda$ *such that* $a(sx) = a(x) + \ell$ *for all* $x \in X$.

If $x_i \in X_i$ $(i = 0,1)$ then $[x_0, x_1] = [x_0, \varepsilon_0) \amalg [x_1, \varepsilon_1)$ and $s[x_0, x_1] = [sx_0, s\varepsilon_0) \amalg [sx_1, s\varepsilon_1) = [sx_0, sx_1] = [sx_0, \varepsilon_0) \amalg [sx_1, \varepsilon_1)$, whence $s\varepsilon_i = \varepsilon_i$ $(i = 0,1)$. Now by (2.34) there is a unique $\ell \in \Lambda$ such that $\beta(x) := a(sx)$ and $\gamma(x) := a(x) + \ell$ agree for all $x \in X_0$. By the uniqueness part of (2.35), $\beta = \gamma$ on X.

We close this section with an elaboration of Theorem (2.28).

(2.37) PROPOSITION. *Let* $(A_s)_{s \in \Gamma}$ *be a family of closed subtrees of* X *such that* $A_s \cap A_t \neq \emptyset$ *for all* $s, t \in \Gamma$, *and put* $A = \cap_s A_s$. *Let* Y *be a closed subtree of* X *such that* $A_s \cap Y \neq \emptyset$ *for all* $s \in \Gamma$.

(a) *If* $A \neq \emptyset$ *then* $A \cap Y \neq \emptyset$.

Suppose now that $A = \emptyset$.

(b) *If* ε *is an (open) end of* X *common to all* A_s *then* ε *belongs to* Y; *if* $y \in Y$ *then* $[y, \varepsilon) \subset Y$.

(c) *Suppose that there is a cut* $X = X_0 \amalg X_1$ *of* X *making the open ends* ε_i *of* X_i $(i = 0,1)$ *meet and such that* $A_s \cap X_i \neq \emptyset$ $(i = 0,1)$ *for all* $s \in \Gamma$. *Put* $Y_i = Y \cap X_i$ $(i = 0,1)$. *Then* $Y_i \neq \emptyset$ $(i = 0,1)$, ε_i *belongs to* Y_i, $Y = Y_0 \amalg Y_1$ *is an open cut making the ends* ε_i *of* Y_i $(i = 0,1)$ *meet, and* $A_s \cap Y_i \neq \emptyset$ $(i = 0,1)$ *for all* $s \in \Gamma$.

Proof. (a) Choose $a \in A$ and let $[a,p]$ be the bridge from a to Y. If $s \in \Gamma$ then $a \in A_s$ and $A_s \cap Y \neq \emptyset$ so $[a,p] \subset A_s$. Thus $p \in A_s \cap Y$ for all $s \in \Gamma$, i.e. $p \in A \cap Y$.

(b) If ε does not belong to Y consider the bridge $[y_\varepsilon, \varepsilon)$ from Y to ε (see Proposition (2.24) (e)). If $s \in \Gamma$ and $y \in A_s \cap Y$ then $[y, \varepsilon) \cap Y = [y, y_\varepsilon] \subset A_s \cap Y$. Hence $y_\varepsilon \in A_s \cap Y$ for all $s \in \Gamma$, contrary to the assumption that $\cap_s A_s = \emptyset$.

(c) Suppose that $Y \cap X_1 = \emptyset$. Choose $x_1 \in X_1$. For $y \in Y$, $[y, x_1] \cap Y$ is a closed segment (Y is closed). Since $[y, x_1] = [y, \varepsilon_0) \amalg [x_1, \varepsilon_1)$ we have $[y, x_1] \cap Y = [y, \varepsilon_0) \cap Y = [y, y_{\varepsilon_0}]$ ((2.24) (e)). If $s \in \Gamma$, $y \in A_s \cap Y$, and $x_1 \in A_s \cap X_1$ then $[y, y_{\varepsilon_0}] = [y, x_1] \cap Y \subset A_s \cap Y$. Hence $y_{\varepsilon_0} \in A_s \cap Y$ for all $s \in \Gamma$, contrary to the assumption that $\cap_s A_s = \emptyset$. Thus $Y_i = Y \cap X_i \neq \emptyset$ ($i = 0, 1$). Choose $y_i \in Y_i$ ($i = 0, 1$). Then Y contains $[y_0, y_1] = [y_0, \varepsilon_0) \amalg [y_1, \varepsilon_1)$, so ε_i belongs to Y_i ($i = 0, 1$) and $Y = Y_0 \amalg Y_1$ is an open cut making the ε_i meet. If $s \in \Gamma$ choose $x_i \in A_s \cap X_i$ ($i = 0, 1$). Then $[x_i, \varepsilon_i)$ and $[y_i, \varepsilon_i)$ meet inside $A_s \cap Y_i$ ($i = 0, 1$) so $A_s \cap Y_i \neq \emptyset$ ($i = 0, 1$). This completes the proof.

3. A Chiswell style construction of rooted Λ-trees

By a *rooted Λ-tree,"* we mean a Λ-tree with a distinguished base point, "the root," $x_0 \in X$. Given $x, y \in X$ put

$$x \wedge_{x_0} y = d(Y(x, x_0, y), x_0)$$

This function

$$\wedge_{x_0} : X \times X \to \Lambda$$

satisfies the following conditions, for all $x, y, z \in X$:

(3.1) $$\begin{cases} \text{RT 0.} & x \wedge_{x_0} y \geq 0 \\ \text{RT 1.} & x \wedge_{x_0} y = y \wedge_{x_0} x \\ \text{RT 2.} & x \wedge_{x_0} z \geq \min(x \wedge_{x_0} y, y \wedge_{x_0} z). \end{cases}$$

RT 0 is obvious, RT 1 follows from the Y-Proposition (2.11), and RT 2 follows from the H-Proposition (2.15). The Y-Proposition further implies that

(3.2) $$\begin{cases} d(x,y) = d(x,x_0) + d(y,y_0) - 2x \wedge_{x_0} y \\ d(x,x_0) = x \wedge_{x_0} x. \end{cases}$$

Thus the metric d and the function $\wedge_{x_0}$ determine each other. In principle therefore we could axiomatize (rooted) Λ-trees in terms of $(X, \wedge_{x_0})$ rather than (X,d). In fact we shall show that (3.1) furnishes such a set of axioms; see Theorem (3.17) below. This is a consequence of a more general construction, which has other useful applications.

The idea is simply this. Suppose that $X' \subset X$ *spans* X, in the sense that X is the smallest subtree of X containing X'. Then X is the union of the intervals $[x_0, x]$ $(x \in X')$, and $[x_0, x]$ is isometric to $[0,d] \subset \Lambda$, where $d = d(x_0, x) = x \wedge_{x_0} x$. Further, knowing $x \wedge_{x_0} y$, we know $[x_0, x] \cap [x_0, y]$. We can thus reconstruct X and d from the disjoint union of intervals $[0, x \wedge_{x_0} x]$ by identifying $[0, x \wedge_{x_0} y] \subset [0, x \wedge_{x_0} x]$ with $[0, x \wedge_{x_0} y] \subset [0, y \wedge_{x_0} y]$ $(x,y \in X')$. We shall in fact do this axiomatically, using only properties (3.1). This procedure is like that used by Chiswell [C1] in the case $\Lambda = \mathbf{R}$.

We give ourselves now "*rooted Λ-tree data*" $(X, \wedge)$, consisting of a set X and a function $\wedge : X \times X \to \Lambda$ satisfying, for all $x,y,z \in X$,

(3.1)
$$\begin{aligned}&\text{RT 0.} && x \wedge y \geq 0 \\ &\text{RT 1.} && x \wedge y = y \wedge x \\ &\text{RT 2.} && x \wedge z \geq \min(x \wedge y, y \wedge z).\end{aligned}$$

(3.3) REMARKS. 1. Put $|x| = x \wedge x$ for $x \in X$. Then we have $|x| = x \wedge x \geq \min(x \wedge y, y \wedge x) = x \wedge y$, i.e.

$$|x| = x \wedge x \geq x \wedge y$$

for $x,y \in X$.

2. Axiom RT 2 can be restated as follows:

RT 2′: For all $x,y,z \in X$ and $a \in \Lambda$,

$$x \wedge y \geq a \text{ and } y \wedge z \geq a \implies x \wedge z \geq a.$$

In the presence of RT 1, another equivalent form is:

RT 2″: For all $x,y,z \in X$, the smallest two of $x \wedge y$, $y \wedge z$, $x \wedge z$ are equal.

Given $(X, \wedge)$ as above we form

$$\Sigma = \Sigma(X, \wedge) = \{(x,t) \in X \times \Lambda \mid 0 \leq t \leq |x|\},$$

and define

$$d : \Sigma \times \Sigma \longrightarrow \Lambda$$

by

$$d((x,s), (y,t)) = \begin{cases} |s-t| & \text{if } s \text{ or } t \text{ is } \leq x \wedge y \\ |s-x \wedge y| + |t-x \wedge y| & \text{if } s \text{ or } t \text{ is } \geq x \wedge y. \end{cases}$$

Note that this definition is consistent in the overlapping cases. For example if $s \leq x \wedge y \leq t$ then $|s-x \wedge y| + |t-x \wedge y| = (x \wedge y - s) + (t - x \wedge y) = t - s = |t-s|$. (Cf. [C1], §7.)

The following properties show that d is a "pseudo-Λ-metric." Properties (3.4) and (3.5) are clear.

(3.4) $$d((x,s),(y,t)) \geq 0 ,$$

and

$$d((x,s),(y,t)) = 0 \text{ iff } s = t \leq x \wedge y$$

(3.5) $$d((x,s),(y,t)) = d((y,t),(x,s)) .$$

The triangle inequality in Λ shows that

(3.6) $$d((x,s),(y,t)) \geq |s-t| .$$

(3.7) LEMMA. *Given* $(x,u),(y,s),(z,t)$ *in* Σ, *we have*

$$d((x,u),(z,t)) \leq d((x,u),(y,s)) + d((y,s),(z,t)) .$$

Proof. If $d((x,u),(z,t)) = |u-t|$ then the lemma follows from (3.6) and the triangle inequality in Λ, $|u-v| \leq |u-s| + |s-t|$. So we may assume that $u, t > x \wedge z$, and $d((x,u),(z,t)) = (u-x \wedge z) + (t-x \wedge z) = u+t - 2x \wedge z$. By RT 2″ we have one of the following cases:

(i) $x \wedge y = y \wedge z \leq x \wedge z$

(ii) $x \wedge y = x \wedge z \leq y \wedge z$

(iii) $x \wedge z = y \wedge z \leq x \wedge y$.

In case (i),

$$d((x,u),(y,s)) = (u-x \wedge y) + |s-x \wedge y| \geq u - x \wedge z$$

and

$$d((y,s),(z,t)) = |s-y \wedge z| + (t-y \wedge z) \geq t - x \wedge z ,$$

whence the lemma in this case.

Consider case (ii). Then

$$d((x,u),(y,s)) = (u - x \wedge z) + |s - x \wedge z|$$

and

$$d((y,s),(z,t)) \geq |t-s| \qquad \text{(by (3.6))}$$

so $d((x,u),(y,s)) + d((y,s),(z,t)) \geq (u - x \wedge z) + |s - x \wedge z| + |t-s| \geq (u - x \wedge z) + |t - x \wedge z| = d((x,u),(z,t))$.

Case (iii) follows from case (ii) by symmetry.

(3.8) COROLLARY. *The relation,*

$$(x,s) \sim (y,t) \quad \text{iff} \quad d((x,s),(y,t)) = 0$$

is an equivalence relation.

It is clearly reflexive and symmetric; transitivity follows from (3.7).
We put

$$T = T(X, \wedge) = \Sigma/\sim,$$

and denote the class in T of $(x,s) \in \Sigma$ by $<x,s>$. It follows further from (3.7) that if $(x,s) \sim (x',s')$ and $(y,t) \sim (y',t')$ then $d((x',s'),(y',t')) = d((x,s),(y,t))$. Indeed $|d(x,s),(y,t)) - d((y,t),(y',t'))| \leq d((x,s),(y',t')) \leq d((x,s),(y,t)) + d((y,t),(y',t'))$ so $d((x,s),(y',t')) = d((x,s),(y,t))$, etc.
Consequently we can define

$$d : T \times T \longrightarrow \Lambda$$

by

$$d(<x,s>,<y,t>) = d((x,s),(y,t)),$$

and the properties (3.4), (3.5), (3.7) above show now that (T, d) is a Λ-metric space. The case analysis in the next proof resembles the arguments of [C1] and [AM].

(3.9) THEOREM. $T = T(X, \wedge)$ *is a Λ-tree.*

The points $<x,0>$ $(x \in X)$ coincide; we shall denote this point by 0, or O_T if there is need to avoid confusion. We shall also write $\bar{x}$ for $<x,|x|>$ $(x \in X)$. The map $a_x : [0, |x|] \to T$, $a_x(t) = <x,t>$, is easily seen to be an isometry $[0, |x|] \to [0, \bar{x}]$.

Let $<x,s>$, $<y,t> \in T$ and $d = d(<x,s>, <y,t>)$. We shall construct a metric morphism $a : [0,d] \to T$ such that $a(0) = <x,s>$ and $a(d) = <y,t>$. Case 1: $t \leq x \wedge y$. Then $<y,t> = <x,t>$ and we obtain a easily from a_x; if $s \leq t$ then $a(r) = a_x(r+s)$ $(0 \leq r \leq t-s)$. We similarly obtain a from a_y when $s \leq x \wedge y$. Case 2: $s, t > x \wedge y = v$. Then we construct a by concatenating the parametrized paths just constructed from $<x,s>$ to $<x,v> = <y,v>$ and from $<y,v>$ to $<y,t>$. If $s \geq s' \geq v$ and $v \leq t' \leq t$ then $d(<x,s'>, <y,t'>) = (s'-v) + (t'-v)$, so it follows that a is a metric morphism. We shall write $[<x,s>, <y,t>]$ for the image of a, isometric to $[0,d]$. If $<z,u> \in [<x,s>, <y,t>]$ then

(3.10) $d(<x,s>, <y,t>) = d(<x,s>, <z,u>) + d(<z,u>, <y,t>)$.

We claim, conversely, that this *property (3.10) characterizes the points of* $[<x,s>, <y,t>]$. Once this is shown the uniqueness of a (as required in the definition of a Λ-tree) follows, since any other parametrization must have the same image, hence differ from a by a metric automorphism of $[0,d]$ fixing 0 and d, and such an automorphism is the identity, by (2.6).

So consider (z,u) satisfying (3.10). Put $v = x \wedge y$, $v_x = x \wedge z$, and $v_y = y \wedge z$, and recall (RT 2″) that the smallest two of v, v_x, v_y are equal.

Case 1. $s \leq v$ or $t \leq v$.

Say $t \leq v$, so that $<y,t> = <x,t>$, and we may assume that $x = y$. If $u > v_x$ then (3.10) equates

$$d(<x,s>, <x,t>) = |s-t|$$

with

$$d(<x,s>,<z,u>) + d(<z,u>,<x,t>)$$
$$= (|s-v_x| + (u-v_x)) + ((u-v_x) + |t-v_x|)$$
$$= |s-v_x| + |v_x-t| + 2(u-v_x) > |s-t| ;$$

contradiction. It follows that $u \leq v_x$, hence $<z,u> = <x,u>$, and (3.10) takes the form $|s-t| = |s-u| + |u-t|$, whence $u \in [s,t]$ in Λ, and the claim follows.

Case 2. $s > v$ and $t > v$.

Then $d(<x,s>,<y,t>) = (s-v) + (t-v) = s+t - 2v$. If $s \leq v_x$ and $t \leq v_y$ then we have $v < \min(v_x, v_y)$, which is impossible, by RT 2. Thus we have one of three cases:

(i) $s \geq v_x$ and $t \geq v_y$; (ii) $s \geq v_x$ and $t < v_y$; or
(iii) $s < v_x$ and $t \geq v_y$.

(i) We have $d(<x,s>,<z,u>) + d(<z,u>,<y,t>) = ((s-v_x) + |u-v_x|) + ((t-v_y) + |u-v_y|) = s+t - v_x - v_y + |u-v_x| + |u-v_y|$, so (3.10) implies that

(*) $$2v = v_x + v_y - |u-v_x| - |u-v_y| .$$

We have:

(a) $v_x = v_y \leq v$;
(b) $v_x = v \leq v_y$; or
(c) $v_y = v \leq v_x$.

Assuming (a), (*) implies that $v = v_x - |u-v_x| \leq v_x$ and so $v = v_x = u$, whence $<z,u> = <x,v> \in [<x,s>,<y,t>]$.

Assuming (b), (*) implies that $v_y - v = |v_y-u| + |u-v|$, whence $u \in [v,v_y]$ in Λ, and so $<z,u> = <y,u>$ with $u \geq v$. We have $v \leq u \leq v_y$ and $v_y \leq t$ (case (i)), whence $u \leq t$, so $<z,u> = <y,u> \in [<x,s>,<y,t>]$, as claimed.

The case (c) is equivalent, by symmetry, to case (b). This concludes case (i).

(ii) $s \geq v_x$ and $t < v_y$. Since $v < t$ we must have $v = v_x$. Now

$$d(<x,s>,<z,u>) + d(<z,u>,<y,t>)$$
$$= ((s-v_x)+|u-v_x|) + |t-u| = s-v + |v-u| + |u-t|$$

equals $s+t-2v$, so $t-v = |t-u| + |u-v|$, whence $u \in [v,t]$ in Λ, and, since $u \leq t < v_y$ we have $<z,u> = <y,u> \in [<x,s>,<y,t>]$, as claimed.

(iii) $s > v_x$ and $t \leq v_y$. This case follows by symmetry from case (ii). Thus we have proved that (3.10) characterizes the points of $[<x,s>,<y,t>]$, and so also established the uniqueness of the isometric path in T from $<x,s>$ to $<y,t>$. This gives axiom (2.9) (a) for a Λ-tree. It remains to establish:

(2.9) (b) $\qquad [<x,s>,<y,t>] \cap [<x,s>,<y',t'>]$

$$= [<x,s>,<z,u>] \text{ for some } <z,u> \in T ;$$

and

(2.9) (c) If $[<x,s>,<y,t>] \cap [<y,t>,<z,u>] = \{<y,t>\}$ then $[<x,s>,<y,t>] \cup [<y,t>,<z,u>] = [<x,s>,<z,u>]$.

To prove (2.9) (b) put $v = x \wedge y$ and $v' = x \wedge y'$. We have $[<x,s>,<y,t>] = [<x,s>,<x,t>]$ if $t \leq v$, and otherwise $[<x,s>,<y,t>]$
$= [<x,s>,<x,v>] \cup [<y,v>,<y,t>]$, where $[<x,s>,<x,v>] = [<x,s>,<y,t>]$
$\cap [0,\bar{x}]$, and $[<y,v>,<y,t>] \cap [0,\bar{x}] = \{<x,v>\} = \{<y,v>\}$. Similarly for $[<x,s>,<y',t'>]$. If $t \geq v$ and $t' \leq v'$ we have the equivalent of an intersection of intervals $[s,t]$ and $[x,t']$ in Λ, and the conclusion is obvious. Suppose that $t > v$ but $t' \leq v'$. Then $[<x,s>,<y,t>] \cap [<x,s>,<y',t'>] = [<x,s>,<y,t>] \cap [<x,s>,<x,t'>] = [<x,s>,<x,v>] \cap [<x,s>,<x,t'>]$, and we have the previous case again. Similarly when $t \leq v$ and $t' > v'$.

Suppose finally that $t > v$ and $t' > v'$. Then $[<x,s>,<y,t>] \cap [<x,s>,<y',t'>]$ is the union of $[<x,s>,<x,v>] \cap [<x,s>,<x,v'>]$ and $[<y,v>,<y,t>] \cap [<y',v'>,<y',t'>]$. The first of these is a closed segment, by the previous cases. Further $<y,r> = <y',r'>$ iff $r = r' \leq y \wedge y'$ so that

$[<y,v>,<y,t>] \cap [<y',v'>,<y',t'>] = \{<y,r> | \max(v,v') \le r \le \min(t',y \wedge y')\}$, which is either empty, or a closed segment. The latter happens only if $v = v' \le y \wedge y'$ (RT 2) in which case $[<x,s>,<y,t>] \cap [<x,s>,<y',t'>] = [<x,s>,<x,v>] \cup [<y,v>,<y,w>] = [<x,s>,<y,w>]$, $w = \min(t',y \wedge y')$.

Finally, to prove (2.9) (c), put $v_x = x \wedge y$ and $v_z = y \wedge z$. If $s \le v_x$ and $u \le v_z$ then $<x,s> = <y,s>$ and $<z,u> = <y,u>$. The assumption of (2.9) (c) implies then that, in Λ, we have $[s,t] \cap [t,u] = \{t\}$, hence $[s,t] \cup [t,u] = [s,u]$, and so $[<y,s>,<y,t>] \cup [<y,t>,<y,u>] = [<y,s>,<y,u>]$, as claimed.

Next suppose that $s \le v_x$ and $u > v_z$. Then $<x,s> = <y,s>$ and $[<y,t>,<z,u>] \cap [0,\bar{y}] = [<y,t>,<y,v_z>]$. Our assumption then implies that, in Λ. $[s,t] \cap [t,v_z] = \{t\}$, so $t \in [s,v_z]$. Further we have $[<x,s>,<z,u>] = [<y,s>,<z,u>] = [<y,s>,<y,v_z>] \cup [<z,v_z>,<z,u>] = [<y,s>,<y,t>] \cup [<y,t>,<y,v_z>] \cup [<z,v_z>,<z,u>] = [<x,s>,<y,t>] \cup [<y,t>,<z,u>]$.

The case $s > v_x$, $u \le v_z$ follows by symmetry from the preceding one.

Suppose finally that $s > v_x$ and $u > v_z$. The assumption implies that, in Λ, we have $[v_x,t] \cap [t,v_z] = \{t\}$ so that $t \in [v_x,v_z]$. Since the smallest two of v_x, v_z, and $v = x \wedge z$ are equal we have either

(i) $v = v_x \le t \le v_z < u$;

(ii) $v = v_z \le t \le v_x < x$; or

(iii) $v_x = t = v_z \le v$.

In case (i), $<y,t> = <z,t>$ and $[<x,s>,<y,t>] \cup [<y,t>,<z,u>] = [<x,s>,<x,v>] \cup [<z,v>,<z,t>] \cup [<z,t>,<z,u>] = [<x,s>,<x,v>] \cup [<z,v>,<z,u>] = [<x,s>,<z,u>]$, as claimed.

Case (ii) follows from case (i) by symmetry.

Consider finally case (iii). Put $w = \min(s,v,u)$. Then we have $v_x = t = v_z \le w \le v$ so $<x,w> = <z,w> \in [<x,s>,<y,t>] \cap [<y,t>,<z,u>] = \{<y,t>\}$. Hence $w = t$. But $s > v_x = t$ and $u > v_z = t$ (by assumption), so we must have $v = t = v_x = v_z$. Then $[<x,s>,<y,t>] \cup [<y,t>,<z,u>] = [<x,s>,<x,v>] \cup [<z,v>,<z,u>] = [<x,s>,<z,u>]$, as claimed.

This concludes the proof of Theorem (3.9).

The description above of $[<x,s>,<y,t>]$ (at the start of the proof of (3.9)) shows the following:

(3.11) $\begin{cases} Y(<x,s>,0,<y,t>) = <x,m> = <y,m>, \text{ where} \\ \\ m = \min(s,t,x \wedge y) = d(Y(<x,s>,0,<y,t>),0) . \end{cases}$

In other words,

(3.12) $\qquad <x,s> \wedge_0 <y,t> = \min(s,t,x \wedge y) .$

When $s = |x|$ and $t = |y|$, both $\geq x \wedge y$, we find that

(3.13) $\qquad d(Y(\bar{x},0,\bar{y}),0) = x \wedge y$, i.e.

$$\bar{x} \wedge_0 \bar{y} = x \wedge y .$$

To formulate the universal property of T we equip T with the base point $0 = 0_T$, and with the function $\phi : X \to T$, $\phi(x) = \bar{x}$, satisfying (3.13).

(3.14) *Universal Property of* $T = T(X, \wedge)$. *Let* (W, w_0) *be a rooted Λ-tree and let* $\psi : X \to W$ *be a map satisfying*

(3.15) $\qquad \psi(x) \wedge_{w_0} \psi(y) = x \wedge y$

for all $x, y \in X$. *Then there is a unique metric morphism* $\bar{\psi} : T \to W$ *such that* $\bar{\psi}(0) = w_0$ *and*

$$\begin{array}{ccc} & X & \\ \phi \swarrow & & \searrow \psi \\ T & \xrightarrow{\bar{\psi}} & W \end{array}$$

commutes.

For $w \in W$ put $|w| = d(w,w_0) = w \wedge_{w_0} w$, and let $a_w : [0, |w|] \to W$ be the metric morphism such that $a_w(0) = w_0$, $a_w(|w|) = w$. If $v, w \in W$, $s \in [0, |v|]$, and $t \in [0, |w|]$ then it is easily seen that

$$a_v(s) \wedge_{w_0} a_w(t) = \min(s, t, v \wedge_{w_0} w).$$

Now define $\bar{\psi} : T \to W$ by $\bar{\psi}(<x,s>) = a_{\psi(x)}(s)$. If $<x,s> = <y,t>$ then $s = t \leq x \wedge y = \psi(x) \wedge_{w_0} \psi(y)$ and so $a_{\psi(x)}(s) = a_{\psi(y)}(t)$; thus $\bar{\psi}$ is well defined. Further

$$\bar{\psi}(<x,s>) \wedge_{w_0} \bar{\psi}(<y,t>) = a_{\psi(x)}(s) \wedge a_{\psi(y)}(t)$$

$$= \min(s, t, \psi(x) \wedge_{w_0} \psi(y))$$

$$= \min(s, t, x \wedge y)$$

$$= <x,s> \wedge_0 <y,t>.$$

Since $\bar{\psi}(0) = w_0$ it follows thus from (3.2) that $\bar{\psi}$ is an isometry.

Finally the uniqueness of $\bar{\psi}$ follows since $\bar{\psi}$ is prescribed on $\phi(X) \cup \{0\}$, which manifestly spans T, and clearly a metric morphism of Λ-trees is determined on a closed segment once it is known on the end points.

(3.16) PROPOSITION. *Let (W, w_0) be a rooted Λ-tree, and let X be a subset of W, equipped with the function $\wedge_{w_0} : X \times X \to \Lambda$. Form the Λ-tree $T = T(X, \wedge_{w_0})$ with root 0_T and canonical map $\phi : X \to T$. The inclusion $\psi : X \to W$ (tautologically) satisfies (3.15) and so defines a metric morphism $\bar{\psi} : T \to W$ as in (3.14). The image of $\bar{\psi}$ is the subtree of W spanned by $X \cup \{w_0\}$. In particular $\bar{\psi}$ is an isomorphism iff $X \cup \{w_0\}$ spans W.*

This is obvious from the foregoing discussion.

(3.17) THEOREM. *Let (X,d) be a Λ-metric space with a base point $x_0 \in X$. Define $\wedge : X \times X \to \frac{1}{2}\Lambda$ by (cf. [C1], [L]),*

$$x \wedge y = \frac{1}{2}(d(x,x_0) + d(y,y_0) - d(x,y)).$$

Then X is isometric to a subspace of a Λ-tree iff the following conditions hold.

(0) $x \wedge y \in \Lambda$ *for all* $x,y \in X$.

(1) *For all* $x,y,z \in X$,

$$x \wedge z \geq \min(x \wedge y, y \wedge z).$$

Moreover X is a Λ-tree iff we have conditions (0), (1), and

(2) *For all $x \in X$ there is a metric morphism $a_x : [0, |x|] \to X$ such that $a_x(0) = x_0$ and $a_x(|x|) = x$, where $|x| = d(x,x_0)$.*

If X is a subspace of a Λ-tree then (0) follows from (3.2) and (1) follows from (2.15). Moreover (2) clearly holds if X is a Λ-tree.

Suppose conversely that (0) and (1) hold. Of the axioms RT 0, RT 1, RT 2 of (3.1) for $\wedge : X \times X \to \Lambda$, RT 2 is just (1), RT 1 (symmetry) is obvious (from the symmetry of d), and RT 0 (positivity) follows from the triangle inequality for d. Thus we can form the Λ-tree $T = T(X, \wedge)$, with base point 0, and the map $\phi : X \to T$, $\phi(x) = \bar{x} = \langle x, |x| \rangle$, where $|x| = x \wedge x = d(x,x_0)$, which satisfies $\bar{x} \wedge_0 \bar{y} = x \wedge y$. We have $d(x,y) = x \wedge x + y \wedge y - 2x \wedge y$, by definition, and $d(\bar{x},\bar{y}) = \bar{x} \wedge_0 \bar{x} + \bar{y} \wedge_0 \bar{y} - 2\bar{x} \wedge_0 \bar{y}$ in the Λ-tree T by (2.11) (a) (iii). It follows that ϕ is a metric morphism, hence an isometry $X \to \phi(X) \subset T$. It remains only to show, assuming (2), that ϕ is surjective. A typical point of T has the form $\langle x,s \rangle$, $x \in X$, $0 \leq s \leq |x|$. Put $y = a_x(s)$, where a_x is as in (2). Since a_x is a metric morphism we have $|y| = d(y,x_0) = s$ and $d(x,y) = |x| - s$. Hence

$x \wedge y = \frac{1}{2}(|x|+|y|-d(x,y)) = \frac{1}{2}(|x|+s-(|x|-s)) = s$. By definition of the equivalence relation defining T (cf. (3.4) and (3.8)) we have $<x,s> = <y,s> = <y,|y|> = \bar{y} = \phi(y)$, whence the surjectivity of ϕ.

(3.18) REMARKS. 1. Let (X,d) and x_0 satisfy (0) and (1) of (3.17), whence an isometric embedding $\phi: X \to T = T(X, \wedge)$ sending x_0 to $0 \in T$. Suppose that $\psi: X \to W$ is an isometric embedding of X into a Λ-tree W. Put $w_0 = \psi(x_0)$. Then we have $\psi(x) \wedge_{w_0} \psi(y) = x \wedge y$ for $x, y \in X$ so, by the universal property (3.16), there is a unique metric morphism $\bar{\psi}: T \to W$ such that $\psi = \bar{\psi} \circ \phi$ and (hence) $\bar{\psi}(0) = w_0$. If $\psi(X)$ spans W then $\bar{\psi}$ must be an isomorphism of Λ-trees. This gives a strong uniqueness condition on the Λ-tree into which X embeds in (3.17). It thus makes sense, given (X,d) (and x_0) satisfying (0) and (1) of (3.17), to speak of *the* Λ-tree spanned by X.

2. Suppose that (X,d) and x_0 satisfy (0) and (1) of (3.17), whence an embedding $X \subset T = T(X, \wedge)$. Let $x_1 \in X$ be another choice of base point. Then $x \wedge_1 y = \frac{1}{2}(d(x,x_1)+d(y,x_1)-d(x,y))$ satisfies (0) and (1), by the converse part of (3.17), so we have another embedding $\phi_1: X \to T_1 = T(X, \wedge_1)$ sending x_1 to the base point 0_1 of T_1. Put $w_0 = \phi_1(x_0)$. Then by Remark 1 we have a unique isomorphism $\bar{\phi}_1: T \to T_1$ such that $\phi_1 = \bar{\phi}_1 \circ \phi$ (and so $\bar{\phi}_1(0) = w_0$). In this sense, conditions (0) and (1) of (3.17) are independent of the choice of base point, as is the Λ-tree spanned by X.

3. Suppose that $\Lambda = \mathbf{R}$. Suppose that a real metric space (X,d) is isometric to a subspace of an $\mathbf{R}$-tree. Let $(\bar{X}, \bar{d})$ be the metric completion of (X,d). Choose $x_0 \in X$. Condition (0) of (3.17) is trivial when $\Lambda = \mathbf{R}$. Condition (1) is clearly inherited by $(\bar{X}, \bar{d})$. Hence, we have $\mathbf{R}$-trees $X \subset T(X)$ and $\bar{X} \subset T(\bar{X})$ and, in view of Remark 1, a commutative diagram

$$X \quad \subset \quad \overline{X}$$
$$\cap \quad \quad \cap$$
$$T(X) \quad \subset \quad T(\overline{X}).$$

Since X spans $T(X)$, $\overline{X}$ spans $T(\overline{X})$, and X is dense in $\overline{X}$, it follows from (2.16) (b) that $T(X)$ is dense in $T(\overline{X})$; in fact that $T(\overline{X}) = T(X) \cup \overline{X}$. Indeed given $z \in T(\overline{X})$, there exist $\overline{x}, \overline{y} \in \overline{X}$ such that $z \in [\overline{x}, \overline{y}]$. Suppose that $z \neq \overline{x}$ and $z \neq \overline{y}$. Choose $x, y \in X$ such that $d(x, \overline{x}) + d(y, \overline{y}) < \min(d(\overline{x}, z), d(\overline{y}, z))$, whence also $d(x, \overline{x}) + d(y, \overline{y}) < d(\overline{x}, \overline{y})$.

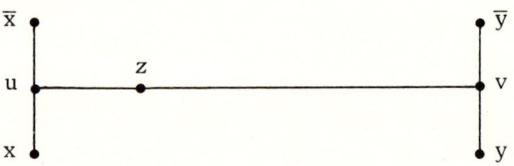

Then by (2.16) (b) (letting $\overline{x}$ play the role there of x_0), we have $[x,y] \cap [\overline{x}, \overline{y}] = [u,v]$, where $u = Y(\overline{x}, x, \overline{y})$ and $v = Y(y, \overline{x}, \overline{y})$. Since $d(\overline{x}, u) \leq d(\overline{x}, x) < d(\overline{x}, z)$ and $d(\overline{y}, v) \leq d(\overline{y}, y) < d(\overline{y}, z)$, we have $z \in [u,v] \subset [x,y] \subset T(X)$. This shows, as claimed, that $T(\overline{X}) = \overline{X} \cup T(X)$.

Suppose that X is an **R**-tree, so that $X = T(X)$. Then $X \subset \overline{X} \subset T(\overline{X})$ and, by the last observation, X is dense in $T(\overline{X})$, whence $\overline{X} = T(\overline{X})$. In other words, *the metric completion of an* **R***-tree is an* **R***-tree*. (Cf. [CM], Corollary II.1.10, [AM], §1, and [I1], Theorem 2).

4. In the context of (3.14), let (W, w_0) be a rooted Λ-tree and let $\psi : X \to W$ be a map satisfying the following weaking of (3.15):

(3.15′) $\begin{cases} (0) \; \psi(x) \wedge_{w_0} \psi(x) = x \wedge x \\ (1) \; \psi(x) \wedge_{w_0} \psi(y) \geq x \wedge y \end{cases}$

for all $x,y \in X$. Then there is a unique map $\bar{\psi} : T \to W$ such that $\bar{\psi}(0) = w_0$, $\psi = \bar{\psi} \circ \phi$, and $\bar{\psi}$ is Λ-metric on each segment $[0, \phi(x)]$ $(x \in X)$. Moreover $\bar{\psi}$ is distance non-increasing on T.

The proof follows that of (3.14), showing, for $p, q \in T$, that $p \wedge q \leq \bar{\psi}(p) \underset{w_0}{\wedge} \bar{\psi}(q)$. Then (3.2) shows that $d(\bar{\psi}(p), \bar{\psi}(q)) \leq d(p,q)$.

A. Appendix. The tree X_v of a valuation v

As an application of Theorem (3.17) we give a simplified presentation of the Morgan-Shalen construction of the tree attached to a (not necessarily discrete or rank 1) valuation. See Chiswell [C4] for still another approach, in the case $\Lambda = \mathbb{Z}$.

Let Λ be an ordered abelian group. Let F be a field with multiplicative group $F^x = F - \{0\}$. A Λ-*valuation* on F is a surjective homomorphism $v : F^x \to \Lambda$ such that, putting $v(0) = \infty$ (an upper bound for Λ), we have $v(a+b) \geq \min(v(a), v(b))$ for all $a, b \in F$. Thus $A = \{a \in F \mid v(a) \geq 0\}$ is a subring of F, called the *valuation ring*, with unique maximal ideal $m = \{a \in F \mid v(a) > 0\}$. The ideals of A are totally ordered by inclusion, with $J \subset J'$ iff $v(J) \subset v(J')$. Every finitely generated ideal is principal: $Aa_1 + \cdots + Aa_n = Aa_i$, where $v(a_i) = \min(v(a_1), \cdots, v(a_n))$.

NOTATION. For elements $x_1, \cdots, x_n$ in an A-module we shall write $\langle x_1, \cdots, x_n \rangle$ for the submodule $Ax_1 + \cdots + Ax_m$ they generate.

(A.1) PROPOSITION. *Let $\mathfrak{M}$ denote the class of finitely presented torsion A-modules. There is a unique function $| \ | : \mathfrak{M} \to \Lambda$ such that*

(a) $|A/Aa| = v(a)$ if $a \in A$, $a \neq 0$.

(b) If $0 \to M' \to M \to M'' \to 0$ is an exact sequence of modules in $\mathfrak{M}$ then $|M| = |M'| + |M''|$.

Proof. If $M = <x_1, \cdots, x_n>$, put $M_i = <x_1, \cdots, x_i>$. Then M_i/M_{i-1} is finitely presented and cyclic, hence isomorphic to A/Aa_i for some a_i. We put $|M| = v(a_1 \cdots a_n) = v(a_1) + \cdots + v(a_n)$.

In case $M = A/Aa$ we must have $M_i = Ab_i/Aa$ with $Aa \subset Ab_1 \subset \cdots \subset Ab_n = A$; say $b_n = 1$. Writing $b_{i-1} = a_i b_i$ (with $b_0 = a$) we have $M_i/M_{i-1} \cong A/Aa_i$. Moreover $a = a_1 b_1 = a_1 a_2 b_2 = \cdots = a_1 a_2 \cdots a_n$. Thus $|M|$ agrees with (a) when M is cyclic. In general $|M|$ is well defined therefore, by the Jordan-Holder-Zassenhaus Refinement Theorem. To verify (b) we simply choose generators so that $M' \subset M$ equals one of the M_i; then (b) is evident.

Let V denote the two dimensional F-module F^2. By an A-lattice $L \subset V$ we mean a finitely generated A-module that spans V as F-module. If L' is another such lattice we write $L \sim L'$ if $L' = aL$ for some $a \in F^x$. This is evidently an equivalence relation. We write $\bar{L}$ for the equivalence class of L, and X $(= X_V)$ for the set of such equivalence classes. Our aim is to give X a Λ-metric making it a Λ-tree.

(A.2) PROPOSITION. *Let L, L_1 be A-lattices in V.*

(a) *There is an A-basis e, f of L, and elements $a, b \in F^x$, such that $L_1 = <ae, bf>$.*

(b) *Say $v(b) \leq v(a)$, i.e. $a = bc$ with $c \in A$. Then $L_1 = bL_1'$ where $L_1' = <ce, f> \subset L$, and $L/L_1' \cong A/Ac$, a cyclic module.*

(c) *Let H be an A-lattice in V such that $H \sim L_1$ and $H \subset L$. The following conditions are equivalent.*

(i) $H = L_1'$

(ii) L/H *is cyclic*

(iii) H *is maximal subject to the conditions $H \sim L_1$ and $H \subset L$.*

Proof. Assertion (a) is classical (cf. [MS], II.3.2). Assertion (b) is obvious. To prove (c), write $L_1 = \beta H$, so that $H = \langle ace, af \rangle$, where $a = b/\beta$. Since $H \subset L$ we have $a \in A$. Thus $H = aL_1' \subset L_1'$ and $L/H \cong A/acA \oplus A/aA$. Now it is clear that (i), (ii), and (iii) are equivalent to the condition, $a \in A^\times$.

(A.3) DEFINITIONS. If L, L' are A-lattices in V write $L' \leq L$ if $L' \subset L$ and L/L' is a cyclic A-module. Note then that if L'' is an intermediate lattice, $L' \subset L'' \subset L$, then $L' \leq L'' \leq L$. If L, L_1 are A-lattices in V then, by (A.2), there is a unique lattice L_1' such that $L_1' \sim L_1$ and $L_1' \leq L$. We put

$$d(L, L_1) = |L/L_1'| \in \Lambda.$$

Thus $d(L, L_1) = v(c)$, with c as in (A-2) (b).

(A.4) PROPOSITION. *Let L_0, L_1, L_2 be A-lattices in V.*
  (a) $d(a_0 L_0, a_1 L_1) = d(L_0, L_1)$ *for all* $a_0, a_1 \in F^\times$, *and* $d(sL_0, sL_1) = d(L_0, L_1)$ *for all* $s \in GL(V)$.
  (b) $d(L_0, L_1) \geq 0$, *and* $d(L_0, L_1) = 0$ *iff* $L_0 \sim L_1$.
  (c) $d(L_0, L_1) = d(L_1, L_0)$.
  (d) $d(L_0, L_1) \leq d(L_0, L_2) + d(L_2, L_1)$.

Proof. (a) $d(sL_0, sL_1) = d(L_0, L_1)$ by "transport of structure" via s. Therefore $d(a_0 L_0, a_1 L_1) = d(L_0, aL_1)$ where $a = a_1/a_0$, and it is clear from the definition that $d(L_0, aL_1) = d(L_0, L_1)$.

(b) is clear from the definition.

(c) Suppose that $L_1' = aL_1 \leq L_0$, $a \in F^\times$, $L_0 = \langle e, f \rangle$, and $L_1' = \langle ce, f \rangle$. Then $cL_0 = \langle ce, cf \rangle \subset L_1'$ and $L_1'/cL_0 \cong A/Ac$. Thus $d(L_1, L_0) = d(L_1', L_0) = |L_1'/cL_0| = v(c) = d(L_0, L_1)$.

(d) Choose $L_i' \sim L_i$ such that $L_i' \leq L_2$ ($i = 0, 1$), and put $L = L_0' + L_1' \leq L_2$. Then

$$d(L_0, L_1) = d(L_0', L_1') \quad \text{(part (a))}$$
$$= |L/L_0'| + |L/L_1'| \quad ((A.5) \text{ below})$$
$$\leq |L_2/L_0'| + |L_2/L_1'|$$
$$= d(L_0, L_2) + d(L_2, L_1).$$

(A.5) LEMMA. *Let L_0, L_1, L be Λ-lattices in V such that $L_i \leq L$ ($i = 0, 1$) and $L_0 + L_1 = L$. Then $d(L_0, L_1) = |L/L_0| + |L/L_1|$.*

Proof. If $L_0 = L$ or $L_1 = L$ this is clear. Suppose that $L_i \neq L$ ($i = 0, 1$). Then $L/L_0 \cap L_1 \cong (L/L_0) \oplus (L/L_1)$ is not cyclic, whence $L_0 \cap L_1 \subset mL$, so elements that generate L modulo $L_0 \cap L_1$ actually generate L (Nakayama's Lemma). Choose $e_{1-i} \in L_i$ so that $L = Ae_0 + L_0 = L_1 + Ae_1$. Then e_0, e_1 generate L modulo $L_0 \cap L_1$ so $L = <e_0, e_1>$. Since $e_1 \in L_0$ we have $L_0 = <a_0 e_0, e_1>$ for some a_0. Similarly $L_1 = <e_0, a_1 e_1>$ for some a_1. Then $a_0 L_1 = <a_0 e_0, a_0 a_1 e_1> \subset L_0$, and $L_0/a_0 L_1 \cong A/a_0 a_1 A$. Thus $d(L_0, L_1) = v(a_0 a_1) = v(a_0) + v(a_1) = |L/L_0| + |L/L_1|$.

In view of (A.4), d defines a Λ-metric,

$$d(\overline{L}_0, \overline{L}_1) = d(L_0, L_1)$$

on the set X of equivalence classes $\overline{L}$ of Λ-lattices $L \subset V$. Moreover $GL(V)$ acts on X as a group of Λ-isometries.

Fix now a lattice $L_0 \subset V$, and put $x_0 = \overline{L}_0$. If $x \in X$, put $|x| = d(x, x_0)$, and let L_x denote the unique lattice such that $\overline{L}_x = x$ and $L_x \leq L_0$; e.g. $L_{x_0} = L_0$. There is a basis e_x, f_x of L_0 and an $a_x \in A$ such that $L_x = <a_x e_x, f_x>$, and thus $|x| = |A/Aa_x| = v(a_x)$.

(A.6) LEMMA. *Let $x \in X$. For each $\lambda \in [0, |x|] \subset \Lambda$ choose $a_\lambda \in A$ so that $v(a_\lambda) = \lambda$, and put $L_\lambda = <a_\lambda e_x, f_x>$. Then $a : [0, |x|] \to X$, $a(\lambda) = \overline{L}_\lambda$, is a Λ-metric morphism such that $a(0) = x_0$ and $a(|x|) = x$.*

If $0 \le \lambda \le \lambda' \le |x|$ then $Aa_{\lambda'} \subset Aa_\lambda$ and $L_{\lambda'} \le L_\lambda$ with $L_\lambda/L_{\lambda'} \cong Aa_\lambda/Aa_{\lambda'} \cong A/A \cdot (a_{\lambda'}/a_\lambda)$, so $d(L_\lambda, L_{\lambda'}) = v(a_{\lambda'}/a_\lambda) = \lambda' - \lambda$; whence the lemma.

For $x, y \in X$ put

$$x \wedge y = \tfrac{1}{2}(d(x,x_0) + d(y,x_0) - d(x,y)).$$

To prove that X is a Λ-tree it suffices, in view of Theorem (3.17), to show, for all $x, y, z \in X$:

(0) $x \wedge y \in \Lambda$

(0) $x \wedge z \ge \min(x \wedge y, y \wedge z)$

These properties follows from the next two lemmas.

(A.7) LEMMA. $x \wedge y = |L_0/L_x + L_y| \in \Lambda$.

(A.8) LEMMA. *Two of* $L_x + L_y$, $L_x + L_z$, $L_y + L_z$ *are equal and contain the third. Hence the smaller two of* $x \wedge y$, $x \wedge z$, $y \wedge z$ *are equal.*

Proof of (A.7). Applying (A.5) to $L_x, L_y \subset L_x + L_y$, we have

$$2(x \wedge y) = |L_0/L_x| + |L_0/L_y| - (|L_x+L_y/L_x| + |L_x+L_y/L_y|)$$
$$= 2|L_0/L_x+L_y|.$$

Proof of (A.8). The submodules of the cyclic module L_0/L_y are totally ordered, so one of $L_x + L_y$ and $L_y + L_z$ contains the other, say $L_y + L_z \subset L_x + L_y$. Thus $L_x + L_y = L_x + L_y + L_z$. Similarly one of $L_x + L_z$ and $L_y + L_z$ equals $L_x + L_y + L_z$; whence the lemma.

This proves:

(A.9) THEOREM. *The Λ-metric space (X,d) is a Λ-tree, on which the natural action of $GL(V)$ is by Λ-isometries.*

(A.10) PROPOSITION. *If* $s = \begin{bmatrix} a & b \\ c & d \end{bmatrix} \in GL_2(F)$ *put* $v(s) = \min(v(a), v(b), v(c), v(d))$. *Let* $L_0 = A^2 \subset V = F^2$. *Then*

$$d(sL_0, L_0) = v(\text{Det}(s)) - 2v(s).$$

Suppose that $\text{Det}(s) \in A^\times$ (e.g. that $s \in SL_2(F)$). Then $d(sL_0, L_0) = -2v(s) \in 2\Lambda$, so s *is not an inversion* ((6.3) (d)). *Further* $d(sL_0, L_0) = 0$ *iff* $s \in GL_2(A)$ *iff* $sL_0 = L_0$.

Proof. Write $s = at$ with $a \in F^\times$ and $v(a) = v(s)$, so that $v(t) = 0$. Thus $sL_0 = aH$ where $H = tL_0 \subset L_0$ and, by (A.2) (c) (iii), $H \leq L_0$. (H is spanned by the column vectors of t.) Thus $d(sL_0, L_0) = |L_0/H|$. By (A.2) (a), $t = udw$ where $u, w \in GL_2(A)$ and $d = \begin{bmatrix} c & 0 \\ 0 & 1 \end{bmatrix}$, where $L_0/H \cong A/Ac$. Thus $d(sL_0, L_0) = v(c) = v(\text{Det}(t)) = v(\text{Det}(a^{-1}s)) = v(a^{-2}\text{Det}(s)) = v(\text{Det}(s)) - 2v(a) = v(\text{Det}(s)) - 2v(s)$. The final assertions of the proposition now follow easily from this formula.

For further results on the action of $GL(V)$ on X see Morgan-Shalen ([MS], Ch. II), or Appendix B to Section 6 below.

4. Base change

We fix a homomorphism $h: \Lambda \to \Lambda'$ of ordered abelian groups such that $a \geq 0 \Longrightarrow h(a) \geq 0$. This amounts to saying that $\text{Ker}(h)$ is a convex subgroup of Λ, and that the embedding $\Lambda/\text{Ker}(h) \to \Lambda'$ preserves order.

We propose to construct a functor $X \to X \otimes_\Lambda \Lambda'$ from Λ trees to Λ'-trees.

Let X be a Λ-tree. Choose $x_0 \in X$, and consider $\wedge_{x_0} : X \times X \to \Lambda$, as in Section 3:

$$x \wedge_{x_0} y = \frac{1}{2}(d(x, x_0) + d(y, y_0) - d(x, y)).$$

It satisfies (cf. (3.17)):

(0) $x \wedge_{x_0} y \in \Lambda$ for all $x, y \in X$

(1) For all $x, y, z \in X$,

$$x \wedge_{x_0} z \geq \min(x \wedge_{x_0} y, y \wedge_{x_0} z).$$

We have a canonical identification then of X with $T = T(X, \wedge_{x_0})$. (See (3.17) and Remark (3.17) 1.) Now form

$$\wedge' = h \circ \wedge_{x_0} : X \times X \to \Lambda'$$

$$x \wedge' y = h(x \wedge_{x_0} y).$$

This clearly satisfies

(1′) For all $x, y, z \in X$,

$$x \wedge' z \geq \min(x \wedge' y, y \wedge' z).$$

Hence we may form the Λ'-tree $T' = T(X, \wedge')$ with a canonical map $\phi' : X \to T'$, $\phi'(x) = x'$, satisfying $\phi'(x_0) = 0'$, the base point of T', and

(4.1) $$x' \wedge_{0'} y' = h(x \wedge_{x_0} y)$$

for all $x, y \in X$.

(4.2) LEMMA. *Let W be a Λ'-tree, $\psi : X \to W$ a map, and $w_0 = \psi(x_0)$. The following conditions are equivalent.*

(a) $\psi(x) \wedge_{w_0} \psi(y) = h(x \wedge_{x_0} y) \ (= x' \wedge y')$ *for all* $x, y \in X$.

(b) $d(\psi(x), \psi(y)) = h(d(x, y))$ *for all* $x, y \in X$.

Proof. We have

$$x \wedge_{x_0} y = \frac{1}{2}(d(x, x_0) + d(y, y_0) - d(x, y))$$

and

$$d(x, y) = x \wedge_{x_0} x + y \wedge_{x_0} y - x \wedge_{x_0} y,$$

and similarly for $\psi(x)$ and $\psi(y)$ in W. The lemma follows easily.

The lemma applied to $\phi': X \to T'$ and (4.1) yields

(4.3) $\qquad d(x',y') = h(d(x,y))$ for all $x,y \in X$.

Now let $\psi: X \to W$ be as in (4.2), satisfying (a) and (b). Then the universal property (3.14) of $T' = T(X, \Lambda')$ yields a unique Λ'-metric morphism $\overline{\psi}: T' \to W$ such that $\psi = \overline{\psi} \circ \phi'$ and (hence) $\overline{\psi}(0') = w_0$. We have thus proved:

(4.4) PROPOSITION. *Let* $h: \Lambda \to \Lambda'$ *be a homomorphism of ordered abelian groups satisfying* $a \geq 0 \implies h(a) \geq 0$. *Let* X *be a* Λ-*tree. There is a* Λ'-*tree* T' *and a map* $\phi': X \to T'$ *satisfying*

(4.3) $\qquad d(\phi'(x), \phi'(y)) = h(d(x,y))$

for all $x,y \in X$. *If* W *is a* Λ'-*tree and if* $\psi: X \to W$ *is a map satisfying*

(4.2) (b) $\qquad d(\psi(x), \psi(y)) = h(d(x,y))$

for all $x,y \in X$ *then there is a unique* Λ'-*metric morphism* $\overline{\psi}: T' \to W$ *such that* $\psi = \overline{\psi} \circ \phi'$. (*Cf.* [MS], *Theorem II.1.9.*)

It follows that T' is determined up to a unique Λ'-isometry. This permits us to write

$$T' = X \otimes_\Lambda \Lambda'.$$

It is independent of the base point x_0 used in its construction. We shall denote $\phi': X \to T'$ by

$$\phi_X: X \to X \otimes_\Lambda \Lambda'.$$

(4.5) COROLLARY. *Let* $a: X \to Z$ *be a* Λ-*metric morphism of* Λ-*trees. There is a unique* Λ'-*metric morphism* $a \otimes_\Lambda \Lambda': X \otimes_\Lambda \Lambda' \to Z \otimes_\Lambda \Lambda'$ *such that* $(a \otimes_\Lambda \Lambda') \circ \phi_X = \phi_Z \circ a$. *This makes* $\otimes_\Lambda \Lambda'$ *a functor from* Λ-*trees to* Λ'-*trees.*

Proof. In the commutative diagram

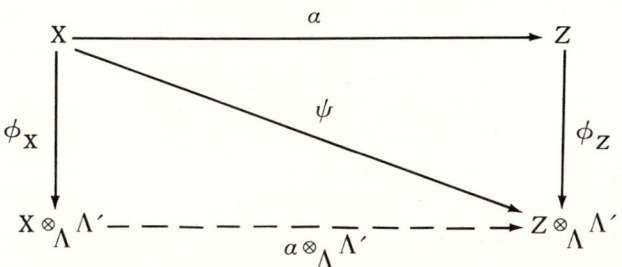

we apply (4.4) to $W = Z \otimes_\Lambda \Lambda'$ and $\psi = \phi_Z \circ a$ to get $\bar{\psi} = a \otimes_\Lambda \Lambda'$. Functoriality is similarly a standard argument.

(4.6) EXAMPLES. 1. If $a \leq b$ in Λ and $[a,b]_\Lambda = \{x \in \Lambda | a \leq x \leq b\}$ then $[a,b]_\Lambda \otimes_\Lambda \Lambda' = [h(a), h(b)]_{\Lambda'} = \{x' \in \Lambda' | h(a) \leq x' \leq h(b)\}$. More generally, if A is a subtree of Λ then $A \otimes_\Lambda \Lambda'$ is the sub Λ'-tree of Λ' spanned by $h(A)$.

2. If $\Lambda = Z$, so that the Z-tree X is a simplicial tree, then $X \otimes_Z \frac{1}{2} Z$ corresponds to the barycentric subdivision of X, while $X \otimes_Z R$ is the geometric realization of X.

CHAPTER II. TREE ACTIONS AND LENGTH FUNCTIONS

5. *Lyndon length functions*

We fix an ordered abelian group Λ.

Let Γ be a group.

(5.1) DEFINITION. By a (Λ-*valued*) *Lyndon length function* on Γ we understand a function $L: \Gamma \to \Lambda$ satisfying the following conditions, where we put

(5.2) $$\delta(s,t) = \frac{1}{2}(L(s) + L(t) - L(s^{-1}t))$$

for $s, t \in \Gamma$. (Cf. [L], [C1], [H].)

L0. $L(1) = 0$.

L1. $L(s) = L(s^{-1})$ for all $s \in \Gamma$.

L2. For all $r, s, t \in \Gamma$,

$$\delta(r,t) \geq \min(\delta(r,s), \delta(s,t)).$$

L3. $\delta(s,t) \in \Lambda$ for all $s, t \in \Gamma$.

(5.3) EXAMPLE. Suppose that Γ acts on a Λ-tree X. Such actions are assumed to be by Λ-isometries. Choose $x_0 \in X$, and define

$$L = L_{x_0} : \Gamma \to \Lambda \text{ by } L(s) = d(sx_0, x_0).$$

Then properties L0 and L1 are evident. Moreover

$$\begin{aligned}\delta(s,t) &= \tfrac{1}{2}(L(s) + L(t) - L(s^{-1}t)) \\ &= \tfrac{1}{2}(d(sx_0,x_0) + d(tx_0,x_0) - d(s^{-1}tx_0,x_0)) \\ &= \tfrac{1}{2}(d(sx_0,x_0) + d(tx_0,x_0) - d(tx_0,sx_0)) \\ &= sx_0 \wedge_{x_0} tx_0,\end{aligned}$$

by (3.2), whence properties L2 and L3, by (3.1).

The aim of this section is to show that every Lyndon length function arises in this way, in an essentially unique way. The lack of uniqueness arises from the fact that the Λ-tree X_0 spanned by the orbit Γx_0 may not be all of X, yet L_{x_0} relative to X_0 and to X coincide.

(5.4) THEOREM. *Let* $L : \Gamma \to \Lambda$ *be a Lyndon length function, and define* δ *as in (5.2). Then there is a Λ-tree* $T = T(\Gamma, \delta)$ *with base point* $0 \in T$, *and an action of* Γ *on* T *(by Λ-isometries), with the following properties.*

(a) *For all* $s \in \Gamma$, $L(s) = d(s \cdot 0, 0)$, *i.e.* L *is the length function* L_0 *arising from the Γ-action on* T.

(b) *Suppose that Γ acts on a Λ-tree X and that $L = L_{x_0}$ for some $x_0 \in X$. Then there is a unique Γ-equivariant Λ-metric morphism $\psi: T \to X$ sending 0 to x_0. The image of ψ is the subtree of X spanned by the orbit Γx_0.*

We begin by observing that (Γ, δ) constitute rooted Λ-tree data, in the sense that δ satisfies the properties of (3.1): For all $r, s, t \in \Gamma$

RT0 $\delta(s,t) \geq 0$

RT1 $\delta(s,t) = \delta(t,s)$

RT2 $\delta(r,t) \geq \min(\delta(r,x), \delta(s,t))$.

Property RT2 is just L2, and RT1 follows from the definition of δ and L1. To check RT0 note first that

$$\delta(1,s) = 0 \text{ and } \delta(s,1) = 0,$$

the first by L0, and the second by L0 and L1. It follows, using L2, that $\delta(s,t) \geq \min(\delta(s,1), \delta(1,t)) = 0$, whence RT0.

Now by Theorem (3.9) we can form the Λ-tree $T = T(\Gamma, \delta)$, with base point $0 = <1,0>$, and the map $\phi: \Gamma \to T$, $\phi(s) = \bar{s} = <s, |s|>$, where $|s| = \delta(s,s) = L(s) = d(\bar{s}, 0)$; in particular $\phi(1) = \bar{1} = 0$. We further have

(5.5) $$\bar{s} \wedge_0 \bar{t} = \delta(s,t),$$

by (3.13).

Let $s \in \Gamma$ and define $\phi_s: \Gamma \to T$ by $\phi_s(t) = \overline{st}$. We claim that, for $s, t, u \in \Gamma$, we have

(5.6) $$\overline{st} \wedge_{\bar{s}} \overline{su} = \bar{t} \wedge_0 \bar{u} \quad (= \delta(t,u)).$$

Suppose that this has been shown. Then it follows from the universal property (3.14) of $T(\Gamma, \delta)$ that there is a unique Λ-isometry ρ_s making

the diagram

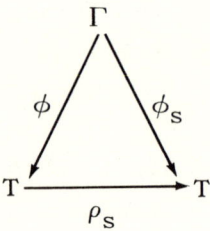

commute. From the uniqueness it is clear that $\rho_1 = 1_T$ and that $\rho_s \circ \rho_t = \rho_{st}$. Thus we have defined an action of Γ on T so that $\rho_s(\bar{t}) = \overline{st}$. In particular $\rho_s(0) = \rho_s(\bar{1}) = \bar{s}$, and $d(\bar{s},0) = \delta(s,s) = L(s)$. Thus assertion (a) is established once we confirm (5.6) above.

To that end we calculate, using (3.2):

$$\overline{st} \wedge_{\bar{s}} \overline{su} = \frac{1}{2}[d(\overline{st},\bar{s}) + d(\overline{su},\bar{s}) - d(\overline{st},\overline{su})]$$

$$= \frac{1}{2}[|st| + |s| - 2\delta(st,s)$$

$$+ |su| + |s| - 2\delta(su,s)$$

$$- |st| - |su| + 2\delta(st,su)]$$

$$= |s| - \delta(st,s) - \delta(su,s) + \delta(st,su),$$

where, as observed above, $|s| = L(s) = d(\bar{s},0)$. Now $\delta(st,s) = \delta(s,st) = \frac{1}{2}(L(st) + L(s) - L(s^{-1}st)) = \frac{1}{2}(L(st) + L(s) - L(t))$. Similarly $\delta(su,s) = \frac{1}{2}(L(su) + L(s) - L(u))$ and $\delta(st,su) = \frac{1}{2}(L(st) + L(su) - L(t^{-1}u))$. Thus

$$st \wedge_{\bar{s}} su = L(s)$$

$$- \frac{1}{2}(L(st) + L(s) - L(t))$$

$$- \frac{1}{2}(L(su) + L(s) - L(u))$$

$$+ \frac{1}{2}(L(st) + L(su) - L(t^{-1}u))$$

$$= \frac{1}{2}(L(t) + L(u) - L(t^{-1}u))$$

$$= \delta(t,u),$$

whence (5.6).

To prove (b) suppose that Γ acts on a Λ-tree X, $x_0 \in X$, and $L(s) = d(sx_0,x_0)$ for all $s \in \Gamma$. Define $\psi_0 : \Gamma \to X$ by $\psi_0(s) = sx_0$. We claim that

(5.7)
$$sx_0 \wedge_{x_0} tx_0 = \delta(s,t)$$

for all $s,t \in \Gamma$. Indeed, by (3.2),

$$sx_0 \wedge_{x_0} tx_0 = \frac{1}{2}(d(sx_0,x_0) + d(tx_0,x_0) - d(tx_0,sx_0))$$

$$= \frac{1}{2}(L(s) + L(t) - L(s^{-1}t))$$

$$= \delta(s,t) .$$

It follows now from (5.7) and the universal property (3.14) of $T = T(\Gamma, \delta)$ that there is a unique Λ-metric morphism $\psi : T \to X$ such that $\psi(\bar{s}) = \psi_0(s) = sx_0$ for $s \in \Gamma$; in particular $\psi(0) = \psi(\bar{1}) = x_0$. To show that ψ is Γ-equivariant note first that, for $s,t \in \Gamma$, $\psi(\rho_s \bar{t}) = \psi(\overline{st}) = stx_0 = s\psi(\bar{t})$. Thus $\psi \circ \rho_s$ and $s_X \circ \psi$ (where s_X is the action of s on X) are Λ-metric morphisms $T \to X$ which agree on the set $\phi(\Gamma) = \{\bar{t} | t \in \Gamma\}$, which spans T, and so they agree. Finally it is clear that $\psi(T)$ is the subtree of X spanned by $\psi(\phi(\Gamma)) = \Gamma x_0$.

(5.7) COROLLARY. *Suppose that Γ acts on a Λ-tree X. Let $x_0 \in X$, and consider the Lyndon length function $L_{x_0}(s) = d(sx_0,x_0)$. Let Λ_0 be a subgroup of Λ containing $\delta_{x_0}(\Gamma)$. Then there is a unique Γ-invariant subspace X_0 of X containing x_0 which is a Λ_0-tree and spanned as a Λ_0-tree by $\Gamma \cdot x_0$. In fact (X_0, x_0) is Γ-equivariantly isomorphic to the rooted Λ_0-tree $(T(\Gamma, \delta_{x_0}), 0)$ associated to $\delta_{x_0} : \Gamma \to \Lambda_0$. The sub Λ-tree X' spanned by Γx_0 is Γ-equivariantly isomorphic to $X_0 \otimes_{\Lambda_0} \Lambda$, as defined in (4.5).*

6. Automorphisms of Λ-trees; hyperbolic length

We here study an automorphism s of a Λ-tree X. (Cf. [T2], [Tg], [S], [MS].) The group generated by s is denoted by <s>.

(6.1) PROPOSITION. *Suppose that s has a fixed point. Let A_s denote the set of fixed points.*

(a) (i) A_s *is a closed subtree of X (cf. (2.10))*.

(ii) A_s *meets every <s>-invariant subtree of X.*

(iii) A_s *is contained in every subtree with the property described in (ii)*.

(b) *Let $x \in X$ and let $[x,p]$ be the bridge from x to A_s (cf. (2.17)). Then $p = Y(x,a,sx)$ for any $a \in A_s$, and p is the mid point of $[x,sx]$:*

(6.2) $\quad d(x,sx) = 2d(x,p) = 2d(x,A_s).$

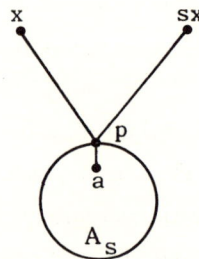

We have $[x,sx] \cap A_s = \{p\}$ *and* $[x,p] \subset [s^{-1}x,x] \cap [x,sx]$.

Clearly A_s is a subtree of X and satisfies (a) (iii). Let $a \in A_s$. If $x \in X$ and $0 \leq m \leq d(x,a)$ then s carries the unique point at distance m from a on $[a,x]$ to the unique point at distance m from a on $[a,sx]$. It follows that $[a,x] \cap [a,sx] = [a,x] \cap A_s$; the left side is just $[a,p_x]$, where $p_x = Y(x,a,sx)$. Moreover $[x,sx] = [x,p_x,sx]$ and $[p_x,sx] = s[p_x,x]$, so $d(x,sx) = 2d(x,p_x)$. Since $[x,p_x] \cap A_s = \{p_x\}$, $[x,p_x]$ is the bridge from x to A_s, and $d(x,p_x) = d(x,A_s)$. Since $[x,sx] = [x,p_x,sx]$ we have $[s^{-1}x,x] = [s^{-1}x,p_x,x]$, so $[x,p_x] \subset [s^{-1}x,x] \cap [x,sx]$.

To show that A_s is closed, suppose that a closed segment $[x,y]$ contains some point $a \in A_s$. Then we must have $p_x \in [x,a]$ and $[x,a] \cap A_s = [p_x,a]$. Similarly $[y,a] \cap A_s = [p_y,a]$, where $p_y = Y(y,a,sy)$. Hence $[x,y] \cap A_s = ([x,a] \cap A_s) \cup ([a,y] \cap A_s) = [p_x,a] \cup [a,p_y] = [p_x,p_y]$. This shows that A_s is closed, and (b) now follows from the preceding paragraph.

(6.3) PROPOSITION. *The following conditions are equivalent.*

(a) *s has no fixed points, but stabilizes a closed segment.*

(b) *There is a closed segment* [x,y] *of "odd" length* $d(x,y) \notin 2\Lambda$ *such that* sx = y, sy = x.

(c) *s has no fixed points, but* s^2 *has one.*

(d) s^2 *has a fixed point, and* d(x,sx) *is "odd" (i.e.* $\notin 2\Lambda$ *) for all* $x \in X$.

(e) *s has no fixed point, and if* Λ' *is an ordered abelian group containing* Λ *such that* $d(x,sx) \in 2\Lambda'$ *for some* $x \in X$ *then* $d(x,sx) \in 2\Lambda'$ *for all* $x \in X$, *and the automorphism* $s' = s \otimes_\Lambda \Lambda'$ *of* $X' = X \otimes_\Lambda \Lambda'$ *has a unique fixed point.*

(a) $\Longrightarrow$ (b). Let [x,y] be stabilized by s. Since s has no fixed points it must exchange the end points x and y. Moreover [x,y] can't have a mid point (s would fix it) so $d(x,y) \notin 2\Lambda$.

(b) $\Longrightarrow$ (c). From (b) and (6.1)(b) we see that s has no fixed points, whereas $s^2 x = x$.

(c) $\Longrightarrow$ (e). If $x, y \in X' = X \otimes_\Lambda \Lambda'$ write [x,y]' for the Λ'-segment from x to y in X'. If t' is a Λ'-automorphism of X' write $A'_{t'}$ for its fixed points in X'.

Suppose that $x \in X$ and $d(x,sx) \in 2\Lambda'$. Let [x,y] be the bridge from x to A_{s^2}. Then $[y,sy] \subset A_{s^2}$ so $[x,y] \cap [y,sy] = \{y\}$. Applying s we further have $[sx,sy] \cap [sy,y] = \{sy\}$. Hence [x,sx] = [x,y,sy,sx], so d(x,sx) = 2d(x,y) + d(y,sy). Thus d(y,sy), like d(x,sx), is in $2\Lambda'$, so [y,sy]' has a mid point, p_x, which is also the mid point of [x,sx]', and since s' stabilizes [y,sy]', $s' p_x = p_x$. Now (6.1)(b) implies that $d(x', s'x') \in 2\Lambda'$ for every $x' \in X'$. It remains to show that s' has a unique fixed point. We first show that p_x is the same for all $x \in X$. Since p_x is the mid point of [y,sy]' above we have $p_x = p_y$, with $y \in A_{s^2}$. Hence it suffices to show that if $x, y \in A_{s^2}$ then $[x, sx]'$ and $[y, sy]'$ have a common mid point. First note that $[x, sx] \cap [y, sy] \neq \emptyset$. Otherwise the bridge from $[x, sx]$ to $[y, sy]$ would be stabilized by s, and s would fix each of its end points, contrary to assumption. Now $[x, sx] \cap [y, sy]$ is, by (2.16) (a), a (nonempty) closed segment; it is stable under the "inversions" s performs on $[x, sx]$ and $[y, sy]$, so it is centrally situated in each of $[x, sx]$ and $[y, sy]$. Thus $[x, sx]'$ and $[y, sy]'$ have a common mid point, as claimed. This shows that the segments $[x, sx]'$ ($x \in X$) have a common mid point $p \in A'_s$. Let $q \in A'_s$. Then $q \in [x_0, x_1]'$ for some $x_0, x_1 \in X$. We have $[x_0, x_1]' \subset [x_0, p]' \cup [p, x_1]'$, and the reverse inclusion holds since $[x_0, x_1]'$ meets A'_s and $[x_i, p]'$ is the bridge from x_i to A'_s (6.1) (b). Hence for $i = 0$ or 1 we have $q \in [x_i, p]' \cap A'_s = \{p\}$.

(e) $\Longrightarrow$ (d). Apply (e) with $\Lambda' = \frac{1}{2} \Lambda$.
If $x \in X$ then s' fixes the midpoint of $[x, sx]'$, which cannot belong to X, so $\frac{1}{2} d(x, sx) \notin \Lambda$. Put $y = Y(x, sx, s^2 x)$. By (6.1) (b) the fixed point p of s' belongs to $[x, sx]' \cap [s^2 x, sx]'$ so $y \in [x, p]' \cap [s^2 x, p]'$. Now $s^2 y \in [s^2 x, p]'$ and $d(s^2 y, p) = d(y, p)$, whence $s^2 y = y$. This proves (d).

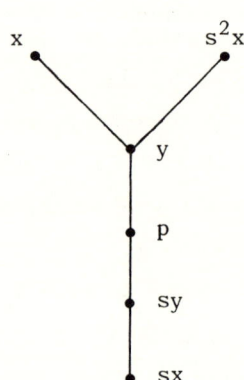

(d) $\Longrightarrow$ (a). If $d(x, sx)$ is "odd" we can't have $sx = x$. If $s^2 x = x$ then s stabilizes $[x, sx]$. Whence (a).

(6.4) DEFINITION. Under the conditions of (6.3) we call s an *inversion*.

(6.5) REMARKS. 1. If $\Lambda = 2\Lambda$ then inversions don't exist.

2. Morgan-Shalen [MS] call them "phantom inversions." When $\Lambda = \mathbf{Z}$ our definition agrees with Serre's notion of "inversion" [S].

3. The main substance of the following theorem is due to Tits [T2] for $\Lambda = \mathbf{Z}$, to Imrich [I1], Theorem 3, for $\Lambda = \mathbf{R}$, and to Morgan-Shalen [MS], Theorem II.2.3 in general.

(6.6) THEOREM. *Suppose that the automorphism* s *of* X *is not an inversion.*

(a) $\ell(s) = \min_{x \in X} d(x,sx)$ *exists, and*

$$A_s = \{p \in X | d(p,sp) = \ell(s)\}$$

is equal to $\{p \in X | [s^{-1}p,p] \cap [p,sp] = \{p\}\}$.

(b) (i) A_s *is a closed* $\langle s \rangle$-*invariant subtree of* X.

(ii) A_s *meets every* $\langle s \rangle$-*invariant subtree of* X.

(iii) A_s *is contained in every* $\langle s \rangle$-*invariant subtree with the property described in (ii).*

(c) *Let* $x \in X$ *and let* $[x,p]$ *be the bridge from* x *to* A_s. *Then* $[x,sx] \cap A_s = [p,sp]$, *a closed segment of length* $\ell(s)$ *centrally situated in* $[x,sx]$. *We have*

(6.7) $\qquad d(x,sx) = \ell(s) + 2d(x,p) = \ell(s) + 2d(x,A_s)$.

(d) *Suppose that* $\ell(s) > 0$. *Then* $p = Y(s^{-1}x,x,sx)$. *Moreover* A_s *is a linear tree and* $s | A_s$ *is equivalent to a translation,* $a \mapsto a + \ell(s)$.

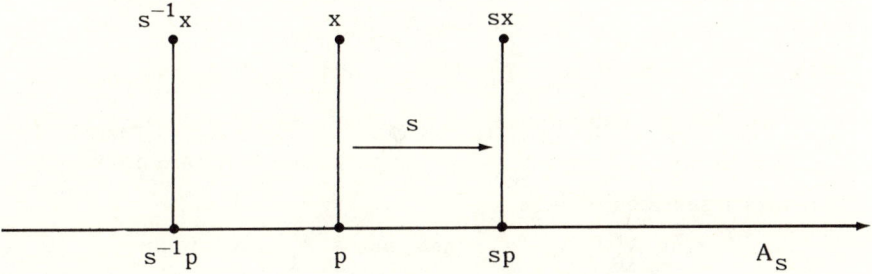

(6.8) DEFINITION. We call $\ell(s)$ the *hyperbolic length* of s and A_s the *characteristic subtree* of s. When s is an inversion we agree to put $\ell(s) = 0$ and $A_s = \emptyset$. If $\ell(s) > 0$ we call s a *hyperbolic automorphism* and A_s the *s-axis*. If $\ell(s) = 0$ then A_s is just the fixed point set of s (empty if s is an inversion).

Proof of Theorem (6.6). When s has a fixed point we have $\ell(s) = 0$ and Theorem (6.6) reduces to Proposition (6.1). So *assume now that s has no fixed points.*

Put $A = \{p \in X | [s^{-1}p,p] \cap [p,sp] = \{p\}\}$

$= \{p \in X | p \in [s^{-1}p,sp]\}$ ((2.9)(c)).

Clearly A is $<s>$-invariant.

Let $p \in A$. Then we have $[s^{n-1}p, s^n p] \cap [s^n p, s^{n+1}p] = \{s^n p\}$ for all $n \in \mathbf{Z}$. It follows then from the Piecewise Geodesic Proposition (2.14)(c) that, for all $n, m \in \mathbf{Z}$, $m \geq 0$,

$$[s^n p, s^{n+m} p] = [s^n p, s^{n+1} p, s^{n+2} p, \cdots, s^{n+m} p]$$

and

$$d(s^n p, s^{n+m} p) = d(p, s^m p) = \sum_{i=1}^{m} d(s^{i-1}p, s^i p) = m \ell_p ,$$

where $\ell_p = d(p, sp)$. It follows that any three of the points $s^n p$ ($n \in \mathbf{Z}$) are colinear, so the tree A_p they span is linear, by Proposition (2.20). Evidently A_p is $<s>$-invariant, and $s|A_p$ is equivalent to a translation of amplitude ℓ_p (cf. (2.19)). Putting $A_{p+} = \bigcup_{n \geq 0} [p, s^n p]$ and $A_{p-} = \bigcup_{n \geq 0} [p, s^{-n}p]$, we have $A_p = A_{p-} \cup A_{p+}$, and $A_{p-} \cup A_{p+} = \{p\}$. If J is a subtree of A_p containing p then clearly: $sJ \subset J \Longrightarrow A_{p+} \subset J$; and $J \subset sJ$ (i.e. $s^{-1}J \subset J$) $\Longrightarrow A_{p-} \subset J$. In particular there are no proper $<s>$-invariant subtrees of A_p.

If $p, q \in A$ and $A_p \cap A_q \neq \emptyset$, then, since $A_p \cap A_q$ is an $<s>$-invariant subtree, it contains both A_p and A_q. Thus:

(6.9) $\qquad A_p \cap A_q \neq \emptyset \Longrightarrow A_p = A_q .$

Let $x \in X$ and put $p = Y(s^{-1}x, x, sx)$, so that $[x,p] = [x,sx] \cap [x, s^{-1}x]$. Hence $[sx, sp] \subset [sx, x]$, so $[p, sp] \subset [x, sx]$. We have two possibilities.

(1) $[x,sx] = [x,sp,p,sx]$

or

(2) $[x,sx] = [x,p,sp,sx]$.

Assume (1). We have $[p,sp] \subset [sp,sx]$ and $[p,sp] \subset [p,x]$ so $[sp,s^2p] \subset [sp,sx]$. Thus p and s^2p both lie in $[sp,sx]$, at the same distance from sp, whence $s^2p = p$, contradicting our assumption that s is not an inversion.

Hence (2) holds, and

$$d(x,sx) = d(x,p) + d(p,sp) + d(sp,sx) = d(p,sp) + 2d(x,p).$$

We have $[s^{-1}p,p] \cap [p,sp] \subset [s^{-1}x,x] \cap [x,sx] = [x,p]$, so $[s^{-1}p,p] \cap [p,sp] \subset [x,p] \cap [p,sp]$, which, by (2), is just $\{p\}$. Thus $p \in A$. If $y \in [x,p]$ then $[y,sy] = [y,p,sp,sy]$ meets $[s^{-1}y,y] = [s^{-1}y,s^{-1}p,p,y]$ in $[y,p]$, so $y \in A$ iff $y = p$. It follows easily that $[x,sx] \cap A = [p,sp]$, that $[x,p]$ is the bridge from x to A, and that

(6.10) $\qquad d(x,sx) = \ell_p + 2d(x,p) = \ell_p + 2d(x,A)$.

(6.11) LEMMA. *Let* $p,q \in A$ *be such that* $A_p \cap A_q = \emptyset$
 (a) *Either* $p \notin [sp,q]$ *or* $q \notin [sq,p]$.
 (b) *If* $p \notin [sp,q]$ *then* $[p,sq] = [p,sp,q,sq]$ *contains* q, $[p,q] \cap A_p = A_{p+}$, *and* $[p,q] \cap A_q = A_{q-}$.
 (c) *If* $q \notin [sq,p]$ *then* $[sp,q] = [sp,p,sq,q]$ *contains* p, $[p,q] \cap A_p = A_{p-}$, *and* $[p,q] \cap A_q = A_{q+}$.

It suffices, by symmetry, to prove (a) and (b). Let $[p_1,q_1]$ be the bridge from $[p,sp]$ to $[q,sq]$ (2.17), with $[p_1,q_1] \cap [p,sp] = \{p_1\}$ and $[p_1,q_1] \cap [q,sq] = \{q_1\}$. We claim that $p_1 = p$ or sp. Suppose not; then $[p_1,q_1] \cap A_p = \{p_1\}$. For suppose, on the contrary, that there is an $r \neq p_1$ in $[p_1,q_1] \cap A_p$. Since r must be outside $[p,sp]$ and since A_p is linear, $[p_1,r]$ must contain p or sp, thus forcing p or sp to lie in $[p_1,q_1] \cap [p,sp] = \{p_1\}$; contradiction. Having shown now that $[p_1,q_1] \cap$

$A_p = \{p_1\}$ we have $[sp_1, sq_1] \cap A_p = \{sp_1\}$ as well. By the Bridge Proposition (2.17) (a), $[p_1, q_1] \cap [sp_1, sq_1]$ must contain a point $p' \in A_p$. But this forces $p_1 = p' = sp_1$; contradiction.

This proves, as claimed, that $p_1 = p$ or sp; say $p_1 = sp$. We similarly have $q_1 = q$ or sq. If $q_1 = sq$ we have $[p,q] = [p, sp, sq, q]$, thus forcing the impossibility $d(sp, sq) < d(p, q)$. Hence $q_1 = q$, and we have $[p, sq] = [p, sp, q, sq]$. It follows that $p \notin [sp, q]$ and $q \in [sp, q]$.

If $p_1 = p$ we similarly reason that $[q, sq] = [q, sq, p, sp]$, so $p \in [sp, q]$ and $q \notin [sp, q]$. This proves (a).

In proving (b) we must be in the case

$$(*) \qquad [p, sq] = [p, sp, q, sq],$$

above. Write $[p,q] = J_p \cup J' \cup J_q$ where $J_p = [p,q] \cap A_p$, $J_q = [p,q] \cap A_q$, and $J' = [p,q] - (J_p \cup J_q)$. From $(*)$ we see that $[sp, sq] = J'_p \cup J' \cup J'_q$, where $J_p = [p, sp] \cup J'_p$, and $J_q \cup [q, sq] = J'_q$. On the other hand $[sp, sq] = sJ_p \cup sJ' \cup sJ_q$ with $sJ_p \subset A_p$, $sJ_q \subset A_q$, and sJ' disjoint from $A_p \cup A_q$. It follows that $sJ_p = J'_p \subset J_p$ and $sJ_q = J'_q \supset J_q$, so $A_{p+} \subset J_p$ and $A_{q-} \subset J_q$. Since J_p contains sp, and p as end point, we have $J_p \subset A_{p+}$; hence $J_p = A_{p+}$. Similarly $J_q = A_{q-}$. This proves (b), and completes the proof of (6.11).

For $p, q \in A$ we now define

$$p < q \iff p \notin [sp, q].$$

If $A_p = A_q$ this corresponds to the order on the linear tree A_p for which s translates in the positive direction. If $A_p \neq A_q$ then $A_p \cap A_q = \emptyset$, by (6.9), so, by Lemma (6.11), $p < q$ or $q < p$.

Say $p < q$. Lemma (6.11) tells us further that $[p, sq] = [p, sp, q, sq]$ and $[p,q] \cap A_p = A_{p+}$. If $p' \in A_{p+}$ then clearly $p' < q$. Hence $s^n p < q$ for all $n \geq 0$. If $p' \in A_{p-}$ then $[p', p] \cap [p, q] = \{p\}$ so $[p', q] = [p', p, q]$ contains $[p', p] \cup A_{p+} = A_{p+}$, whence $p' < q$. Similarly $p < q'$ for all $q' \in A_q$. Whence:

(6.12) If $A_p \cap A_q = \emptyset$ and $p < q$ then $p' < q'$ for all $p' \in A_p$, $q' \in A_q$. Further, any three points of $A_p \cup A_q$ are collinear.

We claim finally that any three points p, q, r of A are collinear. In view of the foregoing discussion it remains only to treat the case when A_p, A_q, and A_r are pairwise disjoint. In view of Lemma (6.11) we have $p < q$ and $q < r$, or else this holds after permuting the roles of p, q, and r. So suppose that $p < q$ and $q < r$. We claim then that $[p,r] = [p,q,r]$. Otherwise $[p,q] \cap [q,r] = [q,u]$ with $u \neq q$. Since $[p,q] \cap A_q = A_{q-}$ and $[q,r] \cap A_q = A_{q+}$ meet only in $\{q\}$ we must have $u \notin A_q$, so $A_u \cap A_q = \emptyset$. By (6.11) either $u \in [su,q]$ or $q \notin [sq,u]$. In the first case $A_{q-} = [u,q] \cap A_q \subset [q,r] \cap A_q = A_{q+}$; contradiction. In the second case $A_{q+} = [u,q] \cap A_q \subset [q,p] \cap A_q = A_{q-}$; contradiction. Thus $[p,r] = [p,q,r]$ as claimed, and so any three points of A are collinear. It follows from Proposition (2.20) that A is a linear tree. It is clearly $\langle s \rangle$-invariant, so $s|A$ is a translation (Remark (2.19)), and $d(p,sp) = \ell_p$ is the same for all $p \in A$. From (6.10) we see then that $\ell_p = \ell(s)$, $A = A_s$, and $d(x,sx) = \ell(s) + 2d(x,A_s)$ for all $x \in X$. If $x, y \in X$ and $[x,y]$ meets A_s then clearly $[x,y] \cap A_s = [p,q]$, where $[x,p]$ and $[y,q]$ are the bridges from x and y, respectively, to A_s (cf. (6.10)). Thus A_s is a closed subtree of X. Evidently A_s meets every $\langle s \rangle$-invariant subtree. Let B be another subtree with this property. If $p \in A_s$ then $A_p \cap B$ is nonempty and $\langle s \rangle$-invariant, hence equals A_p. Thus $A_p \subset B$ for all $p \in A_s$, whence $A_s \subset B$. This completes the proof of Theorem (6.6).

(6.13) COROLLARY. *Let* s, t *be automorphisms of* X.

(a) $\ell(tst^{-1}) = \ell(s)$, *and* $tA_s = A_{tst^{-1}}$. *If* $ts = st$ *then* $tA_s = A_s$.

(b) *For any* $n \in \mathbb{Z}$ *we have* $\ell(s^n) = |n| \ell(s)$, *and* $A_s \subset A_{s^n}$, *with equality if* $\ell(s) > 0$ *and* $n \neq 0$.

(c) *Let* $x \in X$ *and put* $L_x(s^n) = d(x, s^n x)$ *for* $n \in \mathbb{Z}$. *Then*

(6.14) $$\ell(s) = \max(L_x(s^2) - L_x(s), 0).$$

(d) *If Y is an $<s>$-invariant subtree of X then $\ell(s|Y) = \ell(s)$.*

Since A_s is intrinsically defined by s ((6.6)(b)), part (a) follows by "transport of structure" from the commutative diagram

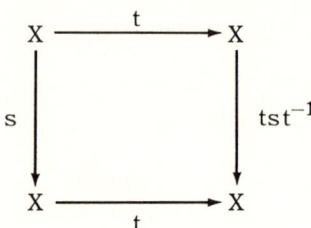

If $\ell(s) = 0$ then A_s is the fixed point set of s, and (b) is clear, as it is also when $n = 0$. So assume that $\ell(s) > 0$ and $n \neq 0$. Then s^n is a translation of amplitude $|n|\ell(s)$ on A_s, whence $A_s \subset A_{s^n}$, by (6.6)(a), and $\ell(s^n) = |n|\ell(s)$. By part (a), A_{s^n} is $<s>$-invariant and, since A_{s^n} is linear, s must induce a translation on A_{s^n} (see (2.19)); hence $A_{s^n} \subset A_s$, thus proving (b).

To prove (c) suppose first that s is not an inversion. If $x \in X$ then from (6.7) and (b) above we have $L_x(s^n) = d(x, s^n x) = \ell(s^n) + 2d(x, A_{s^n}) = |n|\ell(s) + 2d(x, A_{s^n})$. Hence $L_x(s^2) - L_x(s) = \ell(s) + 2d(x, A_s) - d(x, A_{s^2}))$. If $\ell(s) > 0$ then $A_{s^2} = A_s$ (cf. (b) above) so $L_x(s^2) - L_x(s) = \ell(s)$. If $\ell(s) = 0$ then, since $A_s \subset A_{s^2}$, $L_x(s^2) - L_x(s) = 2(d(x, A_{s^2}) - d(x, A_s)) \leq 0$.

Finally, to prove (c) when s is an inversion we embed X in $X \otimes_\Lambda \frac{1}{2}\Lambda$, where $s' = s \otimes_\Lambda \frac{1}{2}\Lambda$ acquires a fixed point, and apply (6.14) to s'. Since $\ell(s) = 0 = \ell(s')$, $L_x(s) = L_x(s')$, and $L_x(s^2) = L_x(s'^2)$, (6.14) for s follows from its validity for s', proved above.

If Y is an $<s>$-invariant subtree and s is not an inversion then $Y \cap A_s \neq \emptyset$ by (6.6)(b)(ii), so $\ell(s|Y) = \ell(s)$. If s is an inversion then $A_s = \emptyset$ and $A_{s^2} \neq \emptyset$, so $A_s \cap Y(= A_{s|Y}) = \emptyset$ and $A_{s^2} \cap Y \neq \emptyset$, by (6.6)(b)(ii) again, so $s|Y$ is an inversion, and $\ell(s) = 0 = \ell(s|Y)$; whence (d). This completes the proof of (6.13).

(6.15) LEMMA. *Let s be an automorphism of* X. *Let* $h: \Lambda \to \Lambda'$ *be a homomorphism of ordered groups such that* $a \geq 0 \implies h(a) \geq 0$. *Let* s' *denote the automorphism* $s \otimes_\Lambda \Lambda'$ *of the* Λ'*-tree* $X' = X \otimes_\Lambda \Lambda'$. *Then* $\ell(s') = h(\ell(s))$ *and the canonical map* $\Phi: X \to X'$ *maps* A_s *into* $A_{s'}$ (cf. (4.4)).

If $\ell(s) = 0$ then A_s is the fixed point set of s, so $\phi(A_s)$ consists of fixed points of s'. Since s^2 has a fixed point, $\ell(s'^2) = 0$, so $\ell(s') = 0$ by (6.13)(b).

If $\ell(s) > 0$ then, by Corollary (4.5) and Example (4.6), $A_s \otimes_\Lambda \Lambda'$ is a linear sub-Λ'-tree of X' on which s' induces a translation of amplitude $h(\ell(s))$, whence $\ell(s') = h(\ell(s))$.

(6.16) COROLLARY. (a) *If s is not an inversion then neither is* $s' = s \otimes_\Lambda \Lambda'$.

(b) *Suppose that* Λ *is a subgroup of* Λ'. *Let* Γ *be a group acting on* X. *Then* Γ *acts without inversions on* $X \otimes_\Lambda \Lambda'$, *iff, for some* $x \in X$, $d(x,sx) \in 2\Lambda'$ *for all* $s \in \Gamma$ *such that* $\ell(s) = 0$.

(a) follows since $A_{s'}$ contains $\phi(A_s) \neq \emptyset$.

(b) The "only if" assertion follows from (6.7). The converse follows from (a) and (6.3)(e).

(6.17) PROPOSITION. *Let s be an automorphism of* S. *The open ends of* X *fixed by s are precisely the open ends of* A_s.

Clearly s fixes the ends of A_s, and, since A_s is a closed subtree, any open end of A_s is also an (open) end of X. For the converse, suppose first that s is not an inversion, and let ε be an open end of X fixed by s. Let $x \in A_s$. If $\ell(s) = 0$ then $s[x,\varepsilon) = [x,\varepsilon)$ so s fixes all points of $[x,\varepsilon)$, and ε is an end of A_s. Suppose now that $\ell(s) > 0$. If $[x,\varepsilon) \not\subset A_s$ then $[x,\varepsilon) \cap A_s = [x,y]$ for some y, and $[y,\varepsilon) \cap A_s = \{y\}$. Since $[sy,\varepsilon) \cap A_s = \{sy\}$ we see

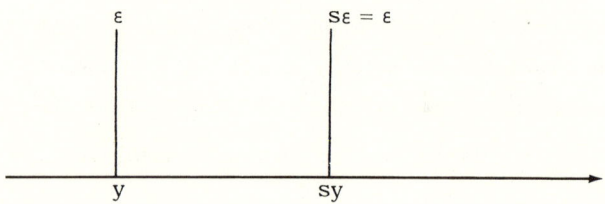

that [y,ε) and [sy,ε) are disjoint; contradiction. Thus [x,ε) ⊂ A_s, as required. Finally, if s is an inversion we pass from X to $X' = X \otimes_\Lambda \tfrac{1}{2}\Lambda$, where s has a unique fixed point, hence no fixed open ends, by the discussion above. Since X and X' have the "same" open ends, s fixes no open ends of X.

(6.18) COROLLARY. *If a group* Γ *acts on* X *and fixes an open end* ε *then* Γ *is without inversions, and* ε *belongs to* A_s *for all* s ∈ Γ.

B. Appendix. Hyperbolic length of s ∈ $GL_2(F)$

Consider, as in Appendix A of Section 3, a field F with valuation $v: F^x \to \Lambda$, valuation ring A, and the tree X_v (Theorem (A.9)) of classes $\bar{L}$ of A-lattices L ⊂ V = F^2. The group $GL_2(F)$ acts on X_v. Given s ∈ $GL_2(F)$ we propose to calculate its hyperbolic length $\ell(s)$.

(B.1) LEMMA. $Tr(s^2) = Tr(s)^2 - 2 Det(s)$.

Proof. Put t = Tr(s) and d = Det(s). Then $s^2 - ts + d \cdot I = 0$ so $Tr(s^2)$ = Tr(ts−dI) = tTr(s) − 2d.

(B.2) LEMMA. *The following conditions are equivalent.*

  (a) s *is conjugate to an element of* $GL_2(A)$.
  (b) sL = L *for some A-lattice* L ⊂ V.
  (c) Tr(s) ∈ A *and* Det(s) ∈ A^x.

Clearly (a) ⟺ (b) ⟹ (c). Assuming (c), for any A-lattice L, s stabilizes L' = L + sL, since s^2 − Tr(s)s + Det(s) I = 0. Since Det(s) ∈ A^x we must have sL' = L'.

(B.3) PROPOSITION. *The following conditions are equivalent.*
  (a) s *has a fixed point in* X_v
  (b) $2\,v(\text{Tr}(s)) \geq v(\text{Det}(s)) \in 2\Lambda$.

Proof. (a) $\Longrightarrow$ (b). If $s\bar{L} = \bar{L}$ then $asL = L$ for some $a \in F^x$, so
((B.2)) $\text{Tr}(as) = a\text{Tr}(s) \in A$ and $\text{Det}(as) = a^2 \cdot \text{Det}(s) \in A^x$. Therefore

$$\frac{\text{Tr}(s)^2}{\text{Det}(s)} = \frac{(a\,\text{Tr}(s))^2}{a^2\,\text{Det}(s)} \in A,$$

whence $v(\text{Det}(s)) \leq v(\text{Tr}(s)^2) = 2\,v(\text{Tr}(s))$. Moreover $0 = v(a^2 \text{Det}(s)) = 2v(a) + v(\text{Det}(s))$, so $\text{Det}(s) \in 2\Lambda$.

  (b) $\Longrightarrow$ (a). Choose $a \in F^x$ so that $v(\text{Det}(s)) + 2v(a) = 0$. Then $\text{Det}(as) \in A^x$. Moreover $v(\text{Tr}(as)) = v(a) + v(\text{Tr}(s)) \geq -\frac{1}{2} v(\text{Det}(s)) + \frac{1}{2} v(\text{Det}(s)) = 0$. By (B.2), as fixes some A-lattice $L \subset V$, hence $s\bar{L} = \bar{L}$.

(B.4) COROLLARY. *If* $\text{Det}(s) \in A^x$ *and* $s\bar{L} = \bar{L}$ *then* $sL = L$.

Proof. In the proof of (a) $\Longrightarrow$ (b) above we have $a^2 \text{Det}(s) \in A^x$ so $a \in A^x$, hence $sL = L$.

(B.5) COROLLARY. *The following conditions are equivalent.*
  (a) $\ell(s) = 0$
  (b) $v(\text{Tr}(s^2)) \geq v(\text{Det}(s))$
  (c) $2v(\text{Tr}(s)) \geq v(\text{Det}(s))$.

Proof. By (6.3) and (6.13), $\ell(s) = 0$ iff s^2 has a fixed point, iff $2v(\text{Tr}(s^2)) \geq v(\text{Det}(s^2)) \in 2\Lambda$, by (B.3). Since $v(\text{Det}(s^2)) = 2v(\text{Det}(s))$, the latter condition is equivalent to (b), whence (a) $\Longleftrightarrow$ (b). From (B.1) we have $\frac{\text{Tr}(s^2)}{\text{Det}(s)} = \frac{\text{Tr}(s)^2}{\text{Det}(s)} - 2$. Thus

$$\text{(b)} \Longleftrightarrow \left[\frac{\text{Tr}(s^2)}{\text{Det}(s)} \in A\right] \Longleftrightarrow \left[\frac{\text{Tr}(s)^2}{\text{Det}(s)} \in A\right] \Longleftrightarrow \text{(c)}.$$

(B.6) COROLLARY. *The following conditions are equivalent.*

(a) s *is an inversion.*

(b) $2v(\operatorname{Tr}(s)) \geq v(\operatorname{Det}(s)) \notin 2\Lambda$.

Proof. (a) $\iff$ [$\ell(s) = 0$ and s has no fixed point], whence (a) $\iff$ (b) by (B.5) and (B.3).

(B.7) PROPOSITION. $\ell(s) = \max(v(\operatorname{Det}(s)) - v(\operatorname{Tr}(s^2)), 0)$

$$= \max(v(\operatorname{Det}(s)) - 2v(\operatorname{Tr}(s)), 0).$$

Proof. Put $t = \operatorname{Tr}(s)$, $t' = \operatorname{Tr}(s^2)$, and $d = \operatorname{Det}(s)$. By (B.1) we have $t'/d = (t^2/d) - 2$. By (B.5) we have $\ell(s) = 0 \iff v(d) \leq v(t') \iff v(d) \leq 2v(t)$. If $v(d) > v(t')$ then $v(t'/d) < 0$, i.e. $t'/d \notin A$, so $v(t^2/d) = v((t'/d)+2) = v(t'/d)$; hence $2v(t) = v(t')$ and $v(d) - v(t') = v(d) - 2v(t) > 0$. It suffices to show in this case that $\ell(s) = v(d) - 2v(t)$. Note that $t \neq 0$, since $v(0) = \infty$. Replacing s by s/t, which clearly affects neither side of the formula in the proposition, we further reduce to the case where $t = 1$, and hence $v(d) > 0$. We must now show that $\ell(s) = v(d)$. Take any lattice $L_0 \subset V$, and set $L = L_0 + sL_0$. We have $s^2 - s + d \cdot I = 0$ so $sL \subset L$ and, putting $u = I - s$,

$$s + u = I, \quad su = s - s^2 = dI, \quad u = ds^{-1}.$$

Thus we have the "lattice of lattices"

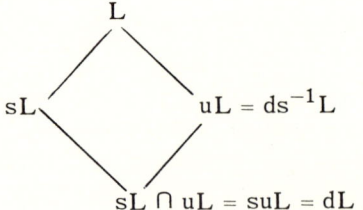

$sL \cap uL = suL = dL$.

(If $x = sy = uz \in sL \cap uL$ then $x = (s+u)x = suz + usy \in suL$, whence $sL \cap uL = suL$.) Thus $L/dL = L/sL \oplus L/uL$, whence $L/sL \cong L/uL \cong A/dA$. From (A.5) we now conclude that, in X_V,

$$[s\bar{L}, s^{-1}\bar{L}] = [\overline{sL}, \overline{s^{-1}L}] = [\overline{sL}, \overline{uL}]$$

$$= [\overline{sL}, \bar{L}, \overline{uL}].$$

It follows therefore from (6.6) that $\bar{L} \in A_s$ and so $\ell(s) = d(\bar{L}, s\bar{L}) = |L/sL| = |A/dA| = v(d)$.

7. The main theorems

We fix an ordered abelian group Λ, and a group Γ. By a Γ-(Λ-tree) we mean a Λ-tree X with an action of Γ (by Λ-isometries). We have the obvious notion of Γ-(Λ-morphism). We say that X is *without inversions* if the action of Γ on X is without inversions (6.4).

Associated to a Γ-(Λ-tree) X is its *hyperbolic length function* $\ell_X : \Gamma \to \Lambda$. The following result recapitulates (6.13), (6.15), and (6.16).

(7.1) PROPOSITION. *Let X be a Γ-(Λ-tree).*

(a) $\ell_X(tst^{-1}) = \ell_X(s)$ and $tA_s = A_{tst^{-1}}$ for all $s, t \in \Gamma$.

(b) For all $n \in \mathbb{Z}$ and $s \in \Gamma$, $\ell_X(s^n) = |n|\ell(s)$, and $A_s \subset A_{s^n}$, with equality if $\ell(s) > 0$ and $n \neq 0$.

(c) Putting $L_X(s) = d(x, sx)$ for all $x \in X$, $s \in \Gamma$, we have

(7.2) $$\ell_X(s) = \max(L_X(s^2) - L_X(s), 0).$$

(d) If Y is a Γ-invariant subtree of X then $\ell_Y = \ell_X$.

(e) Let $h : \Lambda \to \Lambda'$ be a homomorphism of ordered abelian groups such that $a \geq 0 \Longrightarrow h(a) \geq 0$. Then $h \circ \ell_X : \Gamma \to \Lambda'$ is the hyperbolic length function of the Γ-(Λ'-tree) $X \otimes_\Lambda \Lambda'$.

(f) If Λ' is an ordered abelian group containing Λ then the Γ-(Λ'tree) $X \otimes_\Lambda \Lambda'$ is without inversions iff, for some (and hence every) $x \in X$, we have $d(x, sx) \in 2\Lambda'$ whenever $s \in \Gamma$ and $\ell_X(s) = 0$. In particular Γ acts without inversions on $X \otimes_\Lambda \frac{1}{2}\Lambda$.

(7.3) DEFINITION. Let X be a Γ-(Λ-tree). Define $\beta_X : \Gamma \times \Gamma \to \Lambda$ by

$$\beta_X(s,t) = \ell_X(st) - \ell_X(s) - \ell_X(t) .$$

Call ℓ_X *abelian* if $\beta_X(s,t) \leq 0$ for all $s,t \in \Gamma$. Otherwise call ℓ_X *non-abelian*. (This terminology was proposed by John Morgan.) Call ℓ_X *dihedral* if ℓ_X is non-abelian but $\beta_X(s,t) \leq 0$ whenever $\ell_X(s) > 0$ and $\ell_X(t) > 0$. If ℓ_X is neither abelian nor dihedral we say that ℓ_X is of a *general type*.

The following theorem will be proved in Section 8, Theorem (8.4)(a). Meanwhile we shall use it here to derive further conclusions.

(7.4) THEOREM. *Let X be a Γ-(Λ-tree). If $s,t \in \Gamma$ are not inversions then*

$$d(A_s, A_t) = \max\left(\frac{1}{2} \beta_X(s,t), 0\right).$$

Thus if X is without inversions then ℓ_X is abelian iff $A_s \cap A_t \neq \emptyset$ for all $s,t \in \Gamma$, while ℓ_X is dihedral iff $A_s \cap A_t \neq \emptyset$ for all hyperbolic $s,t \in \Gamma$, but $A_s \cap A_t = \emptyset$ for some $s,t \in \Gamma$.

The terminology in the next theorem is taken from (2.23), (2.25), and (2.27).

(7.5) THEOREM. *Let X be a Γ-(Λ-tree) with abelian hyperbolic length function $\ell = \ell_X$.*

(a) *Suppose that Γ acts with an inversion. Then $\ell = 0$ and $X^\Gamma = \emptyset$. More generally, let Λ' be an ordered abelian group containing Λ. Then $(X \otimes_\Lambda \Lambda')^\Gamma = \emptyset$ if $d(x, sx) \notin 2\Lambda'$ for some $x \in X$, $s \in \Gamma$; otherwise $(X \otimes_\Lambda \Lambda')^\Gamma$ consists of a single fixed point.*

(b) *Suppose that Γ acts without inversions. Then exactly one of the following cases occurs.*

Case ($\cap A_s \neq \emptyset$). *The intersection $B = \cap_{s \in \Gamma} A_s$ is not empty. If $\ell = 0$ then $B = X^\Gamma$. If $\ell \neq 0$ then $X^\Gamma = \emptyset$ and B is a Γ-invariant linear subtree on which Γ acts as a non-trivial group of translations.* (We say here that X is of *core type*.)

Case (ε). We have $\bigcap_{s \in \Gamma} A_s = \emptyset$ and Γ fixes an end ε of X. Then ε is unique, it is an open end, and for all $s \in \Gamma$ and all $x \in A_s$, $[x, \varepsilon) \subset A_s$. (We say here that X is of *end type*.)

Case ($X = X_0 \amalg X_1$). We have $\bigcap_{s \in \Gamma} A_s = \emptyset$ and Γ fixes no end of X. Then there is a unique open cut $X = X_0 \amalg X_1$ of X such that, for all $s \in \Gamma$, $A_s \cap X_i \neq \emptyset$ ($i = 0, 1$). Γ leaves each X_i invariant and fixes the ends ε_i of X_i ($i = 0, 1$) that meet at the cut; thus the Γ-(Λ-trees) X_i ($i = 0, 1$) are each of end type. (We say here that X is of *cut type*.)

If $\Lambda = \mathbb{Z}$ or $\mathbb{R}$ then cut type cannot occur.

Proof of (7.5). (a). Since Γ contains an inversion, $X^\Gamma = \emptyset$. Let $s \in \Gamma$ be such that $\ell(s) = 0$, e.g. an inversion. Then, by (6.3)(d), s acts as an inversion on $X' = X \otimes_\Lambda \Lambda'$ if $d(x, sx) \notin 2\Lambda'$ for some $x \in X$. Suppose, on the contrary, that $d(x, sx) \in 2\Lambda'$ for all $x \in X$ and all $s \in \Gamma$ such that $\ell(s) = 0$. Let S denote the set of inversions in Γ. Then, by (6.3)(e), each $s \in S$ has a unique fixed point $x_s \in X'$; $A_s = \{x_s\}$. If $s, t \in S$ and $x_s \neq x_t$ then, by Theorem (7.4) above, $\beta_{X'}(s, t) = 2d(x_s, x_t) > 0$. However since $\ell_{X'} = \ell_X$ (by (7.1)(e)) we have $\beta_{X'} = \beta_X$, so, since ℓ_X is abelian, $\beta_{X'}(s, t) \leq 0$; contradiction. Thus the elements of S have a common unique fixed point $x_0 \in X'$. Since S is invariant under conjugation in Γ, Γ must fix x_0, hence $X'^\Gamma = \{x_0\}$ and $\ell_X = \ell_{X'} = 0$.

(b). Assume that Γ acts without inversions, and put $B = \bigcap_{s \in \Gamma} A_s$. Suppose first that $B \neq \emptyset$. Then B is a Γ-invariant subtree. If $\ell = 0$ then $B = X^\Gamma$. If $\ell \neq 0$ then B is contained in every hyperbolic axis so B is a linear subtree on which Γ acts as a group of translations.

Suppose now that $B = \emptyset$. Then, by Theorem (2.28), either:

(ε) There is an end ε of X common to all A_s; it is unique, it is an open end, and, for all $s \in \Gamma$, $[x, \varepsilon) \subset A_s$ for all $x \in A_s$; or

($X = X_0 \amalg X_1$) There is a unique cut $X = X_0 \amalg X_1$ of X such that, for all $s \in \Gamma$, $A_s \cap X_i \neq \emptyset$ ($i = 0, 1$). The ends ε_i of X_i ($i = 0, 1$) meeting at the cut are open. For all $s \in \Gamma$ and all $x \in A_s \cap X_i$, $[x, \varepsilon_i) \subset (A_s \cap X_i)$.

Moreover the latter case cannot occur if $\Lambda = \mathbf{Z}$ or $\mathbf{R}$.

Since Γ preserves $\{A_s | s \epsilon \Gamma\}$ ($tA_s = A_{tst^{-1}}$) we see that Γ preserves ϵ in case (ϵ) and Γ preserves $\{X_0, X_1\}$ in case ($X = X_0 \amalg X_1$). We claim, in the latter case, that Γ leaves each X_i invariant. Let $s \epsilon \Gamma$ and put $A_{si} = A_s \cap X_i \neq \emptyset$. If $\ell(s) = 0$ then s fixes the points of each A_{si}, hence $sX_i = X_i$ ($i = 0,1$). Suppose that $\ell(s) > 0$. Then $A_s = A_{s0} \amalg A_{s1}$ is an open cut of the s-axis. (E.g. apply (2.37) with $Y = A_s$.) Since s is equivalent to a translation on A_s it cannot exchange A_{s0} and A_{s1}. Hence again $sX_i = X_i$ ($i = 0,1$). Now from (2.26)(b), Γ must fix the unique ends ϵ_i of X_i ($i = 0,1$) that meet at the cut.

To complete the proof of Theorem (7.5), we recall from (6.18) that if Γ fixes an open end ϵ of X then, for all $s \epsilon \Gamma$, ϵ belongs to A_s.

The next theorem gives an approximate geometric picture of Γ-(Λ-trees) with abelian hyperbolic length functions.

(7.6) THEOREM. *Let X be a Γ-(Λ-tree) with abelian hyperbolic length function $\ell = \ell_X$.*

(a) *There is a homomorphism $h: \Gamma \to \Lambda$, unique up to a factor ± 1, such that $\ell(s) = |h(s)|$ for all $s \epsilon \Gamma$.*

(b) *Suppose that Γ acts without inversions. Choose h as in (a) and let Γ act on Λ by $s(a) = a + h(s)$ for $s \epsilon \Gamma$, $a \epsilon \Lambda$. Then there is a Γ-equivariant map $\alpha: X \to \Lambda$ with the following properties.*

(i) *α is Λ-metric on every hyperbolic axis A_s ($s \epsilon \Gamma$, $\ell(s) > 0$).*

(ii) *If X is of end type, with Γ fixing the end ϵ then α is Λ-metric on $[x, \epsilon)$ for all $x \epsilon X$.*

(iii) *If X is of cut type, corresponding to the cut $X = X_0 \amalg X_1$, then α is Λ-metric on $[x_0, x_1]$ for all $(x_0, x_1) \epsilon X_0 \times X_1$.*

If X is of end type or cut type then α is unique modulo $\text{Aut}_{\text{metric}}(\Lambda)$, and even unique up to a translation of Λ if $\ell \neq 0$. If x is of core type this uniqueness holds for the restriction of α to the union of the hyperbolic axes A_s ($s \epsilon \Gamma$, $\ell(s) > 0$).

Proof of (7.6). We first note that the uniqueness property of h follows from (1.4). If Γ acts with an inversion then $\ell = 0$ by (7.5)(a). For the rest of the proof we may therefore assume that Γ *acts without inversions.* Then by (7.5) we have one of the following cases:

Core type: $B = \bigcap_{s \in \Gamma} A_s \neq \emptyset$

End type: $B = \emptyset$ and Γ fixes an end ε of X.

Cut type: $B = \emptyset$, Γ fixes no end of X, and there is an open cut $X = X_0 \amalg X_1$ of X, making ends ε_i of X_i ($i = 0,1$) meet, and Γ leaves each X_i and ε_i invariant.

For core type we have $\ell_X = \ell_B$ ((7.1)(d)). If $\ell_X = 0$, any Γ-equivariant map $a: X \to \Lambda$ will do. If $\ell_X \neq 0$ then B is a linear tree on which Γ acts as a non-trivial group of translations. Thus, if we embed B in Λ, $a_B: B \to \Lambda$, then there is a unique homomorphism $h: \Gamma \to \Lambda$ so that $a_B(sx) = a_B(x) + h(s)$ for all $x \in B$ and all $s \in \Gamma$. Then we have $|h(s)| = \ell_B(s) = \ell_X(s)$ for all $s \in \Gamma$. By (2.5)(b) and (2.31) there is an extension of a_B to a map $a: X \to \Lambda$ that is Λ-metric on every linear subtree containing B. We can choose a to be Γ-equivariant outside the union of such linear subtrees. If $s \in \Gamma$ then $a(sx)$ and $a(x) + h(s)$ have the same property and agree on B, hence they agree on all linear subtrees containing B by (2.31). Suppose $a': X \to \Lambda$ is another Γ-equivariant map that is Λ-metric on every hyperbolic axis $A_s (s \in \Gamma$, $\ell(s) > 0)$. By (2.5)(b) $a'|_B = \phi \circ a_B$ for some $\phi \in \text{Aut}(\Lambda)$. Now ϕ is a translation or a reflection, and a reflection would convert the action of Γ on $a_B(B)$ from h to $-$h; hence, since $h \neq 0$, ϕ must be a translation. Thus, modifying a' by a translation, we may assume that $a'|_B = a_B$. Then, by (2.30)(a), a' and a coincide on every hyperbolic axis. This proves the theorem when X is of core type.

If X is of end type then (2.33) gives us an $a: X \to \Lambda$ that is Λ-metric on each X-ray $[x, \varepsilon)$. If $s \in \Gamma$ then $s\varepsilon = \varepsilon$ by (7.5) so, by (2.34), $a(sx) = a(x) + h(s)$ for a unique $h(s) \in \Lambda$. For $s, t \in \Gamma$ we have $a(stx) = a(tx) + h(s) = a(x) + h(s) + h(t)$, so $h(st) = h(s) + h(t)$. Thus $h: \Gamma \to \Lambda$ is a homomorphism, and $a: X \to \Lambda$ is Γ-equivariant relative to the translation action on Λ defined by h. By considering $a|A_s$ we see easily that $\ell(s) = |h(s)|$ for $s \in \Gamma$. Finally, if $a': X \to \Lambda$ is also Γ-equivariant relative to h and if a' is Λ-metric on $[x, \varepsilon)$ for all $x \in X$ then $a' = \phi \circ a$ for a unique $\phi \in \text{Aut}(\Lambda)$, by (2.33) and ϕ can't be a reflection if $\ell \neq 0$, since it would then convert the action defined by h into that defined by $-h$. Thus ϕ is a translation if $\ell \neq 0$. This proves the theorem when X is of end type.

If X is of cut type then (2.35) gives us an $a: X \to \Lambda$ that is Λ-metric on $[x_0, x_1]$ whenever $x_i \in X_i$ $(i = 0,1)$. If $s \in \Gamma$ then $sX_i = X_i$ $(i = 0,1)$ by (7.5) so, by (2.36), $a(sx) = a(x) + h(s)$ for a unique $h(s) \in \Lambda$. As in the case of end type we now conclude that h is a homomorphism, a is Γ-equivariant for the translation action defined by h, that any other such map a' is of the form $a' = \phi \circ a$ for a unique $\phi \in \text{Aut}(\Lambda)$, and that, if $\ell \neq 0$, ϕ must be a translation. This concludes the proof of (7.6).

(7.7) REMARKS. 1. Let X be a Γ-(Λ-tree) without inversions and abelian hyperbolic length function. Then $\ell = \ell_X$ can be interpreted also as a kind of geometric Lyndon length function, as follows.

Core type: Choose $x \in \bigcap_{s \in \Gamma} A_s$. Then it is clear from (6.7) that $L_x(s) = \ell(s)$ for all $s \in \Gamma$.

End type: For $x, y \in X$ define $d_\varepsilon(x, y) = |d(x, z) - d(y, z)|$ for any $z \in [x, \varepsilon) \cap [y, \varepsilon)$. This is easily seen to be independent of the choice of z. Moreover one can deduce from (6.6)(c) that for all $s \in \Gamma$, $\ell(s) = $ "$L_\varepsilon(s)$" is the common value of $d_\varepsilon(x, sx)$ for all $x \in X$.

Cut type: Let $\mu = (X_0, X_1)$ define the cut. For $x, y \in X$ define $d_\mu(x, y) = d(x, y)$ if x, y are not both in one of the X_i's. If $x, y \in X_0$

define $d_\mu(x,y) = |d(x,z)-d(y,z)|$ for any $z \in X_1$, and analogously if $x,y \in X_1$. Then, as above, one can show that d_μ is well defined and that, for all $s \in \Gamma$, $\ell(s) = "L_\mu(s)"$ is the common value of $d_\mu(x,sx)$ for all $x \in X$.

2. Let $(A_i)_{i \in I}$ be any family of closed subtrees of a Γ-(Λ-tree) X such that $A_i \cap A_j \neq \emptyset$ for all $i,j \in I$, and put $A = \bigcap_i A_i$. Assume that Γ leaves the family invariant. If $A \neq \emptyset$ it is a Γ-invariant subtree. If $A = \emptyset$ we have one of the following cases, by (2.28).

End case. There is a unique open end ε common to all A_i. Clearly ε is fixed by Γ. Hence, by (6.18) and (7.4) Γ acts without inversions and ℓ_X is abelian.

Cut case. There is a unique open cut $X = X_0 \amalg X_1$ such that $A_{ij} = A_i \cap X_j \neq \emptyset$ $(j=0,1)$ for all $i \in I$. It follows that Γ permutes $\{X_0, X_1\}$. Let Γ_t denote the subgroup of index ≤ 2 stabilizing each X_j. Then Γ_t fixes the open ends ε_j of X_j $(j=0,1)$ that meet at the cut. By (6.18) and (7.4) again, Γ_t acts without inversions and $\ell_X|\Gamma_t$ is abelian. The argument at the end of the proof of (7.5) shows further that Γ_t contains all non-inversions of Γ, in particular, all hyperbolic elements. It follows therefore that if $\Gamma_t \neq \Gamma$ then ℓ_X is dihedral and every $s \notin \Gamma_t$ is an inversion which exchanges X_0 and X_1.

That the latter situation can really arise is seen in the following example. Let $X = \Lambda = Z^2$, with the lexicographic ordering. Then we have the open cut $X = X_0 \amalg X_1$ where $(a,b) \in X_j$ iff $a \leq 0$ for $j = 0$, and $a \geq 1$ for $j = 1$. For $n, m \in Z$ put $A_{n,m} = [(0,n),(1,m)]$. Then $A_{n,m} \cap A_{n',m'} \neq \emptyset$ always, $\bigcap_{n,m} A_{n,m} = \emptyset$, and $A_{n,m} \cap X_j \neq \emptyset$ $(j=0,1)$ for all n,m. For $p \in Z$ define $t_p, r_p \in \text{Aut}(X)$ by $t_p(a,b) = (a, b+p)$ and $r_p(a,b) = (1-a, p-b)$. Then $\Gamma = \{t_p, r_p | p \in Z\}$ is a group, each t_p is a translation stabilizing X_0 and X_1, and each r_p is an inversion exchanging X_0 and X_1. Moreover $t_p A_{n,m} = A_{n+p, m+p}$ and $r_p A_{n,m} = A_{p-m, p-n}$.

Writing $Y' = Y \otimes_\Lambda \frac{1}{2}\Lambda$ for a Λ-tree Y, we have $X' = X_0' \amalg L \amalg X_1'$, where $L = \left(\frac{1}{2}, \frac{1}{2} Z\right)$. Thus the cut is filled not by a point, as one might expect, but by an entire interval L. Moreover L is the unique minimal Γ-invariant sub $\left(\frac{1}{2}\Lambda\right)$-tree of X'. It contains the fixed points of the inversions r_p.

We now proceed to the non-abelian case.

(7.8) THEOREM. *Let X be a Γ-(Λ-tree) without inversions and with nonabelian hyperbolic length function ℓ_X. Choose $u, v \in \Gamma$ with $\beta_X(u,v) > 0$, and let $[x_u, x_v]$ be the bridge $[A_u, A_v]$ from A_u to A_v (see (2.17)). (Note that $A_u \cap A_v = \emptyset$ by (7.4).) Then the Lyndon length function $L_{x_u}(s) = d(x_u, sx_u)$ is given by*

(7.9) $\quad L_{x_u}(s) = \ell_X(s) + 2d(A_s, x_u)$

$\qquad = \ell_X(s) + 2 \max(d(A_s, A_u), d(A_s, A_v) - d(A_u, A_v))$

$\qquad = \ell_X(s) + \max(\beta_X(s,u), 0, \beta_X(s,v) - \beta_X(u,v))$

$\qquad = \max(\ell_X(su) - \ell_X(u), \ell_X(s), \ell_X(sv) - \ell_X(v) - \beta_X(u,v))$.

Recall formula,

(7.2) $\qquad \ell_X(s) = \max(L_{x_u}(s^2) - L_{x_u}(s), 0) \qquad (s \in \Gamma)$

(7.10) COROLLARY. *In the setting of Theorem (7.8), formulas (7.9) and (7.2) express each of L_{x_u} and ℓ_X in terms of the other. In particular $L_{x_u}(\Gamma)$ and $\ell_X(\Gamma)$ generate the same subgroup of Λ.*

(7.11) REMARK. Which functions $\ell : \Gamma \to \Lambda$ are hyperbolic length functions? In principle we can now answer this, at least when ℓ is "nonabelian," in the sense that $\ell(uv) - \ell(u) - \ell(v) > 0$ for some $u, v \in \Gamma$. For we can then define $L : \Gamma \to \Lambda$ by the analogue of (7.9), with L, ℓ in the roles of L_{x_u}, ℓ_X, respectively. If ℓ is to be a hyperbolic length function then L must be a Lyndon length function, i.e. L must satisfy axioms L0, L1, L2, L3 of (5.1). Conversely, if L is a Lyndon length function then, by Theorem (5.4), $L = L_{x_0}$, the Lyndon length function of some Λ-tree X on which Γ acts, relative to a base point $x_0 \in X$. Thus we obtain axioms for (nonabelian) hyperbolic length functions as follows. First use (7.9) to translate conditions L0, L1, L2, L3 into conditions on ℓ. Then add further conditions, if necessary, to assure that $\ell(s) = \max(L(s^2) - L(s), 0)$ for all $s \in \Gamma$. In view of (7.2), it will then follow that $\ell = \ell_X$, the hyperbolic length function of the tree X with Γ-action constructed above from L.

If this program is carried out by direct assault it leads, unfortunately, to an unmanageably large and complicated set of axioms for ℓ. It does show, however, that hyperbolic length functions are characterized by conditions in the form of "conditional inequalities." In Section 9 below we attempt to derive a more economical set of axioms for ℓ.

For the proof of Theorem (7.8) we require:

(7.12) LEMMA. *In the setting of (7.8), we have for all $s \in \Gamma$,*

$$d(A_s, x_u) = \max(d(A_s, A_u), \; d(A_s, A_v) - d(A_u, A_v)).$$

Proof of (7.12). Let $[x_u, x_s]$ be the bridge from x_u to A_s; $x_u = x_s$ if $x_u \in A_s$. Put $p = Y(x_s, x_u, x_v)$. Then $d(A_s, x_u) = d(x_s, x_u) \geq d(A_s, A_u)$, clearly. Moreover $d(x_s, x_u) = d(x_s, p) + d(p, x_u) \geq d(x_s, p) - d(p, x_u) = (d(x_s, p) + d(p, x_v)) - (d(x_u, p) + d(p, x_v)) = d(x_s, x_v) - d(x_u, x_v) = d(x_s, x_v) - d(A_u, A_v) \geq d(A_s, A_v) - d(A_u, A_v)$. It suffices therefore to establish the following:

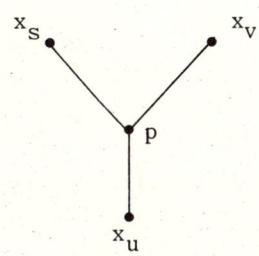

(a) $p \neq x_u \implies d(x_s, x_u) = d(A_s, A_u)$

(b) $p = x_u \neq x_s \implies d(x_s, x_u) = d(A_s, A_v) - d(A_u, A_v)$

(c) $p = x_u = x_s \implies d(x_s, x_u) = 0 = d(A_s, A_u)$.

Assertion (c) is immediate. To prove (b) we need only show that the two inequalities in the derivation above are equalities when $p = x_u \neq x_s$. The first one is an equality since $d(p, x_u) = 0$. For the second we must show that $d(x_s, x_v) = d(A_s, A_v)$, i.e. that $[x_s, x_v] = [x_s, x_u, x_v]$ is the bridge from A_s to A_v. Since $[x_s, x_v] \cap A_v$ is a segment containing x_v but no other point of $[x_u, x_v]$, and since $x_u \neq x_v$, we must have $[x_s, x_v] \cap A_v = \{x_v\}$. Similarly $[x_s, x_v] \cap A_s$ is a segment containing x_s but no other point of $[x_s, x_u]$, and $x_s \neq x_u$ by hypothesis, so $[x_s, x_v] \cap A_s = \{x_s\}$; whence the claim.

Suppose finally, to prove (a), that $p \neq x_u$. Then $[x_s, x_u] \cap A_u$ is a segment containing x_u but no other point of $[x_u, p] \subset [x_u, x_v]$, with $x_u \neq p$, and so $[x_s, x_u] \cap A_u = \{x_u\}$. It follows that $[x_s, x_u]$ is the bridge from A_s to A_u, so $d(x_s, x_u) = d(A_s, A_u)$, as claimed.

Proof of (7.8). We have, with $\beta = \beta_X$ and $\ell = \ell_X$,

$$L_{x_u}(s) \overset{\text{def}}{=} d(x_u, sx_u) = \ell(s) + 2d(x_u, A_s) \qquad (6.7)$$

$$= \ell(s) + 2 \max(d(A_s, A_u), d(A_s, A_v) - d(A_u, A_v)) \qquad (7.12)$$

$$= \ell(s) +$$
$$2 \max\left(\max\left(\tfrac{1}{2}\beta(s,u), 0\right), \max\left(\tfrac{1}{2}\beta(s,v), 0\right) - \max\left(\tfrac{1}{2}\beta(u,v), 0\right)\right) \quad (7.4)$$

$$= \ell(s) + \max(\beta(s,u), 0, \max(\beta(s,v), 0) - \beta(u,v)) \qquad (\beta(u,v) > 0)$$

$$= \ell(s) + \max(\beta(s,u), 0, \beta(s,v) - \beta(u,v)) \qquad (\beta(u,v) > 0)$$

$$= \max(\ell(s) + \beta(s,u), \ell(s), \ell(s) + \beta(s,v) - \beta(u,v))$$

$$= \max(\ell(su) - \ell(u), \ell(s), \ell(sv) - \ell(v) - \beta(u,v)). \qquad (7.3)$$

(7.13) THEOREM. *Let X be a Γ-(Λ-tree) without inversions and with nonabelian hyperbolic length function ℓ_X.*

(a) There is a unique minimal Γ-invariant subtree $X_{\min}$ of X, and $\ell_{X_{\min}} = \ell_X$. If $u,v \in \Gamma$, $A_u \cap A_v = \emptyset$, and $[x_u, x_v]$ is the bridge from A_u to A_v then $X_{\min}$ is the subtree spanned by the orbit $\Gamma \cdot x_u$. (We call X minimal if $X_{\min} = X$.)

(b) If Y is a Γ-(Λ-tree) without inversions and $\ell_Y = \ell_X$ then there is a unique Γ-(Λ-isomorphism) $Y_{\min} \to X_{\min}$.

(c) Let $\Lambda_X = \langle \ell_X(\Gamma) \rangle$, the subgroup of Λ generated by $\ell_X(\Gamma)$. Let Λ_0 be a subgroup of Λ such that $\Lambda_X \cap 2\Lambda \subset 2\Lambda_0$. There is a subspace X_0 of $X_{\min}$ which is a Λ_0-tree, invariant under Γ, such that the inclusion induces a Γ-(Λ-isomorphism) $X_0 \otimes_{\Lambda_0} \Lambda \to X_{\min}$ (cf. Section 4).

Before proving (7.13) we discuss some simple examples.

(7.14) EXAMPLES. Let $X = \Lambda$. Recall from (2.4) and (2.5) that $\text{Aut}(X)$ consists of translations, $t_a(x) = a + x$, and reflections $r_a(x) = a - x$ ($a \in \Lambda$), with multiplication table $t_a t_b = t_{a+b}$, $t_a r_b = r_{a+b}$, $r_a t_b = r_{a-b}$, $r_a r_b = t_{a-b}$. In particular $r_a^2 = 1$, $r_a t_b r_a^{-1} = t_{-b} = t_b^{-1}$, and $t_a r_b t_a^{-1} = r_{b+2a}$. Moreover r_a is an inversion iff $a \notin 2\Lambda$; otherwise it has the unique fixed point $a/2$. We have $\ell(t_a) = |a|$ and $\ell(r_a) = 0$.

Fix a subgroup Λ_0 of Λ, and put $\Gamma = t_{\Lambda_0} \rtimes \langle r \rangle$, the semi-direct product of the group t_{Λ_0} of translations t_a ($a \in \Lambda_0$) with a group of order two generated by r such that $r t_a r^{-1} = t_{-a} = t_a^{-1}$ for $a \in \Lambda_0$. Given $c \in \Lambda$ let Γ act on X with the natural action of t_{Λ_0} and so that r acts via r_c; denote the resulting Γ-(Λ-tree) X_c.

1. We have $\ell_{X_c}(t_a) = |a|$ and $\ell_{X_c}(t_a r) = 0$ for all $a \in \Lambda_0$. Thus ℓ_{X_c} is independent of c; denote it by ℓ.

2. X_c is without inversions iff $c + \Lambda_0 \subset 2\Lambda$ iff $\Lambda_0 \subset 2\Lambda$ and $c \in 2\Lambda$.

3. If $\Lambda_0 = 0$ then $\ell = 0$. If $\Lambda_0 \neq 0$ then $\ell \neq 0$ and, in fact, ℓ is nonabelian. For if $a \neq 0$ in Λ_0 then $\ell((t_a r) \cdot r) - \ell(t_a r) - \ell(r) = \ell(t_a) - 0 - 0 = |a| > 0$.

4. X_c and X_d are isomorphic as Γ-(Λ-trees) iff $c \equiv d \mod 2\Lambda$. In fact if $d = c + 2b$ then $t_b : X_c \to X_d$ is a Γ-(Λ-isomorphism). Conversely if r_c and r_d are conjugate in $\text{Aut}(X)$, say $r_d = u r_c u^{-1}$, then, by replacing u by $u r_c$, if necessary, we can assume that u is a translation, t_b, and then $t_b r_c t_b^{-1} = r_{c+2b}$.

5. Suppose that $0 \neq \Lambda_0 \subset 2\Lambda$ and $c \notin 2\Lambda$. Then X_c and X_0 furnish Γ-(Λ-trees) with the same nonabelian hyperbolic length function such that Γ acts with inversions on X_c but without inversions on X_0. We shall see in (7.16) that these are essentially the only such examples.

6. Suppose that X_c is without inversions and that ℓ is nonabelian. Then $\Lambda_0 \neq 0$, by 3, and $X_c \cong X_0$ by 2 and 4. It is easy to see that $(X_0)_{\min} = \overline{\Lambda}_0$, the convex subgroup of Λ generated by $\Lambda_0 = \ell(\Gamma)$.

7. From 6 one can easily deduce the following: let Γ be any group. Let X be a Γ-(Λ-tree) without inversions and with nonabelian ℓ_X. Suppose further that X is a *linear* tree. Then $\Lambda_X = \ell_X(\Gamma)$ is a nonzero subgroup of Λ, and $X_{\min} \cong \overline{\Lambda}_X$, the convex closure of Λ_X.

Proof of (7.13). Let $u, v \in \Gamma$ and $x_u, x_v \in X$ be as in (a). Define $X_{\min}$ to be the Λ-tree spanned by $\Gamma \cdot x_u$; clearly $X_{\min}$ is Γ-invariant. Let X' be a Γ-invariant subtree of X. By Theorem (6.6)(b)(ii), X' meets A_u and A_v, and so, by (2.17), X' contains the bridge $[A_u, A_v] = [x_u, x_v]$, whence X' contains $X_{\min}$. Moreover, $\ell_{X_{\min}} = \ell_X$ by (7.1)(d). This proves (a).

Consider the Lyndon length function $L = L_{x_u} : \Gamma \to \Lambda$ and $\delta = \delta_{x_u}$ defined by $\delta(s,t) = \frac{1}{2}(L(s) + L(t) - L(s^{-1}t))$. According to Corollary (7.10) L takes values in $\Lambda_X = \langle \ell_X(\Gamma) \rangle$, so δ takes values in $\frac{1}{2}\Lambda_X \cap \Lambda$, by (5.3). Let Λ_0 be a subgroup of Λ such that $\Lambda_X \cap 2\Lambda \subset 2\Lambda_0$; then $\frac{1}{2}\Lambda_X \cap \Lambda \subset \Lambda_0$ so δ takes values in Λ_0. Let $T = T(\Gamma, \delta)$ be the rooted Λ_0-tree with Γ-action and root $0 \in T$ associated by Theorem (5.4) to $\delta: \Gamma \to \Lambda_0$. We have $L_0 = L(-L_{x_u}): \Gamma \to \Lambda_0 \subset \Lambda$. It follows from Corollary (5.7) that there is a Γ-(Λ_0-isomorphism) α from T onto a Γ-invariant sub Λ_0-tree X_0 of X_{min}, spanned as Λ_0-tree by $\Gamma \cdot x_u$, such that $\alpha(0) = x_u$. Further α induces a Γ-(Λ-isomorphism) $T \otimes_\Lambda \Lambda \to X_{min}$. This proves (c).

To prove (b), let Y be a Λ-(Γ-tree) without inversions such that $\ell_Y = \ell_X$. Let $[y_u, y_v]$ be the bridge from A_u to A_v in Y. (Note that $A_u \cap A_v = \emptyset$ in Y, by (7.4), since $\beta_Y(u,v) = \beta_X(u,v) > 0$.) Then by (7.9) we have $L_{y_u} = L_{x_u}$, hence $\delta_{y_u} = \delta_{x_u}$, so $T = T(\Gamma, \delta_{x_u}) = T(\Gamma, \delta_{y_u})$. Now by (c) Y_{min} and X_{min}, being both Γ-(Λ-isomorphic) to $T \otimes_{\Lambda_0} \Lambda$, are Γ-(Λ-isomorphic) to each other.

For uniqueness, suppose that $\alpha, \beta: Y_{min} \to X_{min}$ are Γ-(Λ-isomorphisms). Then they differ by a Γ-(Λ-automorphism) γ of X_{min}. Since γ preserves each A_s ($s \in \Gamma$), γ preserves the bridge $[A_u, A_v] = [x_u, x_v]$, hence γ fixes x_u. The fixed points of γ then form a Γ-invariant subtree of X_{min}, which must equal X_{min}, whence $\gamma = \text{Id}$ and $\alpha = \beta$.

(7.15) PROPOSITION. *Let X be a Γ-(Λ-tree) without inversions.*

(a) *If X contains a linear Γ-invariant subtree then ℓ_X is either abelian or dihedral.*

(b) *If ℓ_X is non-abelian the following conditions are equivalent.*

   (i) *X_{min} is a linear tree.*

   (ii) *ℓ_X is dihedral.*

Moreover, in this case $X_{min} \cong \overline{\Lambda}_X$, the convex subgroup of Λ generated by $\Lambda_X = <\ell_X(\Gamma)>$.

Proof. If $Y \subset X$ is linear and Γ-invariant and if $s \in \Gamma$, $\ell(s) > 0$, then $Y \subset A_s$. Hence $A_s \cap A_t \neq \emptyset$ if $\ell(s) > 0$ and $\ell(t) > 0$ so, by (7.4), ℓ_X is either abelian or dihedral.

In (b), the implication (i) $\Longrightarrow$ (ii) follows from (a), and $X_{min} \cong \overline{\Lambda}_X$ by Examples (7.14)7. It remains to show that X_{min} is linear when ℓ_X

is dihedral. Put $H = \{s \in \Gamma | \ell_X(s) > 0\} \neq \emptyset$. Since H is invariant under conjugation in Γ, $(A_s)_{s \in H}$ is a Γ-invariant family of closed subtrees of X any two of which meet (since ℓ_X is dihedral). Then $A = \bigcap_{s \in H} A_s$ is Γ-invariant. If $A = \emptyset$ then, by Remark (7.7)2, ℓ_X is abelian, contrary to assumption. Thus $A \neq \emptyset$; if $s \in H$ then $A \subset A_s$ so A is linear. Hence $X_{\min} \subset A$ is linear, as claimed.

(7.16) PROPOSITION. *Let X be a Γ-(Λ-tree) without inversions. Let*

$$\Gamma_1 = \{s \in \Gamma | \beta_X(s,t) \leq 0 \text{ for all } t \in \Gamma\}$$
$$= \{s \in \Gamma | A_s \cap A_t \neq \emptyset \text{ for all } t \in \Gamma\}.$$

(a) *Γ_1 is a normal subgroup of Γ, and $\ell_X | \Gamma_1$ is abelian.*

(b) *If $\ell_X | \Gamma_1 \neq 0$ then ℓ_X is either abelian or dihedral, according as $[\Gamma : \Gamma_1] = 1$ or 2, respectively.*

Since Γ_1 is evidently invariant under conjugation the family $(A_s)_{s \in \Gamma_1}$ of closed subtrees of X is Γ-invariant, hence so also is $A = \bigcap_{s \in \Gamma_1} A_s$.

If $A = \emptyset$ then, by Remark (7.7)2, ℓ_X is abelian, so $\Gamma_1 = \Gamma$.

Suppose that $A \neq \emptyset$. Then each $A_u (u \in \Gamma)$ meets A, by (6.6)(b)(ii).

Case 1. $\ell_X | \Gamma_1 = 0$.
Then $s | A = \text{Id}$ for all $s \in \Gamma_1$, so

$$\Gamma_1 \subset N = \text{Ker}(\Gamma \xrightarrow{\text{restr.}} \text{Aut}(A)).$$

If $s \in N$ then $A \subset A_s$ so $A_s \cap A_u \neq \emptyset$ for all $u \in \Gamma$, i.e. $s \in \Gamma_1$; thus $\Gamma_1 = N$.

Case 2. $\ell_X(s) > 0$ for some $s \in \Gamma_1$.

Then $A \subset A_s$ is linear so, by (7.15), either ℓ_X is abelian, and $\Gamma_1 = \Gamma$, or ℓ_X is dihedral, and then Γ_1 is the normal subgroup of index 2 consisting of elements whose restriction to A is a translation.

(7.17) PROPOSITION. *Let X be a Γ-(Λ-tree) without inversions and with ℓ_X of general type, i.e. non-abelian and non-dihedral (cf. (7.3)). Let Y be a Γ-(Λ-tree) such that $\ell_Y = \ell_X$. Then Y is also without inversions.*

By hypothesis there exist $u,v \in \Gamma$ such that, with $\ell = \ell_X = \ell_Y$ and $\beta = \beta_X = \beta_Y$, we have $\ell(u) > 0$, $\ell(v) > 0$, and $\beta(u,v) > 0$. Let $[x_u, x_v]$ (resp., $[y_u, y_v]$) be the bridge from A_u to A_v in X (resp., in Y). From (7.9) we have $L_{y_u} = L_{x_u}$. If $s \in \Gamma$ and $\ell(s) = 0$ then, by (7.1)(f), $L_{x_u}(s) \in 2\Lambda$, since X is without inversions. Thus $L_{y_u}(s) \in 2\Lambda$, so again by (7.1)(f), Y is without inversions.

(7.18) PROPOSITION. *Let X be a Γ-(Λ-tree) without inversions. The following conditions are equivalent.*

(a) *The subgroup $\Lambda_X = \langle \ell_X(\Gamma) \rangle$ of Λ generated by $\ell_X(\Gamma)$ is cyclic.*

(b) *There is a Γ-(Z-tree) Y (i.e. a simplicial tree action) and a homomorphism $\phi : Z \to \Lambda$ such that $\ell_X = \phi \circ \ell_Y$. If ℓ_X is non-abelian then we further have a unique Γ-(Λ-isomorphism) $Y_{min} \otimes_Z \Lambda \to X_{min}$.*

(b) $\Longrightarrow$ (a) since $\ell_X(\Gamma) \subset \phi(Z)$.

(a) $\Longrightarrow$ (b). If ℓ_X is abelian then, by (7.6)(a), $\ell_X(s) = |h(s)|$ for some homomorphism $h : \Gamma \to \Lambda$. Let $\phi : Z \to \Lambda$ map $1 \in Z$ onto the generator ≥ 0 of $h(\Gamma)$, so that $h(s) = \phi(h_1(s))$ with $h_1 : \Gamma \to Z$ a homomorphism. Let Γ act on $Y = Z$ by translations, $s : x \mapsto x + h_1(s)$. Then $\ell_Y(s) = |h_1(s)|$ so $\ell_X(s) = |h_1(s)| = |\phi(h_1(s))| = \phi(|h_1(s)|) = \phi(\ell_Y(s))$.

If ℓ_X is non-abelian put $\Lambda_0 = \frac{1}{2}(\Lambda_X \cap 2\Lambda)$. In view of (a) this is a cyclic group, and $\Lambda_X \cap 2\Lambda \subset 2\Lambda_0$. According to (7.13)(c) there is a Λ_0-tree $Y \subset X_{min}$, stable under Γ, such that the inclusion induces a Γ-(Λ-isomorphism) $Y \otimes_{\Lambda_0} \Lambda \to X_{min}$. Uniqueness of the latter follows from (7.13)(b). Identifying the cyclic group Λ_0 with Z we obtain the proposition in this case.

8. The hyperbolic length of a product

We define the *diameter* of a Λ-tree Y to be

$$\text{diam}(Y) = \{d(x,y) | x,y \in Y\}.$$

This is a subtree of Λ consisting of elements ≥ 0, and containing 0. We write $\text{diam}(Y) \leq \text{diam}(Y')$ if $\text{diam}(Y) \subset \text{diam}(Y')$. If $\text{diam}(Y) = [0,d]$ for some $d \in \Lambda$ we write $\text{diam}(Y) = d$. If $d \in \Lambda$, $d \geq 0$, and $\Delta = \text{diam}(Y)$, then "$\Delta + d$" denotes $\cup_{a \in \Delta} [0, a+d]$.

We fix here a group Γ, a Γ-(Λ-tree) X without inversions, and we put $\ell = \ell_X$, $\beta = \beta_X$,

$$\beta(s,t) = \ell(st) - \ell(s) - \ell(t).$$

Clearly $\beta(s,t) = \beta(t,s) = \beta(s^{-1}, t^{-1})$. Given $s, t \in \Gamma$, our object here is to investigate the possibilities for $\ell(st)$ and A_{st}.

(8.1) PROPOSITION. *Suppose that* $A_s \cap A_t = \emptyset$.
  (a) $\ell(st) = \ell(s) + \ell(t) + 2d(A_s, A_t) > 0$, *i.e.* $\beta(s,t) = 2d(A_s, A_t)$.
  (b) A_{st} *contains the bridge* $[A_s, A_t]$, *and* $\text{diam}(A_{st} \cap A_u) = \ell(u)$ *for* $u = s, t$.
  (c) *Suppose that* $\ell(s) > 0$ *and* $\ell(t) > 0$. *Then* $\beta(s^n, t^m) = \beta(s,t)$ *for all non-zero* $n, m \in \mathbb{Z}$.
  (d) *Suppose that* $\ell(s) > 0$ *and* $\ell(t) = 0$. *Then* $\beta(s^n, t^{\pm 1}) = \beta(s,t)$ *for all non-zero* $n \in \mathbb{Z}$.
  (e) *Suppose that* $\ell(s) = 0$ *and* $\ell(t) = 0$. *Then* $0 \leq \beta(s^n, t^m) \leq \beta(s,t)$ *for all* $n, m \in \mathbb{Z}$, *and* $\beta(s^n, t^m) = \beta(s,t)$ *if* $n, m = \pm 1$. *This remains true even if* $A_s \cap A_t \neq \emptyset$, *in which case we evidently have* $\ell(s^n t^m) = 0$ *and* $\beta(s^n, t^m) = 0$ *for all* $n, m \in \mathbb{Z}$.
  (f) *If* $\ell(s) > 0$ *and* $u = st$ *then* $\ell(u) > 0$ *and* $\beta(u, s^2 u s^{-2}) = 2\ell(s) > 0$.

Proof. Once (a) is established, (c), (d), and (e) result from (7.1)(b): $A_s \subset A_{s^n}$ for all $n \in \mathbb{Z}$, with equality if $\ell(s) > 0$ and $n \neq 0$, or if

$\ell(s) = 0$ and $n = \pm 1$. Moreover (f) follows from (a) and (b). In fact $\ell(u) > 0$ by (a). If $D = A_u \cap A_s$ then $\mathrm{diam}(D) = \ell(s)$, by (b). Hence $d(s^2 D, D) = \ell(s)$. Since $A_{s^2 u s^{-2}} \cap A_s = s^2(A_u \cap A_s) = s^2 D$, we see that A_u and $A_{s^2 u s^{-2}}$ are disjoint, and $d(A_u, A_{s^2 u s^{-2}}) = d(D, s^2 D) = \ell(s)$. Thus $\beta(u, s u s^{-2}) = 2\ell(s)$, by (a). To prove (a) and (b) we distinguish the three cases as in (c), (d), (e). We shall denote by $B = [S, T]$ the bridge from A_s to A_t.

Case $\ell(s) > 0$, $\ell(t) > 0$

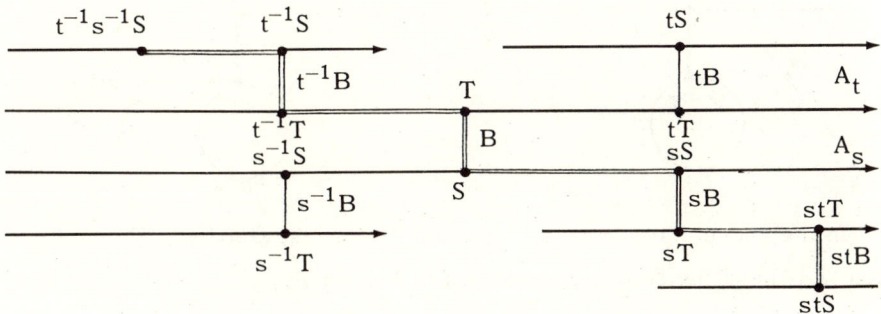

In the above diagram we see from the Piecewise Geodesic Proposition (2.14) and the Bridge Proposition (2.17) that,

$$[t^{-1}s^{-1}S, stS] = [t^{-1}s^{-1}S, t^{-1}S, t^{-1}T, T, S, sS, sT, stT, stS].$$

It follows that $[(st)^{-1}S, S] \cap [S, stS] = \{S\}$, hence $S \in A_{st}$, by (6.6)(a), and so

$$\ell(st) = d(S, stS) = d(S, sS) + d(sS, sT) + d(sT, stT) + d(stT, stS)$$
$$= \ell(s) + d(S, T) + \ell(t) + d(T, S)$$
$$= \ell(s) + \ell(t) + 2d(A_s, A_t).$$

Further $B = [S, T] \subset A_{st}, A_{st} \cap A_t = [t^{-1}T, T]$ has diameter $\ell(t)$, and $A_{st} \cap A_s = [S, sS]$ has diameter $\ell(s)$.

Case $\ell(s) > 0$, $\ell(t) = 0$

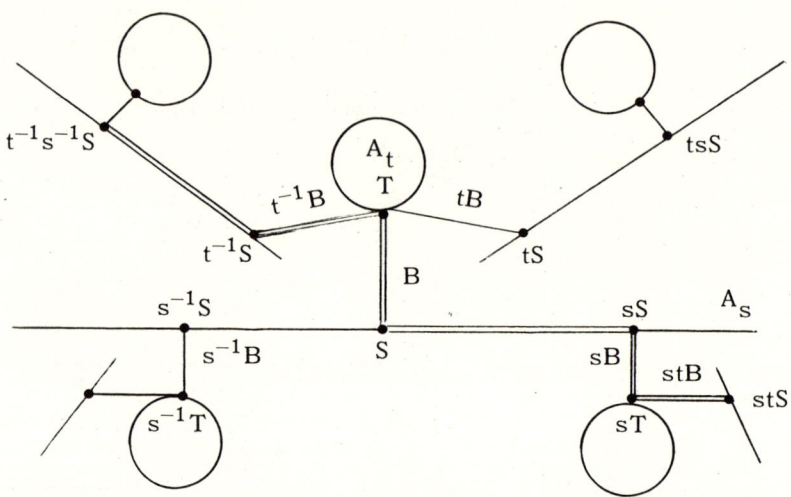

In the above diagram we see from the Piecewise Geodesic Proposition (2.14) and the Bridge Proposition (2.17) that,

$$[t^{-1}s^{-1}S, stS] = [t^{-1}s^{-1}S, t^{-1}S, T, S, sS, sT, stS] \ .$$

It follows that $[(st)^{-1}S, S] \cap [S, stS] = \{S\}$, hence $S \in A_{st}$, by (6.6)(a), and

$$\begin{aligned} \ell(st) = d(S, stS) &= d(S, sS) + d(sS, sT) + d(sT, stS) \\ &= \ell(s) + d(S, T) + d(T, tS) \\ &= \ell(s) + 2d(S, T) \\ &= \ell(s) + \ell(t) + 2d(A_s, A_t) \ . \end{aligned}$$

Further $B = [S, T] \subset A_{st}$, $A_{st} \cap A_t = \{T\}$ has diameter $0 = \ell(t)$, and $A_{st} \cap A_s = [S, sS]$ has diameter $\ell(s)$.

Case $\ell(s) = 0$, $\ell(t) = 0$

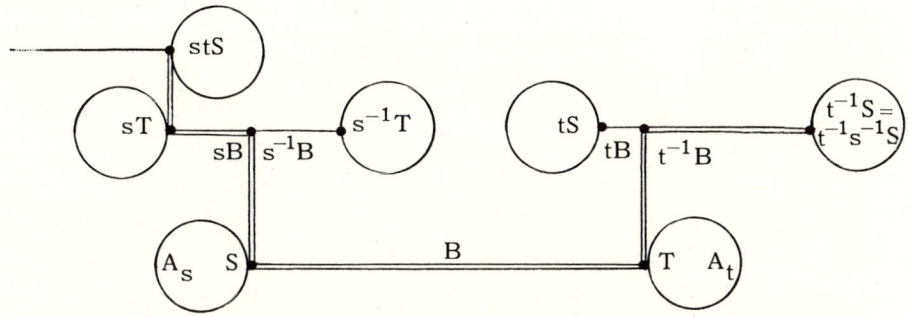

In the diagram above we see from the Piecewise Geodesic Proposition (2.14) and the Bridge Proposition (2.17) that,

$$[t^{-1}s^{-1}S, stS] = [t^{-1}s^{-1}S = t^{-1}S, T, S, sT, stS].$$

It follows that $[(st)^{-1}S, S] \cap [S, stS] = \{S\}$, hence $S \in A_{st}$, by (6.6)(a), and

$$\ell(st) = d(S, sT) + d(sT, stS) = 2d(S, T)$$
$$= \ell(s) + \ell(t) + 2d(A_s, A_t).$$

Further $B = [S, T] \subset A_{st}$, $A_{st} \cap A_t = \{T\}$ has diameter $0 = \ell(t)$, and $A_{st} \cap A_s = \{S\}$ has diameter $0 = \ell(s)$.

(8.2) PROPOSITION. *Assume that* $\ell(s) > 0$ *and that* t *fixes a unique point*, T, *of* A_s. *Let* $P = Y(t^{-1}s^{-1}T, T, stT)$. *Then*

$$\ell(st) = \max(\ell(s) - 2d(P, T), 0) = \text{diam}(A_{st} \cap A_s),$$

so

$$\beta(s, t) = \max(-2(P, T), -\ell(s))$$
$$= -2 \min(d(P, T), \tfrac{1}{2}\ell(s)) \leq 0.$$

Moreover $\beta(s,t) \in 2\Lambda$. If $\ell(st) = 0$ then $A_{st} \cap A_s$ is the midpoint of $[T,sT]$. If $\ell(st) > 0$ then $A_{st} \cap A_s = [P,stP]$. In any case $A_{sts^{-1}} \cap A_s = \{sT\}$ is disjoint from $\{T\}$ so $A_t \cap A_{sts^{-1}} = \emptyset$ ((2.13 (b))) and hence $\beta(t,sts^{-1}) = 2\ell(s) > 0$ ((8.1)(a)).

Proof. If $2d(P,T) \geq \ell(s)$ we have the following picture.

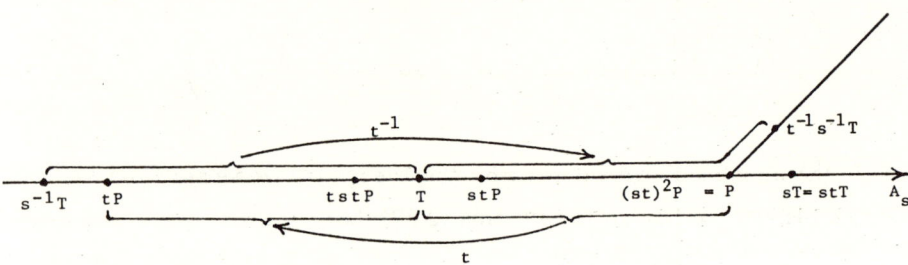

Since $(st)^2 P = P$, and since Γ is without inversions, st fixes the midpoint Q of $[stP,P]$, which is likewise the midpoint of $[T,sT]$, so $\ell(s) = 2d(T,Q) \in 2\Lambda$, and $\beta(s,t) = -\ell(s)$. Clearly st fixes no point of $[stP,P]$ other than Q, so $A_{st} \cap A_s = \{Q\}$ if $P \neq Q$. If $P = Q$ then $st[T,P] = [stT,stP]$, so st fixes only P in $[T,sT]$, and again $A_{st} \cap A_s = \{P\}$.

Now assume that $2d(P,T) < \ell(s)$. Then we have the following picture

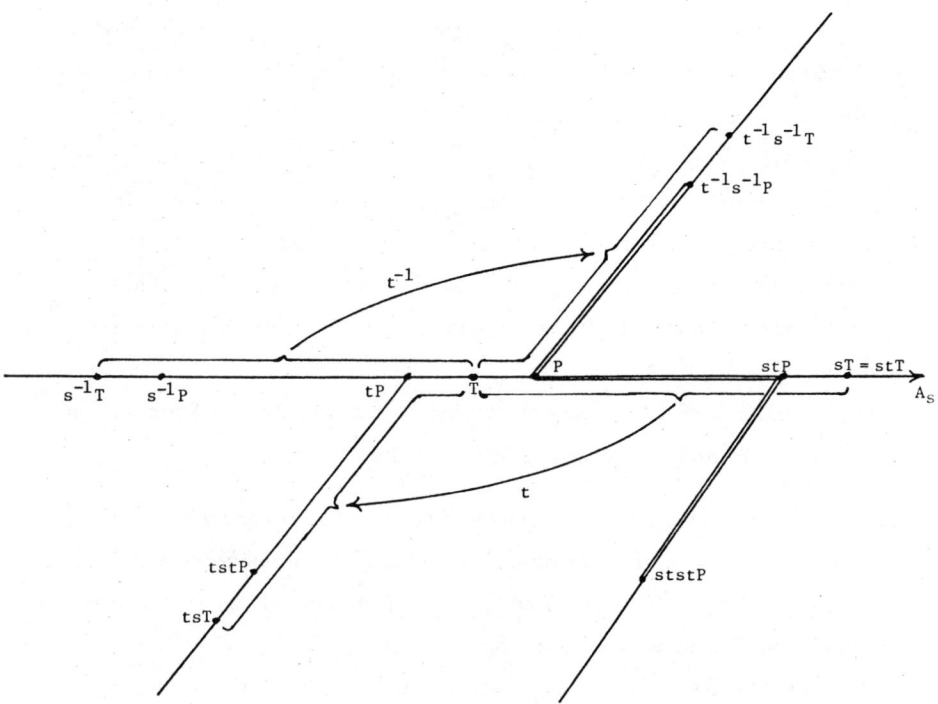

Since $d(tP,P) = 2d(T,P) < \ell(s) = d(tP,stP)$, we see that $st[T,P] = [sT,stP]$ is disjoint from $[T,P]$, so st fixes no point of $[T,P]$. Thus (6.1)(b) implies that $\ell(st) > 0$, and then (6.6)(d) implies that $P \in A_{st}$. Hence $\ell(st) = d(P,stP) = d(T,sT) - d(T,P) - d(stP,sT) = \ell(s) - d(P,T) - d(tP,T) = \ell(s) - 2d(P,T)$. The diagram above shows further that $A_{st} \cap A_s = [P,stP]$, whence the claims in this case.

The final assertion of (8.2) is clear.

The cases not covered by Propositions (8.1) and (8.2) are covered next by Proposition (8.3), for which we introduce the following terminology. We say that s and t *meet* if $A_s \cap A_t \neq \emptyset$, and then put $\Delta(s,t) = \text{diam}(A_s \cap A_t)$. Thus $\Delta(s,t) = \Delta(t,s) = \Delta(s^{\pm 1}, t^{\pm 1}) = \Delta(usu^{-1}, utu^{-1})$ for all $u \in \Gamma$. We say that s and t *meet incoherently* if $\ell(s) > 0$, $\ell(t) > 0$, $\Delta(s,t) > 0$, and s and t translate points of $A_s \cap A_t$ in opposite directions. Otherwise we say that s and t *meet coherently*. Thus if s and t meet incoherently then s and t^{-1} meet coherently (but not conversely).

In Proposition (8.1) st meets s and t coherently. In Proposition (8.2) $\Delta(s,t) = 0$ and st meets s coherently.

(8.3) PROPOSITION. *Assume that* s *and* t *meet coherently and that, if exactly one of* s *and* t *is hyperbolic, then* $\Delta > 0$, *where* $\Delta = \Delta(s,t)$.

(a) $\ell(st) = \ell(s) + \ell(t)$, *i.e.* $\beta(s,t) = 0$. *Moreover* st *meets* s *and* t *coherently, and* $\Delta(st,u) = \Delta + \ell(u)$ *for* u = s,t.

(b) $\ell(st^{-1}) = \ell(s) + \ell(t) - 2m$, *i.e.* $\beta(s,t^{-1}) = -2m$, *where* $m \in \Lambda$ *satisfies*

$$\min(\ell(s), \ell(t), \Delta) \leq m \leq \frac{1}{2}(\ell(s) + \ell(t)).$$

Either $m = \min(\ell(s), \ell(t))$ *or else* $\Delta < \max(\ell(s), \ell(t))$, *say* $\Delta < \ell(s)$, *in which case* $\beta(t, sts^{-1}) > 0$.

(c) *If* $\ell(t) = \min(\ell(s), \ell(t), \Delta)$ *then either* st^{-1} *and* t *don't meet (and* $\ell(s) > \ell(t) = \Delta$ *) or* st^{-1} *and* t *meet coherently. Otherwise* st^{-1} *and* t^{-1} *meet coherently*.

If $\ell(s) = \min(\ell(s), \ell(t), \Delta)$ *then either* st^{-1} *and* s^{-1} *don't meet (and* $\ell(t) > \ell(s) = \Delta$ *) or* st^{-1} *and* s^{-1} *meet coherently. Otherwise* st^{-1} *and* s *meet coherently*.

We first observe that if suffices to treat the case $\ell(s) \geq \ell(t)$. This follows since s^{-1} conjugates st,s into ts,s, and t conjugates st,t into ts,t, and since $ts^{-1} = (st^{-1})^{-1}$. Assume now that $\ell(s) \geq \ell(t)$.

Recall that $\Delta = \Delta(s,t) = \text{diam}(D)$, $D = A_s \cap A_t$. If $\ell(s) > \Delta$ then D is disjoint from $sD = s(A_s \cap A_t) = A_s \cap A_{sts^{-1}}$. It follows easily that $A_t \cap A_{sts^{-1}} = \emptyset$, and so, by (8.1)(a), that $\beta(t,sts^{-1}) = 2d(A_t, A_{sts^{-1}}) = 2d(D,sD) = 2(\ell(s) - \Delta) > 0$.

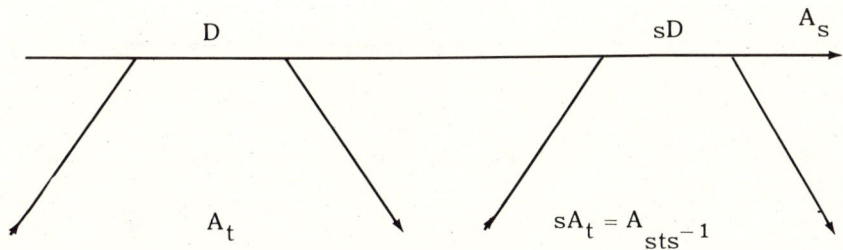

Thus the proposition will follow if we verify (a) and (c) in the cases of the following table, whose content establishes (b).

Case	$\ell(st^{-1})$	$m \in \Lambda$	$\Delta(st^{-1}, s)$	$\Delta(st^{-1}, t)$	$d(A_{st^{-1}}, A_t)$
$\ell(s) = \ell(t) = 0$	0	0	Δ	Δ	0
$\ell(s) > \ell(t) = 0$ ($\Delta > 0$)	$\ell(s)$	0	$\Delta + \ell(s)$	Δ	0
$\ell(s) \geq \ell(t) > \Delta$	$\ell(s) + \ell(t) - 2\Delta$	Δ	$\ell(s) - \Delta$	$\ell(t) - \Delta$	0
$\ell(s) \geq \ell(t)$, $\Delta > \ell(t)$	$\ell(s) - \ell(t)$	$\ell(t)$	$\ell(s) + \Delta - 2\ell(t)$	$\Delta - \ell(t)$	0
$\ell(s) \geq \ell(t) = \Delta \geq 0$, $e \geq d$	$e - d$	$\ell(t) + d$	$e - d$	(0 if $d = 0$)	d
$e \leq d$	0	$\frac{1}{2}(\ell(s) + \ell(t))$	0	(0 if $\ell(s) = \ell(t)$)	$\frac{1}{2}(\ell(s) - \ell(t))$

The notation in the last case has the following meaning: Let $D = [L,R]$ with $R = tL$. Put $P = Y(ts^{-1}R, R, st^{-1}R)$, $d = d(R,P)$, and $d + e = d(R, st^{-1}R)$ $(= \ell(s) - \ell(t))$.

Case 0. $\ell(s) = \ell(t) = 0$

Then st fixes the points of $D = A_s \cap A_t$, so $\ell(st) = 0$, and clearly any two of A_s, A_t, A_{st} intersect in D. The same conclusions apply to s, t^{-1}. Whence the proposition in this case.

Case 1. $\ell(s) > \ell(t) = 0$ and $\Delta > 0$

Since $\Delta > 0$ we can choose $P \in D$ and a $Q \neq P$ in $[s^{-1}P, P] \cap D$.

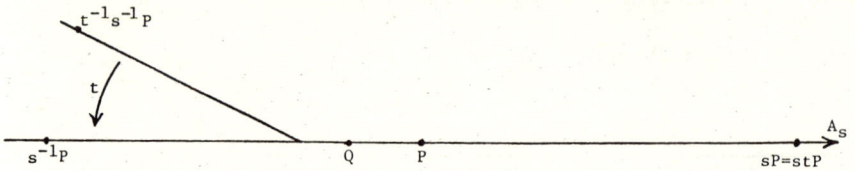

Then $[s^{-1}P, P] = [s^{-1}P, Q, P]$. Applying t^{-1} we have $[t^{-1}s^{-1}P, P] = [t^{-1}s^{-1}P, Q, P]$. Hence

(*) $\qquad [t^{-1}s^{-1}P, P] \cap [P, stP] = [t^{-1}s^{-1}P, Q, P] \cap [P, sP]$.

Since $[Q, P] \cap [P, sP] \subset [s^{-1}P, P] \cap [P, sP] = \{P\}$, and since $Q \neq P$, it follows from (*) that $[t^{-1}s^{-1}P, P] \cap [P, stP] = \{P\}$. By (6.6)(a) we have $P \in A_{st}$, so $\ell(st) = d(P, stP) = d(P, sP) = \ell(s)$. Moreover A_{st} contains $([s^{-1}P, P] \cap D) \cup [P, sP]$ for every $P \in D$ such that $[s^{-1}P, P] \cap D \neq \{P\}$. It follows easily from this that $A_{st} \cap A_s$ is the span of $D \cup sD$, which has diameter $\Delta + \ell(s)$, and that $A_{st} \cap A_t = D$, of diameter Δ. Since s and t^{-1} satisfy the same hypotheses as s and t, the proposition follows in this case.

Case 2. $\ell(s) \geq \ell(t) > \Delta$

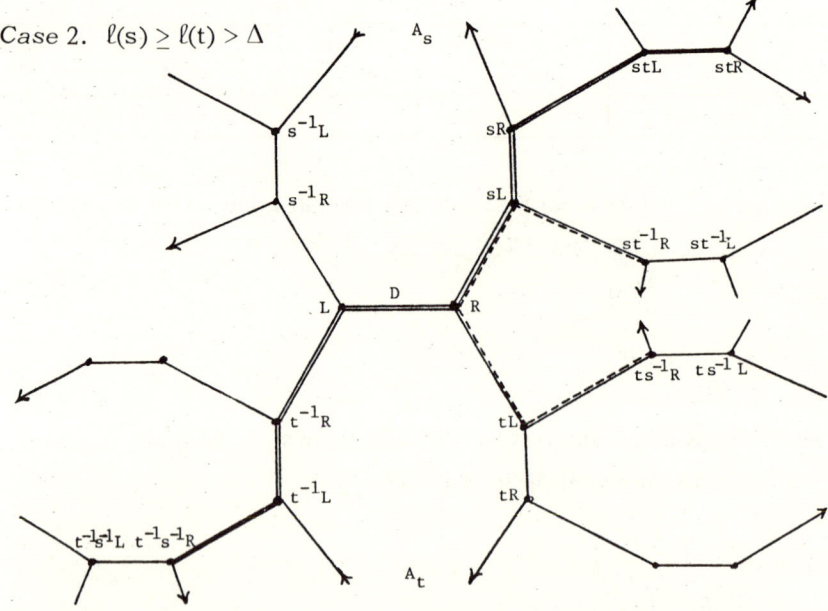

Write $D = [L,R]$ where s and t translate L toward (and past) R. Then $[L,sL] = [L,R,sL], [L,tL] = [L,R,tL]$, $[s^{-1}R,R] = [s^{-1}R,L,R]$, and $[t^{-1}R,R] = [t^{-1}R,L,R]$. The intermediate points here might equal one, but not both, of the endpoints. Thus $[s^{-1}R,tL] = [s^{-1}R,L,R,tL]$ and $[L,tR] = [L,R,tL,tR]$ have union $[s^{-1}R,tR] = [s^{-1}R,L,R,tL,tR]$, whence

$$[R,stR] = [R,sL,sR,stL,stR].$$

Similarly

$$[t^{-1}s^{-1}R,R] = [t^{-1}s^{-1}R, t^{-1}L, t^{-1}R,L,R].$$

We have $[R,sR] \cap A_t = \{R\}$ so $[t^{-1}R,R] \cap [R,sR] = \{R\}$, and hence $[t^{-1}s^{-1}R,R] \cap [R,stR] = \{R\}$. Thus, by (6.6)(a), $R \in A_{st}$ and $\ell(st) = d(R,stR) = d(R,sR) + d(sR,stR) = \ell(s) + \ell(t)$. Moreover we see from the diagram above that $D \subset A_{st}$, that $A_{st} \cap A_s = D \cup [R,sR]$ has diameter $\Delta + \ell(s)$, that $A_{st} \cap A_t = D \cup [t^{-1}L,L]$ has diameter $\Delta + \ell(t)$, and that st meets s and t coherently.

Similar calculations (cf. diagram above) show that $[ts^{-1}R,R] = [ts^{-1}R,tL,R]$ meets $[R,st^{-1}R] = [R,sL,st^{-1}R]$ only in R. Thus, by (6.6)(a) again, $R \in A_{st^{-1}}$, $\ell(st^{-1}) = d(R,st^{-1}R) = d(R,sL) + d(sL,st^{-1}R)$
$= d(L,sL) - d(L,R) + d(L,t^{-1}R) = \ell(s) - \Delta + d(R,t^{-1}R) - d(R,L) = \ell(s) + \ell(t) - 2\Delta$. Moreover $A_{st^{-1}} \cap A_s = [R,sL]$ has diameter $\ell(s) - \Delta$ and $A_{st^{-1}} \cap A_t = [R,tL]$ has diameter $\ell(t) - \Delta$. Further st^{-1} meets s coherently and t^{-1} coherently. Whence the proposition in this case.

Case 3. $\ell(s) \geq \ell(t) > 0$ and $\Delta > \ell(t)$

Let "left to right" denote the direction that s and t translate along D. If D has a left (resp., right) end point denote it by L (resp., R).

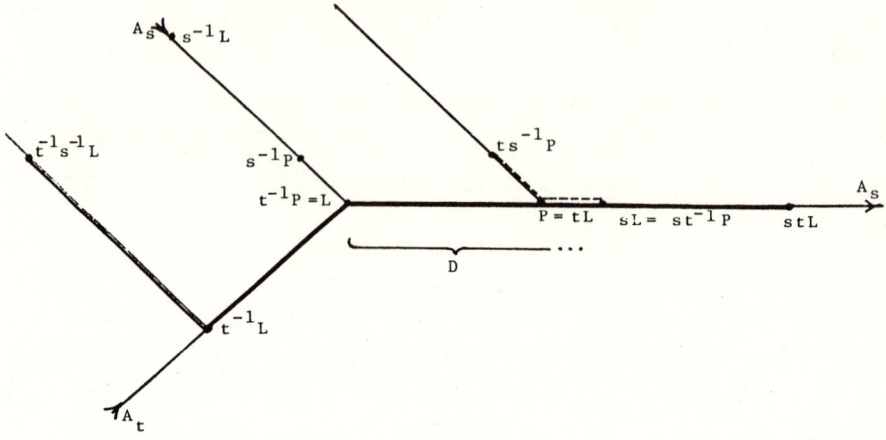

The above diagram, when L exists, shows that $L \in A_{st}$, so $\ell(st) = d(L, stL) = \ell(s) + \ell(t)$, that $A_{st} \cap A_s = [L, ?]$, and that $A_{st} \cap A_t = [t^{-1}L, ?]$. Further $P = tL \in A_{st^{-1}}$, so $\ell(st^{-1}) = d(P, st^{-1}P) = d(tL, sL) = \ell(s) - \ell(t) = \ell(s) + \ell(t) - 2\ell(t)$, and $A_{st^{-1}} \cap A_s = [tL, ?]$ and $A_{st^{-1}} \cap A_t = [tL, ?]$.

In case D has no left endpoint then, since D is a closed subtree ((2.16)(a)), D contains the left ends of both A_s and A_t (cf. (2.24)(e)). We see easily that A_{st} and $A_{st^{-1}}$ do likewise, that $\ell(st) = \ell(s) + \ell(t)$, and that $\ell(st^{-1}) = \ell(s) - \ell(t)$. Thus the length formulas are established and st and st^{-1} both meet s and t coherently in all cases. Further the following trees have common left ends: $A_{st} \cap A_s$ and D; $A_{st} \cap A_t$ and $t^{-1}D$; $A_{st^{-1}} \cap A_s$ and tD; $A_{st^{-1}} \cap A_t$ and tD.

The proposition will follow now once we show that the following trees have common right ends: $A_{st} \cap A_s$ and sD; $A_{st} \cap A_t$ and D; $A_{st^{-1}} \cap A_s$ and $st^{-1}D$; $A_{st^{-1}} \cap A_t$ and D.

If D has no right end point then D contains the right ends of A_s and A_t, and so also do A_{st} and $A_{st^{-1}}$, whence the assertions in this case. So assume that D has a right end point R. The following diagrams then show that $R \in A_{st} \cap A_{st^{-1}}$ and confirm the claims above.

LENGTH FUNCTIONS

(In the second diagram we might instead have $[L,R] = [L, tL, t^{-1}R, R]$, but the argument is unaffected by this.)

$\ell(t) \leq \ell(s) \leq \Delta$

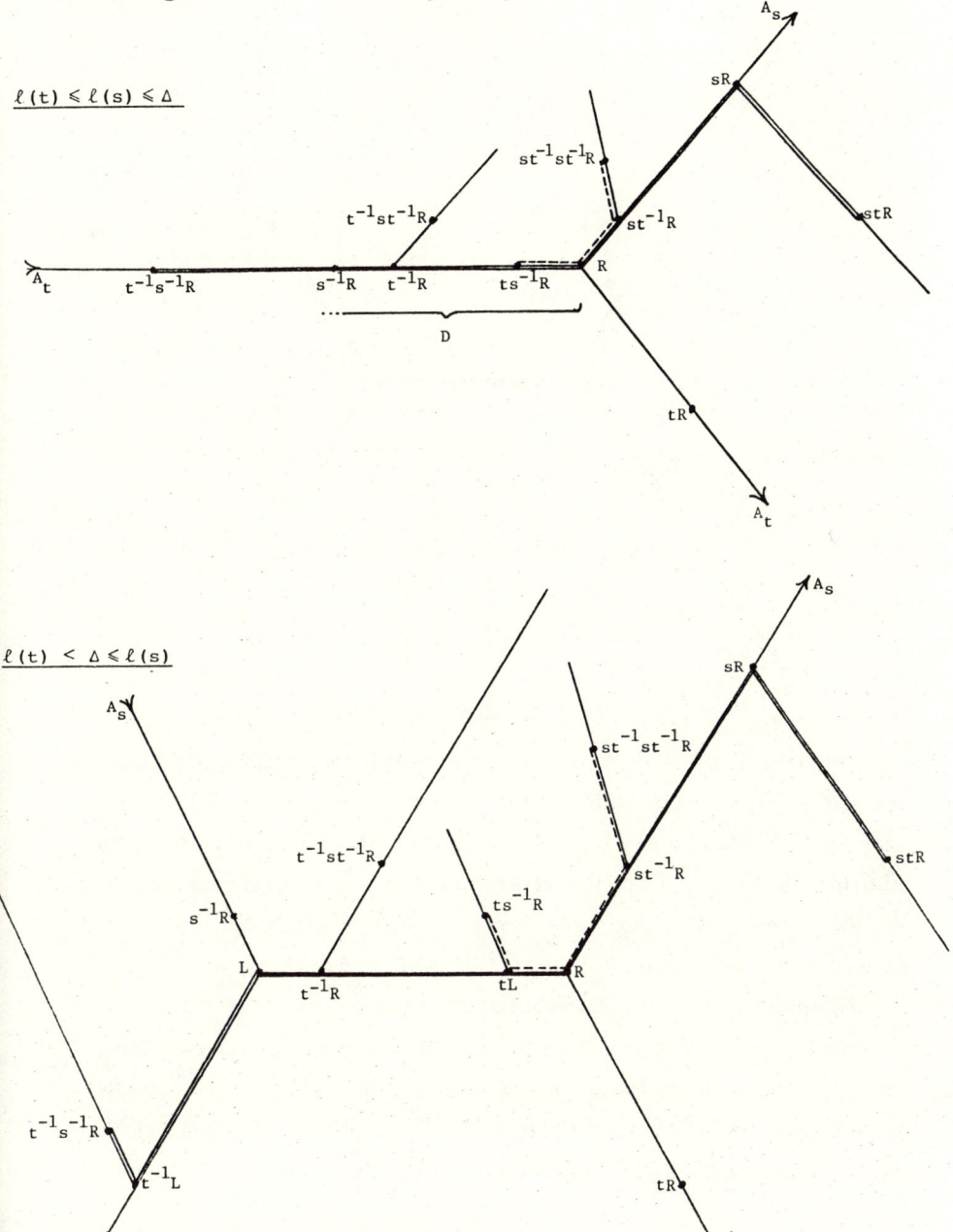

$\ell(t) < \Delta \leq \ell(s)$

Case 4. $\ell(s) \geq \ell(t) = \Delta$

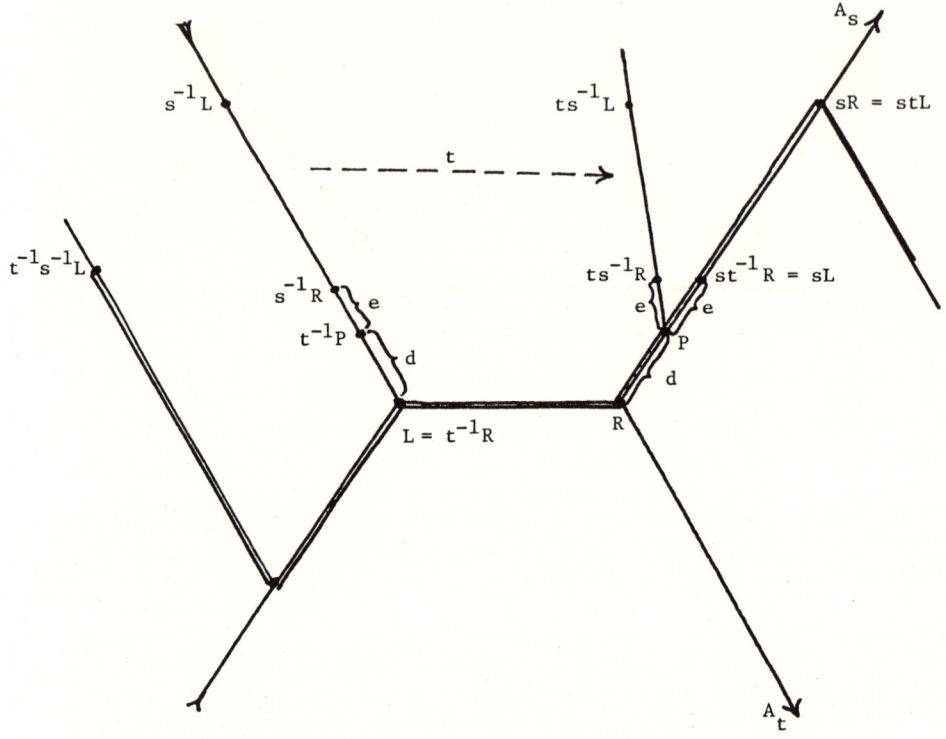

Let $D = [L,R]$ with $R = tL$, and put $P = Y(ts^{-1}R, R, st^{-1}R)$, $d = d(R,P)$, and $d+e = d(R, st^{-1}R) = d(R, sL) = \ell(s) - \ell(t)$. From the diagram it is easy to see that $[t^{-1}s^{-1}L, L] \cap [L, stL] = \{L\}$, hence, by (6.6)(a), $L \in A_{st}$ and $\ell(st) = d(L, stL) = \ell(s) + \ell(t)$. Moreover st meets s and t coherently, $A_{st} \cap A_s$ is the subtree spanned by $D \cup sD$, of diameter $\Delta + \ell(s)$, and $A_{st} \cap A_t = t^{-1}D \cup D$, of diameter $\Delta + \ell(t)$.

To analyze st^{-1} we distinguish the cases $e > d$ and $e \leq d$.

In case $\ell(s) = \ell(t) = \Delta$ then $e = d = 0$ and it is easily seen that $\ell(st^{-1}) = 0$ and R is the unique fixed point of st^{-1} in A_s or in A_t. In particular $\beta(s, t^{-1}) = -(\ell(s) + \ell(t)) = -2\ell(t)$, whence all claims in this case. Henceforth we assume that $\ell(s) > \ell(t) = \Delta$.

Case 4(a). $e > d$

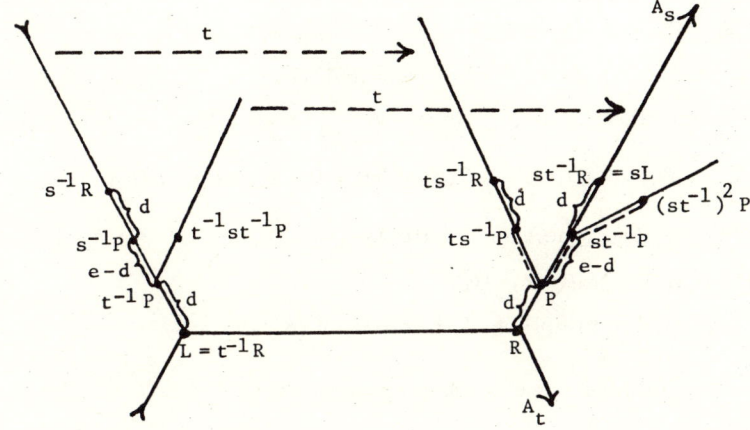

From the diagram we see that $[ts^{-1}P,P] \cap [P,st^{-1}P] = \{P\}$, whence, by (6.6)(a), $P \in A_{st^{-1}}$ and $\ell(st^{-1}) = d(P,st^{-1}P) = e-d = e+d-2d = \ell(s)-\ell(t)-2d = \ell(s) + \ell(t) - 2(\ell(t)+d)$, so $m = \ell(t) + d$. Moreover st^{-1} meets s coherently in $[P,st^{-1}P]$, so $\Delta(st^{-1},s) = e-d$ also, and $d(A_{st^{-1}},A_t) = d(P,R) = d$. If $d = 0$ then $A_{st^{-1}} \cap A_t = \{R\}$ so $\Delta(st^{-1},t) = 0$.

Case 4(b). $e \leq d$

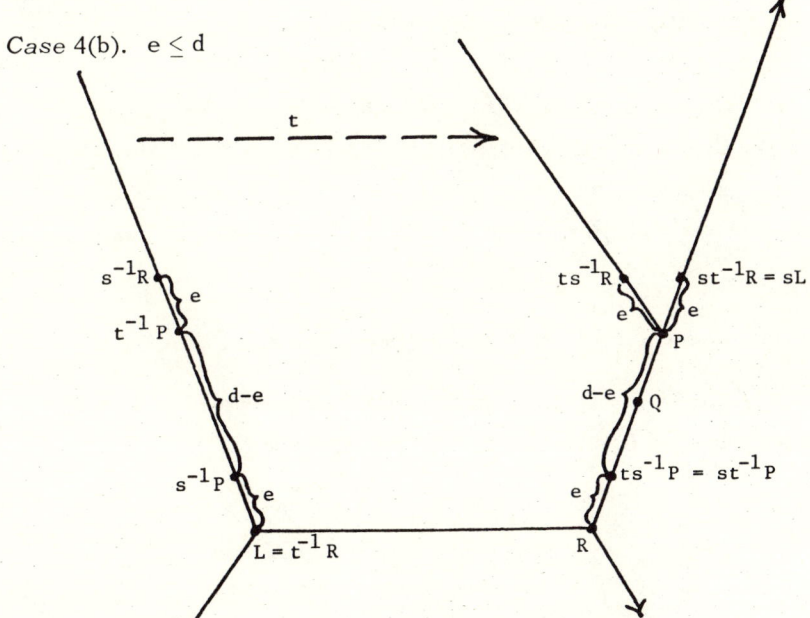

From the diagram we see that $st^{-1}P = ts^{-1}P$. Since, by hypothesis, Γ acts without inversions, we conclude from (6.1)(b) that $\ell(st^{-1}) = 0$ and $[P, st^{-1}P]$ has a mid-point Q (also the mid-point of $[R, st^{-1}R]$) fixed by st^{-1}. It follows further from (6.1)(b) that, in $[R, st^{-1}R]$, st^{-1} fixes no point other than Q. Hence $A_{st^{-1}} \cap A_s = \{Q\}$, so $\Delta(st^{-1}, s) = 0$, and $d(A_{st^{-1}}, A_t) = d(Q, R) = \frac{1}{2}(\ell(s) - \ell(t)) > 0$. Finally, $\beta(s, t^{-1}) = -(\ell(s) + \ell(t)) = -2m$, where $m = \frac{1}{2}(\ell(s) + \ell(t)) \in \Lambda$ because $\ell(s) + \ell(t) = \ell(s) - \ell(t) + 2\ell(t) = 2(d(Q,R) + \ell(t))$.

This concludes the proof of Proposition (8.3).

We now catalogue some basic properties of ℓ.

(8.4) THEOREM. *Let* $s, t \in \Gamma$.

(a) $\beta(s,t) \in 2\Lambda$, *and* $d(A_s, A_t) = \max\left(\frac{1}{2}\beta(s,t), 0\right)$.
  (i) $\beta(s,t) \leq 0 \Longrightarrow \beta(s^2, t) \leq 0$.
  (ii) *If* $\beta(s,t) > 0$ *then* $\beta(s^2, t) \leq \beta(s,t) = \beta(s^{-1}, t)$. *If also* $\ell(s) > 0$ *then* $\beta(s^2, t) = \beta(s,t)$, *and* $\beta(st, s^2(st)s^{-2}) > 0$.
  (iii) $\ell(s) = \ell(t) = 0 \Longrightarrow \beta(s,t) \geq 0$.
  (iv) *If* $u \in \Gamma$ *and* $\ell(x) = 0$ *for* $x = s, t, u, st, tu, us$, *then* $\ell(stu) = 0$.

(c) *If* $\ell(s) > \ell(t) = 0 > \beta(s,t)$ *then* $\beta(t, sts^{-1}) > 0$.

(d) *Suppose that* $\ell(s) > 0$, $\ell(t) > 0$, *and* $\beta(s,t) \leq 0$.
  (i) $\{\beta(s,t), \beta(s, t^{-1})\} = \{0, -2m\}$ *where* $0 \leq m \in \Lambda$.
  (ii) *Either* $m = \min(\ell(s), \ell(t))$ *or if say* $\ell(s) \geq \ell(t)$ *then* $\beta(t, sts^{-1}) > 0$.

(e) *Let* $\Gamma_1 = \{s \in \Gamma \mid \beta(s, u) \leq 0 \text{ for all } u \in \Gamma\}$. *If* $s, t \in \Gamma_1$ *then* $st \in \Gamma_1$.

Note that *assertion (a) establishes Theorem (7.4)*.

Proof. (a) If $A_s \cap A_t = \emptyset$ then $\beta(s,t) = 2d(A_s, A_t)$ by (8.1). In all other cases $\beta(s,t) \leq 0$ and $\beta(s,t) \in 2\Lambda$ by (8.2) and (8.3).

(b)(i). If $\beta(s,t) \le 0$ then, by (a), $A_s \cap A_t \ne \emptyset$. Since $A_s \subset A_{s^2}$ we have $A_{s^2} \cap A_t \ne \emptyset$, so $\beta(s^2,t) \le 0$, by (a) again.

(ii). Using (a) again, if $\beta(s,t) > 0$ then $\beta(s^2,t) \le \beta(s,t) = \beta(s^{-1},t)$, since $A_{s^2} \supset A_s = A_{s^{-1}}$, with equalities if $\ell(s) > 0$. In the latter case also $\beta(st, s^2(st)s^{-2}) > 0$ by (8.1)(f).

(iii). This follows from (8.1)(e).

(iv). The hypotheses imply that $\beta(s,t)$, $\beta(t,u)$, and $\beta(u,s)$ all vanish. Hence, by (a), any pair of A_s, A_t, A_u meet. By (2.13)(b) then $A_s \cap A_t \cap A_u \ne \emptyset$. But any $x \in A_s \cap A_t \cap A_u$ is fixed by s, t, and u, hence by stu, so $\ell(stu) = 0$, as claimed.

(c) Suppose that $\ell(s) > \ell(t) = 0 > \beta(s,t)$. Then $\Delta(s,t) = 0$, for otherwise (8.3) implies that $\beta(s,t) = 0$, contrary to assumption. Hence (8.2) applies to give $\beta(t, sts^{-1}) = 2\ell(s) > 0$.

(d) follows from Proposition (8.3)(b).

(e) follows from Proposition (7.16).

9. *In search of hyperbolic length axioms*

Fix a group Γ and a function $\ell : \Gamma \to \Lambda$. We seek reasonable necessary and sufficient conditions for ℓ to be the hyperbolic length function ℓ_X of a Γ-(Λ-tree) X without inversions.

Our procedure will be to list several conditions which, by the preceding results, are necessary, and then to go as far as we can toward proving sufficiency. In the statements, $s, t, u, \cdots$ vary over Γ.

(H0) $\qquad\qquad\qquad\qquad \ell(s) \ge 0$

(H1)$_n$ $\qquad\qquad\qquad\qquad \ell(s^n) = |n|\ell(s)$

In fact we shall use only

$\qquad\qquad$ (H1)$_0$ $\qquad\quad \ell(1) = 0$

$\qquad\qquad$ (H1)$_{-1}$ $\qquad\; \ell(s^{-1}) = \ell(s)$

$\qquad\qquad$ (H1)$_2$ $\qquad\quad \ell(s^2) = 2\ell(s)$.

(H2) $$\ell(sts^{-1}) = \ell(t), \quad \text{i.e.}$$
$$\ell(st) = \ell(ts).$$

These properties follow from (7.1)(a) and (b). Now define,

$$\beta(s,t) = \ell(st) - \ell(s) - \ell(t)$$

for $s, t \in \Gamma$.

(9.1) LEMMA. *For all* $s, t \in \Gamma$ *we have*:

(a) $\beta(s,t) = \beta(t,s) = \beta(s^{-1}, t^{-1})$

and

(b) $\beta(1,t) = 0 = \beta(t,t)$.

This follows immediately from (H2), $(H1)_{-1}$, $(H1)_0$ and $(H1)_2$.

(H3) $$\beta(s,t) \in 2\Lambda.$$

This is a "non-inversion" condition.

(H4) $\beta(s,t) \leq 0 \Longrightarrow \beta(s^2, t) \leq 0$

(H5) $\beta(s,t) > 0 \Longrightarrow \beta(s^2, t) \leq \beta(s,t) = \beta(s^{-1}, t)$

If also $\ell(s) > 0$, then $\beta(s^2, t) = \beta(s,t)$, and $\beta(st, s^2(st)s^{-2}) > 0$.

(H6) $\ell(s) = \ell(t) = 0 \Longrightarrow \beta(s,t) \geq 0$

(H7) $\ell(s) > \ell(t) = 0 > \beta(s,t) \Longrightarrow \beta(t, sts^{-1}) > 0$

(H8) Suppose that $\ell(s) > 0$, $\ell(t) > 0$, and $\beta(s,t) \leq 0$.
(a) $\{\beta(s,t), \beta(s, t^{-1})\} = \{0, -2m\}$, where $0 \leq m \in \Lambda$.
(b) Either $m = \min(\ell(s), \ell(t))$ or if say $\ell(s) \geq \ell(t)$ then $\beta(t, sts^{-1}) > 0$.

(H9) If $\ell(x) = 0$ for $x = s, t, u, st, tu, us$ then $\ell(stu) = 0$.

(H10) Let $\Gamma_1 = \{s \in \Gamma | \beta(s,u) \leq 0$ for all $u \in \Gamma\}$. Then $s,t \in \Gamma_1 \Longrightarrow$ st $\in \Gamma_1$.

The references for the necessity of these conditions are as follows: (H3), (8.4)(a); (Hn), n = 4,5,6,9, (8.4)(b)(i)-(iv); (H7), (8.4)(c); (H8) (8.4)(d); and (H10), (8.4)(e). It is quite possible that they are not logically independent.

(9.2) DEFINITION. Call ℓ *abelian* if $\beta(u,v) \leq 0$ for all $u,v \in \Gamma$, *dihedral* if ℓ is not abelian but $\beta(u,v) \leq 0$ whenever $\ell(u) > 0$ and $\ell(v) > 0$, and of *general type* if ℓ is neither abelian nor dihedral.

(9.3) PROPOSITION. *Suppose that ℓ is abelian, and assume (H0), (H1)$_0$, (H1)$_{-1}$, (H2), (H6), (H7), and (H8). Then there is a homomorphism $h : \Gamma \to \Lambda$, unique up to a factor ± 1, such that $\ell(s) = |h(s)|$ for all $s \in \Gamma$.*

According to Lemma (1.4) we need only verify the following conditions, for all $s,t \in \Gamma$.

(1.1) $0 = \ell(1) \leq \ell(s) = \ell(s^{-1})$

(1.2) $\ell(sts^{-1}) = \ell(t)$

(1.3) $\{\beta(s,t), \beta(s,t^{-1})\} = \{0, -2m\}$, where $m = \min(\ell(s), \ell(t))$.

Now (1.1) follows from (H1)$_0$, (H0), and (H1)$_{-1}$, while (1.2) is just (H2). To verify (1.3) we may, in view of Lemma (9.1), assume that $\ell(s) \geq \ell(t)$. If $\ell(s) = \ell(t) = 0$ then $\beta(s,t) \geq 0$ by (H6), so $\beta(s,t) = 0$ since ℓ is abelian. Similarly $\beta(s,t^{-1}) = 0$. Suppose that $\ell(s) > \ell(t) = 0$. Since ℓ is abelian it follows from (H7) that $\beta(s,t) = 0$; similarly $\beta(s,t^{-1}) = 0$. Finally assume that $\ell(s) \geq \ell(t) > 0$. Then, since ℓ is abelian, (1.3) follows from (H8).

We assume henceforth that ℓ is non-abelian. Then we can choose $u,v \in \Gamma$ so that
$$\beta(u,v) > 0.$$

Now define $L = L_{u,v} : \Gamma \to \Lambda$ by

(9.4) $\quad L(s) = \ell(s) + \max(\beta(s,u), 0, \beta(s,v) - \beta(u,v))$
$\quad\quad\quad = \max(\ell(su) - \ell(u), \ell(s), \ell(sv) - \ell(v) - \beta(u,v))$.

(9.5) PROPOSITION. *Assuming (H1)$_2$, (H4), and (H5), we have*
$\ell(s) = \max(L(s^2) - L(s), 0)$ *for all* $s \in \Gamma$.

Proof. Put $\lambda(s) = \max(\beta(s,u), 0, \beta(s,v) - \beta(u,v))$. Then $L(s^2) - L(s) = (\ell(s^2) - \ell(s)) + (\lambda(s^2) - \lambda(s))$. By (H1)$_2$, $\ell(s^2) = 2\ell(s)$. Thus we must show that:

(a) $\ell(s) > 0 \Longrightarrow \lambda(s^2) = \lambda(s)$
(b) $\ell(s) = 0 \Longrightarrow \lambda(s^2) \leq \lambda(s)$.

From (H4) and (H5) we have $\beta(s,t) \leq 0 \Longrightarrow \beta(s^2,t) \leq 0$, and $\beta(s,t) > 0 \Longrightarrow \beta(s^2,t) \leq \beta(s,t)$, with equality if $\ell(s) > 0$. Bearing in mind that $\beta(u,v) > 0$, assertions (a) and (b) follow readily.

Now in view of (9.5), (7.1)(c), and (5.4), it suffices, in order to show that ℓ is a hyperbolic length function, to show that L is a Lyndon length function, in the sense of (5.1), i.e. that L satisfies, for all $r, s, t \in \Gamma$:

(L0) $\quad\quad\quad\quad\quad\quad L(1) = 0$.

(L1) $\quad\quad\quad\quad\quad\quad L(s) = L(s^{-1})$.

If $Y(s,t) = \frac{1}{2}(L(s) + L(t) - L(s^{-1}t))$, then,

(L2) $\quad\quad\quad\quad\quad\quad Y(r,t) \geq \min(Y(r,s), Y(s,t))$.

(L3) $\quad\quad\quad\quad\quad\quad Y(s,t) \in \Lambda$.

(9.6) LEMMA. (a) *(L0) follows from (H1)$_0$.*
(b) *(L1) follows from (H5) and (H1)$_{-1}$.*
(c) *(L3) follows from (H1)$_{-1}$ and (H3).*

Proof. (a). $(H1)_0$ implies that $\beta(1,t) = 0$ for all t, whence $L(1) = 0$.

(b). From (H5) we have $\beta(s,t) > 0 \Longrightarrow \beta(s^{-1},t) = \beta(s,t)$. It follows from this that $\beta(s,t) \leq 0 \Longrightarrow \beta(s^{-1},t) \leq 0$. Consequently $\lambda(s^{-1}) = \lambda(s)$, where $L(s) = \ell(s) + \lambda(s)$. Since $\ell(s^{-1}) = \ell(s)$, by $(H1)_{-1}$, we have $L(s^{-1}) = L(s)$.

(c). $Y(s,t) = \frac{1}{2}(L(s) + L(t) - L(s^{-1}t))$

$\qquad = \frac{1}{2}(\ell(s) + \ell(t) - \ell(s^{-1}t) + \lambda(s) + \lambda(t) - \lambda(s^{-1}t))$

$\qquad = \frac{1}{2}(-\beta(s^{-1},t) + \lambda(s) + \lambda(t) - \lambda(s^{-1}t))$ (by $(H1)_{-1}$).

The expression in parentheses lies in the subgroup of Λ generated by all $\beta(x,y)$ $(x,y \in \Gamma)$. By (H3) this subgroup is contained in 2Λ, whence $Y(s,t) \in \Lambda$.

Combining our conclusions thus far, we have the following characterization of hyperbolic length functions, in terms of conditional inequalities.

(9.7) THEOREM. *Let* $\ell : \Gamma \to \Lambda$ *satisfy* (H0)-(H8), *with* $(H1)_n$ *only for* $n = 0, -1, 2$.

(a) *If* ℓ *is abelian, i.e. if* $\beta(u,v) \leq 0$ *for all* $u, v \in \Gamma$, *then there is an action of* Γ *on* $X = \Lambda$ *by translations such that* $\ell = \ell_X$.

(b) *Suppose that* ℓ *is not abelian. Choose* $u, v \in \Gamma$ *such that* $\beta(u,v) > 0$, *and define* $L = L_{u,v} : \Gamma \to \Lambda$ *by* (9.4). *Then* ℓ *is a hyperbolic length function iff* L *satisfies* (L2) *above*.

Proof. If ℓ is abelian then by (9.3), $\ell(s) = |h(s)|$ for some homomorphism $h : \Gamma \to \Lambda$. Now let $s \in \Gamma$ act on $X = \Lambda$ by $x \mapsto x + h(s)$, and we have $\ell = \ell_X$.

Suppose now that $\beta(u,v) > 0$ and define L by (9.4). From (9.6), L satisfies (L0), (L1), and (L3). If L further satisfies (L2) then L is a Lyndon length function. By Theorem (5.4), $L = L_X$ for some Γ-(Λ-tree) X and base point $x \in X$. By (7.1)(c) and (9.5) we have $\ell = \ell_X$.

Conversely if $\ell = \ell_X$ is a hyperbolic length function then L is a Lyndon length function, hence satisfies (L2), by (7.9). This proves the theorem.

Unfortunately (L2), when stated directly in terms of ℓ, or β, is excessively complicated. Hopefully it is already a consequence of our other assumptions.

Note that (L2) is equivalent to the condition that, of $Y(r,s)$, $Y(s,t)$, $Y(r,t)$, the smallest two are equal. Thus (L2) is equivalent to:

(L2′) $\qquad\qquad Y(r,s) > Y(s,t) \Longrightarrow Y(r,t) = Y(s,t)$.

Writing $L(s) = \lambda(s) + \ell(s)$, we noted in the proof of (9.6)(c) that

$$2Y(s,t) = \lambda(s) + \lambda(t) - \lambda(s^{-1}t) - \beta(s^{-1},t).$$

Thus (L2′) becomes "(a) $\Longrightarrow$ (b)," where:

(a) $\lambda(r) + \lambda(s) - \lambda(r^{-1}s) - \beta(r^{-1},s) > \lambda(s) + \lambda(t) - \lambda(s^{-1}t) - \beta(s^{-1},t)$

(b) $\lambda(r) + \lambda(t) - \lambda(r^{-1}t) - \beta(r^{-1},t) = \lambda(s) + \lambda(t) - \lambda(s^{-1}t) - \beta(s^{-1},t)$.

Cancelling $\lambda(s)$ from (a) and $\lambda(t)$ from (b), they take the form

(a′) $y(r,s) > y(t,s)$

(b′) $y(r,t) = y(s,t)$

where

(9.8) $y(p,q) = \lambda(p) - \lambda(p^{-1}q) - \beta(p^{-1},q)$

$\qquad\qquad = \max(\beta(p,u), 0, \beta(p,v) - \beta(u,v))$

$\qquad\qquad - \max(\beta(p^{-1}q,u) + \beta(p^{-1},q), \beta(p^{-1},q), \beta(p^{-1}q,v) + \beta(p^{-1},q) - \beta(u,v))$

Thus, defining $y(p,q)$ by (9.8), we have the condition:

(H11) $\qquad\qquad y(r,s) > y(t,s) \Longrightarrow y(r,t) = y(s,t)$,

which is equivalent to (L2).

In trying to verify (H11) from the other axioms it might be convenient to restrict to the non-dihedral case, where we can choose u,v above so that $\ell(u) > 0$ and $\ell(v) > 0$. This reduction is admissible, by virtue of the next proposition.

(9.9) PROPOSITION. *Assume (H0)-(H8) as in Theorem (9.7), and (H9) and (H10) as well. If ℓ is dihedral (cf. (9.2)) then there is an action without inversions of Γ on $X = \Lambda$ such that $\ell = \ell_X$.*

Proof. Put $\Gamma_1 = \{s \in \Gamma | \beta(s,u) \leq 0 \text{ for all } u \in \Gamma\}$. If $s, t \in \Gamma_1$ then $st \in \Gamma_1$, by (H10). Moreover $\beta(s^{-1}, u) = \beta(s, u^{-1})$, by (9.1)(a), so $s^{-1} \in \Gamma_1$. If $v \in \Gamma$ then $\beta(vsv^{-1}, u) = \beta(s, v^{-1}uv)$, by (H2), so $vsv^{-1} \in \Gamma_1$. Thus Γ_1 is a normal subgroup of Γ. Clearly $\ell | \Gamma_1$ is abelian. Hence, by (9.3), there is a homomorphism $h: \Gamma_1 \to \Lambda$ such that $\ell(s) = |h(s)|$ for all $s \in \Gamma_1$. Put $\Gamma_0 = \text{Ker}(h) = \{s \in \Gamma_1 | \ell(s) = 0\}$. If $s \in \Gamma_0$ and $u \in \Gamma$ then $\ell(usu^{-1}) = \ell(s) = 0$, so Γ_0 is a normal subgroup of Γ. We claim that ℓ factors through Γ/Γ_0, i.e. that $\ell(st) = \ell(s)$ if $s \in \Gamma$ and $t \in \Gamma_0$. Since $\ell(t) = 0$ this is equivalent to the condition, $\beta(s,t) = 0$. Since $t \in \Gamma_0 \subset \Gamma_1$ we have $\beta(s,t) \leq 0$. If $\ell(s) = 0$ then, by (H5), $\beta(s,t) \geq 0$, hence $\beta(s,t) = 0$. Suppose that $\ell(s) > 0$. If $\beta(s,t) < 0$ then (H7) implies that $\beta(t, sts^{-1}) > 0$, contradicting the fact that $t \in \Gamma_0 \subset \Gamma_1$. Thus $\beta(s,t) = 0$ in all cases, as claimed. Now we can replace Γ by Γ/Γ_0 and so reduce to the case when $\Gamma_0 = \{1\}$, i.e. $h: \Gamma_1 \to \Lambda$ is injective.

Now assume further that ℓ is dihedral.

Claim 1. If $\ell(s) > 0$ then $s \in \Gamma_1$.

We must show that $\beta(s,t) \leq 0$ for all $t \in \Gamma$. If $\beta(s,t) > 0$ then $\ell(st) = \ell(s) + \ell(t) + \beta(s,t) > 0$ and, by (H5), $\beta(st, s^2(st)s^{-2}) > 0$, contradicting the fact that ℓ is dihedral.

Claim 2. If $s \in \Gamma_1$, $t \notin \Gamma_1$ then $\beta(s,t) = -\ell(s)$.

In fact, $st \notin \Gamma_1$ so, by Claim 1, $0 = \ell(st) = \ell(s) + \ell(t) + \beta(s,t) = \ell(s) + \beta(s,t)$.

Claim 3. If $s, t \notin \Gamma_1$ then $st \in \Gamma_1$.

For suppose that $st \notin \Gamma_1$, i.e. that $\beta(st, u) > 0$ for some u; evidently $u \notin \Gamma_1$. Using Claim 1 we have

$$\ell(stu) = \ell(st) + \ell(u) + \beta(st,u) = \beta(st,u) > 0$$
$$= \ell(s) + \ell(tu) + \beta(s,tu) = \ell(tu) + \beta(s,tu).$$

Thus $tu \notin \Gamma_1$, for otherwise the last expression vanishes, by Claim 2. Moreover $stu \in \Gamma_1$ since $\ell(stu) > 0$, hence $ust \in \Gamma_1$; but $t \notin \Gamma_1$, hence $us \notin \Gamma_1$. Thus Γ_1 contains none of the elements s,t,u,st,tu,us, so ℓ vanishes on each of them, by Claim 1. Now (H9) applies to give $\ell(stu) = 0$. This contradiction establishes Claim 3.

Now fix an $r \in \Gamma$, $r \notin \Gamma_1$. (Note that $\Gamma \neq \Gamma_1$, otherwise ℓ would be abelian, not dihedral.) Then $\ell(r) = 0$ (Claim 1) so $\ell(r^2) = 0$, by $(H1)_2$. But $r^2 \in \Gamma_1$ by Claim 3. Hence $r^2 = 1$, since $h: \Gamma_1 \to \Lambda$ is injective. It follows now from Claim 3 that $\Gamma = \Gamma_1 \rtimes \langle r \rangle$, the semi-direct product of Γ_1 with $\langle r \rangle = \{1, r\}$. For $s \in \Gamma_1$ we have $|h(rsr^{-1})| = \ell(rsr^{-1}) = \ell(s) = |h(s)|$. By the uniqueness property of h in (9.3), there is an $e = \pm 1$ such that $h(rsr^{-1}) = eh(s) = h(s^e)$ for all $s \in \Gamma_1$. Since h is injective we have $rsr^{-1} = s^e$ for all $s \in \Gamma_1$.

Claim 4. $e = -1$.

Choose s such that $\ell(s) > 0$, hence $s \in \Gamma_1$ (Claim 1). Then $\ell(sr) = \ell(s) + \ell(r) + \beta(s,r) = \ell(s) + \beta(s,r)$, so $\ell(s) > \ell(r) = 0 > \beta(s,r)$. Thus (H7) applies to give $\beta(r, srs^{-1}) > 0$. But if $e = 1$ then $srs^{-1} = r$, whereas $\beta(r,r) = 0$ by (9.1)(b). Thus $e = -1$, as claimed.

Now we can let Γ act on $X = \Lambda$ by $s(x) = x + h(s)$ for $s \in \Gamma_1$ and $r(x) = -x$. This action is without inversions because, for $s \in \Gamma_1$, $\ell(s) = -\beta(s,r)$, as we saw above, and $\beta(s,r) \in 2\Lambda$, by (H3), and so sr fixes $h(s)/2 \in \Lambda$.

ROGER ALPERIN
DEPARTMENT OF MATHEMATICS
UNIVERSITY OF OKLAHOMA
NORMAN, OKLAHOMA 73019

HYMAN BASS
DEPARTMENT OF MATHEMATICS
COLUMBIA UNIVERSITY
NEW YORK, NEW YORK 10027

REFERENCES

[AM] Alperin, R. C. and Moss, K. N., "Complete trees for groups with a real-valued length function," J. London Math. Soc., 2, 31 (1985), 55-68.

[C1] Chiswell, I. M., "Abstract length functions in groups," Math. Proc. Cambridge Phil. Soc. 80 (1976), 451-463.

[C2] _____, "Embedding theorems for groups with an integer-valued length function," Math. Proc. Cambridge Phil. Soc. 85 (1979), 417-429.

[C3] _____, "Length functions and free products with amalgamation of groups," Proc. London Math. Soc. 42 (1981), 42-58.

[C4] _____, "An example of an integer valued length function on a group," J. London Math. Soc. 16 (1977), 67-75.

[CM] Culler, M. and Morgan, J. W., Draft manuscript, (1984).

[CS] Culler, M. and Shalen, P. B., "Varieties of group representations and splittings of 3-manifolds," Annals of Math. 117 (1983), 109-146.

[D] Dress, Andreas W. M., "Trees, tight extensions of metric spaces, and the cohomological dimension of certain groups: A note on combinatorial properties of metric spaces," Adv. in Math. 53 (3) (1984), 321-402.

[H] Harrison, N., "Real length functions in groups," Trans. Amer. Math. Soc. 174 (1972), 77-106.

[Ho1] Hoare, A. H. M., "On length functions and Nielsen methods in free groups," J. London Math. Soc. 14 (1976), 188-192.

[Ho2] _____, "An embedding for groups with length functions," Mathematika 26 (1979), 99-102.

[Ho3] Hoare, A. H. M. and Wilkins, D. I., "On groups with unbounded non-archimedean elements," preprint 1983.

[I1] Imrich, W., "On metric properties of tree-like spaces," *Beitrage zur Graphentheorie und deren Anwendungen*, Sektion MAROK der Technischen Hochschule Ilmenau, pp. 129-156, Oberhof, East Germany, 1979.

[I2] Imrich, W. and Schwarz, G., "Tree and length functions on groups," Ann. Discrete Math. 17 (1982), 347-359.

[L] Lyndon, R. C., "Length functions in groups," Math. Scand. 12 (1963), 209-234.

[MB] Morgan, J. W. and Bass, H., *The Smith Conjecture*, Academic Press, New York (1984).

[MO] Morgan, J. W. and Otal, J. P. Draft manuscript, (1984).

[MS] Morgan, J.W. and Shalen, P.B., "Valuations, trees, and degenerations of hyperbolic structures: I," Annals of Math. (to appear).

[P] Promislow, D., "Equivalence classes of length functions on groups," preprint 1984.

[S] Serre, J-P., "*Trees*," Springer-Verlag, New York, 1980.

[Tg] Tignol, J-P., "Remarque sur le groupe des automorphismes d'un arbre," Ann. Soc. Sci. Bruxelles 93 (1979), 196-202.

[T1] Tits, J., "A 'theorem of Lie-Kolchin' for trees," *Contributions to Algebra: A Collection of Papers Dedicated to Ellis Kolchin*, pp. 377-388, Academic Press, New York, 1977.

[T2] ———, "Sur le groupe des automorphismes d'un arbre," *Essays on Topology and Related Topics* (Mémoires dédies à G. de Rham), pp. 188-211, Springer-Verlag, New York, 1970.

[Wi] Wilkins, D.I., "Length functions and normal subgroups," J. London Math. Soc. 22 (1980), 439-448.

RESIDUAL FINITENESS FOR 3-MANIFOLDS

John Hempel[1]

ABSTRACT. It is shown that every Haken 3-manifold has a residually finite fundamental group. This easily extends to prove the same for the fundamental groups of a large class of 3-manifolds which has been conjectured to include all compact 3-manifolds

1. *Introduction*

A group G is *residually finite* if every non-trivial element of G is mapped nontrivially in some finite quotient group of G. Residually finite groups are of interest for a number of reasons (cf. [Mg]) — they have a solvable word problem, the finitely generated ones satisfy the Hopfian property that every epimorphism $G \to G$ is an isomorphism, and so on. Since much of what is known about the structure of 3-dimensional manifolds reduces to the study of their fundamental groups, it has been natural to ask whether these groups are residually finite. See [H; Chapter 15] for more discussion on this matter. The answer, as stated in Theorem 3.3 of [T], is yes for an important class of 3-manifolds:

1.1. THEOREM. *The fundamental group of a Haken manifold is residually finite.*

Since residual finiteness is preserved under free products and finite index supergroups we have (calling a manifold which is finitely covered by a Haken manifold *virtually Haken*)

---

[1] Supported in part by NSF MCS-8301400.
AMS SUBJECT CLASSIFICATION (1983) 57M05, 20E26.

1.2. COROLLARY. *The fundamental group of a 3-manifold whose prime factors are either virtually Haken or have finite or cyclic fundamental groups is residually finite.*

It is unsolved whether this hypothesis gives the class of all compact 3-manifolds — cf. [T; §6].

The proof of 1.1 goes as follows. It is easy to reduce to the case in which the Haken manifold, M, is closed. The Jaco-Shalen-Johannson decomposition theorem [J,S], [J] gives a splitting of M by incompressible tori into pieces which are either Seifert fibered or, by the work of Thurston [T], have a hyperbolic structure. Correspondingly $\pi_1(M)$ is a graph product whose vertex groups (the fundamental groups of the pieces) are known to be residually finite and whose edge groups are peripheral $Z \oplus Z$'s. The remainder of the proof is in showing that residual finiteness is preserved under this type of graph product. Here some care must be taken since residual finiteness is not preserved under graph products in general — the non-hopfian, non-residually finite groups [B,S]

$$\langle a, b : ab^p a^{-1} = b^q \rangle$$

are graph products with one vertex and one edge group — both infinite cyclic. The proof of 1.1 depends on properties of the placement of the edge groups in the vertex groups which come from topological considerations. Briefly, the key properties are that the vertex groups are linear groups (although the Seifert fibered pieces are treated somewhat differently) and that the edge groups are maximal unipotent subgroups which do not intersect excessively with the commutator subgroups of their vertex groups (a reflection of duality).

The purpose of this paper is to provide the details of this proof and to present some natural generalizations to similar types of graph product

We first note an application:

1.3. COROLLARY. *Suppose* M *and* N *are Haken manifolds with (possibly empty) boundary and that*

$$f : (M, \partial M) \rightleftarrows (N, \partial N) : g$$

are degree one maps. Then f *is homotopic to a homeomorphism.*

The proof, which follows directly from Theorem 15.13 of [H], depends on the fact that $f_* : \pi_1(M) \to \pi_1(N)$ is an epimorphism — hence by the Hopfian property; an isomorphism — and so the theory of Waldhausen [W] applies.

The paper is organized as follows. In the remainder of this section we develop our notation and conventions. Section 2 reviews graph products — of groups, of spaces, and of covering spaces. Section 3 gives a general condition under which a graph product of residually finite groups is residually finite. Section 4 applies this to the case of Haken manifolds, and Section 5 gives some natural generalizations to similar types of graph products with linear groups as vertex groups and maximal unipotent subgroups as edge groups.

We will use covering space theory to translate between group theory and topology. Thus for $G = \pi_1(X)$, G is residually finite if and only if for each homotopically nontrivial loop $\alpha : (I, \partial I) \to (X, *)$ there is a finite sheeted regular covering space $\widetilde{X} \to X$ such that some (hence, by regularity, every) lifting $\widetilde{\alpha} : I \to \widetilde{X}$ of α is not a loop. If G is finitely generated, we may drop the requirement of regularity from this condition; since in this case every finite sheeted cover is covered by a finite sheeted regular cover corresponding to the intersection of the finitely many distinct conjugates of the associated subgroup.

By a *Haken manifold* we mean a compact, orientable, irreducible 3-manifold which contains a 2-sided incompressible surface (cf. [H; Ch. 6]).

By definition, a hyperbolic 3-manifold, M, is a quotient of hyperbolic 3-space, H^3, by a discrete, torsion-free subgroup of its group, $PSL(2, C)$, of isometries. We find it convenient to use the fact [C,S; 3.1.1] that the associated representation $\pi_1(M) \to PSL(2,C)$ lifts to a (discrete, faithful) representation $\pi_1(M) \to SL(2,C)$.

Commuting elements of $SL(2,C)$ have identical eigenspaces and thus can be simultaneously conjugated to triangular form. Thus an abelian subgroup of $SL(2,C)$ is conjugate to one consisting entirely of elements of one of the following forms:

$$(1) \quad \begin{bmatrix} a & 0 \\ 0 & a^{-1} \end{bmatrix} \quad \text{or} \quad (2) \quad \begin{bmatrix} 1 & b \\ 0 & 1 \end{bmatrix}.$$

A subgroup of $SL(2,C)$ of the second type is called a *unipotent* subgroup [Bo]. If the group is also discrete then it has rank at most 1 in case (1) and at most 2 in case (2).

2. Graph products

We will consider a *graph*, Y, as a collection $V = V(Y)$ of vertices and a collection $E = E(Y)$ of oriented edges with each $e \in E$ having initial vertex $i(e) \in V$ and terminal vertex $t(e) \in V$. The oppositely oriented edge is denoted by $\bar{e}$ where $i(\bar{e}) = t(e)$ and $t(\bar{e}) = i(e)$. The same notation will be used for the geometric realization of Y as a 1-dimensional CW complex. We assume that all graphs are connected.

A *graph of groups*, based on a graph, Y, cf. [S], consists of collections:

$$\{G_v : v \in V(Y)\}, \{G_e : e \in E(Y)\}$$

of groups such that $G_{\bar{e}} = G_e$, together with monomorphisms

$$\phi_e : G_e \to G_{t(e)}$$

and hence

$$\phi_{\bar{e}} : G_e \to G_{i(e)}.$$

Similarly, a *graph of* CW *complexes*, based on a graph, Y, consists of collections:

$$\{X_v : v \in V(Y)\}, \{X_e : e \in E(Y)\}$$

of CW complexes together with cellular embeddings

$$f_e : X_e \to X_{t(e)}$$

which induce monomorphisms

$$\pi_1(X_e) \to \pi_1(X_{t(e)}).$$

It is convenient to assume that the images of the f_e's are pairwise disjoint and have collar neighborhoods. This can always be arranged, without changing the homotopy type of any of the complexes involved, by attaching products.

By analogy it is possible to define a graph in a general category as a correspondence (functor) which associates objects to the edges and vertices and monomorphisms from each "edge" object to its terminal "vertex" object. We do not pursue this formalism except for the occasional convenience of the use of categorical language in describing properties such as "naturality."

By the *graph product* of a graph of CW complexes we mean the CW complex, X, obtained from the disjoint union of the X_v's by identifying each subcomplex

$$f_e(X_e) \text{ with } f_{\bar{e}}(X_e) \text{ via } f_{\bar{e}} \circ f_e^{-1}.$$

The underlying graph, Y, can be embedded in X in such a way that each vertex corresponds to the base point of the corresponding vertex complex, and each edge crosses the image of the corresponding edge complex at its base point, and so that the intersection of Y with the interior of each vertex complex is contractible. This provides paths joining the various base points necessary to make the induced maps

$$\pi_1(X) \to \pi_1(X_{t(e)})$$

well defined and converts the collection of fundamental groups to a graph of groups based on the same graph Y. Choice of a maximal tree in Y provides a consistent collection of well-defined maps

$$\pi_1(X_v) \to \pi_1(X).$$

The *graph product* (or fundamental group) of a graph of groups can be defined, as in [S], as a solution to a certain universal mapping problem; however, by the Seifert-Van Kampen theorem, this amounts to:

2.1. THEOREM. *The fundamental group of the graph product of a graph of CW complexes is naturally isomorphic to the graph product of the associated graph of fundamental groups.*

Given a graph

$$\{X_v; v \in V(Y)\}, \ \{X_e; e \in E(Y)\}, \ \{f_e; e \in E(Y)\}$$

of CW complexes, a collection

$$\{P_v : \tilde{X}_v \to X_v; v \in V(Y)\}, \ \{P_e : \tilde{X}_e \to X_e; e \in E(Y)\}$$

of covering spaces is called a *compatible collection of covers* provided that for each $e \in E(Y)$ and for each component $\tilde{X}$ of $p^{-1}(f_e(X_e))$ there is an embedding

$$\tilde{f} : \tilde{X}_e \to \tilde{X}_{t(e)}$$

with image $\tilde{X}$ which covers

$$f_e : X_e \to X_{t(e)}.$$

Note that this requires that for each e that all of the covers over $f_e(X_e)$ are equivalent (which is automatic if $p_{t(e)} : \tilde{X}_{t(e)} \to X_{t(e)}$ is regular) and that each is equivalent to each of the covers over $f_{\bar{e}}(X_e)$.

2.2. THEOREM. *Let X be the graph product of a graph of CW complexes based on a finite graph Y and let*

$$\{p_v : \tilde{X}_v \to X_v; v \in V(Y)\}, \quad \{p_e : \tilde{X}_e \to X_e; e \in E(Y)\}$$

be a compatible collection of finite sheeted covering spaces. Then there is a finite covering space

$$p : \tilde{X} \to X$$

where $\tilde{X}$ is the graph product of a graph whose vertex (edge) complexes are copies of the $\tilde{X}$ ($\tilde{X}_e$) and p is induced by the p_v (p_e).

Proof. If Y has a single vertex, v, and a single edge, e, then it follows that $p_v^{-1}(f_e(X_e))$ and $p_v^{-1}(f_{\bar{e}}(X_e))$ have the same number of components. We can identify these components in pairs by homeomorphisms which cover the identification of $f_e(X_e)$ with $f_{\bar{e}}(X_e)$ to obtain $\tilde{X}$.

If Y has two vertices, v_1 and v_2 and one edge, e, then suppose that $t(e) = v_1$. Let $p_{v_i}^{-1}(f_e(X_e))$ have n_i components. We take $\text{lcm}(n_1,n_2)/n_1$ copies of $\tilde{X}_{v_1}$, $\text{lcm}(n_1,n_2)/n_2$ copies of $\tilde{X}_{v_2}$ and identify each copy of a component of $p_{v_1}^{-1}(f_e(X_e))$ with a copy of a component of $p_{v_2}^{-1}(f_{\bar{e}}(X_e))$ as above to obtain $\tilde{X}$.

The proof then can be completed by induction on the number of edges of Y.

Note that the covering space $\tilde{X} \to X$ given by this construction is definitely not unique; nor is it likely to be regular even if all the component covers are regular. We will say that any cover produced by this construction is an *amalgam* of the corresponding compatible collection of covers.

3. Residual finiteness for certain graph products

The following theorem seems to be more or less well known. A version for amalgamated free products is given in [B] while a version for HNN extensions is given in [H]. We present it here in the form in which it will be used. A collection

$$\{H_v < G_v; v \in V(Y)\}, \{H_e < G_e; e \in E(Y)\}$$

of subgroups of the vertex and edge groups in a graph of groups is called a *compatible collection of subgroups* provided that for each $e \in E(Y)$

$$\phi_e(G_e) \cap H_{t(e)} = \phi_e(H_e).$$

3.1. THEOREM. *Let G be the graph product of a graph of finitely generated groups based on a finite graph Y. Suppose that for each $v_0 \in V(Y)$ and for each $g \in G_{v_0} - \{1\}$ there is a compatible collection $\{H_v\}, \{H_e\}$ of subgroups each normal and of finite index in the corresponding G_v, G_e with $g \notin H_{v_0}$. Suppose further that if for some fixed e_0 with $t(e_0) = v_0$, $g \notin \phi_{e_0}(G_{e_0})$ we can choose the collection so that $g \notin H_{v_0} \cdot \phi_{e_0}(G_{e_0})$.*

Then G is residually finite.

Proof. We can identify G with $\pi_1(X)$ where X is the graph product of a graph of CW complexes based on the same finite graph Y. Since G is finitely generated it suffices to show that for each nontrivial loop

$$\alpha : (I, \partial I) \to (X, *)$$

there is a finite sheeted covering space $\tilde{X} \to X$ such that some lifting

$$\tilde{\alpha} : I \to \tilde{X}$$

of α is not a loop.

There is a collection $\{A_e; e \in E(Y)\}$ of bicollared subcomplexes of X (homeomorphic to the X_e's) which split X into pieces homeomorphic to the X_v's. We may suppose that $\alpha(I)$ meets $\cup A_e$ minimally. We induct on the number, n, of subintervals J of I such that α maps the end points of J to the same A_e and maps neighborhoods in J of its endpoints to the same side of A_e.

We may assume that the graph Y is embedded in X with each vertex in the corresponding vertex space and each edge crossing the corresponding A_e, and that there is a retraction

$$\rho : X \to Y$$

mapping each A_e to the point $A_e \cap Y$. If

$$\rho \circ a : (I, \partial I) \to (Y, *)$$

is homotopically nontrivial, then the conclusion follows from the fact that $\pi_1(Y)$, being free, is residually finite. This must be the situation if $n = 0$.

If $\rho \circ a$ is nullhomotopic and $a(I)$ does not lie in a single vertex space in X, then there is a subinterval J of I such that $a(\partial J)$ lies in a single A_{e_0}, a neighborhood of ∂J in J is mapped to one side of A_{e_0}, and $a(\text{Int}(J))$ lies in a vertex space which we may correspond to X_{v_0} with $v_0 = t(e_0)$. By joining the points of $a(\partial J)$ in A_{e_0} we create a loop which corresponds to an element $g \in \pi_1(X_{v_0}) = G_{v_0}$. Now $g \notin \phi_{e_0}(G_{e_0})$ – otherwise $a|J$ could be homotoped into A_{e_0} then pushed to the opposite side of A_{e_0} to simplify the intersection of $a(I)$ with $\cup A_e$.

Now take the compatible collection $\{H_v\}$, $\{H_e\}$ of normal subgroups of finite index given by hypothesis and consider the corresponding collection of finite sheeted regular covering spaces $\{\tilde{X}_v \to X_v\}$, $\{\tilde{X}_e \to X_e\}$. This is a compatible collection of covering spaces in the sense of Section 2. Moreover the condition that $g \notin H_{v_0} \cdot \phi_{e_0}(G_{e_0})$ implies that the liftings of $a|J$ to $\tilde{X}_{v_0}$ begin and end in different components of $p_{v_0}^{-1}(f_{e_0}(X_{e_0}))$.

By Theorem 2.2 we get a finite covering $p : \tilde{X} \to X$ which is an amalgam of the given collection of covering spaces. Take a lifting

$\tilde{\alpha} : I \to \tilde{X}$ of α. If $\tilde{\alpha}$ is not a loop we are done. If it is a loop then its complexity, $\tilde{n}$, relative to the lifted graph product structure on $\tilde{X}$ is smaller than the complexity, n, of α. Now the hypotheses of the theorem are easily seen to be inherited by the lifted graph product structure on $\tilde{X}$. Thus induction applies.

The case $(n=1)$ in which $\alpha(I)$ lies in a single vertex space is handled in much the same way, but without the extra complications involving the product subgroup.

4. Proof of Theorem 1.1

Let M be a Haken manifold. We will show that $\pi_1(M)$ is residually finite. If ∂M is nonempty then, since residual finiteness is preserved under free products, we may assume that ∂M is incompressible in M. The double of M along its boundary is a closed Haken manifold whose fundamental group contains $\pi_1(M)$ as a subgroup. Since residual finiteness is inherited by subgroups, it suffices to prove the theorem for closed Haken manifolds. Thus we assume that M is closed.

By [J,S], [J] there is a collection $\{T_1, \cdots, T_m\}$ of disjoint, incompressible tori in M which split M into 3-manifolds $M_1, \cdots, M_n$ which are either Seifert fibered or, by the work of Thurston, cf. [T], have a complete hyperbolic structure of finite volume. We assume that this collection is nonempty; however, the proof in the case where M is already Seifert fibered or hyperbolic follows easily from the same techniques and without all of the complications. Now M is the graph product of the graph whose vertex spaces are the M_i and whose edge spaces are the T_j. The edge groups in the corresponding graph of groups are the $\pi_1(T_j) \cong Z \oplus Z$. Observe that the characteristic subgroups of finite index in $Z \oplus Z$ are precisely the subgroups $n(Z \oplus Z)$ with index n^2 and quotient group $Z_n \oplus Z_n$ $(n = 1, 2, \cdots)$. Thus to achieve the compatibility conditions of Theorem 3.1, it suffices to find, for a fixed integer n, normal subgroups N_i of finite index in $\pi_1(M_i)$ such that for every j with $T_j \subset \partial M_i$, $N_i \cap \pi_1(T_j)$ is the characteristic subgroup of index n^2 in $\pi_1(T_j)$.

Thus the proof of Theorem 1.1 will be completed by:

4.1. LEMMA. *Let M be a compact 3-manifold which is either Seifert fibered or whose interior has a complete hyperbolic structure and whose boundary consists of tori* $T_1, \cdots, T_m$. *If* $g \in \pi_1(M) - \{1\}$ *then for almost all primes p there is a normal subgroup, N, of finite index in* $\pi_1(M)$ *such that for each j,* $N \cap \pi_1(T_j)$ *is the characteristic subgroup of index* p^2 *in* $\pi_1(T_j)$. *Moreover if for some boundary component, say* T_1, $g \in \pi_1(M, *) - \pi_1(T_1, *)$, *we can choose N so that* $g \notin N \cdot \pi_1(T_1)$.

Proof. We first consider the case in which M has a hyperbolic structure. We also suppose the case in which $g \in \pi_1(M) - \pi_1(T_1)$. Now for each j,

$$\dim(\text{image}(H_1(T_j; Q) \to H_1(M, Q))) \geq 1 \ .$$

We choose generators μ_j, λ_j for $\pi_1(T_j) = H_1(T_j)$ with the convention that whenever the above image has dimension 1, λ_j maps to 0 in $H_1(M; Q)$.

As discussed in Section 1, the hyperbolic structure on M gives rise to a discrete, faithful representation

$$\rho : \pi_1(M) \to SL(2, C)$$

under which the peripheral subgroups, $\pi_1(T_j)$, are mapped to unipotent subgroups. Thus for each j there is a $P_j \in SL(2, C)$ such that

$$P_j \rho(\mu_j) P_j^{-1} = \begin{bmatrix} 1 & m_j \\ 0 & 1 \end{bmatrix}, \quad P_j \rho(\lambda_j) P_j^{-1} = \begin{bmatrix} 1 & \ell_j \\ 0 & 1 \end{bmatrix}$$

where m_j and ℓ_j are elements of C which are linearly independent over Z. We may assume that $P_1 = I$.

Let A be a finitely generated ring in C which contains all the entries of each element of $\rho(\pi_1(M))$ as well as the entries of each P_j and P_j^{-1}.

It follows that $\pi_1(T_j)$ is a maximal unipotent subgroup of $\pi_1(M)$. This can be argued using hyperbolic geometry (cf. [Mr]) by using the fact that $\pi_1(T_j)$ corresponds to the parabolic subgroup associated to a cusp of M. An alternate argument that $\pi_1(T_j)$ cannot be properly contained in a subgroup $H \cong Z \oplus Z$ goes as follows. Take the covering $\widetilde{M} \to M$ corresponding to H. There is (cf. [H; 8.61]) a compact 3-submanifold, N, of $\widetilde{M}$ such that $\pi_1(N) \to \pi_1(\widetilde{M})$ is an isomorphism. We may assume that N contains a homeomorphic lifting $\widetilde{T}_j$ of T_j in its boundary. By [H; 10.5] and the assumption of orientability, N is a punctured product: $\widetilde{T}_j \times I$ with some 3-cells removed. In particular $\pi_1(\widetilde{T}_j) \to \pi_1(M)$ is an isomorphism.

So if $\rho(g) = \begin{bmatrix} a & b \\ c & d \end{bmatrix}$ then either $c \neq 0$ or $a \neq 1$ and $d \neq 1$ — otherwise $\pi_1(T_1)$ and g are contained together in some larger unipotent subgroup of $\pi_1(M)$.

The finitely generated ring $A \subset C$ is residually a finite field [M] (see Lemma 4.2 below). Thus for each prime p outside of a finite set there is a homomorphism

$$\eta : A \to F$$

where F is a finite field of characteristic p such that ker η contains none of the non-zero elements among:

$$m_j, \ell_j, c, a-1, d-1 .$$

Now η induces a representation

$$\bar\rho : \pi_1(M) \to SL(2,F) .$$

By the construction $\bar\rho(g) \notin \bar\rho(\pi_1(T_1))$. Moreover $\bar\rho$ maps each P_j to an invertible element of SL(2,F). Thus $\bar\rho(\mu_j)$, $\bar\rho(\lambda_j)$ are conjugate in SL(2,F) to upper triangular matrices which clearly have order p. However, $\rho(\pi_1(T_j))$ may be cyclic instead of $Z_p \oplus Z_p$ as desired. To correct for this, let

$$\theta : \pi_1(M) \to H_1(M; Z)/\text{torsion} = H .$$

By the choice of generators we have that for each j either:

i) $\theta|\pi_1(T_j)$ is monic, or

ii) $\theta(\lambda_j) = 0$.

By eliminating another finite set of values of p we may assume that under the induced map

$$\bar{\theta}: \pi_1(M) \to H/H^p$$

$\bar{\theta}(\pi_1(T_j)) \cong Z_p \oplus Z_p$ in case i) and Z_p in case ii). Thus we can take

$$N = \ker\{\bar{\rho} \times \bar{\theta}: \pi_1(M) \to SL(2,F) \times H/H^p\}$$

to finish the proof in the case M is hyperbolic.

Now suppose that M is Seifert fibered. By killing the (normal) subgroup generated by a nonsingular fiber we get a map of $\pi_1(M)$ to a Fuchsian group which must (cf. [H; 12.2]) contain a normal, torsion-free subgroup of finite index. By pulling this subgroup back to $\pi_1(M)$ and taking the corresponding covering space, we obtain a finite sheeted regular covering space $\widetilde{M} \to M$ in which every boundary component of $\widetilde{M}$ projects homeomorphically to M. By [H; 11.10], $\widetilde{M}$ is an S^1-bundle over a compact surface. After passing to another covering, we may suppose that $\widetilde{M}$ is a product $S^1 \times F$ (and preserve the property that boundary components project homeomorphically). It suffices then to prove the lemma for $\widetilde{M}$. For this it suffices, by using product coverings, to show that if $g \in \pi_1(F,*) - \pi_1(J,*)$ for some component J of ∂F, then, for almost all primes p, there is a normal subgroup $N \subset \pi_1(F)$ of finite index such that $g \notin N \cdot \pi_1(J)$ and so that for any component K of ∂F, $N \cap \pi_1(K)$ is the subgroup of index p. This follows just as in the first case regarding $\pi_1(F)$ as a discrete subgroup of $SL(2,R) \subset SL(2,C)$.

For completeness we include a proof of:

4.2. LEMMA. *Let* $A \subset C$ *be a finitely generated ring and* $a_1, \cdots, a_n$ *be nonzero elements of* A. *Then for almost every prime,* p, *there is a finite field*, F, *of characteristic* p *and a homomorphism* $\eta: A \to F$ *such that for each* i, $\eta(a_i) \neq 0$.

Proof. By taking a product it suffices to consider the case of a single element, a. By the Noether normalization lemma (cf. [A,M; pg. 70]), A is integral over the subring

$$B = Z[1/s][x_1, \cdots, x_k]$$

obtained from Z by inverting an integer, s, and adjoining polynomials in independent transcendentals $x_1, \cdots, x_k$. For any prime, p, not dividing s we have a homomorphism

$$Z[1/s] \to Z_p.$$

Any choice of images for the x_i's gives rise to an extension

$$B \to Z_p.$$

For any finite set of elements of B we can choose p large enough and evaluate in such a way that none of these elements are mapped to 0. We apply this to the coefficients of the polynomial over B satisfied by a. This polynomial then maps to a polynomial over Z_p which has a non-trivial root in some extension field of Z_p. Hence we can further extend to a homomorphism

$$B[a] \to Z_p(a')$$

of B[a] to a finite extension field of Z_p in which a is mapped non-trivially. Since A is finitely generated we can, in a similar way, extend inductively to get the desired homomorphism of A to a finite extension of Z_p.

5. Generalizations

In the context of finite graph products of finitely generated groups the arguments used in the proof of Theorem 1.1 depended on three basic points.

(i) The vertex groups were linear groups — the basis of the analysis by matrix arguments. Dimension two was incidental. They could well have been subgroups, say, of $SL(n,C)$.

(ii) An element of a vertex group not in the image of a particular edge group could (after appropriate conjugation) be distinguished by values of certain entries of its matrix — a property which, by Lemma 4.2, can be preserved by mappings to finite fields of almost any characteristic. This was easily established for maximal unipotent subgroups.

(iii) The kernels of the finite representations of the vertex groups coming from (ii) could, after some modification, be made to intersect the images of the adjacent edge groups in characteristic subgroups identifiable by their indices. This was essential in satisfying the compatibility conditions of Theorem 3.1. This depended on the easy recognition of the characteristic subgroups in the (free abelian) edge groups as well as the ability to make the necessary modifications. The latter came from topological facts about the map $H_1(\partial M) \to H_1(M)$ for a 3-manifold M.

To generalize it is natural to take (i) as an assumption. Item (ii) offers many possibilities; however in connection with (iii) we take the most direct. For this we say (following [Bo]) that a subgroup of $SL(n,C)$ is *unipotent* if it is conjugate to a subgroup of

$$N(n,C) = \{T \in SL(n,C) : T \text{ is upper triangular}$$
$$\text{with 1's on the main diagonal}\}.$$

$N(n,C)$ is a torsion-free nilpotent group. Elements of $N(n,C)$ are of the form:

$$I + A \text{ with } A^n = 0.$$

Since $(I+A)^p \equiv I + A^p \mod p$, we see that for F a field of characteristic p that

$N(n,F)$ has exponent p if $p \geq n$.

Now any finitely generated subgroup of $SL(n,C)$ maps (as in the proof of Lemma 4.1) to a finite subgroup of $SL(n,F)$. By the above remarks any unipotent subgroup will map to a group of exponent p. Clearly a matrix not in $N(n,C)$ can be so distinguished by the values of some of its entries. By suitable choices we can keep the image of such a matrix outside of the image of a given unipotent subgroup.

With regard to item (iii) we say that a subgroup, H, of a group, G, is p-*full* in G provided that there is a homomorphism of G to a finite group whose kernel intersects H in the subgroup H^p generated by the p-th powers of elements of H. Note that H^p is a characteristic subgroup of H and either $H^p = H$ or H^p is the smallest normal subgroup of H whose quotient has exponent p. So if $H \to K$ is any other homomorphism to a group, K, of exponent p then

$$H^p = \ker\{H \to H/H^p \times K\}.$$

The arguments for Theorem 1.1 then provide a proof for:

5.1. THEOREM. *Let G be the graph product of a finite graph of finitely generated groups such that each vertex group is a subgroup of $SL(n,C)$ for some n and each edge group maps to a maximal unipotent subgroup of its terminal vertex group which is p-full in this vertex group for almost all primes p. Then G is residually finite.*

Regarding the last condition we note

5.2. LEMMA. *If H is a finitely generated abelian, unipotent subgroup of a subgroup, G, of $SL(n,C)$ such that the image of H in G/G' has rank $\geq$ rank $H-1$, then for almost all p, H is p-full in G.*

Proof. Choose $p \geq n$ and such that p does not divide the image of H in the minimal direct summand of G/G' which contains this image. Then just as in the 3-manifold case we can take the direct product of suitably defined maps $G \to SL(n,F)$ and $G \to G/G' \otimes Z_p$ to produce one whose kernel intersects H in H^p.

5.3. COROLLARY. *Suppose G is a finite graph product of finitely generated groups such that each vertex group is a subgroup of $SL(n,C)$ for some n and such that each edge group, G_e maps to a maximal unipotent subgroup of its terminal vertex group $G_{t(e)}$. Suppose that each G_e is either*

  i) *cyclic or*

  ii) *free abelian of rank 2 and projects to an infinite subgroup of* $G_{t(e)}/G'_{t(e)}$.

Then G is residually finite.

Clearly there are many variations of the above theme. Rather than try to list them in generality, we suggest that it is more important to find other applications (as to the case of Haken manifolds) of the general principles involved and recommend this as a problem for study.

DEPARTMENT OF MATHEMATICS
RICE UNIVERSITY
HOUSTON, TEXAS 77251

REFERENCES

[A,M] M. F. Atiyah and I. G. MacDonald, *Introduction to Commutative Algebra*, Addison Wesley (1969).

[B] G. Baumslag, "On the residual finiteness of the generalized free product of nilpotent groups," Trans. Amer. Math. Soc. 106 (1963), 193-209.

[B,S] G. Baumslag and D. Solitar, "Some two generator, one relator non-Hopfian groups," Bull. Amer. Math. Soc. 68 (1962), 199-201.

[Bo] A. Borel, *Linear Algebraic Groups*, W. A. Benjamin (1969).

[C,S] M. Culler and P. Shalen, "Varieties of group representations and splittings of 3-manifolds," Ann. of Math. 117 (1983), 109-146.

[H]   John Hempel, *3-manifolds*, Annals of Math. Study No. 86, Princeton University Press (1976).

[J,S] W. Jaco and P. Shalen, *Seifert Fibered Spaces in 3-manifolds*, Memoirs of the Amer. Math. Soc. vol. 21 (1979).

[J]   K. Johannson, *Homotopy Equivalences of 3-manifolds with Boundaries*, Lecture Notes in Math. Vol. 761, Springer-Verlag (1979).

[M]   A. I. Mal'cev, "On the faithful representation of infinite groups by matrices," Mat. Sb. 8(50) (1940), 405-422; Translations of the Amer. Math. Soc. (2) 45 (1965), 1-18.

[Mg]   W. Magnus, "Residually finite groups," Bull. Amer. Math. Soc. 75 (1969), 305-315.

[Mr]   A. Marden, "The geometry of finitely generated Kleinian groups," Ann. of Math. 99 (1974), 383-462.

[S]   J. P. Serre, *Trees*, Springer-Verlag (1980).

[T]   W. Thurston, "Three dimensional manifolds, Kleinian groups and hyperbolic geometry," Bull. Amer. Math. Soc. (n.s.) 6 (1982), 357-381.

[W]   F. Waldhausen, "On irreducible 3-manifolds which are sufficiently large," Ann. of Math. 87 (1968), 56-88.

THE NIELSEN-THURSTON THEORY OF SURFACE AUTOMORPHISMS

Steven A. Bleiler

(Based on lectures and notes by A. J. Casson)

Let F be a closed orientable surface of genus g. The fundamental group of F can be expressed in terms of $2g$ generators and a single relator as given below, and is called a *surface group*.

$$\pi_1(F) = <x_1, y_1, \cdots, x_g, y_g | [x_1, y_1][x_2, y_2] \cdots [x_g, y_g]> .$$

An observation of Nielsen allows one to study the automorphisms of surface groups geometrically. In particular:

LEMMA (Nielsen [9]). *If* $g \geq 1$ *then every element of* $\mathrm{Out}(\pi_1(F))$ *is represented by a unique isotopy class of self-homeomorphisms of* F.

There is no difference in considering surface automorphisms up to isotopy or up to homotopy, as homotopic automorphisms of a closed orientable surface are isotopic.

An important subgroup of $\mathrm{Out}(\pi_1(F))$ is the *mapping class group* $\mathfrak{M}_g$. This is the index 2 subgroup of $\mathrm{Out}(\pi_1(F))$ consisting of elements represented by orientation-preserving homeomorphisms. The group $\mathfrak{M}_g$ is finitely generated by *Dehn twists* in simple closed curves on F. In particular, a simple closed curve C on an orientable surface F has a neighborhood A homeomorphic to an annulus which we consider as being parameterized by $\{[r,\theta] | 1 \leq r \leq 2\}$. The *Dehn twist* in C is defined to be the automorphism $T_C : F \to F$ given by the identity off A and by $[r, \theta] \to [r, \theta + 2\pi r]$ on A.

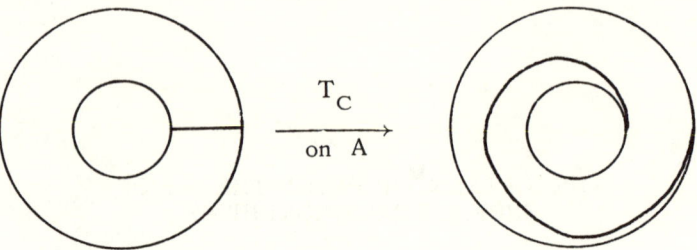

EXAMPLE: The torus T^2.

Identify T^2 with $\mathbb{R}^2/\mathbb{Z}^2$. Then $\pi_1(T^2) \cong \mathbb{Z} \oplus \mathbb{Z}$, $\text{Out}(\pi_1(T^2)) \cong GL_2(\mathbb{Z})$, and $\mathfrak{M}_1 \cong SL_2(\mathbb{Z})$. This is easily seen, as an α in $GL_2(\mathbb{Z})$ maps $\mathbb{Z}^2$ to itself and thus induces a continuous map $h_\alpha : T^2 \to T^2$. As α is in $GL_2(\mathbb{Z})$, h_α is a homeomorphism with inverse $h_{\alpha^{-1}}$ and $h_\alpha^* : \pi_1(T^2) \to \pi_1(T^2)$ has matrix α. This map preserves orientation if and only if $\det(\alpha) = 1$.

Consider α to be in $SL_2(\mathbb{Z})$. If α is the 2×2 matrix $\begin{pmatrix} p & q \\ r & s \end{pmatrix}$ the characteristic polynomial of α, $t^2 - (p+s)t + (ps-rq)$, can be written as $t^2 - (\text{trace}(\alpha))t + 1$, so the eigenvalues λ, λ^{-1} of α are either

1) complex, i.e. $\text{trace}(\alpha) = 0, 1$, or -1

or 2) both ± 1, i.e. $\text{trace}(\alpha) = \pm 2$,

or 3) distinct reals, i.e. $|\text{trace}(\alpha)| > 2$.

In case 1) an easy exercise in the Cayley-Hamilton theorem shows α to be of finite order and that $(h_\alpha)^{12} = 1$.

In case 2) h_α leaves an essential simple closed curve invariant (the image of an integral eigenvector for α under the quotient map $\mathbb{R}^2 \to T^2$) and is a power of a Dehn twist in this curve.

In case 3) suppose that $|\lambda| > 1 > |\lambda^{-1}|$ and that x and x are the corresponding real eigenvectors. In this case h_α has infinite order and leaves no essential simple closed curve invariant. In addition, translating x, x′ yields vector fields $\mathcal{F}, \mathcal{F}'$ which are carried by h_α to $\lambda \mathcal{F}$ and $\lambda^{-1} \mathcal{F}'$ respectively. That is, h_α is a linear homeomorphism which stretches by a factor λ in one direction and shrinks by the same factor in a complimentary direction. When this occurs, h_α is called *Anosov*.

Nielsen-Thurston theory generalizes this classification to all orientable surfaces.

§2. General theorems

The following is a condensation of some of Nielsen's results [9,10,11].

THEOREM 1 (Nielsen). *If* $\alpha \in \mathfrak{M}_g \subset \mathrm{Out}(\pi_1(F))$, *then either*

1) α *has finite order*

or 2) α *is represented by a homeomorphism leaving invariant a disjoint union of essential simple closed curves on* F.

or 3) *for all* $x \in \pi_1(F) - \{1\}$ *and all* $n \neq 0$, $\alpha^n(x)$ *is not conjugate to* x *in* $\pi_1(F)$.

Let us examine these cases in turn.

Case 1). α *is periodic.*

Fenchel [4,5] and Nielsen [12] proved that if $\alpha \in \mathfrak{M}_g$ has finite order then α is represented by a periodic homeomorphism. For prime order elements this can be proved using basic facts about the action of the mapping class group on the Teichmuller space of hyperbolic structures on F. For composite order, one also needs Smith fixed point theory.

More generally, Kerckhoff [7] proved:

THEOREM 2. *If* $h: G \to \mathrm{Out}(\pi_1 F)$ *is a homomorphism of a finite group* G, *then* h *lifts to a homomorphism* $\bar{h}: G \to \mathrm{Homeo}(F)$.

The proof of Theorem 2 uses deep results of Thurston on the compactification of Teichmuller space. A corollary of this theorem is that if Γ is a torsion-free group with a finite index subgroup isomorphic to a surface group, then Γ is a surface group. This corollary can also be proven algebraically [1].

Case 2). α *is reducible*

Let α be represented by $h : F \to F$ and that $h(C) = C$ for a disjoint union C of essential simple closed curves in F. The idea is to 'simplify' h by cutting F along C. The behavior of h can still be quite complicated, as it may permute the components of F-C; note also that h remains reducible after composition with Dehn twists in the components of C.

THEOREM 3 (Nielsen). *If* $\alpha \in \mathfrak{M}_g$ *is reducible then* α *is represented by a homeomorphism* $h : F \to F$ *such that*

1) $h(A) = A$ for $A = A_1 \cup \cdots \cup A_k$ *a disjoint union of essential annuli in* F.

2) *If* $G_1, \cdots, G_j$ *are the components of* F-Int A *and* $h^s(G_i) = G_i$, *then either* $h^s|G_i$ *is periodic or, for every non-peripheral element* x *of* $\pi_1(G_i)$ *and every* $n = 0$, $(h_*^s)^n(x)$ *is not conjugate to* x.

A *non-peripheral* element of $\pi_1(G_i)$ is one which is not conjugate to a loop in the boundary of G_i. These conditions imply that if $h^r(A_i) = A_i$ then $h^{2r}|A_i$ is a power of a Dehn twist.

Case 3). Homeomorphisms satisfying this condition were not well understood until Thurston's work [14]. He termed a homeomorphism representing an α of this type *pseudo-Anosov*, and proved

THEOREM 4. *An element* $\alpha \in \mathfrak{M}_g$ *is represented by a pseudo-Anosov automorphism if and only if* α *is neither periodic nor reducible.*

Before defining exactly pseudo-Anosov automorphisms in §3, consider some of the properties a pseudo-Anosov automorphism enjoys [14].

1) The pseudo-Anosov representative of $\alpha \in \mathfrak{M}_g$ is unique up to conjugacy in Homeo(F).

2) h has a *stretching factor* $\lambda > 1$ such that for any x in $\pi_1(F) - \{1\}$ and for all n,
$$k_x \lambda^{|n|} < \ell(h^n(x)) < K_x \lambda^{|n|}$$

where $\ell(h^n(x))$ is the length of the shortest word conjugate to $h^n(x)$ and k_x and K_x are constants depending on x.

3) The stretching factor λ is an algebraic integer.

4) For a given surface F and constant K there are only finitely many conjugacy classes of pseudo-Anosov automorphisms with stretching factor less than K.

The following examples (Thurston [14]) correctly suggest that 'most' elements of $\mathfrak{M}_g$ are represented by pseudo-Anosov automorphisms.

Let C_1 and C_2 be simple closed curves in F with minimal intersection, that is, C_1 and C_2 intersect transversely and there do not exist arcs A_1, A_2 in C_1, C_2 respectively, having common endpoints such that $A_1 \cup A_2$ is the boundary of a disc in F.

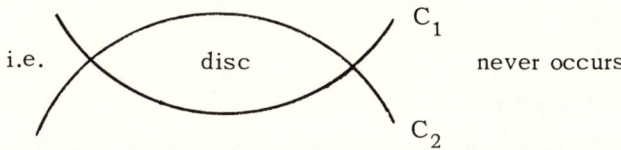

Moreover, suppose that every component of $F - (C_1 \cup C_2)$ is an open disc (C_1, C_2 are said to *fill* F).

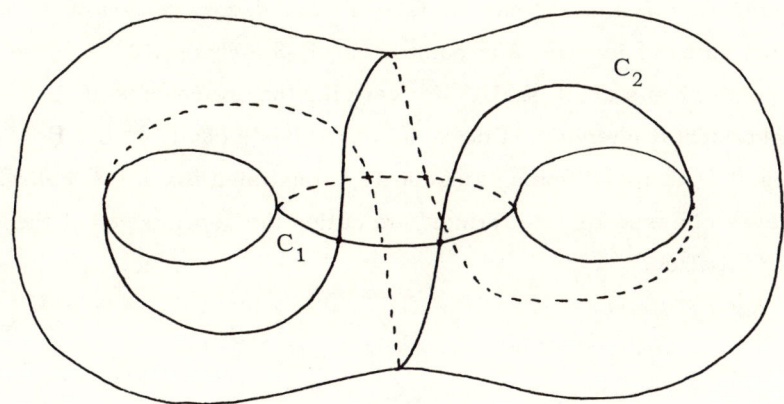

Let T_1, T_2 be Dehn twists in C_1, C_2 and let

$$A_1 = \begin{pmatrix} 1 & n \\ 0 & 1 \end{pmatrix} \qquad A_2 = \begin{pmatrix} 1 & 0 \\ n & 1 \end{pmatrix}$$

where $n = |C_1 \cup C_2|$. If $w(x,y)$ is a word, $w(T_1,T_2)$ is isotopic to a pseudo-Anosov automorphism if and only if $w(A_1,A_2)$ has distinct real eigenvalues. In this case the stretching factor is equal to the absolute value of the larger eigenvalue.

With this construction, the stretching factor is an algebraic integer of degree two. Long [8] and Penner [13] generalized the construction (in different ways), achieving stretching factors of higher degrees. In both generalizations C_1 and C_2 are replaced by disjoint unions of simple closed curves.

§3. Measured foliations

A homeomorphism $h: F \to F$ is *pseudo-Anosov* if there are transverse measured singular foliations $(\mathcal{F}^s, \mu^s), (\mathcal{F}^u, \mu^u)$ on F such that

1) $h_*(\mathcal{F}^s, \mu^s) = (\mathcal{F}^s, \lambda^{-1}\mu^s)$
2) $h_*(\mathcal{F}^u, \mu^u) = (\mathcal{F}^u, \lambda\mu^u)$

where $\lambda > 1$ is the stretching factor.

A *singular foliation* $\mathcal{F}$ on a surface F is a decomposition of F into a disjoint union of *leaves*. Any point x in $F-S$ (where S is a finite set of points) has a chart $\Phi: U \to \mathbf{R}^2$ carrying the components of $U \cap$ leaf to horizontal intervals. For x in S, x has a chart $\Phi: U \to \mathbf{R}^2$ carrying $\mathcal{F} \cap U$ to a 'k-prong singularity,' illustrated for $k = 4$ below. The leaves corresponding to 'prings' are called the *separatrices* of the singular foliation.

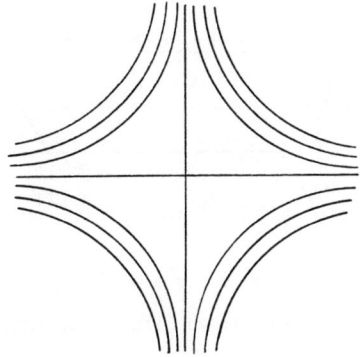

Singular foliations $\mathcal{F}^s, \mathcal{F}^u$ are transverse if they have the same singular set and at regular points the leaves are transverse. Further we require a standard 'prong' model, as illustrated below, at the singular points.

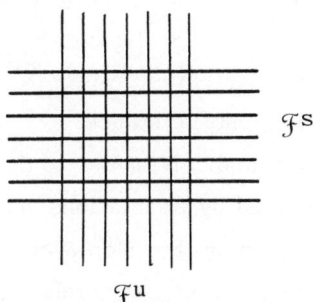
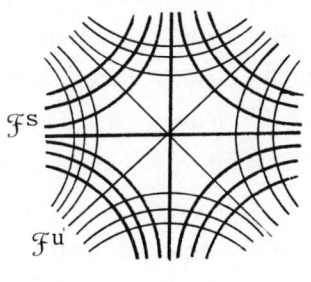

Away from S · · · · · · · · · · · · · · · · Near a point of S

For a_0, a_1 closed arcs transverse to a foliation $\mathcal{F}$, say that $a_0 \sim a_1$ if there exists a homotopy $H: I \times I \to F$ such that $H(I \times 0) = a_0$, $H(I \times 1) = a_1$, $H^{-1}(\mathcal{F}) = A \times I$ where $A \approx a_0 \cap \mathcal{F} \approx a_1 \cap \mathcal{F}$. The map $H_1 H_0^{-1} : a_0 \cap \mathcal{F} \to a_1 \cap \mathcal{F}$ is called *projection along the leaves*.

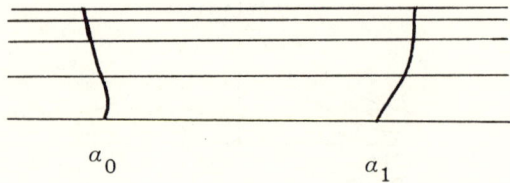

$a_0 \qquad a_1$

A *transverse measure* μ on $\mathcal{F}$ is a collection of finite, non-negative measures on the transverse arcs to $\mathcal{F}$ which is invariant under projection along the leaves and consistent with restriction if $\beta \subset \alpha$. A *measured foliation* is a foliation equipped with a transverse measure. If a homeomorphism h leaves $\mathcal{F}$ invariant, then the measure $h(\mu)$ is the measure defined by $h(\mu)(\alpha) = \mu(h(\alpha))$.

We remark that if $(\mathcal{F}^s, \mu^s), (\mathcal{F}^u, \mu^u)$ are transverse measured foliations with singular set S, then every point x not in S has a 'canonical' chart $\Phi : U \to \mathbf{R}^2$ such that

1) $\mathcal{F}^s$ leaves map to horizontals and μ^s to vertical distance.
2) $\mathcal{F}^u$ leaves map to verticals and μ^u to horizontal distance.

The map Φ is unique up to translation in $\mathbf{R}^2$ and π-rotation, so the transition functions lie in the pseudo-group generated by translations and rotation by angle π, giving a branched flat structure on F. Moreover, the measured foliations $(\mathcal{F}^s, \mu^s), (\mathcal{F}^u, \mu^u)$ determine a C^∞ structure on F, and so a pseudo-Anosov h can be smoothed away from the singular set. Away from the singular set, h is therefore a diffeomorphism with respect to this smooth structure; indeed, h is Anosov away from the singular set. However, a pseudo-Anosov automorphism cannot be smoothed to a diffeomorphism near the singular points.

§4. A sketch of the proof of Theorem 4

Assume that the genus of F is at least two, as the automorphisms of tori have already been classified. In this case, as detailed in [2] or [3], F can be given a hyperbolic metric; the universal cover $\tilde{F}$ of F can be identified with the hyperbolic plane H^2 in such a way that the deck translations are all isometries of H^2. We will represent H^2 by the Poincaré disc model. While the boundary S^1_∞ of the Poincaré disc is disjoint from H^2, it does have significance for hyperbolic geometry; for example, any two disjoint points of S^1_∞ can be regarded as 'ideal' endpoints for a unique geodesic in H^2.

For any homeomorphism $h: F \to F$, a fundamental lemma of Nielsen [9] says that any lift $\tilde{h}: H^2 \to H^2$ of h to the universal cover extends to a unique self-homeomorphism of the disc $H^2 \cup S^1_\infty$. For γ a geodesic in H^2 with endpoints P,Q on S^1_∞, the (not necessarily geodesic) curve $\tilde{h}(\gamma)$ thus stays a bounded distance from the geodesic $\overline{h}(\gamma)$ with endpoints $\tilde{h}(P)$, $\tilde{h}(Q)$.

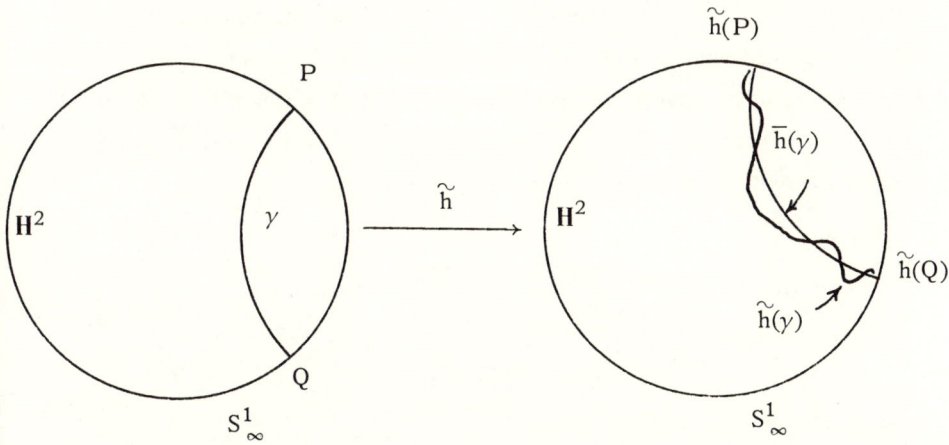

The homeomorphism h can now be seen to induce a self map of the set of geodesics in F. For any closed geodesic C in F, the closed curve $h(C)$ is freely homotopic to a unique closed geodesic $\overline{h}(C)$. By the above we can also 'tighten' $h(C)$ for non-closed geodesics; for C a geodesic extending to infinity in both directions, take γ to be a geodesic in H^2 projecting to C, and define $\overline{h}(C)$ to be the image of the geodesic $\overline{h}(\gamma)$ in F.

Let C be a simple closed geodesic in F. With respect to the Hausdorff metric on the non-empty closed subsets of F (i.e. for non-empty closed subsets A, B say that $d(A,B) < \epsilon$ if $A \in N_\epsilon(B)$ and $B \in N_\epsilon(A)$) the sequence $\bar{h}^n(C)$ (n = 1,2,···) has a subsequence $\bar{h}^{n_i}(C)$ which converges to a non-empty closed subset $K \subset F$. It is not hard to prove that K is a disjoint union of (not necessarily closed) unoriented geodesics in F. Such a set is called a (geodesic) *lamination* in F; the member geodesics are called *leaves*. As the Euler characteristic $\chi(F)$ is negative, there is no continuous line field on all of F, and so every lamination is necessarily a proper subset of F. Indeed, an elaboration of this argument shows that every lamination is locally homeomorphic to the product of an interval with a compact totally disconnected set.

The simplest examples of geodesic laminations are finite disjoint unions of simple closed geodesics. However these are the only laminations which are everywhere locally connected. If h is not periodic, there is a simple closed geodesic C such that $K = \lim \bar{h}^{n_i}(C)$ is not of this type (see, for example, [2, Theorem 2.7]), so $K' = \{x \in K | K$ is not locally connected at $x\}$ is a non-empty lamination.

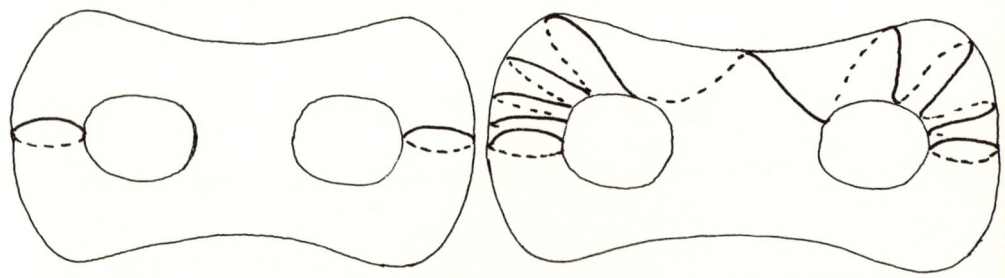

Geodesic laminations

Set $L = \cup_{r \in Z} \bar{h}^r(K')$. To show that L is indeed a lamination, one must check that $\bar{h}^r(K')$ and $\bar{h}^s(K')$ have no transverse intersections. This follows from the easily proved exercise that $\bar{h}^r(K)$ and $\bar{h}^s(K)$ have only finitely many transverse intersections. Clearly, $\bar{h}(L) = L$.

If h is irreducible and non-periodic then each component of the preimage of F-L in the universal cover of F, identified with H^2, is hyperbolically convex. Thus any non-simply connected component of F-L is essential in F, i.e. the fundamental group injects under the inclusion map. If we knew that $h(L) = L$ (rather than $\bar{h}(L) = L$) we could deduce that h is reducible unless all the components of F-L are simply connected. The general case is only slightly more technical.

The preimage $\tilde{L}$ of L in H^2 has many remarkable properties. Let $\pi = \pi_1(F)$ be the group of deck translations. Some of these properties are

1) $\tilde{L}$ is a closed subset of H^2, locally homeomorphic to the product of an interval and a Cantor set.

2) $\tilde{L}$ is a disjoint union of geodesics.

3) $\tilde{L}$ is invariant under π, and $\tilde{h}(\tilde{L}) = \tilde{L}$.

4) There are only finitely many π-orbits among the components of $H^2 - \tilde{L}$.

5) Each component of $H^2 - \tilde{L}$ is a finite-sided polygon with vertices on S^1_∞.

Properties 4) and 5) follow from the fact that each component of $H^2 - \tilde{L}$ (and hence of F-L) has area a non-zero multiple of n bounded by area (F). These properties ensure that for U a component of $H^2 - \tilde{L}$, there is a lift $\tilde{k}$ of a positive power k of h which fixes all the vertices of U. The construction of L from a convergent subsequence of $\{\bar{h}^n(C) | n > 0\}$ suggests that the vertices of U should be sinks for the action of $\tilde{k}$ on S^1_∞. This is indeed the case; in addition, the only other fixed points of $\tilde{k} | S^1_\infty$ are sources, a single source separating any two consecutive sinks.

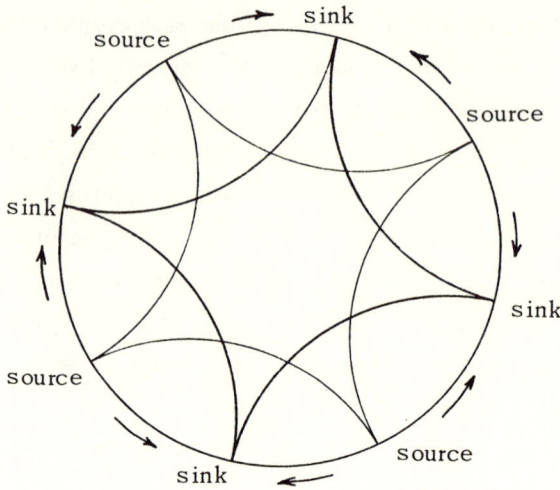

In fact, $\tilde{L}$ can be characterized as the closure of the union of the set of all geodesics joining consecutive sinks of the lift of some positive power of h. This demonstrates that L is independent of all the choices made in its construction. We write L^s for L, and call it the *stable lamination* of h; the *unstable lamination* of h is the stable lamination of h^{-1}. It has a similar characterization, using sources instead of sinks.

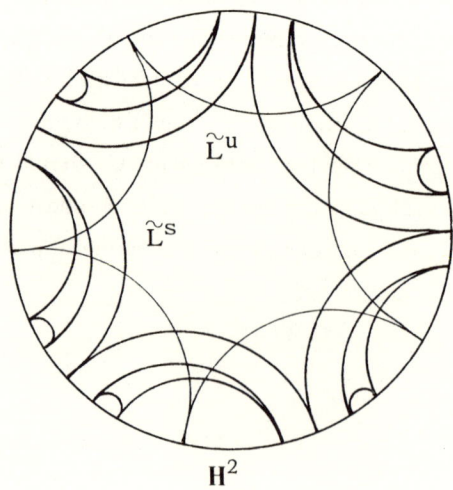

Nielsen [9] arrived at L^s and L^u by using fixed-point theory; a more recent exposition of this approach can be found in [6].

All but finitely many of the components of $F-(L^s \cup L^u)$ are rectangles; the remaining components are 2k-gons with $k > 2$. It is not hard to isotop h so that $h(L^s) = L^s$ and $h(L^u) = L^u$.

Next define an equivalence relation on the points of F by saying that $x \sim y$ if

i)   x and y are in the closure of the same component of $F-(L^s \cup L^u)$

ii)   x and y are in the closure of the same component of $L^s - L^u$

iii)   x and y are in the closure of the same component of $L^u - L^s$

iv) $x = y$.

The quotient $F/\sim$ is a closed surface with a pair of transverse singular foliations $\mathcal{F}^s$, $\mathcal{F}^u$. For $F/\sim$ is Hausdorff and if $\pi: F \to F/\sim$, one can describe neighborhoods of $\pi(x)$ as follows: In a sufficiently small neighborhood U of x in $L^s \cap L^u$ not on a leaf of either lamination that is isolated from one side, the components of $U \cap L^s$ and $U \cap L^u$ incident to x are non-isolated from both sides. Using a 'Cantor collapse' along these geodesic arcs one can obtain a chart $\phi: U \to (-\epsilon, \epsilon) \times (-\epsilon, \epsilon)$ whose composition with π takes the components of $U \cap L^s$ to 'horizontals' and the components of $U \cap L^u$ to 'verticals.' For x in a component of $L^s - L^u$ or $L^u - L^s$ charts to $(-\epsilon, \epsilon) \times (-\epsilon, \epsilon)$ are formed from two 'half-charts,' as follows. For x on a leaf γ in $L^u - L^s$, say, a 'half-chart' to $[0,\epsilon) \times (-\epsilon,\epsilon)$ (or $(-\epsilon,0] \times (-\epsilon,\epsilon)$) is formed on a side of x that is non-isolated from the leaves of L^s as above. On an isolated side, a 'half-chart' is formed using a 'Cantor collapse' along leaves, starting from the intersection of γ with the boundary leaf of L^s nearest x. Similarly, rectangle components of $F-(L^s \cup L^u)$ give rise to charts formed from 'quarter-charts,' and so on for 2k-gons.

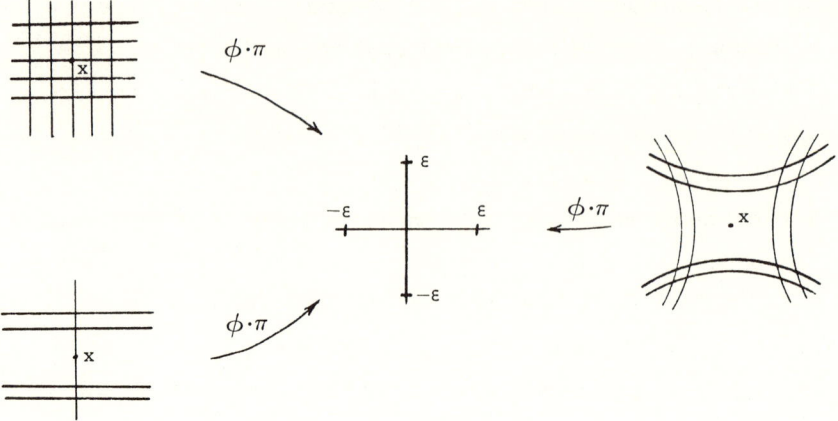

Under π, L^s and L^u become foliations $\mathcal{F}^s$, $\mathcal{F}^u$. The singular points arise from the 2k-gon components of $F - (L^s \cup L^u)$; the associated charts give the standard 'prong' model, as illustrated below.

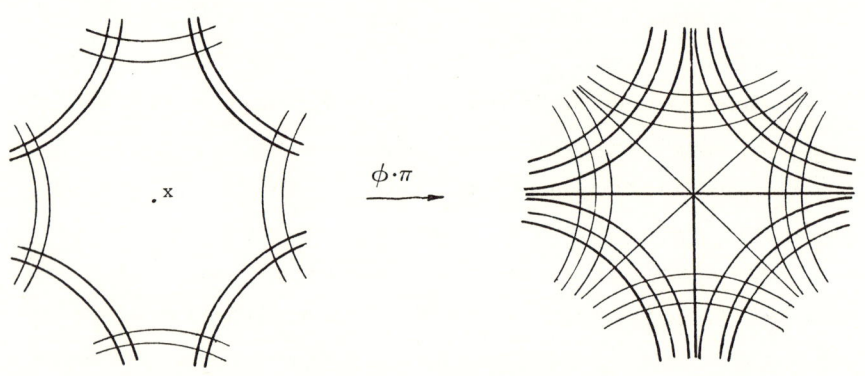

The quotient map π is a homotopy equivalence and so one may replace F by $F/\sim$; alternately, one can show that π is homotopic to a homeomorphism. As h leaves L^s and L^u invariant, it induces a self-homeomorphism $h/\sim$ of $F/\sim$, leaving $\mathcal{F}^s$ and $\mathcal{F}^u$ invariant.

For ease of notation, write F for $F/\sim$ and h for $h/\sim$. In general, h permutes the finitely many separatrices of $\mathcal{F}^s$. If a positive power

h^m of h leaves a separatrix σ invariant, it moves points of σ away from the singularity of σ. For each separatrix σ of $\mathcal{F}^s$, choose a closed arc $a_\sigma \subset \sigma$ with one endpoint at the singularity of σ, such that $h(a_\sigma) \subset a_{h(\sigma)}$, and let $X = \cup a_\sigma$.

Let Y be the set of points in F which can be joined to a singularity or an end-point of X by an arc β of a leaf of $\mathcal{F}^u$ with Int $\beta \subset F - X$. Then $F - (X \cup Y)$ has finitely many components $R_1', \cdots, R_k'$. Each R_j' is an open rectangle, admitting a homeomorphism to $(0,1) \times (0,1)$ carrying the leaves of $\mathcal{F}^s$ and $\mathcal{F}^u$ to horizontal and vertical segments respectively. Let R_j be the closure of R_j', and observe that:

1) $F = \cup R_j = X \cup Y \cup$ Int R_j.
2) $h^{-1}(X) \subseteq X$.
3) $h(Y) \subseteq Y$.

The set $\{R_j\}$ is a *Markov partition* for h.

Note that a Markov partition for h decomposes F into finitely many rectangles whose sides are parallel to $\mathcal{F}^s$ and $\mathcal{F}^u$. Moreover, for rectangles R_i, R_j the intersection of R_i with $h(R_j)$ is the union of a_{ij} disjoint 'horizontal' slices of R_i.

The matrix $A = (a_{ij})$ has the property that there is an n such that every entry of A^n is strictly positive and so the eigenvalue λ with the largest modulus of the eigenvalues of A is real and of multiplicity one. Thus A has a unique (up to multiplication by a positive scalar) positive eigenvector $x = (x_i)$ with eigenvalue λ. One obtains a transverse measure μ^s for $\mathcal{F}^s$ such that $h_*(\mu^s) = \lambda^{-1}\mu^s$ by assigning x_i as the 'height' of the rectangle R_i and extending linearly. Similarly, the λ-eigenvector of A^t gives the transverse measure for $\mathcal{F}^u$.

By using the 'metric' $\mu = ((\mu^s)^2 + (\mu^u)^2)^{1/2}$, the transverse measures give a second method of defining 'arc length' in F. It turns out [3, p. 179] that for each class of loops there are constants k, K such that the ratio of the Riemannian length of a loop α to its 'μ length' is bounded below by k and above by K. The inequality given in the second remark following Theorem 4 follows as an exercise.

§5. Problems

1) Find a good way of constructing all pseudo-Anosov automorphisms, together with stretching factors and foliations.

2) Find a good way of classifying a given automorphism. (Thurston, Penner, and Mosher have given algorithms).

3) Which numbers are realized as stretching factors? For genus g, the stretching factor is an algebraic integer of degree $\leq 6g-6$. (Penner and Long have generalized Thurston's construction to get degree $6g-6$ and others).

DEPARTMENT OF MATHEMATICS
UNIVERSITY OF BRITISH COLUMBIA
VANCOUVER, B.C.
V6T 1Y4

REFERENCES

[1] B. Eckmann, *Poincaré duality groups of dimension 2 are surface groups*, these proceedings.

[2] A. Casson, S. Bleiler, *Automorphisms of surfaces after Nielsen and Thurston*, London Mathematical Society Student Text Series, to appear.

[3] A. Fathi, F. Laudenbach, V. Poenaru, *Travaux de Thurston sur les surfaces*, Asterisque 66-67 (1979).

[4] W. Fenchel, *Estensioni gruppi discontinui e transformaziomi periodiche delle surficie*, Rend. Acc. Naz. Lincei (se.fis., mat. e. nat) 5, 326-329 (1948).

[5] W. Fenchel, *Bemarkongen om endelige gruppen af abbilungsklasser*, Mat. Tidskrift B 1950, 90-95.

[6] M. Handel, W. Thurston, *New proofs of some results of Nielsen*, preprint 1983.

[7] S. Kerckhoff, *The Nielsen realization problem*, Annals of Math. 117 (1983), 235-265.

[8] D. Long, *Cobordism of knots and surface automorphisms*, thesis, Cambridge 1983.

[9] J. Nielsen, *Untersuchungen zur Topologie der geschlossenen zweiseitigen Flachen I*, Acta Math. 50 (1927), 189-358.

[10] _____, *Untersuchunger zur Topologie der geschlossenen zweiseitigen Flachen II*, Acta Math. 53 (1929), 1-76.

[11] _____, *Untersuchungen zur Topologie der geschlossenen zweiseitigen Flachen III*, Acta Math. 58 (1931), 87-167.

[12] _____, *Abbildungsklassen endlicher Ordnung*, Acta Math. (1942), 23-115.

[13] R. Penner, *A computation of the action of the mapping class groups on isotopy classes of curves and arcs in surfaces*, thesis M.I.T. 1982.

[14] W. Thurston, *On the geometry and dynamics of diffeomorphisms of surfaces I*, preprint.

WHITEHEAD GROUPS
OF CERTAIN HYPERBOLIC MANIFOLDS, II

A. J. Nicas and C. W. Stark

Abstract. The Whitehead groups $Wh_j(\pi_1 M)$ $(j = 0,1,2,\cdots)$ are proved to vanish for most of the hyperbolic 3-manifolds obtained from regular hyperbolic polyhedra by face identification. Frobenius induction theorems and well-known results of Waldhausen are used to obtain this vanishing theorem, which follows from computer-assisted homology calculations with the free differentia calculus. We rely on the classification of this family of 3-manifolds by J. S. Richardson and J. H. Rubinstein, which corrects older work of L. A. Best.

Acknowledgments: We wish to thank the Computer Science Department of Brandeis University, especially Jim Storer, for making available to us most of the computing resources this project required, and Bruce Levy at Haverford College for providing additional computing resources and a computer program for arbitrary precision integer matrix reduction.

Introduction

The Whitehead groups $Wh_i(\Gamma)$ $(i = 0,1,2,\cdots)$ are conjectured to vanish if $\Gamma = \pi_1 M$ is the fundamental group of a closed aspherical manifold M. This problem is particularly enticing for geometrically interesting manifolds, i.e. those with special properties of curvature or symmetry. If M carries a flat Riemannian metric then much can be proved, especially for smaller values of i (see [4], [8], [9]). If M admits an effective, low-codimension action by a compact Lie group (necessarily a torus) then information on the orbit space may be promoted to M to give proofs of vanishing results, as in [10], [12], and [14]. In most cases, however, we still know very little about Whitehead groups of aspherical manifolds.

In particular, little is known for closed hyperbolic manifolds (those admitting a Riemannian metric of constant negative sectional curvature), even in dimension three. Our paper [11] considered the Whitehead groups of those closed hyperbolic three-manifolds obtained by face identifications from regular hyperbolic dodecahedra, icosahedra, or hexahedra. The best known member of this family of manifolds is the Seifert-Weber hyperbolic dodecahedral space [17], and Best's paper [1] served us as a reference for the classification of these manifolds up to homeomorphism. Unfortunately, the algorithm for the computer-assisted classification claimed in [1] is not well described and so is hard to verify, and in fact Best's lists contain errors. The classification of these manifolds was re-computed by Richardson and Rubinstein [13], using an explicit algorithm which appears correct. We have not had the opportunity to program and run their classification algorithm but we have verified the group presentations from face identifications given in [13], so while there remains a chance that the classification in [13] is flawed, the groups analyzed below do correspond to hyperbolic three-manifolds.

Our main result is the verification of the vanishing conjecture for $Wh_i(\Gamma)$ for most of the groups in the tables of [13]. Our methods are those of [11]. Frobenius induction theorems for the Whitehead groups and Waldhausen's work on Whitehead groups of graphs of groups [16] are used to reduce the problem to the analysis of certain covering spaces: if each of these covering spaces is a Haken three-manifold then the manifold under study has all its Whitehead groups equal to zero. The most easily detected incompressible surfaces are those derived from essential maps from M to the circle, so we test for this condition by calculating the first homology of our covering spaces with the free differential calculus.

1. *Preliminaries*

Frobenius induction theorems for the functors of K- and L-theory are invaluable tools in topological space-form problems. In their strongest forms these theorems assert that the value of a functor F on a group

G is determined by the values of F on certain subgroups H of G; the model results of this type are induction theorems in the theory of representations of finite groups, which show that any character on a finite group G is a sum of characters induced from the hyperelementary subgroups H of G. Recall that a group H is hyperelementary if there are a prime p, a p-group P, and a finite cyclic group C such that H is an extension

$$1 \to C \to H \to P \to 1.$$

Weaker forms of these results suffice to prove vanishing theorems. Induction theorems and a vanishing result for the structure set S_{TOP} of topological surgery are given in [7]. The following vanishing result for Waldhausen's higher Whitehead groups appears in [10]:

LEMMA 1. *Let A be a group and* $f: A \to G$ *an epimorphism to a finite group, R a subring of the rational numbers, and n a non-negative integer. Suppose that for every hyperelementary subgroup H of G the higher Whitehead groups of* $f^{-1}(H)$ *satisfy* $Wh_j(f^{-1}(H)) \otimes R = 0$ *for* $0 \leq j \leq n$.

Then $Wh_j(A) \otimes R = 0$ *for* $0 \leq j \leq n$.

Recall that $Wh_0(A) = \tilde{K}_0(A)$, $Wh_1(A) = Wh(A)$, and the higher Whitehead groups are constructed in [16], where all $Wh_j(A)$ are shown to vanish if A belongs to a large class of groups which includes the fundamental groups of Haken 3-manifolds. (Recall that a 3-manifold is Haken if it is compact, orientable, irreducible, and possessed of a two-sided incompressible surface.) In particular, Waldhausen's results apply to any closed, orientable, irreducible 3-manifold whose first Betti number is 1 or more.

Any hyperbolic 3-manifold is irreducible but there are many closed, orientable, hyperbolic 3-manifolds which contain no incompressible surfaces. Thurston has made tantalizing conjectures concerning these manifolds [15, p. 380] but even his conjecture that such a hyperbolic

3-manifold M has a finite-sheeted covering space with positive first Betti number does not suffice for a proof using Frobenius induction that the Whitehead groups of M vanish. Hempel [5], [6] has calculated the first homology groups of some covers of the Seifert-Weber hyperbolic dodecahedral space constructed in [17] and [1] to verify this conjecture of Thurston's for the Seifert-Weber space and other hyperbolic 3-manifolds considered in [1]. Hempel's calculations are made with Fox's free differential calculus, which is well suited to the numerous but repetitious computations required to verify that all covers of a 3-manifold M associated to hyperelementary subgroups H of a finite quotient of $\pi_1 M$ have positive first Betti number.

Our treatment of these homology computations follows [6]. Consider the symmetric group S_q as the group of permutations of $\{1,2,\cdots,q\}$, acting from the right so that $a \in S_q$ denotes the function

$$i \to (i)a$$

and if $a, b \in S_q$ then ab is the function

$$i \to ((i)a)b .$$

S is isomorphic to the subgroup of $GL(q,Z)$ which permutes the standard basis elements, where $a \in S_q$ is identified with the q by q matrix

$$(\delta_{(i)a, j}) .$$

If $p: M' \longrightarrow M$ is a path-connected, finite-sheeted covering space of M with q sheets then path lifting defines a homomorphism

$$R: \pi_1 M \to S_q .$$

This homomorphism has a purely algebraic description as well: let $B = p_\#(\pi_1(M',m'))$ and let the right cosets of B in $A = \pi_1(M,m)$ be $B = B, B_2, \cdots, B_q$. Then

$$(i) R(g) = j$$

if and only if

$$B_i g = B_j .$$

Familiar results on covering spaces show that R is well defined once basepoints are selected, that R is well defined up to conjugation without picking basepoints, that the image of R acts transitively on $\{1,2,\cdots,q\}$, that $\pi_1(M',m')$ is the pre-image in $\pi_1(M,m)$ of the stabilizer of 1, and that the correspondence

$$(p : M' \to M) \leftrightarrow R$$

is a one-to-one correspondence between equivalence classes of q-sheeted coverings of M and conjugacy classes of representations of $\pi_1 M$ onto transitive subgroups of S_q. R is called the permutational representation of $\pi_1 M$ associated to $p : M' \to M$. R is called regular if the covering is regular.

Fox showed that if a group A has a presentation $\langle x_1, \cdots, x_n : r_1, \cdots, r_m \rangle$ then the Jacobian matrix of free derivatives

$$J = \left(\frac{\partial r_i}{\partial x_j} \right)$$

with entries in ZA presents a right ZA-module (namely the cokernel of $J : (ZA)^n \to (ZA)^m$)

$$(u_j) \to \left(\sum \frac{\partial r_i}{\partial x_j} u_j \right)$$

which is independent of the presentation given for A. When $A = \pi_1(M)$ and $p : M' \to M$ is a regular covering projection whose group of covering transformations is T, then the image of J over ZT is known to present $H_1(M', p^{-1}(m))$ as a ZT-module. Hempel gives the following variation on this result in [6, Theorem 3.1].

LEMMA 2. *Let* $p: M' \to M$ *be a q-sheeted finite covering of a finite CW complex* M, *let* $R: \pi_1 M \to S_q$ *be the associated permutational representation, and let* $\langle x_1, \cdots, x_n : r_1, \cdots, r_m \rangle$ *be any presentation of* $\pi_1 M$. *Then the* nq *by* mq *matrix of integers* R(J) *is a presentation matrix over* Z *for* $H_1(M') \oplus Z^{q-1}$.

The q by q blocks replacing derivatives $\partial r_i / \partial x_j$ are obtained by regarding S_q as a subgroup of GL(q,Z). Elementary row and column operations over the integers may be used to find the rank and torsion coefficients of the module presented above by reducing the presentation matrix to an equivalent diagonal matrix. Hempel did his example by hand; we used a computer in [11] and for this paper and were successful with some relatively large matrices.

Under some circumstances the permutational representation associated to $p: M' \to M$ is easily determined. Let $A = \pi_1(M,m)$ and suppose that $f: A \to G$ is an epimorphism to a finite group G. If H is a subgroup of G and $p: M' \to M$ is the covering projection associated to $f^{-1}(H)$ then the permutational representation R of p is the composite of f with the right action of G on the right cosets of H. This observation makes the method of [6] relatively economical for our purposes, as we shall use only two finite groups G, namely the alternating groups A_5 and A_6.

The Seifert-Weber hyperbolic dodecahedral space belongs to a family of hyperbolic 3-manifolds obtained by face identifications from regular polyhedra. We shall follow the classification and fundamental group presentations of Richardson and Rubinstein [13] for this interesting family of three-manifolds. Hempel [5] observes that the fundamental group of the Seifert-Weber hyperbolic dodecahedral space admits an epimorphism to A_5 and computes the first homology as in [6] for the cover associated to an A_4 contained in A_5. Re-enacting that calculation served as motivation for our computational scheme.

Consider the situation of Lemma 1: A is a group (the fundamental group of the manifold M) and $f: A \to G$ is an epimorphism to a finite group. In principle, Lemma 1 obliges us to treat $f^{-1}(H)$ for each hyperelementary subgroup H of G, but since conjugate subgroups H and H′ will have conjugate pre-images $f^{-1}(H)$ and $f^{-1}(H')$ (and so the resulting covering spaces of M are equivalent) we only need to verify the vanishing hypotheses for one representative of each conjugacy class of hyperelementary subgroups of G.

If G is a finite simple group and G is not cyclic then G is not hyperelementary. In particular, the alternating group A_n is not hyperelementary as long as n is at least 5. To find the conjugacy classes of hyperelementary subgroups of G one may begin by classifying cyclic subgroups and p-subgroups of G up to conjugacy. Each hyperelementary subgroup H of G containing a non-trivial cyclic normal subgroup C_k lies inside the normalizer $N_G(C_k)$, while the Sylow subgroups of G provide the maximal p-groups contained in G. These observations lead to the next lemma.

LEMMA 3. a) *Each hyperelementary subgroup of* A_5 *is conjugate to a subgroup of*

(i) $C_2 \times C_2$, *consisting of* Id, (12)(34), (13)(24), *and* (14)(23),

(ii) S_3, *generated by* (12)(45) *and* (123), *or*

(iii) $C_5 \rtimes C_2$, *generated by* (12345) *and* (14)(23).

b) *Each hyperelementary subgroup of* A_6 *conjugate to a subgroup of*

(i) D_4, *the dihedral group of order* 8, *generated by* (1234)(56) and (24)(56),

(ii) $C_5 \rtimes C_2$, *generated by* (12345) *and* (14)(23), *or*

(iii) $(C_3 \times C_3) \rtimes C_2$, *generated by* (123), (456), *and* (13)(46). (*Note that this last group is not itself hyperelementary.*)

2. Homology computations and consequences

The main results of this section are computations of the first homology groups of certain covers of the manifolds discussed by Richardson and

Rubinstein [13]. These results are given in tables listing the presentation of each manifold's fundamental group following [13], epimorphisms to alternating groups, and the first homology groups of the covers of the manifold corresponding to the three subgroups of A_5 or A_6 isolated in Lemma 3 above.

These computations were made with the assistance of a computer. We produced files (by hand) recording the Jacobian matrices J of Richardson and Rubinstein's presentations. Right coset decompositions were produced for the three subgroups specified above. Given a subgroup H of G and an epimorphism $f: \pi_1 M \to G$ ($G = A_5$ or A_6) the machine was used to compute the associated permutation representation R by composing f with the right G-action on the right cosets of H. R was then applied to the Jacobian J, by machine, to produce an integral matrix to be reduced to diagonal form by row and column operations over the integers.

The computer was also essential in the search for homomorphisms to finite groups. Here we used exhaustive searches to find and classify homomorphisms up to conjugacy from the fundamental groups of these manifolds to the alternating groups A_5 and A_6. In some cases we made such an analysis for PSL(2,7) and PSL(2,8) as well. The process was accelerated by two observations: we used conjugacy in the finite group to restrict to a small number of possible images for one of the generators and the images of two or more of the generators were determined by the other generators and the relations. Each homomorphism used in the homology computations was checked by hand.

The first table exhibits results of calculations for the eight dodecahedral spaces. These manifolds have fundamental groups presented by six generators u, v, w, x, y, and z, with relations listed in the table. The second table lists our results for the six icosahedral manifolds, whose fundamental groups are presented in terms of ten generators q, r, s, t, u, v, w, x, y, and z. In some instances the tables report that no epimorphism to A_5 or A_6 gave satisfactory results. This occurred when the relevant first homology groups had rank zero.

There is another hyperbolic manifold obtained from a regular hyperbolic polyhedron, the hexahedral space discussed in [1]. We dealt with it in [11].

The following conclusions may be drawn from our homology calculations:

PROPOSITION 1. *All the Whitehead groups* Wh_i $(i = 0,1,2,\cdots)$ *vanish for the first, third, fourth, fifth, sixth, and seventh dodecahedral manifolds, and for the first five icosahedral manifolds.*

This is an immediate consequence of the lemmas of Section 1, Waldhausen's results [16], and the computations of Section 2. Note that the positivity of the first Betti number for the cover corresponding to the third subgroup $(C_3 \times C_3) \rtimes C_2$ we examined in A_6 implies the positivity of the first Betti number for the covers corresponding to its hyperelementary subgroups.

The analogue of the above proposition in surgery is obstructed by Cappell's UNil groups [2], [3], which are presently not known to vanish for incompressible surfaces in 3-manifolds. As these groups vanish away from 2, we obtain the following statement using the above computations and the induction theorems of [7]:

PROPOSITION 2. *Let* M^3 *be any of the manifolds in the above proposition.*

Then $S_{TOP}(M^3 \times D^k, \partial) \otimes Z[1/2] = 0$ *if* $k > 2$.

Table 1. The eight dodecahedral manifolds

(DM1) $uv^{-1}zxw^{-1} = uw^{-1}vyx^{-1} = ux^{-1}wzy^{-1} = 1$

$uy^{-1}xvz^{-1} = uz^{-1}ywv^{-1} = vxzwy = 1$.

The epimorphism to A_5 used was:

$u = (1\,2\,3\,4\,5)$	$v = (1\,2\,4\,5\,3)$	$w = (1\,2\,5\,3\,4)$
$x = (1\,3\,4\,2\,5)$	$y = (1\,4\,2\,3\,5)$	$z = (1\,4\,5\,2\,3)$

Subgroup of A_5	First homology of the corresponding cover	
	Rank	Torsion
$C_2 \times C_2$	14	$(Z/2)^4$
S_3	9	$(Z/2)^2$
$C_5 \rtimes C_2$	5	$(Z/2)^2 + (Z/5)^3$

(DM2) $uv^{-1}xwv^{-1} = uy^{-1}zxy^{-1} = uw^{-1}vzw^{-1} = 1$

$ux^{-1}yvx^{-1} = uz^{-1}wyz^{-1} = vwzyx = 1$.

No epimorphism to A_5 or A_6 gave satisfactory results. (See the text.)

(DM3) $uwv^{-1}xy^{-1} = u^{-1}w^{-1}zyx = uy^{-1}zxv^{-1} = 1$

$u^{-1}zy^{-1}wv = u^{-1}xvwz = vzwx^{-1}y = 1$.

The epimorphism to A_5 used was:

$u = (1\,2\,3)$	$v = (2\,4\,3)$	$w = (1\,5\,3)$
$x = (1\,5\,2)$	$y = (1\,3\,2\,5\,4)$	$z = (1\,4\,5\,3\,2)$

Table 1 (Continued)

Subgroup of A_5	First homology of the corresponding cover	
	Rank	Torsion
$C_2 \times C_2$	3	$(Z/2)^2 + (Z/3)^3 + (Z/8)^3$
S_3	2	$(Z/3)^3 + Z/4 + Z/8 + Z/16$
$C_5 \rtimes C_2$	1	$(Z/3)^2 + Z/4 + Z/16$

(DM4) $ux^{-1}uz^{-1}u = vxv^{-1}w^{-1}x = uyv^{-1}yv = 1$

$w^2z^{-1}wy = uv^{-1}wz^2 = xy^{-1}x^{-1}zy = 1$.

The epimorphism to A_6 used was:

u = (1 2 3)(4 5 6)   v = (1 2 4)(6 3 5)   w = (1 4 6 3 2)
x = (1 3 2 6 5)   y = (2 3)(4 6)   z = (1 2 6 4 3)

Subgroup of A_6	First homology of the corresponding cover	
	Rank	Torsion
D_4	1	
$C_5 \rtimes C_2$	1	$(Z/4)^3 + Z/5 + (Z/7)^3$
		$+ Z/17 + Z/25 + Z/64$
$(C_3 \times C_3) \rtimes C_2$	1	$(Z/4)^3 + Z/64 + Z/119 + Z/125$

(DM5) $uv^{-1}z^{-1}wx = uzyx^{-1}v^{-1} = uw^{-1}vyz = 1$

$u^{-1}wzxy = uxwvy^{-1} = vx^{-1}ywz^{-1} = 1$.

The epimorphism to A_5 used was:

u = (1 2 3 4 5)   v = (1 3 5 4 2)   w = (2 3 5)
x = (1 2)(3 4)   y = (1 3 4)   z = (2 3)(4 5)

Table 1 (Continued)

Subgroup of A_5	First homology of the corresponding cover	
	Rank	Torsion
$C_2 \times C_2$	5	$(Z/2)^5 + (Z/3)^2 + Z/9$
S_3	4	$(Z/2)^3 + Z/3$
$C_5 \rtimes C_2$	1	$(Z/2)^2 + (Z/3)^2 + (Z/5)^2 + Z/9$

(DM6) $uwx^{-1}zw = u^{-1}vy^2z = uxzw^{-1}x = 1$

$u^{-1}y^{-1}xwv = uyv^2z^{-1} = v^{-1}yxz^{-1}w = 1$.

The epimorphism to A_6 used was:

$u = (1\,2\,3\,4\,5)$ $v = (3\,4)(5\,6)$ $w = (3\,4\,5)$
$x = (1\,4)(2\,3\,5\,6)$ $y = (1\,6)(4\,5)$ $z = (1\,2\,3\,5\,6)$

Subgroup of A_6	First homology of the corresponding cover	
	Rank	Torsion
D_4	6	$(Z/2)^2 + (Z/3)^6 + Z/4 + Z/16 + Z/27$
$C_5 \rtimes C_2$	4	$(Z/3)^6 + (Z/4)^2 + (Z/5)^2 + Z/8$
$(C_3 \times C_3) \rtimes C_2$	4	$(Z/3)^5 + (Z/4)^2 + Z/8$

(DM7) $uxzw^{-2} = u^{-1}x^{-1}ywz = uyz^{-1}v^{-1}w = 1$

$u^{-1}zv^{-1}y^2 = u^{-1}vx^2v = vw^{-1}z^{-1}xy = 1$.

The epimorphism to A_6 used was:

$u = (1\,2\,3\,4\,5)$ $v = (1\,3\,5\,2)(4\,6)$ $w = (1\,5\,6)(2\,3\,4)$
$x = (3\,6\,5)$ $y = (1\,3\,4)$ $z = (2\,6\,4)$

Table 1 (Continued)

Subgroup of A_6	First homology of the corresponding cover	
	Rank	Torsion
D_4	8	
$C_5 \rtimes C_2$	7	$(Z/2)^{10} + (Z/3)^3 + (Z/11)^2 + Z/53 + Z/64$
$(C_3 \times C_3) \rtimes C_2$	2	$(Z/2)^6 + (Z/3)^3 + (Z/11)^2 + Z/53 + Z/64$

(DM8) $uv^{-1}xvy = u^{-1}v^{-1}wxz = uzxv^{-1}z = 1$

$vzw^{-1}xw = u^{-2}y^2w = w^{-1}yxyz^{-1} = 1$.

No epimorphism to A_5 or A_6 gave satisfactory results.

Table 2. The icosahedral manifolds

(IM1) $qsy^{-1} = rtz^{-1} = twz = qus^{-1} = r^{-1}xv = 1$

$uzw^{-1} = q^{-1}wu = sxy = vyx^{-1} = rvt^{-1} = 1$.

The epimorphism to A_6 used was:

q = (1 2 3)	r = (1 3 5)(2 4 6)
s = (1 6 4 5 3)	t = (1 6 2 5 4)
u = (2 6 4 5 3)	v = (1 4 5 3 6)
w = (1 3)(2 5 4 6)	x = (1 5 6 2)(3 4)
y = (1 2)(3 6 4 5)	z = (1 3 4 2)(5 6)

Subgroup of A_6	First homology of the corresponding cover	
	Rank	Torsion
D_4	6	$(Z/2)^6 + (Z/3)^4 + (Z/11)^2$
$C_5 \rtimes C_2$	5	$(Z/2)^4 + (Z/3)^2 + (Z/11)^2$
$(C_3 \times C_3) \rtimes C_2$	1	$(Z/2)^4 + (Z/3)^2 + (Z/11)^2$

Table 2 (Continued)

(IM2) $qst = rvt^{-1} = tw^{-1}x = qwr^{-1} = svu^{-1} = 1$

$u^2w = qyv^{-1} = sz^2 = x^2y = rz^{-1}y = 1$.

The epimorphism to A_6 used was:

$q = (1\,2\,3)$	$r = (1\,2\,4\,3\,5)$
$s = (1\,6)(3\,5)$	$t = (1\,6\,3\,5\,2)$
$u = (1\,3\,5\,4)(2\,6)$	$v = (1\,2\,6\,3\,4)$
$w = (1\,5)(3\,4)$	$x = (1\,3\,4\,6)(2\,5)$
$y = (1\,4)(3\,6)$	$z = (1\,3\,6\,5)(2\,4)$

Subgroup of A_6	First homology of the corresponding cover	
	Rank	Torsion
D_4	4	$(Z/2)^2 + (Z/3)^2 + (Z/9)^3$
$C_5 \rtimes C_2$	4	$Z/3 + Z/4 + (Z/9)^2$
$(C_3 \times C_3) \rtimes C_2$	4	$Z/3 + Z/4 + (Z/9)^2$

(IM3) $qsu = rts = uz^{-1}y = qxr^{-1} = syt^{-1} = 1$

$v^{-1}xw = qzw^{-1} = t^{-1}vx = vyz = ru^{-1}w = 1$.

The epimorphism to A_6 used was:

$q = (1\,2\,3)$	$r = (1\,3\,4\,5\,2)$
$s = (1\,2\,3\,5\,6)$	$t = (2\,3\,6\,5\,4)$
$u = (1\,6\,5\,2\,3)$	$v = (1\,3\,6\,4\,5)$
$w = (1\,6\,4\,3\,2)$	$x = (1\,4\,5\,2\,3)$
$y = (1\,5\,6\,4\,2)$	$z = (1\,2\,6\,4\,3)$

Table 2 (Continued)

Subgroup of A_6	First homology of the corresponding cover	
	Rank	Torsion
D_4	4	$(Z/2)^6 + (Z/3)^4 + Z/81$
$C_5 \rtimes C_2$	2	$(Z/3)^4 + (Z/4)^3 + Z/81$
$(C_3 \times C_3) \rtimes C_2$	2	$(Z/3)^2 + (Z/4)^3 + Z/9$

(IM4) $qr^{-1}s = s^{-1}ty = tw^{-1}u = q^2w^{-1} = su^{-1}z = 1$

$w^{-1}xy = rux = t^{-1}v^2 = xz^2 = ryv = 1$.

The epimorphism to A_6 used was:

$q = (1\,2\,3\,4\,5)$ $r = (2\,6\,5\,3\,4)$
$s = (1\,5\,2\,6\,4)$ $t = (1\,2)(3\,6)$
$u = (1\,6\,3\,5)(2\,4)$ $v = (1\,3\,2\,6)(4\,5)$
$w = (1\,3\,5\,2\,4)$ $x = (1\,6)(2\,3)$
$y = (1\,6\,3\,4)(2\,5)$ $z = (1\,2\,6\,3)(4\,5)$

Subgroup of A_6	First homology of the corresponding cover	
	Rank	Torsion
D_4	6	$(Z/2)^2 + (Z/3)^2 + Z/4$ $+ Z/5 + Z/7$
$C_5 \rtimes C_2$	4	$(Z/2)^2 + Z/3 + (Z/4)^3 +$ $+ (Z/5)^2 + Z/7$
$(C_3 \times C_3) \rtimes C_2$	4	$(Z/2)^2 + Z/3 + (Z/4)^3 + Z/7$

(IM5) $q^2t = rwx^{-1} = tu^{-1}v = q^{-1}wz = s^{-1}uv = 1$

$u^{-1}yz = rzs^{-1} = sxw = v^{-1}xy = rty = 1$.

Table 2 (Continued)

The epimorphism to A_6 used was:

$q = (1\,2\,3\,4\,5)$ $\quad\quad$ $r = (2\,5\,4\,6\,3)$
$s = (1\,6\,3\,5\,2)$ $\quad\quad$ $t = (1\,4\,2\,5\,3)$
$u = (1\,4\,6\,2)(3\,5)$ $\quad\quad$ $v = (2\,3)(4\,6)$
$w = (1\,4\,2\,3\,6)$ $\quad\quad$ $x = (1\,4)(2\,5)$
$y = (1\,6\,4)(2\,5\,3)$ $\quad\quad$ $z = (1\,6\,4\,2\,5)$

Subgroup of A_6	First homology of the corresponding cover	
	Rank	Torsion
D_4	2	$(Z/2)^9 + Z/3 + Z/31 + Z/331$
$C_5 \rtimes C_2$	1	$(Z/2)^8 + Z/3 + Z/27 + Z/29$
$(C_3 \times C_3) \rtimes C_2$	1	$(Z/2)^4 + (Z/3)^2 + Z/29$

(IM6) $\quad qtv = rux^{-1} = t^{-1}vz = q^2w^{-1} = stu^{-1} = 1$

$\quad\quad\quad v^{-1}wy = r^{-1}sy = sz^{-1}u = wy^{-1}x = rxz = 1$.

No epimorphism to A_5 or A_6 gave satisfactory results.

ANDREW J. NICAS
DEPARTMENT OF MATHEMATICS
UNIVERSITY OF TORONTO
TORONTO M5S 1A1
CANADA

C. W. STARK
DEPARTMENT OF MATHEMATICS
UNIVERSITY OF FLORIDA
GAINESVILLE, FLORIDA 32611

REFERENCES

[1] L. A. Best, On torsion-free discrete subgroups of PSL(2,C) with compact orbit space, Can. J. Math. 23 (1971), 451-460.

[2] S. E. Cappell, Unitary nilpotent groups and Hermitian K-theory, Bull. Amer. Math. Soc. 80 (1974), 1117-1122.

[3] ———, Manifolds with fundamental group a generalized free product. I, Bull. Amer. Math. Soc. 80 (1974), 1193-1198.

[4] F. T. Farrell and W.-C. Hsiang, The Whitehead group of poly-(finite or cyclic) groups, J. London Math. Soc. (2) 24 (1981), 308-324.

[5] J. Hempel, Orientation reversing involutions and the first Betti number for finite coverings of 3-manifolds, Invent. Math. 67 (1982), 133-142.

[6] _____, Homology of coverings, Pacific J. Math. 112 (1984), 83-113.

[7] A. J. Nicas, Induction theorems for groups of homotopy manifold structures, Memoirs Amer. Math. Soc. 267 (1982).

[8] _____, On Wh_2 of a Bieberbach group, Topology 23 (1984), 313-321.

[9] _____, On Wh_3 of a Bieberbach group, Math. Proc. Camb. Phil. Soc. 95 (1984), 55-60.

[10] A. J. Nicas and C. W. Stark, Higher Whitehead groups of certain bundles over Seifert manifolds, Proc. Amer. Math. Soc. 91 (1984), 1-5.

[11] _____, Whitehead groups of certain hyperbolic manifolds, Math. Proc. Camb. Phil. Soc. 95 (1984), 299-308.

[12] _____, K-theory and surgery of codimension-two torus actions on aspherical manifolds, J. London Math. Soc. 31 (1985), 173-183.

[13] J. S. Richardson and J. H. Rubinstein, Hyperbolic manifolds from regular polyhedra (preprint).

[14] C. W. Stark, Structure sets vanish for certain bundles over Seifert manifolds, Trans. Amer. Math. Soc. 285 (1984), 603-615.

[15] W. Thurston, Three dimensional manifolds, Kleinian groups and hyperbolic geometry, Bull. Amer. Math. Soc. (N.S) 6 (1982), 357-381.

[16] F. Waldhausen, Algebraic K-theory of generalized free products, Ann. of Math. 108 (1978), 135-256.

[17] C. Weber and H. Seifert, Die beiden Dodekaederraume, Math. Z. 37 (1933), 237-253.

A CHARACTERIZATION OF FINITE SUBGROUPS OF THE MAPPING-CLASS GROUP

Jane Gilman

Abstract. The main result of this paper is that a subgroup, G, of the mapping-class group of a surface of genus $g \geq 2$ is finite if and only if it contains only elements of finite order. As a consequence, we are able to characterize finite subgroups by the Nielsen types of the elements of L(G), the lifts of the elements of G to the unit disc. L(G) turns out to be a so-called abstract Fuchsian group. There is a natural way to assign a group of Möbius transformations to an abstract Fuchsian group. This gives an explicit construction for the solution to the Nielsen realization problem when there is a realization.

§1. Introduction

Let S be a compact Riemann surface of genus $g \geq 2$. The purpose of this paper is to characterize finite subgroups of M(S), the mapping-class group of S (also known as the Teichmüller modular group). In Section 2 we show that a subgroup G of M(S) is finite if and only if G contains no elements of infinite order. In one direction this is, of course, obvious. This result has also been obtained independently by Frank Raymond [R].

As a consequence of the main result we are able to characterize finite subgroups of M(S) by the Nielsen types in the group of lifts, L(G), of elements of G to the upper half-plane. This appears in Section 3 as Corollaries 3.2 and 3.3. L(G) turns out to be what is termed an abstract Fuchsian group. There is a natural way to associate to L(G) a group of Möbius transformations, which is denoted Möb(G) and is called the Möbius group associated to G. This appears in Section 4.

If $S = U/F$ where U is the unit disc and F is a Fuchsian group, then Möb(G) is an extension of F. Let $\mathcal{G}(S)$ be the set of all groups

of homeomorphisms of S whose image in M(S) is G. We show
(Theorem 4.2) that if $\mathcal{G}(S)$ contains a group, $\underline{G}_0$ of conformal automorphisms of S, then F is a normal subgroup of Möb(G) and $\underline{G}_0$ =
Möb(G)/F. Kerckhoff's solution to the Nielsen realization problem (see
[K]) says that there exists some surface S for which $\mathcal{G}(S)$ contains
such a conformal group $\underline{G}_0$. Theorem 4.2 shows exactly how $\underline{G}_0$ is constructed. Section 4 contains other results about the Möbius group of G
as well as some open questions.

§2. *Finite subgroups of the mapping-class group*

G will always denote a subgroup of M(S). Our main result is:

THEOREM 2.1. *G is a finite subgroup of* M(S) *if and only if every element of* G *is of finite order.*

Proof. Since the forward implication is obvious, assume that every element of G is of finite order. Let $\tau : M(S) \to Sp(2g, \underline{Z})$ be the map that
assigns to each element, g, of M(S) the matrix representation of the
action on a canonical homology basis of any homeomorphism in the
homotopy class of g. For an element g of finite order, $\tau(g)$ is the
identity if and only if g is the homotopy class of the identity. To see
this note that if g is of finite order, then there is a homeomorphism q
of finite order whose image in M(S) is g, and, in fact, there is a surface
S′ and a map $f : S \to S'$ where fqf^{-1} is homotopic to a conformal map.
The matrix representation of q on a canonical homology basis for S is
the identity if and only if that for fqf^{-1} is on the image of that basis
under f. By Hurwitz's Theorem [A], fqf^{-1} induces the identity on
homology if and only if fqf^{-1} is the identity. Since the kernel of the
restriction of τ to G is the identity, G is isomorphic to $\tau(G)$. Every
element of $\tau(G)$ has order dividing $84(g-1)!$ since every element of G
comes from a (conjugate of a) conformal map and thus has order less
than or equal to $84(g-1)$. Thus $\tau(G)$ is a periodic subgroup of

$GL(2g, \underline{C})$ of finite exponent. By 9.1ii p. 112 of [W] $\tau(G)$ is completely reducible and by 9.1iii $\tau(G)$ is of finite order. Thus G is of finite order.

Equivalent formulations of Theorem 2.1 are:

COROLLARY 2.2.

(a) G *is a finite subgroup of* $M(S)$ *if and only if* G *contains no elements of infinite order.*

(b) *Every periodic subgroup of* $M(S)$ *is finite.*

(c) $M(S)$ *contains no infinite subgroups of bounded exponent.*

§3. *The lifts of a group of mapping-classes*

In this section we characterize finite subgroups, G, of $M(S)$ by the Nielsen types in the group of lifts, $L(G)$, of elements of G to the unit disc, U. First we fix terminology and notation to make this concept well defined.

Let π be the projection from the group of homeomorphisms of S onto $M(S)$. For any fixed subgroup G of $M(S)$, (G not necessarily finite), let G_0 denote any group of homeomorphisms of S with $\pi(G_0) = G$. Let $L(G_0)$ be all lifts to U of elements of G_0. Each $t \in L(G_0)$ determines a pair of integers (v_t, u_t) called *the Nielsen type of* t (see the appendix to this paper (pp. 441-442) or Section 4 of [G] for details). We want to be able to talk about the lifts of G. Define $T(L(G_0))$, *the Nielsen types of* G_0, by $T(L(G_0)) = \{(v_h, u_h) | h \in L(G_0)\}$. We begin with:

LEMMA 3.1. *If two groups of homeomorphisms of* S, G_0 *and* G_0' *have the same image in* $M(S)$, *then there is a 1-1 function from* $T(L(G_0))$ *onto* $T(L(G_0'))$.

Proof. First for $h \in G_0$, let $\langle h \rangle$ be the cyclic group generated by h. $L(\langle h \rangle)$ and $L(h)$ have the obvious meaning. Note that $L(h)$ is not a group. If h and h' are any two elements of G_0 and G_0' respectively with $\pi(h) = \pi(h')$, then Theorem 12.1 of [G] establishes a 1-1 function σ from $L(<h>)$ onto $L(<h'>)$ such that for each $t \in L(<h>)$, $(v_t, u_t) = (v_{\sigma(t)}, u_{\sigma(t)})$. The proof of that theorem actually proceeds by first establishing for each i, σ_i, a 1-1 function from $L(h^i)$ onto $L((h')^i)$, where σ_i preserves Nielsen types. For any $g \in G$, pick $h \in G_0$ and $h' \in G_0'$ with $\pi(h) = \pi(h') = g$. Let σ_g be the function from $L(h)$ onto $L(h')$ which is 1-1 and preserves Nielsen types. Define $\Phi : L(G_0)$ onto $L(G_0')$ by $\Phi(h) = \sigma_g(h)$ when $\pi(h) = g$. Clearly Φ induces the desired mapping from $T(L(G_0))$ to $T(L(G_0'))$.

We say that $T(L(G_0))$ is equivalent to $T(L(G_0'))$ if there is a 1-1 function from one onto the other. Lemma 3.1 says that $T(L(G_0))$ is equivalent to $T(L(G_0'))$ when $\pi(G_0) = \pi(G')$. Therefore, it makes sense to talk about $T(L(G))$, *the Nielsen types of the lifts of* G, where G is any subgroup of $M(S)$. For a subgroup G of $M(S)$, we also talk about *the type of an element* $h \in L(G)$. We say that $h \in L(G)$ is of type (v_t, u_t) when we mean that every representative of the equivalence class of $T(L(G))$ contains a triple $(t', v_{t'}, u_{t'})$ where $\pi(t') = h$ and $(v_t, u_t) = (v_{t'}, u_{t'})$.

We refer to Section 4 of [G] or to the appendix (p. 441) for a detailed discussion of Nielsen types. In particular, recall that a lift of type $(1,0)^{**}$ has two fixed points on the boundary of U, one attracting and one repelling. Since this is the same as the action on ∂U as that of a hyperbolic fractional linear transformation, we call such a lift an *abstract hyperbolic*. An element of type $(0,0)$ has no fixed points on the boundary of U but moves all points in the same direction (either clockwise or counter-clockwise). Some power of such a map is not of type $(0,0)$. If the power which is not of type $(0,0)$ is the identity, then the lift has the same action on the boundary of U as an elliptic fractional linear transformation, and we call such a lift an *abstract elliptic* element.

In terms of Nielsen types, Theorem 2.1 can be reformulated as

COROLLARY 3.2.

(a) G *is a finite subgroup of* M(S) *if and only if every element of* L(G)-{identity} *is either of type* (0,0) *or of type* (1,0)**, *and equivalently,*

(b) G *is a finite subgroup of* M(S) *if and only if* L(G) *contains only abstract hyperbolic and abstract elliptic elements.*

Proof. This follows directly from Theorem 5.2i of [G]. An element g in G is of finite order if and only if every element of $T(L(<g>))$ is of type (0,0) or (1,0)** unless it is the identity.

If $L(G_0)$ contains only abstract hyperbolic elements and abstract elliptic elements it is termed an *abstract Fuchsian group*, and we also refer to L(G) as an abstract Fuchsian group, even though strictly speaking it is not a group.

Another equivalent formulation of Corollary 3.2 is

COROLLARY 3.3. *A subgroup* G *of* M(S) *is of infinite order if and only if there is a* t *in* L(G), $t \neq$ id., *with* t *of type* (v_t, u_t) *and* $v_t + u_t > 1$.

Proof. G is of infinite order if and only if it contains an element h of infinite order. This happens if and only if $L(<h>)$ satisfies the condition 5.1ii, 5.1iii, or 5.1iv of Theorem 5.1 of [G]. It is easy to check that one of these three conditions holds if and only if there is a t as required.

§4. *Abstract Fuchsian extensions and some open questions*

If F is a Fuchsian group of the first kind where U/F = S is a compact Riemann surface of genus $g \geq 2$, then a group, E, of homeomorphisms of U is called an *abstract Fuchsian extension of* F if F is a normal subgroup of E and every element of E is either abstract hyperbolic or abstract elliptic or the identity. Corollary 3.2 says that given a subgroup G of M(S) and a group G_0 of homeomorphisms of S with

$\pi(G_0) = G$, then G is a finite subgroup of $M(S)$ if and only if $L(G_0)$ is an abstract Fuchsian extension of F where $U/F = S$.

Assume from now on that G is a finite subgroup of $M(S)$. If $h \in L(G_0)$ is an abstract hyperbolic, then there is an integer n such that $h^n = f \in F$ and Nielsen theory tells us that $f \neq id$. We call h an abstract n^{th} root of f, and we let $f^{1/n}$ be the hyperbolic fractional linear transformation with the same fixed points as f but with $1/n$ times the translation length. The map $h \mapsto f^{1/n}$ thus associates a hyperbolic fractional linear transformation to a given abstract n^{th} root.

LEMMA 4.1. *$L(G_0)$ can be generated by adjoining to F a finite number of abstract roots of hyperbolic elements of F.*

Proof. It suffices to show that this is true for each $g_0 \in G_0$ and for the cyclic extension $L(<g_0>)$ of F. By using Smith's fixed point theorem to obtain the Nielsen realization in the cyclic case, one knows that the pair $\{L(<g_0>), F\}$ is conjugate to a pair of Fuchsian groups $\{L', F'\}$. Every Fuchsian group can be generated by hyperbolic elements (see Sibner [S]). Thus L' can be obtained by adjoining hyperbolic elements to F'. Since L' is a finite extension of F', L' must be obtained by adjoining hyperbolic roots of elements of F'. Since conjugation preserves Nielsen types, the lemma follows.

Given an abstract Fuchsian extension E of F, let E be obtained by adjoining abstract roots of the elements $f_1, \cdots, f_r$ of F. Assume that the roots are of order $m_1, \cdots, m_r$ respectively and that no f_i is the identity. Let E_F be the group of fractional linear transformations generated by F and by $\{f_i^{1/m_i} | i = 1, \cdots, r\}$. We call E_F a *Möbius group associated to* E. If we obtain E as $L(G_0)$ where $\pi(G_0) = G$ is a finite subgroup of $M(S)$, we also call E_F a *Möbius group associated to* G and write Möb(G) for E_F. Note that there may be more than one Möbius group associated to a given E or G since this depends upon a choice of generators.

Several interesting questions remain to be answered about the Möbius group associated to a group G. For example (1) When are all the possible Möbius groups associated to a given group G the same? (2) When is a Möbius group associated to G discrete? (3) When is Möb(G) a normal extension of F? (4) When is Möb(G) a finite extension of F? (5) When is the map from $L(G_0)$ or $L(G)$ to Möb(G) a group homomorphism? Finally what is the relationship among all of the above questions and can one actually find necessary and sufficient conditions for (1), (2), (3), (4) or (5) to occur? We have some partial answers.

THEOREM 4.2. *If $\underline{G}_0$ is a group of conformal homeomorphisms of S with $\pi(\underline{G}_0) = G$, then F is a normal subgroup of Möb(G) and $\underline{G}_0 = $ Möb(G)/F.*

Proof. By Lemma 4.1, $L(\underline{G}_0)$ is obtained by adjoining abstract hyperbolic roots to F. However, since $\underline{G}_0$ is conformal, these abstract hyperbolic roots are already hyperbolic roots. Thus by construction $L(\underline{G}_0) = $ Möb(G).

More generally, if S is a surface for which $\mathcal{G}(S)$ contains a conformal group, then all five questions have affirmative answers for S. If $\mathcal{G}(S)$ does not necessarily contain a conformal group, we obtain the results below. In particular, if F is a normal subgroup, the realization problem has a positive solution. If F is of finite index in Möb(G), there is a modified realization on a covering of S.

THEOREM 4.3.

(a) *If F is a normal subgroup of Möb(G), then it is of finite index, Möb(G) is discrete, and $\mathcal{G}(U/F)$ contains a group of conformal automorphisms.*

(b) *If F is of finite index in Möb(G), then Möb(G) is discrete and there is a finite sheeted covering of S onto which Möb(G) projects as a group of conformal automorphisms.*

Proof. If F is a normal subgroup of Möb(G) then since Möb(G) consists of fractional linear transformations, Möb(G)/F acts as a group of conformal automorphisms of $S = U/F$, from which it follows that Möb(G)/F is a finite group. Möb(G) will be discrete if and only if it contains no elliptic elements of infinite order. By the way in which Möb(G) is constructed it follows that if $|\text{Möb}(G):F|$ is finite, Möb(G) is discrete. Finally, if $|\text{Möb}(G):F| < \infty$, let F^σ be the intersection of all conjugates of F by elements of Möb(G). F^σ is a normal subgroup of finite index both in Möb(G) and in F, so that $\text{Möb}(G)/F^\sigma$ and F/F^σ act conformally on $U/F^\sigma = S^\sigma$. Further $S = S^\sigma/(F/F^\sigma)$ making S^σ a finite-sheeted covering of S.

Acknowledgment. I would like to thank Professor A. Marden for some helpful conversations and for suggesting the terms abstract hyperbolic and abstract elliptic.

DEPARTMENT OF MATHEMATICS
RUTGERS UNIVERSITY
NEW BRUNSWICK, NEW JERSEY 08903

REFERENCES

[A] Accola, R.D., "Automorphisms of Surfaces," J. D'Anal. Math. 18 (1967), 1-5.

[G] Gilman, J., "On the Nielsen Type and the Classification for the Mapping-class Group," Advances in Math., 40 (1981), 68-96.

[K] Kerckhoff, Steven, "The Nielsen realization problem," Ann. of Math., 117 (1983), 235-265.

[R] Raymond, Frank, Oral communication.

[S] Sibner, R., "Hyperbolic generators for Fuchsian groups," Proc. Amer. Math. Soc. 17 (1966), 963-968.

[W] Wehrfritz, B.A.F., "Infinite Linear Groups," Springer-Verlag, New York-Berlin (1973).

APPENDIX

Definition of the Nielsen Type and Explanation of its Relation to Thurston Theory

Recall that S is a compact Riemann surface of genus $g \geq 2$, $S = U/F$ where U is the unit disc and F the Fuchsian group that uniformizes S. If G_0 is a group of homeomorphisms of S, $L(G_0)$ is the group of all lifts of all elements of G_0 to U.

In his papers [N], Nielsen shows how to assign to each lift g in $L(G_0)$ a pair of integers, (v_g, u_g) called the Nielsen type of the lift g. (The precise definition is given below.) This pair of integers describes the dynamics of the action of the extension of g to the boundary of U. The important connection between Nielsen theory and Thurston theory arises from the fact that for a given homeomorphism h of S the totality of the Nielsen types of the elements of $L(H)$, where H is the cyclic group generated by h, determines the Thurston type of (the isotopy class of) h (e.g. pseudo-Anosov, etc.). Roughly speaking whether or not the integer v_g is always zero governs whether or not h is reducible and whether or not u_g is always zero governs whether or not h (or any of its component maps) are pseudo-Anosov. The precise details are much more complicated and are worked out in [G]. Here we merely summarize the facts necessary to define (v_g, u_g), the Nielsen type of g.

The following facts are established in [N]. (See also [G] for a summary.):

1. Every $g \in L(G_0)$ extends in a natural way to the boundary of U.

2. The fixed element subgroup of g, $N(g) = \{f \in F | gfg^{-1} = f\}$ is finitely generated. We let v_g denote the minimum number of generators and let $v_g = 0$ when $N(g) = \{id\}$. (Clearly, the reducibility of h is related to whether or not v_g is nonzero for some g in $L(H)$. This immediately suggests a connection between Nielsen Theory and Thurston theory.)

3. The isolated fixed points of the extension of g to the boundary of U are either attracting, repelling or neutral depending upon whether in the intervals adjacent to the fixed point g moves points on the boundary towards the fixed point, away from the fixed point, or on one side towards the fixed point and on the other side away from the point.

4. With the exception of the case where $v_g = 1$ and the extension of g to the boundary of U has precisely two fixed points, the isolated fixed points of g are alternately repelling and attracting and the number of N(g) orbits of attracting fixed points is finite. This number is denoted u_g.

5. In the case where the pair $(v_g, u_g) = (0,0)$, there is some power, n, of g for which the pair $(v_{g^n}, u_{g^n}) \neq (0,0)$.

6. If $v_g = 1$, then $N(g) = <f>$ where f is some hyperbolic element of F. On the boundary of U, f has two fixed points, the end points of the hyperbolic line fixed by f. (This is known as the axis of f.) If g fixes no other points on the boundary of U, then we say that g is of Nielsen type (1,0). Further g is of type $(1,0)^{**}$ if one of the fixed points is attracting and the other repelling while g is of type $(1,0)^*$ if both fixed points are neutral.

To each lift g in $L(G_0)$ we have now associated a pair of integers called the Nielsen type of g. The symbol (v_g, u_g) denotes the Nielsen type of g and is defined in general by #2 and #4 with the exceptional cases described by #6.

[G] Gilman, J., On the Nielsen Type and the Classification of the Mapping Class Group, Advances in Math., 40 (1981), 68-96.

[N] Nielsen, J., Untersuchungen zur Topologie der geschlossenen zweiseitigen Flachen, Acta Math., 50 (1927), 189-358, 53 (1929), 1-76, und 58 (1932), 87-167.

———, Surface Transformation Classes of Algebraically Finite Type, Math.-Fys. Medd. Danske Vid. Selsk, XXI 2 (1944), 1-89.

A SEQUENCE OF PSEUDO-ANOSOV DIFFEOMORPHISMS

L. Neuwirth and N. Patterson

The class of knots K_g pictured below are fibered, with fiber a surface Σ_g of genus g and characteristic map f_g (which we will see is pseudo-Anosov) inducing the automorphism ϕ_{2g} of $\pi_1(\Sigma_g) \approx F_{2g}$ (free group of rank 2g). The matrix M_{2g} defined below describes the induced automorphism of $H_1(\Sigma_g)$. $P_{2g}(x)$ is the characteristic polynomial of M_{2g} (as well as being the Alexander polynomial of K_g) and the largest real eigenvalue of M_{2g} is $2g$. We will see that the η_{2g} are the dilatation factors of the pseudo-Anosov maps f_g and form a monotone increasing sequence converging to $3 + 2\sqrt{2}$.

The knots K_g are formed from an array of 2g discs stacked above one another and joined to one another by pairs of twisted bands. The boundary of the surface Σ_g so obtained is the knot and the surface is of minimal genus spanning the knot. The maps of the fundamental group of Σ_g to the fundamental group of the complement of Σ_g defined by pushing loops off each side are both isomorphisms. Thus by [St] the knots fiber.

THEOREM 1. *The characteristic map f_g is pseudo-Anosov, with dilatation η_{2g}.*

Proof. Menasco's Corollary 1 in [M] asserts that a non-split prime alternating link is simple (without companions). Our knots K_g are non-split, alternating and prime by Theorem 1 in the same paper. The simplicity of our knots implies that no element of $\pi_1(\Sigma_g)$ is left fixed by ϕ_{2g} other than that corresponding to the boundary.

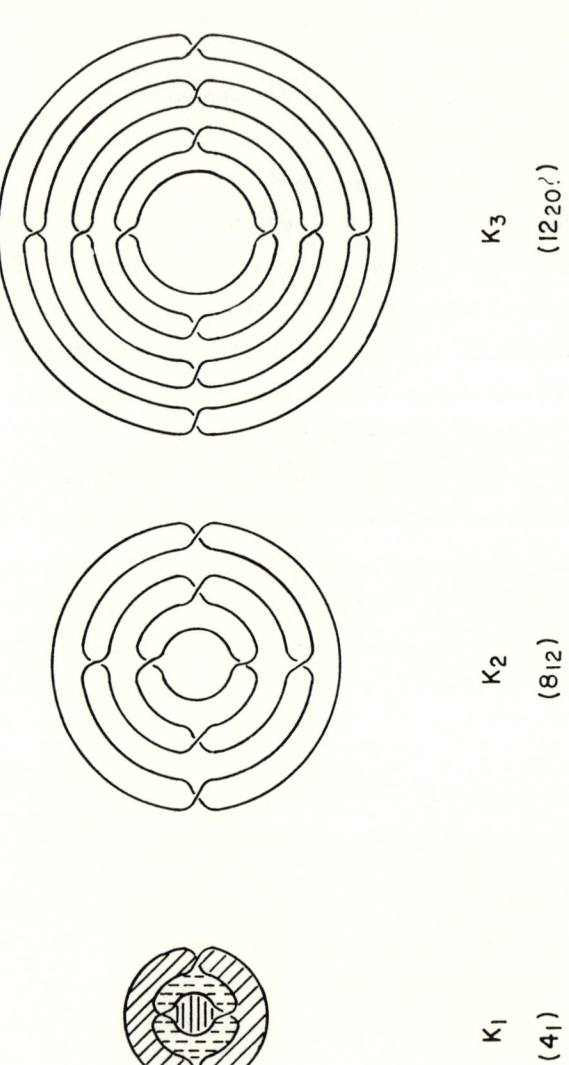

According to Thurston [Th 1], under these circumstances f_g is periodic or pseudo-Anosov. Since M_{2g} has real eigenvalues not equal to ± 1 f_g is not periodic.

According to J. Birman and C. Series [B] (to appear) if the image of each positive generator under ϕ_{2g} has no negative exponents in the reduced word representing the image in F_{2g} then the appropriate lamination (train tracks) are oriented, so that by Thurston [Th 2], the largest eigenvalue η_{2g} is the dilatation for f_g.

ϕ_1

$X_1 \to X_1 X_2$
$X_2 \to X_1 X_2 X_2$

ϕ_2

$X_1 \to X_1 X_2$
$X_2 \to X_1 X_2 X_3 X_2$
$X_3 \to X_1 X_2 X_3 X_4 X_3 X_2$
$X_4 \to X_1 X_2 X_3 X_4 X_4 X_3 X_2$

ϕ_3

$X_1 \to X_1 X_2$
$X_2 \to X_1 X_2 X_3 X_2$
$X_3 \to X_1 X_2 X_3 X_4 X_3 X_2$
$X_4 \to X_1 X_2 X_3 X_4 X_5 X_4 X_3 X_2$
$X_5 \to X_1 X_2 X_3 X_4 X_5 X_6 X_5 X_4 X_3 X_2$
$X_6 \to X_1 X_2 X_3 X_4 X_5 X_6 X_6 X_5 X_4 X_3 X_2$

$\ldots$

$$\begin{bmatrix} 1 & 1 \\ 1 & 2 \end{bmatrix} \begin{bmatrix} 1 & 1 & 1 & 1 \\ 1 & 2 & 2 & 2 \\ 0 & 1 & 2 & 2 \\ 0 & 0 & 1 & 2 \end{bmatrix} \begin{bmatrix} 1 & 1 & 1 & 1 & 1 & 1 \\ 1 & 2 & 2 & 2 & 2 & 2 \\ 0 & 1 & 2 & 2 & 2 & 2 \\ 0 & 0 & 1 & 2 & 2 & 2 \\ 0 & 0 & 0 & 1 & 2 & 2 \\ 0 & 0 & 0 & 0 & 1 & 2 \end{bmatrix} \ldots$$

$P_2(x) = x^2 - 3x + 1$

$P_4(x) = x^4 - 7x^3 + 13x^2 - 7x + 1$

$P_6(x) = x^6 - 11x^3 + 41x^4 - 63x^3 + 41x^2 - 11x + 1$

$$\eta_2 = \frac{3+\sqrt{5}}{2} \cong 2.62^+$$

$$\eta_4 = \frac{\sqrt{35+19\sqrt{5}}}{2\sqrt{2}\sqrt{\sqrt{5}}} + \frac{\sqrt{5}}{4} + \frac{7}{4} \cong 4.39^+$$

$$\eta_6 \cong 5.04^+ .$$

Define M_{2i-1} in the obvious way as a $(2i-1) \times (2i-1)$ matrix with all 1's in the first row, 1 and then 2's in the second row, then each row the preceding shifted right by 1, with zero inserted in the empty far left position $(M_0 = \emptyset)$.

PROPOSITION. *If P_i is the characteristic polynomial of M_i, $Q_i = (-1)^i P_i$ satisfies the recursion $Q_i = (x-1)Q_{i-1} - xQ_{i-2}$ for $i > 1$, $(Q_0 = 1)$.*

Proof. Let R_i denote the determinant of $N_i = (M_i - Ix)$ with 2 instead of $2-x$ in the last row

$$N_2 = \begin{pmatrix} 1-x & 1 \\ 1 & 2 \end{pmatrix}$$

$$N_3 = \begin{pmatrix} 1-x & 1 & 1 \\ 1 & 2-x & 2 \\ 0 & 1 & 2 \end{pmatrix} \quad \text{etc.}$$

By expanding $|M_i - Ix|$ and $|N_i|$ along their last rows one obtains

(1)$_i$
$$P_i = (2-x)P_{i-1} - R_{i-1}$$
$$R_i = 2P_{i-1} - R_{i-1}$$

subtracting gives

(2)$_i$
$$P_i = R_i - xP_{i-1} .$$

From (1)$_i$
$$P_i = (1-x)P_{i-1} + P_{i-1} - R_{i-1} .$$

Substituting (2)$_{i-1}$ gives
$$P_i = (1-x)P_{i-1} + R_{i-1} - xP_{i-2} - R_{i-1} .$$

Thus
$$P_i = (1-x)P_{i-1} - xP_{i-2}$$

or
$$Q_i = (x-1)Q_{i-1} - xQ_{i-2} \qquad \text{Q.E.D.}$$

We are interested in the largest real root η_i of Q_i, particularly when i is even.

LEMMA 1. $\{\eta_i\}$ is monotone increasing.

Proof. We proceed by induction; notice first if $x > \eta_i$, $Q_i(x) > 0$, also $\eta_3 = 2 + \sqrt{3} > \eta_2$.

Suppose $\eta_{i-1} > \eta_{i-2}$, then $Q_{i-2}(\eta_{i-1}) > 0$. By substituting η_{i-1} for x in Proposition 1 we obtain

$$Q_i(\eta_{i-1}) = 0 - \eta_{i-1}Q_{i-2}(\eta_{i-1}) .$$

Since $\eta_{i-1} > \eta_{i-2}$ by induction, and η_{i-2} is the largest real root of Q_{i-2}, $Q_{i-2}(\eta_i - 1) > 0$. Since $\eta_2 = \dfrac{3+\sqrt{5}}{2} > 0$, $\eta_{i-1} > 0$, so $Q_i(\eta_{i-1})$ is negative, hence $\eta_i > \eta_{i-1}$ and the proof is complete.

To study this sequence η_i of roots it is useful to study an arbitrary sequence of numbers $h_0 = 1$, $h_i = Q_i(\lambda)$ for (almost) any real λ.

Clearly the h_i satisfy the recursion $u_i = (\lambda-1)u_{i-1} - \lambda u_{i-2}$.

The roots of the characteristic equation of this recursion $y^2 = (\lambda-1)y - \lambda$ are

$$\tfrac{1}{2}[(\lambda-1) \pm (\lambda^2 - 6\lambda + 1)^{1/2}].$$

Let $(\lambda^2 - 6\lambda + 1)^{1/2} = \mu$. We shall suppose $\mu \neq 0$.

The roots of the characteristic equation are

$$\theta_1 = \tfrac{1}{2}(\lambda - 1 + \mu)$$

and

$$\theta_2 = \tfrac{1}{2}(\lambda - 1 - \mu).$$

The point of this is that any h_i may be written as

$$h_i = \alpha \theta_1^i + \beta \theta_2^i.$$

Now we solve for α, β by setting $i = 0, 1$

$$1 = h_0 = \alpha + \beta$$

$$\lambda - 1 = h_1 = \alpha \theta_1 + \beta \theta_2.$$

Solving gives $\alpha = \theta_1/\mu$, $\beta = -\theta_2/\mu$ so

$$h_i = \frac{1}{\mu}[\theta_1^{i+1} - \theta_2^{i+1}].$$

We are interested in the case when some h_i is zero, for then $Q_i(\lambda) = 0$, but if $\lambda > 3 + 2\sqrt{2}$, μ is real, $\neq 0$, and $\theta_1 > \theta_2$ so h_i is monotone increasing and $h_0 = 1$. So no h_i is zero, hence for any i the largest real root η_i of Q_i must be $\leq 3 + 2\sqrt{2}$.

Suppose, on the other hand, $\lambda < 3 + 2\sqrt{2}$. Taking polar coordinates, we may write

$$\theta_1 = re^{\sqrt{-1}\emptyset}$$

$$\theta_2 = re^{-\sqrt{-1}\emptyset}$$

$$\mu = 2r\sqrt{-1}\sin\emptyset = \theta_1 - \theta_2,$$

hence,

$$h_i = r^i \frac{\sin(i+1)\emptyset}{\sin\emptyset},$$

but $\emptyset \neq k\pi$ so that h_i is negative infinitely often, thus $Q_i(\lambda)$ is negative for $\lambda < 3 + 2\sqrt{2}$ for arbitrarily large i, so the largest root for those i must exceed λ. This, along with Lemma 1, proves

THEOREM 2. *The sequence η_i converges monotonically to $3 + 2\sqrt{2}$.*

We would like to thank Joan Birman and Len Charlap for their help.

IDA-CRD
THANET ROAD
PRINCETON, NEW JERSEY 08540

BIBLIOGRAPHY

[Th 1] Thurston, W. On the Geometry and Dynamics of Diffeomorphisms of Surfaces I (to appear).

[Th 2] _____. Diffeomorphisms of Surfaces, Princeton University, 1978.

[M] Menasco, W. Closed Incompressible Surfaces in Alternating Knot and Link Complements, *Topology*, Vol. 23, No. 1, pp. 37-44 (1984).

[B] Birman, J. and Series, C. Algebraic Linearity in the Mapping Class Group of a Surface, (to appear in J. of Applied Algebra).

[St] Stallings, J. On Fibering Certain 3-Manifolds, University of Georgia Conference on Topology of Manifolds, 1961.

DEHN'S ALGORITHM REVISITED, WITH APPLICATIONS TO SIMPLE CURVES ON SURFACES

Joan S. Birman[*] and Caroline Series

Abstract. Let Γ be a Fuchsian group acting on the Poincaré disc, with orbit space M. We study the relationship between the path of a closed geodesic C on M and shortest words (in generators for Γ) representing the free homotopy class of C. The results are applied to the study of words which represent geodesics having no self-intersections.

Introduction

Let Γ be a finitely generated Fuchsian group which acts on the Poincaré disc, with orbit space $M = U/\Gamma$, $U \subseteq \text{Int } D$, where D is the closed unit disc. In this paper we solve the problem of determining all shortest representatives of conjugacy classes in Γ, and give applications to the study of geodesics on M, for a class of metrics on M.

In the spirit of the foundational work of M. Dehn [D], we will see that for certain special metrics on M there is a remarkable relationship between the geometric data of the path followed by a geodesic on M and the algebraic data of its representation as a product of generators. As described in detail in Section 1, the lift to the Poincaré disc of a geodesic C on M cuts through a sequence $g_1 R, g_2 R, \cdots$ of copies of a fixed fundamental domain R for $\pi_1 M$. If we choose generators Γ_R for $\pi_1 M$ which pair sides of R, then $g_i^{-1} g_{i+1} \in \Gamma_R$ for each i, and if the geodesic C is closed then the product $\prod_{i=1}^{k-1} g_i^{-1} g_{i+1}$, where the sequence runs over exactly one period of a lift of C, represents the free homotopy class $[\gamma]$

---

[*]Partially supported by NSF Grant #MCS 79-04715.

of γ in $\pi_1 M$. In Section 2 we will prove the key fact that for appropriate choices of R and Γ_R this representation of $[\gamma]$ is cyclically shortest, and shortest in its conjugacy class. Any other shortest word in the conjugacy class may be obtained from this one in a simple way (Theorems 2.8 and 2.12). While this circle of ideas, which relates to Dehn's famous algorithm, may seem familiar and thoroughly investigated (e.g. see Chapter V of [L-S]) there are surprises in our work. For example, see Theorem 2.12(c), Example 2.9 and Remark 2.13 below. Our characterization of shortest words gives a slightly sharpened version of Dehn's algorithm which although undoubtedly known to the experts we have not found in print.

In Section 3 we apply these results to the problem of studying "simple words" in closed surface groups. Our main result is to give a precise way to reduce the problem of deciding when a word in a closed surface group corresponds to a simple curve on the closed surface to the corresponding problem on the surface punctured at one point. This easier problem was studied in the paper [Bi-S]. The same problem was also studied for closed surfaces in [R], also for both closed and punctured surfaces simultaneously in [Ch] and closed surfaces only in [Z1] and [Z2]. Our results imply that there is, rather surprisingly, a delicate dependence on the particular generating set in the results of Chillingworth and Zieschang when the surface is closed and the words concerned contain "half of the defining relator."

The authors are grateful to Martin Lustig for pointing out an error in Section 3, in an early version of this paper.

§1. *Preliminaries*

Let Γ be a finitely generated Fuchsian group acting on the unit disc D by isometries of the hyperbolic metric $ds = \dfrac{2|dz|}{1-|z|^2}$. Let L be the limit set of Γ, i.e. the closure on ∂D (the "circle at infinity") of the set of accumulation points of Γ, and let U be the convex hull of L in D. Then $U \subseteq \text{Int } D$ and U/Γ is a surface M, with covering space projection the natural map $\pi: U \to U/\Gamma = M$. We allow Γ to have elliptic elements, so π could be a branched covering. The surface M inherits a metric of constant negative curvature from the Poincaré metric on D. Geodesics in U are Euclidean circles orthogonal to ∂D and their images on M are geodesics on M.

Let R be a finite-sided geodesic polygon which is a fundamental domain for the action of Γ on D. We allow R to have vertices interior to D, or vertices on ∂D (these determine *cusps at infinity* on M) or sides on ∂D (these determine *funnels at infinity* on M). In the passage from R to M the vertices of R interior to D are allowed to fall into several equivalence classes, covering distinct points on M. The sides of R interior to D are identified in pairs by elements of Γ; the set of these elements is a symmetric set of generators for Γ which we denote Γ_R.

Label each side of R by the generator corresponding to it, putting the label inside R. Let N be the net of images of ∂R under Γ. Each oriented side of N is labelled by the same generator as the corresponding side of R. With this convention, if gR, hR are adjacent along a side s, then the side of s which is interior to hR is labelled $g^{-1}h$, and that which is interior to gR is labelled $h^{-1}g$.

Without loss of generality we may suppose that $O \in \text{Int } R$ and that O is not a fixed point of any element of Γ. Let N^* be the dual net to N, obtained by joining gO to hO, $g, h \in \Gamma$, whenever $g^{-1}h \in \Gamma_R$. A path in N^* will be called an *edge path*. The oriented edge of N^* joining gO to hO clearly cuts across the common side of adjacent regions gR, hR and is also labelled $g^{-1}h$. Thus a path in N^* corresponds to a word $e_1 \cdots e_k$, $e_i \in \Gamma_R$, where we write the e_i's in the order they occur from left to right. Since O is not fixed by any $g \in \Gamma$, the map $\Gamma \to \Gamma O$ is an injection and N^* may be regarded as the Cayley graph of Γ. The length of a path E in N^* is the number of edges it contains, written $|E|$, and we write $|g|$ for the length of a shortest path in N^* joining O to gO; equivalently $|g|$ is the length of g in the word metric defined by the generators Γ_R. We shall be only interested in *reduced* paths; that is ones in which a sequence ee^{-1}, $e \in \Gamma_R$, does not occur. *From now on all edge paths which we describe will be taken to be reduced.*

The tiling of D defined by N^* is made up of polygons dual to the polygons in N. A circuit round the edge of one of these polygons corresponds to one of the defining relations in Γ. Any sequence of edges which follows consecutive sides of one of these polygons we call a *cycle*. Suppose that an even number, $2n(v)$, of copies of R meet at the vertex v of R. Then the dual polygon in N^* surrounding v has $2n(v)$ sides. We call a cycle containing $n(v)$ edges a *half-cycle*. A *long* cycle is a cycle which contains more than $n(v)$ edges.

Suppose that P,Q are polygons in N^* with common edge AB, and suppose that C_1, C_2 are cycles lying in $\partial P, \partial Q$ respectively so that C_1 ends and C_2 begins at B, and A is not in C_1 or C_2. Then we call C_1 and C_2 *consecutive*. A sequence of consecutive cycles we call a *chain*. A sequence of polygons in N (or N^*), each meeting the next in a common edge, is a *polygonal chain* in N (or N^*). A chain is *long* if it consists of cycles of lengths $n(v_1), n(v_2)-1, n(v_3)-1, \cdots, n(v_{k-1})-1, n(v_k)$, where $v_1, \cdots, v_k$ are the vertices of N dual to the corresponding polygonal chain. We shall usually include long cycles in the term long chain; this should be clear from the context. (If we wish to discuss either one separately we will say so explicitly.) To each cycle or chain is associated a *complementary* cycle or chain: this complementary chain is the opposite boundary of the corresponding polygonal chain, and represents the same element of Γ. It is clear that the complement of a long cycle or chain is a path of strictly shorter length. If E is an edge path in N^* we denote by $P(E)$ the corresponding polygonal chain in N^*.

Let γ be any oriented arc in D which does not pass through a vertex of N and with $\partial \gamma \cap N = \emptyset$. Then γ cuts through a well-defined sequence of adjacent copies $g_1 R, g_2 R, \cdots, g_k R$ of R and thus defines a polygonal chain in N, denoted $P(\gamma)$. The path in N^* joining $g_1 O, g_2 O, \cdots, g_k O$ is called the *edge path* of γ, denoted $E(\gamma)$. The associated *edge path word* is $e_1 \cdots e_{k-1}$, where $e_i = g_i^{-1} g_{i+1}$. In particular, a path defined by a geodesic arc in this way is called a *geodesic edge path* and corresponding word is a *geodesic word*. If γ projects to a closed curve on $M = D/\Gamma$ and if $g_1 R, g_k R$ contain adjacent equivalent points on γ then it is clear that $e_1 \cdots e_k$ represents the free homotopy class of γ. Notice that a geodesic word is always reduced, for otherwise γ would cut some side of some gR twice.

Suppose now that γ passes through a vertex v of N but is not coincident with a side of N. Then we deform γ in a neighbourhood of this vertex to pass around one or other side (Figure 1a). Notice that the two possible edge paths we obtain correspond to the two complementary halves of the relation in Γ corresponding to v. If γ coincides with a net edge we deform as shown in Figure 1b.

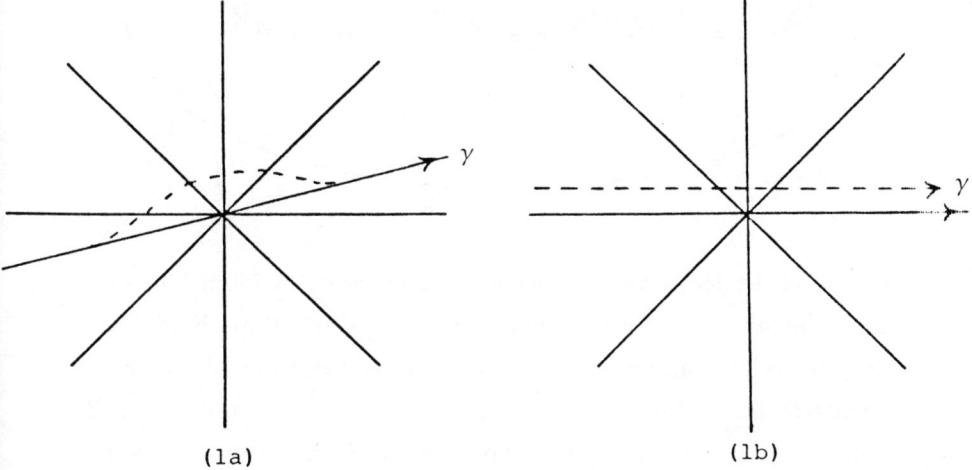

Figure 1

If $w = e_1 e_2 \cdots e_k$ is a word in the symbols of Γ_R, we will say that w contains a cycle, chain, half-cycle, long cycle, long chain, if the edge-path $E(w)$ joining $O, e_1 O, e_1 e_2 O, \cdots, e_1 e_2 \cdots e_k O$ contains a cycle, chain, half-cycle, long cycle or long chain respectively.

We say that a fundamental region R has *even corners* if the net N is a union of complete geodesics in **D**. This means that the extension of a side of R through a vertex is a side of some copy gR of R. If M is a surface and if Γ does not contain elliptic elements then such fundamental regions R always exist: if $\partial M \neq \emptyset$ one may suppose that all vertices of R lie on $\partial \mathbf{D}$ so that the condition is trivial; if M is a closed surface of genus g then cutting M with 2g geodesic loops arranged to intersect in pairs as illustrated below renders the

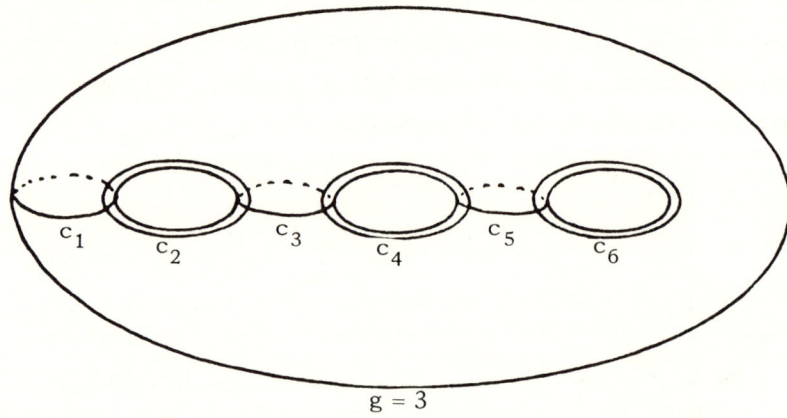

g = 3

complement of the loops simply connected, giving a fundamental region R with 8g–4 sides. Since the extension of any geodesic through an intersection point is still a side of R, it is clear that R has even corners. (This construction is due to Koebe [K].) For groups Γ of the first kind containing elliptic elements it was shown in [Bo-S] that for each abstract Γ which occurs one can find an embedding $\Gamma \to \text{Isom}(\mathbf{D})$ and a fundamental region R with this property. (The result is only proved in [Bo-S] for groups of the first kind, but the extension is obvious to any nonelementary group Γ (cf. [Bo-S] and [Be], Th. 10.4.2).) Fundamental regions with even corners frequently occur in specific cases, for example the symmetric 4g-gon representing a closed surface of genus g and the standard fundamental region $|z| > 1$, $|\text{Re} z| < \frac{1}{2}$ for $SL(2, \mathbf{Z})$.

Throughout, we denote g^{-1}, $g \in \Gamma$, by $\bar{g}$ or g^{-1}, using whichever is more convenient.

§2. *Geodesic paths and shortest words*

In this section we establish the correspondence between geodesic paths and shortest words described in the introduction. The main results of this section are Theorems 2.8 and 2.12. For validity of these results, we have to introduce a slight additional restriction on the fundamental region R.

HYPOTHESIS 2.1.
 (a) If R has three sides, then not all vertices of R lie in Int D.
 (b) If R has four sides, and if all vertices of R lie in Int D, then at least three geodesics in N cross at each vertex of R.

Notice that if D/Γ is a surface without branch points (so that Γ contains no elliptic elements) then both of the above conditions hold. If R is a triangle then any glueing pattern will force a branch point at any vertex of R in Int D. If R has four sides, and all vertices lie in Int D, then the only glueing pattern which produces an orientable surface without branch points identifies opposite sides to give a torus. Since a torus is not covered by the hyperbolic disc, this last case is impossible.

LEMMA 2.2. *A geodesic lying in N cuts a geodesic edge path at most once.*

Proof. Let C be a geodesic in N. Suppose that C intersects the geodesic edge path $E(\gamma)$ joining vertices $g_1 O, \cdots, g_k O$ between $g_1 O$ and $g_2 O$ and again between $g_{k-1} O$ and $g_k O$. Then C must include as sub-arcs $g_1 R \cap g_2 R$ and $g_{k-1} R \cap g_k R$ because C is in the net N of Γ-translates of ∂R. Now γ runs from $g_1 R$ into $g_2 R$ and from $g_{k-1} R$ into $g_k R$; thus γ cuts C twice, which is impossible. ‖

LEMMA 2.3. *Let s and s′ be non-adjacent sides of R. Then the geodesics which extend s and s′ do not intersect.*

458   JOAN S. BIRMAN AND CAROLINE SERIES

Proof. This is Lemma 2.2, [Bo-S]. ‖

LEMMA 2.4. *Let γ be a geodesic intersecting two copies R_1, R_2 of R which have a common vertex v. Suppose also that $v \notin \gamma$. Then $E(\gamma)$, the geodesic edge path of γ, contains exactly one of the two cycles of edges corresponding to the two cycles of fundamental regions meeting at v and connecting R_1 with R_2.*

Proof. Suppose, if possible, that this was not the case. Let the sides of R_1, R_2 meeting at v be extended to half lines λ_1, λ_1' and λ_2, λ_2', which meet at v, labelled so that λ_1 and λ_2 are adjacent. See Figure 2. Then γ certainly intersects either λ_1 and λ_2 or λ_1' and

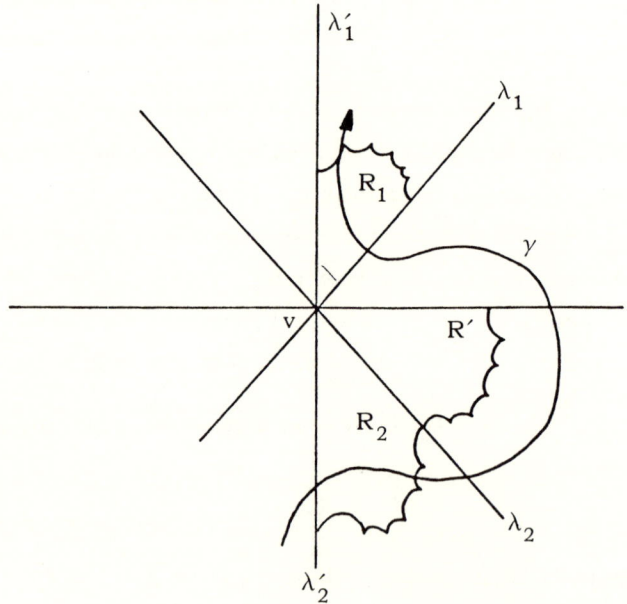

Figure 2

λ_2', suppose the first. Moreover there is a region R' with one vertex at v lying in the sector at v between λ_1 and λ_2, and with $\gamma \cap R' = \emptyset$. If R' has more than four sides then by Lemma 2.3 we can find a side whose geodesic extension C does not intersect λ_1 or λ_2. Then γ intersects C twice, which is impossible.

If R' has four sides then by Hypothesis 2.1 we may choose a geodesic $C \subset N$ through the vertex w of R' which lies on neither λ_1 nor λ_2 and so that $R' \cap C = \{w\}$. By Lemma 2.3 again C lies entirely in the sector cut off by λ_1 and λ_2 so that again γ cuts C twice. Finally if R' is a triangle, then by assumption it has a vertex at infinity and so γ cannot possibly have avoided R'. ‖

Suppose that E is an edge path in N^* which contains a cycle at a vertex v of N. The total sectorial angle at v formed by the regions cut by edges in the cycle we call the *vertex angle* of E at v. If $2n(v)$ regions gR meet at v we denote the vertex angle by π^-, π or π^+ respectively if $n(v)-1$, $n(v)$ or $n(v)+1$ regions make up the vertex angle.

LEMMA 2.5. *The vertex angles along any path E which is either geodesic or which contains no long cycles are at most π^+. If the angle at v is π^+ then we may replace the cycle at v by its complementary cycle and obtain another path of the same length.*

Proof. If a geodesic path contained a vertex angle greater than π^+, then the corresponding geodesic would cut some geodesic in N twice. The remaining statements are obvious. ‖

LEMMA 2.6. *Let E be an edge path which is either geodesic or which contains no long chains. Then the occurrence of a sequence of vertex angles $\pi^+, \pi, \pi, \cdots, \pi, \pi^+$ along E is impossible.*

Proof. The existence of such a sequence implies that E contains a long chain. In the case where E is geodesic, the corresponding geodesic cuts a side of N twice, which is impossible. ‖

LEMMA 2.7. *Let E_1 and E_2 be edge paths containing no long chains and with coincident initial and final points. Then there are no copies*

gR of R *lying strictly inside the region bounded by the polygonal chains* $P(E_1), P(E_2)$ *defined by* E_1 *and* E_2.

Proof. Suppose that R' were such a region. We claim that there is at least one geodesic $C \subset N$ which cuts one of E_1 or E_2 twice. Suppose on the contrary that each extended side of R' cut each of E_1, E_2 at most once. Let s_1, s_2 be those sides whose extensions cut E_1 closest to the endpoints v_0, v_n of E_1 and E_2. If these two sides did not intersect in a common vertex of R' then any side of R' lying between s_1 and s_2 on the side of E_2 would have an extension which cut E_2 twice (see Figure 3), for otherwise it would cut E_1 between v_0 and s_1

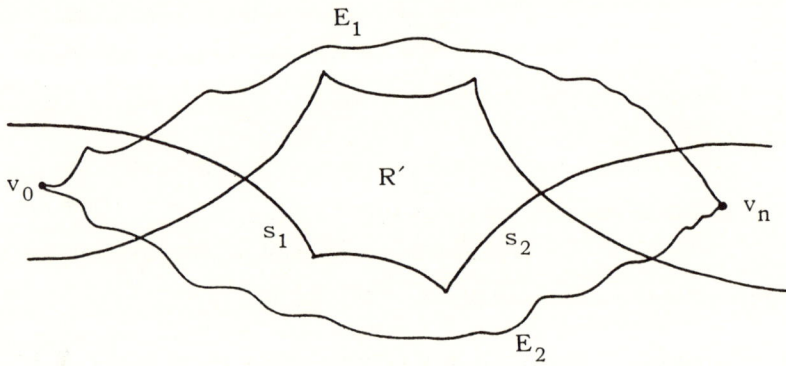

Figure 3

or between v_n and s_2. Thus $s_1 \cap s_2$ is a vertex of R'. If R' has more than four sides, then the extension of any side not adjacent to s_1 or s_2 must, by Lemma 2.3, cut E_1 twice. If R' has four sides, the same is true of any geodesic in N which passes through the vertex of R' not contained in $s_1 \cup s_2$, and which is not a side of R'. Finally the whole figure is impossible if R' has a vertex at infinity.

Now among geodesics $C \subset N$ which cut E_1 twice, choose one where the segment E which C cuts off on E_1 has minimal length. Choose any region gR which meets C (possibly only at a vertex) between these two intersection points and on the same side of C as E. Suppose that gO ∉ E. Then the extension of any side of gR which does not have a point in common with C cuts off a shorter segment than E on E_1, contradicting the choice of C.

Suppose that there is a vertex hO ∈ E such that hR ∩ C = ∅. Choose a vertex v of hR which lies inside the region bounded by E and C. The extension of any side of N through v either cuts E twice, which is impossible by minimality, or cuts C once. Take two geodesics in N which contain v. Since both intersect C we have a triangle formed by sides of N, which is impossible by Hypothesis 2.1.

We have shown that the vertices of E consist, besides its endpoints, precisely in the vertices of all copies of R which meet C on the side of E between the two intersection points of C with E. The two endpoints of E lie on the other side of C. Thus E contains a long chain, contrary to hypothesis. ∥

We are now ready to prove our first theorem.

THEOREM 2.8. *Assume that* R *satisfies the conditions of Hypothesis 2.1. Then*

(a) *An edge path in* N^* *joining* $g_1 O, g_2 O, \cdots, g_k O$, $g_i^{-1} g_{i+1} \in \Gamma_R$, *is a shortest path if and only if it contains no long cycles or long chains.*

(b) *If* γ *is a geodesic arc which begins in* $g_1 R$ *and ends in* $g_k R$, *and if* $E(\gamma)$ *is the corresponding geodesic edge path, then* $E(\gamma)$ *is a shortest edge path in* N^* *from* $g_1 O$ *to* $g_k O$.

(c) *Any shortest path* F *in* N^* *can be converted to a geodesic path* E *in* N^* *with the same end points by successively replacing half-cycles with their complements. These cycle changes may be made at any vertex* v *of* N *which lies in the intersection* $P(E) \cap P(F)$ *of the associated polygonal chains, such that the angle on* $P(F)$ *is* π^-.

Proof. We have already seen that shortest paths cannot contain long chains. Assume inductively that all remaining statements of the theorem hold for paths of length less than n.

Suppose that E is a path of length n containing no long chains. Let F be the geodesic path corresponding to the geodesic joining the endpoints of E. By Lemma 2.7 there are no copies of R between P(E) and P(F). Consider a vertex v of N on the common boundary of P(E), P(F). The sum of the vertex angles on E, F at v is 2π; by Lemma 2.5 the angle on both sides can only be π^-, π or π^+. If the angle on E is π^+ then we may by 2.5 replace the cycle by its complement and divide the new path E' into two segments which meet at the centre g0 of some region $gR \subset P(E') \cap P(F)$ which has v as a vertex. By the inductive hypothesis one can change each of these two paths into the corresponding segments of F by replacement of half-cycles by their complements at vertices of angle π^+ on F; all these vertices also lie on P(E); moreover $|E| = |F|$.

There remains the case when the vertex angle along P(F) is everywhere π or π^+. In particular, the angle on P(F) at the initial and final vertices of $P(F) \cap P(E)$ must be π^+. By Lemma 2.6, this situation is impossible. Hence every path of length n containing no long cycles may be changed into a geodesic path of the same length in the manner asserted.

In particular, it easily follows that geodesic paths are shortest. ∥

Theorem 2.8 has some obvious analogues in terms of shortest words, however we defer the statement until we can improve it. A word in the generators Γ_R is *cyclically shortest* if it and each of its cyclic permutations is a shortest word in Γ_R. Our first question is whether two arbitrary cyclically shortest words in the same conjugacy class have the same length. Unfortunately, this is not always the case. A phenomenon we must rule out in order to be able to establish a positive result is illustrated by the following example.

EXAMPLE 2.9. Take M to be the closed surface of genus 2 whose fundamental region is the symmetric octagon but whose side pairings are as shown in Figure 4. One verifies that these pairings satisfy Poincaré's condition for generation of a discrete group and that the quotient surface is indeed M. The relation in this group is $\bar{d}c\bar{b}dab\bar{a}\bar{c} = 1$. Then if $w = c\bar{b}$, $v = ca\bar{b}a$ both w and v are cyclically shortest and $w = dv\bar{d}$.

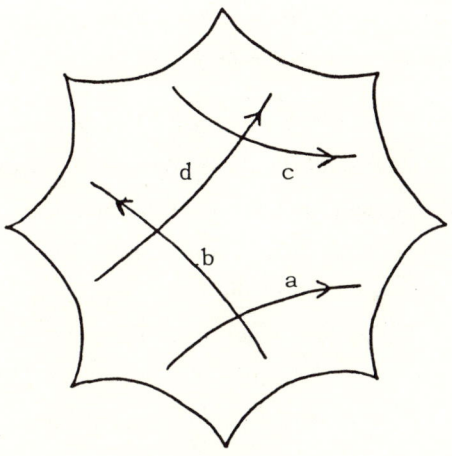

Figure 4

We therefore introduce a property of the edge pairings on R which ensures that such a situation does not arise. This property only makes sense when R has an even number of sides, as for example when R has even corners and Γ_R contains no elliptic elements of order 2.

DEFINITION 2.10. The generating set Γ_R is *alternating* if there is a map $\rho : \Gamma_R \to \{1,-1\}$ so that $\rho(x) = \rho(x^{-1})$ and so that $\rho(e) = -\rho(f)$ whenever $e, f \in \Gamma_R$ are adjacent terms in a cycle.

Clearly the generating set Γ_R of example 2.9 is not alternating. On the other hand, the standard generating set $\Gamma_R = \{a_i, b_i : i = 1, \cdots, g\}$ whose vertex cycle is the commutator product $\prod_{i=1}^{g} [a_i, b_i]$ is alternating, as are the standard generators for $SL(2,\mathbf{Z})$ corresponding to the fundamental region $|z| > 1$, $|\mathrm{Re}\, z| < \frac{1}{2}$.

LEMMA 2.11. *Let* R *be as in Theorem 2.8. Assume, further, that* Γ_R *is alternating. Suppose that* w *is a cyclically shortest word. Then*

$w^n = w w \cdots w$ *is also a cyclically shortest word in its conjugacy class, unless* w *is elliptic.*

Proof. Suppose Lemma 2.11 is false. Then, by Theorem 2.8(a) the word w^n must contain a long chain. After cyclic permutation if necessary, we may assume that the long chain occurs at the beginning of w^n. Assume first that the long chain is not a long cycle. Let $w = e_1 e_2 \cdots e_k$, and let the ρ-value of the first side of N which is cut by the edge path $O, e_1 O, e_1 e_2 O, \cdots$ be $+1$. Then the first cycle in w^n cuts in order sides of N labelled $+ - + - + \cdots + -$. The next cycle cuts sides labelled $- + - \cdots + -$, which pattern repeats until the last cycle which is $- + - + \cdots - +$, as in Figure 5. Now this sequence S generated by the long chain

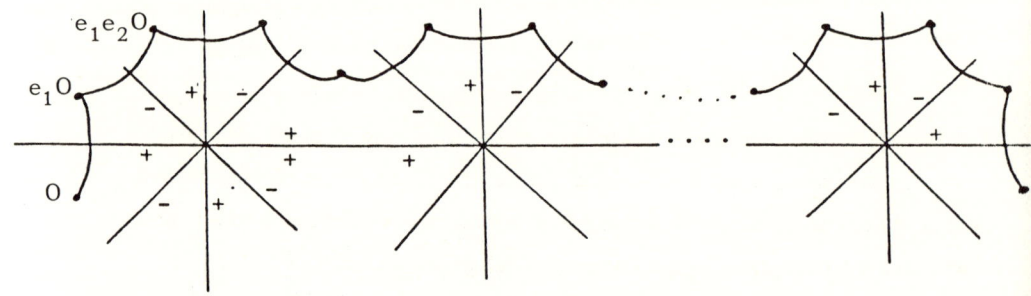

Figure 5

occupies more than one period in the periodic word $\cdots www \cdots$, for w is cyclically shortest. The second occurrence of w in S certainly again begins with the symbol $+$. Examination of S shows that every occurrence of $+$ is preceded by at least one element in the same cycle; moreover the initial cycle in w is a half cycle. These two portions of cycles concatenated produce a long cycle in a cyclic permutation of w or some power w^k. We may therefore assume that w^k contains a long cycle.

If all the elements in the long cycle are distinct then clearly the cycle lies in a cyclic permutation of w, which is impossible. Otherwise not all elements are distinct and the basic period of the cycle is elliptic. Since w^k contains a long cycle it follows that w is elliptic. ∥

We are finally able to establish the theorem which describes all of the cyclically shortest words in a conjugacy class and relates them to one another.

Recall that a cyclic word w is a *geodesic word* if w is a fundamental period in the infinite edge path $E(\hat{\gamma})$ associated to a complete geodesic $\hat{\gamma}$ which covers a closed, smooth geodesic on M.

THEOREM 2.12. *Suppose that R has even corners and satisfies Hypothesis 2.1. Assume that Γ_R is alternating.*
  (a) *A cyclic word in Γ_R is cyclically shortest if and only if it contains no long cycles or long chains.*
  (b) *Geodesic words are cyclically shortest words.*
  (c) *If w,z are cyclically shortest words in the same conjugacy class, then w,z have the same length. Also, they either agree up to a finite number of half-cycle switches and cyclic permutations, or $w = u_1 u_2 \cdots u_k$, $z = v_1 v_2 \cdots v_k$ are chains of consecutive cycles each of length $n(v)-1$, also $z = ewe^{-1}$ in Γ for some $e \in \Gamma_R$ and z, ewe^{-1} are complementary chains. This last case occurs exactly when w represents the projection of some side of N on M.*

Proof. Assertion (a) is an immediate consequence of Theorem 2.8(a).

To prove (b), note that each cyclic permutation of w arises from the edge-path $E(\gamma)$ associated to a finite geodesic arc γ which is a sub-arc of the complete geodesic $\hat{\gamma}$. But then, by Theorem 2.8(b), that cyclic permutation of w is a shortest word, hence w is a cyclically shortest word.

Now for the proof of (c). Let $\hat{\gamma}$ be the complete geodesic in D joining the ideal points $\lim_{n \to \infty} w^{-n} 0$ and $\lim_{n \to \infty} z^n 0$, so that $\hat{\gamma}$ is the axis of $w \in \Gamma$, and let y be the geodesic word determined by $\hat{\gamma}$.

Let E_y be the edge path $E(\gamma)$. If $w = e_1 e_2 \cdots e_k$, let E_w be the edge path defined by w, i.e. the edge path determined by

$\cdots w^{-n}0, \cdots, w^{-1}0, \cdots, e_k^{-1}e_{k-1}^{-1}0, \; e_k^{-1}0, 0, e_1 0, e_1 e_2 0, \cdots, w0, \cdots, w^n 0 \cdots$.
Clearly E_y, E_w have the same end points at infinity.

Suppose that $g0 \in E_y \cap E_w$. Since E_y, E_w are both invariant under w (because the end point of E_y, E_w is a fixed point of w) we have $w^n g0 \in E_y \cap E_w$ for all $n \in \mathbb{Z}$. The segment of E_w between $g0$ and $wg0$ must be some cyclic permutation w^i of w, which by hypothesis is shortest. The corresponding segment of E_y is a cyclic permutation y^j of y. By part (b) above y^j is cyclically shortest and by Theorem 2.8, $|w| = |w^i| = |y^j| = |y|$. Moreover w^i is obtained from y^j by interchanging half cycles, as required. Thus we are done if $E_w \cap E_y \neq \phi$.

Now suppose that $E_w \cap E_y = \emptyset$, so that E_w lies entirely to one side of E_y. Consider the vertex angles of the infinite chain $P(E_y)$ on the E_w side. Note that if $\tilde{\gamma}$ passes through a vertex v of N but is not contained in N, the vertex angle on $P(E_y)$ at v is π^+.

Suppose that the vertex angle at $v \in N$ on $P(E_y)$ is less than π. Since E_y is periodic, this happens periodically; take two such vertices v, v' at one period distance, so that $w(v) = v'$, and fix corresponding regions $R_1, w(R_1)$ in the chain $P(E_y)$.

By the remark above, $\tilde{\gamma}$ cannot pass through v or v'. Let $\tilde{\gamma}_n$ be the geodesic path joining $w^{-n}0$ to $w^n 0$. As $n \to \infty$, the endpoints of $\tilde{\gamma}_n$ approach the endpoints of $\tilde{\gamma}$ so that we can choose n large enough that $\tilde{\gamma}_n$ passes through R_1, R_1' and defines the same path as $\tilde{\gamma}$ between these regions. Let $E_{w,n}$ be the segment of E_w from $w^{-n}0$ to $w^n 0$. By Theorem 2.8(a) $E_{w,n}$ is a shortest path. Moreover the vertex angle of $P(\tilde{\gamma}_n)$ at v and v' is less than π. Thus by Theorem 2.8, v and v' are also vertices along $P(E_{w,n})$ and we can change cycles on $E_{w,n}$ at v and v' so that the new path coincides with $P(\tilde{\gamma}_n)$, and hence E_y, near v and v'. Each cycle change obviously takes place within some period of w; repeating this periodically we find ourselves back in the previous situation.

It remains to consider the case when all vertex angles along $P(E_y)$ on the side of E_w are at least π. By Lemma 2.6 the angle may be π^+

at most one point. But occurrences of a particular angle are obviously periodic, thus the angle along $P(E_y)$ on the E_w side is everywhere π, in other words the boundary of $P(E_y)$ along the E_w side is a complete geodesic $C \subset N$. Since $\tilde{\gamma}$ is everywhere at a bounded distance from C, $\tilde{\gamma}$ and C coincide.

We have shown that either y,w differ by cyclic permutation and half-cycle switches, or that $\tilde{\gamma}$ is a side C of N. In this last case either E_y, E_w are coincident paths along the same side of C, in which case y,w agree possibly after cyclic permutation, or E_y, E_w are paths along opposite sides of C. Either way, both w and y are sequences of consecutive cycles of lengths $n(v)-1$.

If w,y do not coincide then, again after cyclically permuting we may assume that both w and y begin with an $n(v)-1$ cycle at the same vertex v on C, but in opposite senses, so that if $e \in \Gamma_R$ is the label on the E_w side of that part of the common boundary of E_y, E_w which terminates at v (with the orientation given to C by w), then ew and $e^{-1}y$ begin with opposite $n(v)$ cycles. Following E_w for a full period of w, we see that $ewe^{-1} = y$.

Now let $\tilde{\gamma}''$ be the geodesic joining the points $\lim_{n\to\infty} z^{-n}0$, $\lim_{n\to\infty} z^n 0$, and let y' be the corresponding geodesic word. Since z and w are conjugate in Γ, the geodesics $\tilde{\gamma}$ and $\tilde{\gamma}''$ project to the same curve on M. Thus the geodesic words y,y' agree up to cyclic permutation, and $\tilde{\gamma}$ is a side of N if and only if the same is true of $\tilde{\gamma}''$. The words y',z are related in the same way as y,w. It follows that if $\tilde{\gamma}$ is not a side of N, then z and w agree up to cyclic permutation and cycle switches. If $\tilde{\gamma}$ is a side of N then after cyclically permuting, either y' = z or $y' = e_1 z e_1^{-1}$, $e_1 \in \Gamma_R$. But e_1 is such that $e_1 y' = e_1 y$ begins with an $n(v)$ cycle, and so $e_1 = e$. Thus either $z = w$ or $z = ewe^{-1}$. ‖

REMARK 2.13. Theorem 2.12 above does not agree with the solution to the same problem given in [Z-V-C], page 131, Theorem 4.9.5. We have two counter-examples to Theorem 4.9.5 of [Z-V-C]. The first was given earlier in Example 2.9, and shows that some restriction on presentation is needed. The second uses the standard presentation $<a,b,c,d;\ aba^{-1}b^{-1}cdc^{-1}d^{-1}>$ for a closed surface of genus 2. Let $w_1 = aba^{-1}a^{-1}b^{-1}c$, $w_2 = cd^{-1}c^{-1}a^{-1}dc$. Theorem 4.9.5 of [Z-V-C] implies that w_1, w_2 cannot be conjugate, because neither contain any half-cycles and both are shortest words. However, $d^{-1}w_1 d = (d^{-1}aba^{-1})(a^{-1}b^{-1}cd)$, which after two half-cycle switches becomes $(cd^{-1}c^{-1}b)(b^{-1}a^{-1}dc) = (cd^{-1}c^{-1})(a^{-1}dc) = w_2$.

§3. *An application to simple curves on closed surfaces*

Let $M = U/\Gamma$ be as in Sections 1 and 2, but assume Γ has no elliptic elements, so that π is a covering. In an earlier work [Bi-S] the authors considered the problem of characterizing *simple words*, i.e. those words w in the generators of a given presentation for $\pi_1 M$ which are represented by an embedded loop. It was assumed in that work that M is not a closed surface. In this section we extend the results of [Bi-S] to the case when M is closed, in the following seemingly transparent way: each curve on a closed surface determines a curve on a punctured surface when one removes a point from M which is not in the curve. Thus one need only lift w "in the right way" to a free group and study the problem there. Our work in Section 2 is ideally suited for this: we have established a precise connection between algebra and geometry, which we now put to use to determine from the word w exactly how representative curves lie on M with relation to a fixed point v and so to determine a "right way" to lift w to $\pi_1(M-v)$. This problem is solved in Theorem 3.2.

The main tool in the proof of Theorem 3.2 is Theorem 3.1 which may be of interest in its own right. It gives necessary conditions on the form of a word in $\pi_1(M)$ for it to be simple. Many other necessary conditions are given in [Bi-S].

Theorem 3.3 shows that the simple word problem is very strongly dependent on the choice of the presentation. See, in particular, Example 3.4 and the remark which follows it.

Although it will be clear that our method will apply to any fundamental region R with even corners and Γ_R alternating, we shall for simplicity assume for the remainder of this section that $\Gamma = \pi_1(M)$ for a closed surface M of genus g and that R is the symmetrical 4g-gon of vertex angle $\frac{\pi}{2g}$, with side pairings chosen so that all vertices of R project to one point v and so that Γ_R is alternating. Let $X = \{x_1,\cdots,x_{2g}\}$ be the set of transformations which pair the sides of R, for this particular choice of Γ and R. Let $\mathcal{P} = \langle X; \mathcal{R} \rangle$ be the corresponding presentation of Γ. Call a presentation $\mathcal{P}$ satisfying all the conditions just listed an *admissible* presentation.

We denote by ϕ the permutation of $\Gamma_R = X^{\pm}$ which carries a generator to the next generator in the cycle around the vertex v, in clockwise order. Then
$$\mathcal{R} = x\phi(x)\phi^2(x)\cdots\phi^{4g-1}(x), \quad x \in X^{\pm}.$$

We note that the "standard presentations":

$$\mathcal{P}_1 = \mathcal{P}_1(g) = \langle x_1,\cdots,x_{2g}; x_1 x_2 x_1^{-1} x_2^{-1} \cdots x_{2g-1} x_{2g} x_{2g-1}^{-1} x_{2g}^{-1} \rangle$$

$$\mathcal{P}_2 = \mathcal{P}_2(g) = \langle x_1,\cdots,x_{2g}; x_1 x_2 \cdots x_{2g} x_1^{-1} x_2^{-1} \cdots x_{2g}^{-1} \rangle$$

are both admissible, but that while for genus 2 these are the only such presentations the number of possibilities increases with g.

A cyclic word w in the generators of an admissible presentation $\mathcal{P}$ contains a *generalized cycle* if there is a cycle $C = x\phi(x)\phi^2(x)\cdots\phi^r(x)$ in such that for each $j = 0, 1,\cdots, r-1$ the 2-letter sequence $\phi^j(x)\phi^{j+1}(x)$ or its inverse occurs in w. For example, in the standard admissible presentation $\mathcal{P}_1(2)$ the cyclic word $w = x_2^{-1} x_3 x_2^{-1} x_1^{-1} x_2 x_1^{-1}$ contains the generalized 5-cycle $x_1 x_2 x_1^{-1} x_2^{-1} x_3$ because each of the 2-letter sequences $x_1 x_2$, $x_2 x_1^{-1}$, $x_1^{-1} x_2^{-1}$, $x_2^{-1} x_3$ or its inverse occurs in w.

THEOREM 3.1. *Let w be a cyclically shortest or a geodesic word in an admissible presentation* $<X; \mathcal{R}>$. *Then the following are necessary conditions for w to be simple:*

(a) *w contains no generalized cycle of length* $> 2g$.

(b) *If w contains generalized cycles* C_1, C_2 *of length* $2g$, *then* C_1 *and* C_2 *or* C_2^{-1} *are identical or complementary.*

(c) *w contains no consecutive cycles* u_1, u_2 *of length* $2g-1$ *with* $u_1 \neq u_2$.

Proof. Suppose first that $w = w_g$ is geodesic. Suppose that w_g is simple, and let A_g be the (possibly deformed) smooth geodesic corresponding to w. As remarked before A_g is simple and hence $\pi^{-1}(A_g) \cap R$ is a collection of disjoint arcs. It is clear that there is an oriented arc in this family joining the sides of R labelled e, f^{-1} corresponding to each occurrence of the sequence $(ef)^{\pm}$ in w_g. Suppose that this sequence occurs $n(e,f)$ times. The set of integers $n(e,f)$, $e, f \in \Gamma_R$ determine the cyclic word w, for if we construct $n(e,f)$ parallel arcs joining the sides e, f^{-1} of R for each pair $e, f \in \Gamma_R$ then when we identify corresponding sides of F to reconstruct M there is a *unique* way to join these arcs to form a simple closed curve, specified by the order in which they arrive on the sides of R.

Let ψ be the 4g-cycle acting on $X^{\pm}$ which records the sequence of labels encountered on the sides of R as one moves about ∂R anticlockwise. This permutation is related to the 4g-cycle ϕ used earlier as follows: Let τ be the involution on $X^{\pm}$ which takes each $x \in X^{\pm}$ to x^{-1}. Then (see Figure 6) $\psi = \tau \phi$.

Call a component of $\pi^{-1}(A_g) \cap R$ a *corner arc* if it joins adjacent sides of R and if among all of the arcs joining that pair of sides it is closest to the vertex. If the sides in question are labelled $(e, \psi(e))$ then by the above such an arc occurs if and only if w_g contains the 2-letter sequence $(e, \overline{\psi(e)})^{\pm 1}$. Since $\psi = \tau \phi$, this is equivalent to the condition that w_g contain the cycle $(e, \phi(e))^{\pm 1}$ of length 2.

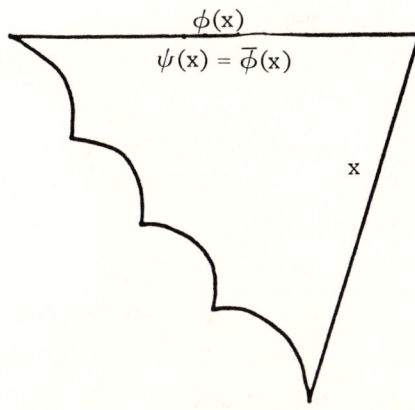

Figure 6

Suppose, now, that w_g contains a generalized 3-cycle $\phi^{-1}(e), e, \phi(e)$, $e \in X^{\pm}$. This means that the 2-letter sequences $(e, \phi(e))^{\pm 1}$ and $(\phi^{-1}(e), e)^{\pm 1}$ occur in w_g implying as above that there are corner arcs joining the sides of R labelled $e, \psi(e)$ and $\phi^{-1}e, \psi\phi^{-1}e$. This last pair is of course the same as $\phi^{-1}e^{-1}, e^{-1}$. But then, when the sides of R labelled e and e^{-1} are identified under the projection $\pi: \mathbf{D} \to M$ these corner arcs (being closest to the common vertex) must join up to form a connected sub-arc of A_g, which shows that there must have been a true 3-cycle (not just a generalized 3-cycle) in w_g. Note that we are using the fact that A_g is a *simple* curve in an essential way.

Continuing in this way, we see that if w_g contains a generalized cycle of length $\geq 2g+1$, then in fact it contains a *cycle* of length $\geq 2g+1$. However, by hypothesis w_g is geodesic, and so contains no cycles of length $\geq 2g+1$ (Theorem 2.12). Therefore no generalized cycle of length $\geq 2g+1$ can occur, and part (a) is proved.

In fact our proof shows more. Suppose that w_g contained distinct cycles C_1, C_2 of length $2g$. Since R has length $4g$, if C_1 and $C_2^{\pm 1}$ are not complementary cycles, then these cycles fit together to give a generalized $(2g+1)$-cycle in w_g, however by part (a) this is impossible, hence (b) is proved.

It remains to prove (c). Suppose that w_g contains adjacent cycles u_1, u_2 of lengths $2g-1$, $2g-1$. Let $u_1 = e_1 e_2 \cdots e_{2g-1}$, $R = e_1 e_2 \cdots e_{2g-1} \cdots e_{4g}$, and let $u_2 = f_1 f_2 \cdots f_{2g-1}$. Then since u_1, u_2 are adjacent, we may extend u_1, u_2 to $2g$-cycles u_1', u_2', with $u_1' = e_1 e_2 \cdots e_{2g-1} e_{2g}$, $u_2' = e_{2g}^{-1} f_1 f_2 \cdots f_{2g-1}$, so that $u_1' u_2' = u_1 u_2$ after cancellation of the adjacent pair $e_{2g} e_{2g}^{-1}$ at the interface. By the alternating property of admissible presentations, $\rho(e_1) \neq \rho(e_{2g}) = \rho(e_{2g}^{-1})$, so $e_{2g}^{-1} \in \{e_2, e_4, \cdots, e_{2g-2}, e_{2g+2}, \cdots, e_{4g}\}$. However, if $e_{2g}^{-1} \in \{e_2, e_4, \cdots, e_{2g-2}\}$, then $u_1 u_2$ includes the $(2g-1)$-cycle $e_1 e_2 \cdots e_{2g-1}$ (from u_1) and the 3-cycle $e_{2g-1} e_{2g} e_{2g+1}$ (from u_2) and hence the generalized $(2g+1)$-cycle $e_1 e_2 \cdots e_{2g+1}$, contradicting (a). Similarly, if $e_{2g}^{-1} \in \{e_{2g}, e_{2g+2}, \cdots, e_{4g-2}\}$, then $u_1 u_2$ contains the generalized $(2g+1)$-cycle $e_{4g-1} e_{4g} e_1 e_2 \cdots e_{2g-1}$, which is also impossible. Finally, if $e_{2g}^{-1} = e_{4g}$, then $u_1 = u_2$, an excluded case.

We now consider the case when w is cyclically shortest rather than geodesic. Since w_g cannot contain a pair of adjacent $(2g-1)$-cycles, it follows that w_g has no long chains. But then, by Theorem 2.12, w and w_g agree up to a finite number of half-cycle switches.

Now suppose that w contains a generalized $(2g+1)$-cycle, and that w' is obtained from w by a single half-cycle switch. Let the half-cycle be $x\phi(x)\phi^2(x) \cdots \phi^{2g-1}(x)$. Without loss of generality we may assume that the generalized $(2g+1)$-cycle in w is $\{\phi^k(x)\phi^{k+1}(x))^{\pm 1}$: $k = 0, 1, \cdots, 2g-1\}$. Observe that the 2-cycle $(\phi^{2g-1}(x)\phi^{2g}(x))^{\pm 1}$ in w must occur in a part of w which is unaltered in the passage to w', for if not w would not be cyclically shortest. It then follows that w' contains the generalized $(2g+1)$-cycle $\{(\phi^k(x)\phi^{k+1}(x))^{\pm 1}$: $k = 2g-1, \cdots, 4g-2\}$. Continuing in this way, we conclude that any word obtained from w by finitely many half-cycle switches contains a generalized $(2g+1)$-cycle. However, that is impossible for w_g. Hence w contains no generalized $(2g+1)$-cycle. But then the entire proof applies equally well to geodesic and to cyclically shortest words, and the proof of Theorem 3.1 is complete. ‖

We now use Theorem 3.1 to give the promised connection between simple words in $\pi_1(M)$ and simple words in $\pi_1(M-v)$. For $x_i \in \Gamma_R$, let $A(x_i)$ be the axis of x_i, $1 \leq i \leq 2g$. The point $w = p(O)$, where O is the origin of D, will be our base point for $\pi_1 M$. If $A(x_i)$ is not a diameter of D, let α_i be the oriented radial arc from O to $A(x_i)$ which is orthogonal to $A(x_i)$. Then $\pi(\alpha_i \cup A(x_i) \cup \alpha_i^{-1})$ is a canonical w-based loop X_i in the homotopy class of x_i, with $v \notin X_i$. This loop represents our generator, as an w-based curve on M.

Let $i : (M-v) \to M$ be the inclusion map. A curve $\hat{A}$ will be said to be the *canonical lift* of a curve A on M if $v \notin A$ and if $i(\hat{A}) = A$. Let $\hat{X}$ be the canonical lift of X, let $\hat{x}$ be the homotopy class of $\hat{X}$ in $\pi_1(M-v, w)$, and let $\hat{\mathcal{P}}$ be the presentation $\langle \hat{x}_1, \cdots, \hat{x}_{2g} \rangle$ of $\pi_1(M-v, w)$. The map i induces a *canonical homomorphism* $i_* : \pi_1(M-v, O) \to \pi_1(M, O)$ defined by $i_*(\hat{x}_i) = x_i$, $1 \leq i \leq 2g$. The word $\hat{w} = \hat{x}_{i_1} \hat{x}_{i_2} \cdots \hat{x}_{i_n}$ in the generators of $\hat{\mathcal{P}}$ is the *canonical lift* of the word $w = x_{i_1} x_{i_2} \cdots x_{i_n}$ in the generators of an admissible presentation $\mathcal{P}$.

THEOREM 3.2. *Let* w *be a cyclically shortest word in the generators of an admissible presentation* $\mathcal{P}$. *Then*

(a) *If* w *contains cycles* C_1, C_2 *of length* $2g$, *a necessary condition for* w *to be simple is that* C_1 *and* $C_2^{\pm 1}$ *be identical or complementary.*

(b) *Assume* w *satisfies (a). Let* w' *be the word obtained from* w *by half-cycle switches chosen so that all half-cycles in* w' *are* $C_1^{\pm 1}$. *Then* w' *is simple if and only if its canonical lift to* $\pi_1(M-v, w)$ *is simple.*

REMARK. To decide whether a word in $\pi_1(M-v, w)$ is simple, one may apply the algorithm of [Bi-S].

Proof. If $\hat{w}$ is a simple word in $\pi_1(M-v, w)$ then it is trivial that w is simple in $\pi_1(M, O)$, because any simple representative $\hat{A}$ of $\hat{w}$ determines a simple representative $A = i(\hat{A})$ of w.

Assume that w is simple. Assertion (a) is just Theorem 3.1(b). Construct the bi-infinite word $\cdots www \cdots = \cdots e_{-3}e_{-2}e_{-1}e_1e_2e_3 \cdots$, $e_i \in X^{\pm 1}$. Then w determines an edge path $E(w)$ in D, which projects to a closed loop A on M which represents the free homotopy class of w because $\cdots e_{-3}e_{-2}e_{-1}e_1e_2e_3 \cdots$ has fundamental period w. We call A *the loop determined by* w. Note that $v \notin A$, because an edge path lies in the net N^*, and $v \notin N^*$.

Now let A_g be the smooth geodesic representative of w, deformed slightly as in Section 1, Figure 1, if $v \in A_g$. Let w_g be the edge path word corresponding to a lift of one period of A_g to D, so that w_g represents the conjugacy class of w. As observed by Poincaré [P], the conjugacy class of w has a simple representative if and only if A is simple.

By Theorem 2.12(b), w_g is a cyclically shortest word. By Theorem 2.12(c), $|w| = |w_g|$ and either w, w_g agree to up half-cycle switches or $w = u_1 u_2 \cdots u_k$, $w_g = v_1 v_2 \cdots v_k$ as in 2.12(c).

Suppose that w, w_g agree up to half-cycle switches. It is clear that we may assume that A and A_g agree outside a small regular neighborhood N of v on M. The components of A and A_g may be assumed to have the same endpoints on N, also arcs in A and A_g with the same endpoints correspond to the pairs of switched cycles in w and w_g. Corresponding arcs differ only in that they pass to one or other side of v (Figure 7). Moreover all arcs in $A \cap N$ pass to the same side of v. Since A_g is simple the arcs in $A_g \cap N$ are pairwise disjoint and hence their endpoints do not separate each other on ∂N. Thus the arcs in $A \cap N$ are also pairwise disjoint and hence A is a simple loop on M whose canonical lift to $M - v$ is also simple, in other words the canonical lift of w to $\pi_1(M-v, w)$ is simple.

It remains to consider the case $w = u_1 u_2 \cdots u_k$, $w_g = v_1 v_2 \cdots v_k$. Assume $k > 1$. If $u_1 = u_2$ or $v_1 = v_2$ then $w = u_1^k$ or $w_g = v_1^k$, because all cycles in a chain of $(2g-1)$-cycles coincide if any two adjacent ones do, however this is impossible because by hypothesis w

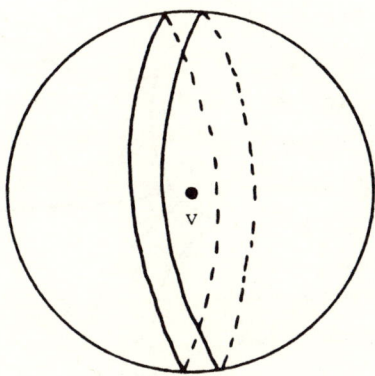

Figure 7

(and hence w_g) is simple while a word which is a proper power is never simple. But then w and w_g each contain two distinct consecutive (2g–1)-cycles, contradicting Theorem 3.1, part (c).

The case $k = 1$ remains. Then $w = u_1$ and $w_g = v_1$ are (2g–1)-cycles, also w and $e w_g e^{-1}$ are complementary cycles for some $e \in \Gamma_R$. This case occurs precisely when a lift of A_g is a geodesic in the net N. Pushing this geodesic off N to either side and projecting to M we obtain two loops A and A' which are the boundaries of a (possibly singular) annulus on M, with A_g as its core. Clearly A' and A'' are simple if and only if A_g is simple. Therefore it will not matter whether we lift u_1 or v_1 to the free group $\pi(M-v, w)$. This completes the proof of Theorem 3.2. ∥

To motivate our last result, we return to Theorem 3.1 and note that the excluded case $u_1 = u_2$ in 3.1(c) does occur. Here is an example. Consider the standard presentation $\mathcal{P}_2(g)$. Then for every cyclic permutation of $\mathcal{R}$ we have $e_{2g}^{-1} = e_{4g}$, hence adjacent (2g–1)-cycles are always identical. Let $g = 2$, $u_1 = u_2 = x_1 x_2 x_3$. We see from Figure 8 that the cyclic word $w = (x_1 x_2 x_3)^2 x_4^{-1}$ is simple. The need to exclude this case motivates our next definition.

A presentation $\mathcal{P} = <X; \mathcal{R}>$ is *strongly admissible* if it is admissible and if, in addition, $\phi^{2g}(x) \neq x^{-1}$ for every $x \in X^{\pm}$. This implies that adjacent $(2g-1)$-cycles begin with distinct letters. Note that the standard presentation $\mathcal{P}_1(g)$ is strongly admissible, but $\mathcal{P}_2(g)$ is not. In fact, in $\mathcal{P}_2(g)$ we have $x^{-1} = \phi^{2g}(x)$ for every $x \in X^{\pm}$.

THEOREM 3.3. *Let w be a simple cyclic word which contains no long cycles, in the generators of a strongly admissible presentation for $\pi_1(M)$. Then w is cyclically shortest. In particular, w contains no long chains.*

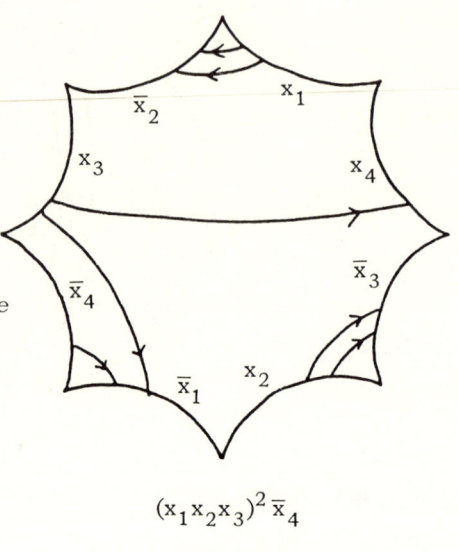

$(x_1 x_2 x_3)^2 \bar{x}_4$

Figure 8

Proof. Suppose that w contains $t \geq 1$ long chains, and that $w = b u_1 u_2 \cdots u_s$ where $u_1 u_2 \cdots u_s$ is a long chain, $s \geq 2$. Then u_1 and u_s are coherently oriented half-cycles. Note that since $\mathcal{P}$ is strongly admissible $u_1 u_2 \cdots u_s$ always contains a pair of distinct consecutive $(2g-1)$-cycles.

Let $w_1 = b v_1 v_2 \cdots v_s$ be the cyclic word obtained from w by replacing $u_1 u_2 \cdots u_s$ by its complement $v_1 v_2 \cdots v_s$. Now, $v_1 v_2 \cdots v_s$ cannot contain a long cycle or long chain because each v_j is a $(2g-1)$-cycle. Also, free cancellation cannot occur at the interface between b and v_1, for if so w would have contained a long cycle from the last symbol in b together with the half-cycle u_1 (a picture in **D** makes this obvious). Similarly, there cannot be a half-cycle in w_1 which includes $2g-1$ symbols from b together with the first symbol in v_1, or a half $\mathcal{R}^{-1}$ cycle which includes the last symbol in b together with the $(2g-1)$-cycle v_1, for if so w would have adjacent symbols $x_1 x_1^{-1}$ at the interface between b and v_1. For exactly the same reasons, there is no free cancellation at the interface between v_s and b, also there cannot be a half-cycle at the interface.

Thus we have proved that if w contains $t \geq 1$ long chains, then w_1 contains t–1 long chains and a pair of consecutive, distinct (2g–1)-cycles. By induction on t we conclude that w is equivalent to a cyclically shortest word w_t which contains a pair of consecutive distinct (2g–1)-cycles. But then w_t (and hence w) is not simple. ‖

EXAMPLE 3.4. We illustrate Theorem 3.3 with an example. Consider the presentation $\mathcal{P}_2(2)$, and let $w = (x_4 x_3 x_2)(x_1 x_2 x_3 x_4)(x_2 x_3 x_4 x_1^{-1}) x_1^{-1}$. This word contains the long chain $(x_1 x_2 x_3 x_4)(x_2 x_3 x_4 x_1^{-1})$, with complementary chain $(x_4 x_3 x_2)^2 x_1^{-1}$. One may verify (Figure 9) that $\hat{w}$ is not

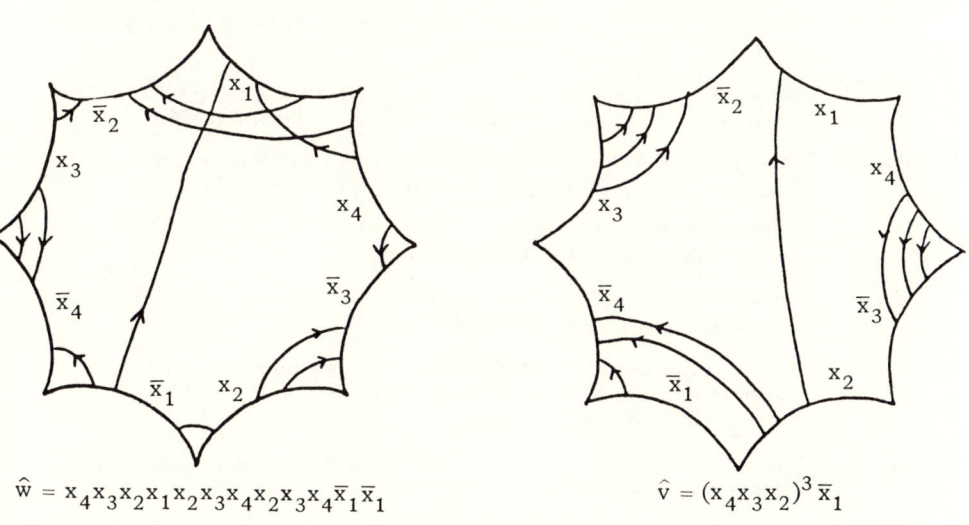

$\hat{w} = x_4 x_3 x_2 x_1 x_2 x_3 x_4 x_2 x_3 x_4 \bar{x}_1 \bar{x}_1$ $\qquad$ $\hat{v} = (x_4 x_3 x_2)^3 \bar{x}_1$

Figure 9

simple, while $\hat{v}$ is simple. Therefore v, and hence w are simple words in $\mathcal{P}_2$. The point we are making in Theorem 3.3 is that such examples do not exist using $\mathcal{P}_1$ or any other strongly admissible presentation.

REMARK 3.5. We found Theorem 3.3 enlightening for the following reasons. In our earlier work [Bi-S] we developed an algorithm for deciding when a word is simple for a surface with boundary. In extending our work to the closed surface case, we were surprised to see that the algorithm in [Ch] did not involve the removal of long chains. Theorem 3.3 explains this fact, since the algorithm of [Ch] relates only to the presentation $\mathcal{P}_1$, which is strongly admissible.

JOAN S. BIRMAN
DEPARTMENT OF MATHEMATICS
COLUMBIA UNIVERSITY
NEW YORK, NEW YORK 10027

CAROLINE SERIES
MATHEMATICS INSTITUTE
UNIVERSITY OF WARWICK
COVENTRY CV4 7AL
U.K.

REFERENCES

[B] Beardon, A., *The Geometry of Discrete Groups*, Springer-Verlag (1983).

[Bi-S] Birman, J.S. and Series, C., "An algorithm for simple curves on surfaces," J. London Math. Soc. (2), 29 (1984), 331-342.

[Bo-S] Bowen, R. and Series, C., "Markov maps associated with Fuchsian groups," Pub. Math. IHES #50 (1979), 401-418.

[Ch] Chillingworth, D., "Simple closed curves on surfaces," Bull. London Math. Soc. 1 (1969), 310-314.

[D] Dehn, M., "Transformation der Kurven auf zweiseitigen Flächen," Math. Ann. 72 (1912), 413-421.

[K] Koebe, P., Riemannsche Manigfaltigkeiten und nichteuklidische Raumformen, IV, Sitzung berichte der Preussischen Akad. der Wissenschaften (1929), 414-457.

[L-S] Lyndon, R. and Schupp, P., *Combinatorial Group Theory*, Springer-Verlag (1977).

[P] Poincaré, H., "Analysis Situs," J. Ecole Polytechn. (2), 1 (1985), 1-121.

[R] Reinhart, B.L., "Algorithms for Jordan curves on compact surfaces," Annals Math 75 (1962), 209-222.

[Z, 1] Zieschang, H., "Algorithmen fur einfache kurven auf Flächen," Math. Scad. 17 (1965), 17-40.

[Z, 2] _____, Ibid II, Math. Scand. 25 (1969), 49-58.

[Z-V-C] Zieschang, Vogt and Coldewey, *Surfaces and Planar Discontinuous Groups*, Springer Verlag Lecture Notes #835 (1980).

PATHS OF GEODESICS AND GEOMETRIC INTERSECTION NUMBERS: I

Marshall Cohen[*] and Martin Lustig[*]

§0. Introduction

Suppose M^2 is a compact, connected orientable 2-manifold with negative Euler characteristic. In this paper (I) we assume that $\partial M^2 \neq \emptyset$. We give a simple combinatorial algorithm which determines for each pair of words V, W in a natural basis of $\pi_1 M^2$, the minimal number $|V \cap W|$ of geometric intersections of curves on M^2 in the free homotopy classes represented by V and W respectively. Similarly, given a single word V, our algorithm determines the self-intersection number $|V|$.

In Part II (the next paper in these proceedings) the second-named author extends this algorithm to the case of closed 2-manifolds. The algorithm in the closed case reduces to the algorithm given here when the words do not involve half of any cyclic permutation of either the defining relator or its inverse.

The algorithms and theorems given are completely combinatorial. ("Given words $V, W, \cdots$.") The method of proof is to impose a hyperbolic structure on M^2 and then to construct and study edge-paths in the Poincaré disk which traverse the same sequence of fundamental domains in the tesselation as do the geodesics corresponding to the given words. The construction of such an edge path for a given word is straightforward when the fundamental group is free, but it becomes the matter of central interest when (Part II) $\partial M^2 = \emptyset$. (In the closed case many very different-

---

[*]The first author was partially supported by an NSF Research Contract. The second author was supported by the Studienstiftung des Deutschen Volkes.

looking words represent the same free homotopy class, so one cannot expect a given word to directly indicate the path of the geodesic in its free homotopy class.)

The outline of this paper is

§1. Background and statement of the theorem
§2. The algorithm
§3. Edge paths and geodesics
§4. Proof of the theorem
§5. Some intersection numbers.

This work grew out of our reading of [B-S 1]. The authors are grateful to Professor Joan Birman for her encouragement and helpful comments.

§1. *Background and statement of the theorem*

We suppose that the orientable manifold M^2 is given as a 2-disk with handles, gotten by choosing pairwise identifications of 2q disjoint closed arcs on the boundary of the disk. Let G be the group of covering transformations of the universal cover. G is freely generated by transformations $a_1, a_2, \cdots, a_q$ which identify the paired edges of a fundamental domain F; say $a_i(\bar{a}_i) = a_i$ ($1 \leq i \leq q$). Thus $G \cong \pi_1 M^2$, with a_i corresponding to the loop which runs over the handle it determines. (See Figure 1.)

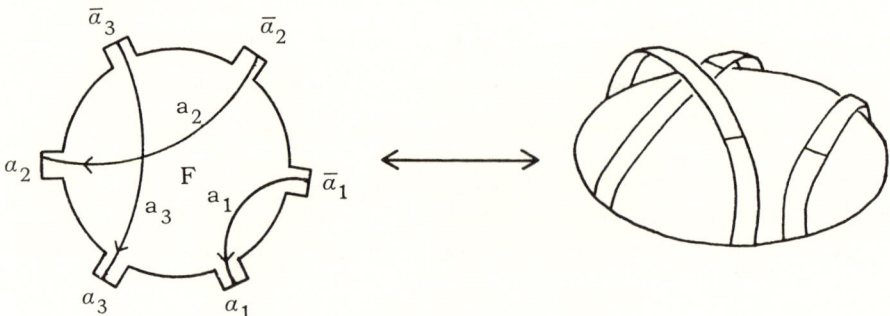

Figure 1

MAIN THEOREM. *Suppose* $V = v_1 v_2 \cdots v_n$ *and* $W = w_1 w_2 \cdots w_m$ *are cyclically reduced words (in the generators* $a_1, a_1^{-1}, \cdots, a_q, a_q^{-1}$*) which are not powers of other words. Then*

(A) *the geometric intersection number* $|V \cap W|$ *is the cardinality of the set of equivalence classes*

$$\{[j,k] \mid (j,k) \text{ is a linking pair}, 1 \leq j \leq n, 1 \leq k \leq m\}$$

where equivalence of pairs is the relation generated by

(1) $(j,k) \sim (j+1, k+1)$ *if* $v_j = w_k$

(2) $(j+1, k) \sim (j, k+1)$ *if* $v_j = w_k^{-1}$

with $j+1, k+1$ *taken modulo* n, m *respectively.*

(B) $|V| = (1/2) |V \cap V| = \text{card } \{[j,k] \mid (j,k) \text{ is a linking pair}, 1 \leq j < k \leq n\}$

where equivalence of pairs is the relation generated by (1) and (2) with the interpretation that (a,b) *is replaced by* (b,a) *if* $a > b$.

An application to the free group $F(a,b)$ *of rank 2*

Identify $F(a,b)$ with $\pi_1(T_0^2)$, where T_0^2 is the punctured torus. As noted in [B-S 1], §5, every automorphism of $F(a,b)$ can be realized by a homeomorphism of T_0^2. Thus words V, W (or pairs (V_1, V_2), (W_1, W_2)) will be equivalent under an automorphism of $F(a,b)$ only if $|V| = |W|$ (or $|V_1| = |W_1|$, $|V_2| = |W_2|$ and $|V_1 \cap V_2| = |W_1 \cap W_2|$.) Using the Main Theorem, this powerful necessary condition for equivalence can be quickly checked (see §2).

Definitions of the terms used

Definition. Let $\alpha, \beta : S^1 \to M^2$ be immersed loops in Int M^2 such that all intersections and self-intersections are transverse. Then $|\alpha \cap \beta|$ is the number of intersections of small arcs of α with small arcs of β. Similarly $|\alpha|$ is the number of transverse intersections of pairs of small arcs of α.

Caution. We allow a curve to pass through a point several times. (For example, this can happen when the curve is a geodesic [B-S 3].) Thus when we say "let $x \in M^2$ be an intersection" we mean that locally a well-defined pair of arcs through x is specified. If s arcs of α and t arcs of β cross at x, this contributes $s \cdot t$ to $|\alpha \cap \beta|$ and $s(s-1)/2$ to $|\alpha|$.

Definition. If V and W are words in $a_1, a_1^{-1}, \cdots, a_q, a_q^{-1}$ then the *intersection number* of V and W —written $|V \cap W|$ —is the minimum of $|\alpha \cap \beta|$, taken over all pairs of transverse, self-transverse maps $\alpha, \beta : S^1 \to \text{Int } M^2$ representing the free homotopy classes determined by V and W respectively. Similarly the *self-intersection number* $|V|$ is the minimum of $|\alpha|$, taken over all self-transverse maps $\alpha : S^1 \to \text{Int } M^2$ representing the free homotopy class determined by V.

The assertion in the Main Theorem that (j,k) *is a linking pair* can be understood hyperbolically, lexicographically or graphically as follows. Let
$$V_j = v_j v_{j+1} \cdots v_n v_1 \cdots v_{j-1} \quad (1 \leq j \leq n)$$
$$W_k = w_k w_{k+1} \cdots w_m w_1 \cdots w_{k-1} \quad (1 \leq k \leq m).$$

Hyperbolically. If M^2 is given a hyperbolic structure and the a_i are realized as Moebius transformations (see §3) then (j,k) is a linking pair if Axis V_j and Axis W_k meet transversely. (I.e., $\text{Fix}(V_j)$ and $\text{Fix}(W_k)$ link in ∂D^2; D^2 = the closed unit disk.)

Lexicographically. Form the infinite words $X^\infty = XX \cdots$ for $X = V_j, V_j^{-1}$, W_k or W_k^{-1}. Then (j,k) is a linking pair if, in the cyclic lexicographic ordering of Birman and Series [B-S 1],
$$V_j^\infty < (W_k^\varepsilon)^\infty < (V_j^{-1})^\infty < (W_k^{-\varepsilon})^\infty$$
holds cyclically for $\varepsilon = \pm 1$. (The definition of this cyclic lexicographic ordering is reviewed in A3 of §2.)

The "geodesic and edge-path" proof we give of the Main Theorem leads to the combinatorial algorithm of §2 because of the following fundamental proposition.

PROPOSITION ([B-S 1], Theorem A). *(j,k) is a linking pair hyperbolically if and only if it is a linking pair lexicographically.* □

Graphically. We shall see (§4) that in the graph Γ (naturally embedded in the Poincaré disk) of the group G appropriate edge paths corresponding to bi-infinite words $\cdots V_j V_j V_j \cdots$ and $\cdots W_k W_k W_k \cdots$ will meet along an arc whose vertices represent an equivalence class (Figure 2). Then

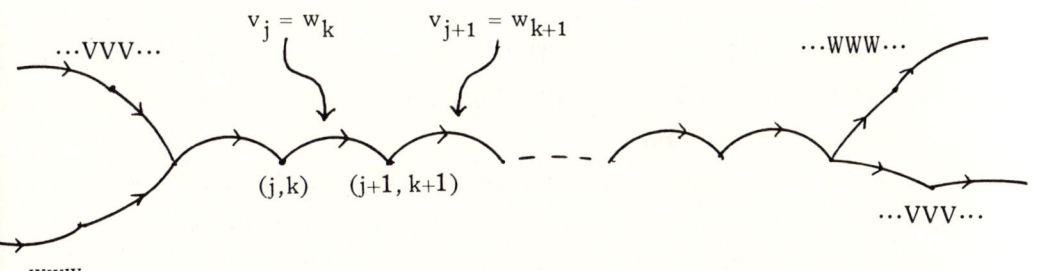

Figure 2

(j,k) is a linking pair if at the two ends of this arc the edge paths diverge in an alternating fashion. (In practice this is often the quickest and most intuitive way to decide whether (j,k) is a linking pair. It gives a direct geometric translation of the lexicographic definition.)

Historical Note. Much of the work on these matters is based on Poincaré ([P], p. 465-475) who pointed out that, upon imposing a hyperbolic structure, the crux is whether axes of conjugates of the word(s) in question cross each other. Reinhart made Poincaré's program algorithmic by demonstrating that the only conjugates which need be considered are cyclic permutations of the word(s) given. (In our terminology he proved that all intersections come from linking pairs (j.k) .) He used a representation into $PSL(2,\mathbb{C})$ and some elementary algebra to check whether axes cross. Work in a similar vein was later done by Călugareanu [Că]. Recently Hempel [He] elegantly streamlined the Reinhart analysis, giving a trace criterion for deciding linking and a representation into $PSL(2,\mathbb{Z}[\sqrt{3}])$ which only requires integer computations. Also Lehner-Sheingorn [L-S] give an algorithm for finding self-intersections of geodesics (given as axes of hyperbolic transformations) in which all computations take place in a single fundamental region of the modular surface.

Birman and Series [B-S 1] use cyclic lexicographic order to avoid representations and give a completely combinatorial algorithm for deciding whether a curve is simple. On the other hand, combinatorial algorithms which do not depend on Poincaré's viewpoint had previously been given by Zieschang [Z 1,2], who used Whitehead's work on automorphisms of free groups, and by Chillingsworth [Ch 1,2,3] who gives an algorithm involving winding numbers.

Of the above-mentioned authors only Reinhart[*] and Lehner-Sheingorn give methods for counting the number of intersections or self-intersections; the other papers deal just with whether there are any such intersections or self-intersections. Our results give the first combinatorial algorithm for counting these. Simultaneously C. Frohman [Fr] has proved results similar to ours using Nielsen's theory of boundary developments.

Turaev [T] and Turaev-Viro [T-V] give a (non-algorithmic) analysis, and present estimates for the intersection and self-intersection numbers in terms of intersection forms over the integral group ring; these result in characterizations of simple or non-intersecting pairs of loops.

---

[*]Theorem 3 on page 219-220 of [R] is not literally correct. For example, if $W_1 = b$, $W_2 = ab^{10}$ on the punctured torus with meridian a and longitude b, then the "indicating words" $B_1 = b(ab)^{-1}$ and $B_2 = b(ab^3)^{-1}$ (corresponding to our ordered pairs (1,2) and (1,4)) are not equivalent according to the criterion given there, while the reader can easily verify that $|W_1 \cap W_2| = 1$.

The realization of minimal numbers of intersections by geodesics in arbitrary Riemannian 2-manifolds and the geometric patterns in which intersections occur are studied in [F-H-S] and [H-S].

§2. *The algorithm*

A1) An ordering of the alphabet $\{a_1, a_1^{-1}, \cdots, a_q, a_q^{-1}\}$ is chosen to indicate the counterclockwise ordering of the disjoint arcs $a_1, \bar{a}_1, \cdots, a_q, \bar{a}_q$ in ∂F which are identified to form 1-handles. [E.g., in Figure 1, an admissible ordering is $a_2, a_3, a_1, \bar{a}_1, \bar{a}_2, \bar{a}_3$.]

A2) Given words $V = v_1 v_2 \cdots v_n$, $W = w_1 w_2 \cdots w_m$ (cyclically reduced, not proper powers) form the $2m+2n$ cyclic permutations
$V = V_1, V_2, \cdots V_n,\ V_1^{-1}, \cdots V_n^{-1},\ W = W_1, \cdots, W_m, W_1^{-1}, \cdots, W_m^{-1}$.

A3) Form the infinite words $X^\infty = XXX \cdots (X = V_j^{\pm 1}$ or $X = W_k^{\pm 1}$ as in A2) and put these words in "cyclic lexicographic order" according to the following rule [B-S 1]:

$$X^\infty = (x_1 x_2 \cdots) < Y^\infty = (y_1 y_2 \cdots)\ \text{iff}$$

a) $x_1 < y_1$ in the ordered alphabet of A1) or

b) $x_i = y_i$ $(1 \le i \le t)$ and $x_{t+1} < y_{t+1}$ *in the new alphabet* gotten by cyclically permuting the alphabet given in A1) so that x_t^{-1} comes first.

A4) From the ordering constructed in A3) form the set

$$\{(j,k) | (j,k)\ \text{is a linking pair},\ 1 \le j \le n,\ 1 \le k \le m\}$$

where (j,k) is a linking pair iff

$$(V_j)^\infty < (W_k^\varepsilon)^\infty < (V_j^{-1})^\infty < (W_k^{-\varepsilon})^\infty$$

holds cyclically for $\varepsilon = +1$ or $\varepsilon = -1$.

A5) Form the equivalence classes for the equivalence relation generated by the conditions

$$(j,k) \sim (j+1, k+1) \text{ if } v_j = w_k$$
$$(j+1,k) \sim (j, k+1) \text{ if } v_j = w_{k+1}$$

where $j+1, k+1$ are taken modulo m, n respectively.

A6) Let N = the number of equivalence classes. Then
    a) $|V \cap W| = N$
    b) If $V = W$ then $|V| = N/2$.

REMARKS.

I. In finding $|V|$ half the calculations can be eliminated if instead of A6 b) one modifies the algorithm: Use only $V_1, \cdots, V_n, V_1^{-1}, \cdots, V_n^{-1}$ in A2). Use only linking pairs (j,k) such that $1 \leq j < k \leq n$ in A4). Use the equivalence relation in part (B) of the Main Theorem in A5). Then $|V|$ = the number of equivalence classes found.

II. The order of A4) and A5) can be interchanged: We can first form equivalence classes in the set of *all* pairs $\{(j,k) | 1 \leq j \leq n, 1 \leq k \leq m\}$ (or if $V = W$ in $\{(j,k) | 1 \leq j < k \leq n\}$). It will transpire in §4 that (j,k) is a linking pair iff every pair in its equivalence class is a linking pair. Thus we may test one element in each equivalence class according to the rule in A4) to see if this equivalence class should be counted in the final intersection number in A6).

This algorithm has been programmed for us in Pascal.[*] Some sample intersection numbers are given in §5.

§3. *Edge paths and geodesics*

We put a hyperbolic structure on M^2 as in [B-S 1]. Thus, in the notation of §1, we specify as fundamental domain the closed 2-disk

---

[*]We would like to express our appreciation to David J. Cohen for the speed, accuracy and cheerfulness with which this was accomplished.

$F = D^2$ − (the interiors of $2q$ symmetrically placed small open disks bounded by disjoint circles orthogonal to ∂D^2).

Also, we let $a_i : \bar{a}_i \to a_i$ be Moebius transformations (with these circles as isometric circles), generating the group G. (See Figure 3.) Then G

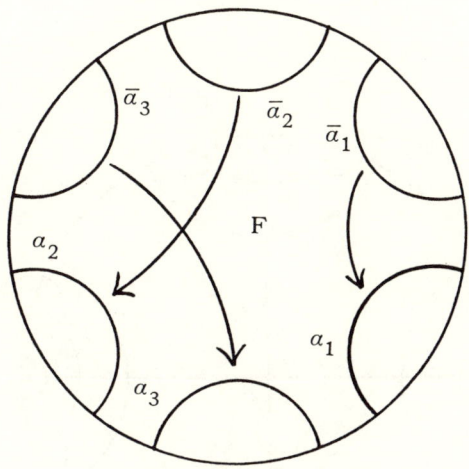

Figure 3

is the group of covering transformations of the universal covering $p : \tilde{M} \to M^2$ where

$$\tilde{M} = \cup \{g(F) | g \in G\} \subset D^2$$

$$\text{Int } \tilde{M} = \text{Int } D^2 .$$

The choice of F implies that G is purely hyperbolic [Be, p. 225]. We will repeatedly use the elementary facts that, if $g, W \in G$,

$$\text{Axis } (gWg^{-1}) = g(\text{Axis } W) ,$$

$$pg(\text{Axis } W) = p(\text{Axis } W) = \gamma(W) ,$$

where $\gamma(W)$ is the unique geodesic in the free homotopy class $[W]$.

We realize Γ (the graph of the group G) as a graph in $\tilde{M}$ as follows: Let 0 be a point in the basic fundamental domain F (say 0 = origin). Then

vertices of $\Gamma = \{g0 | g \epsilon G\}$

edges of $\Gamma = \{\text{geodesic arcs } [g0, ga_i^\epsilon 0] | g \epsilon G, \epsilon = \pm 1\}$.

It is often convenient to direct an arc from $g0$ to $ga_i^\epsilon 0$ and to label it "a_i^ϵ" (Figure 4).

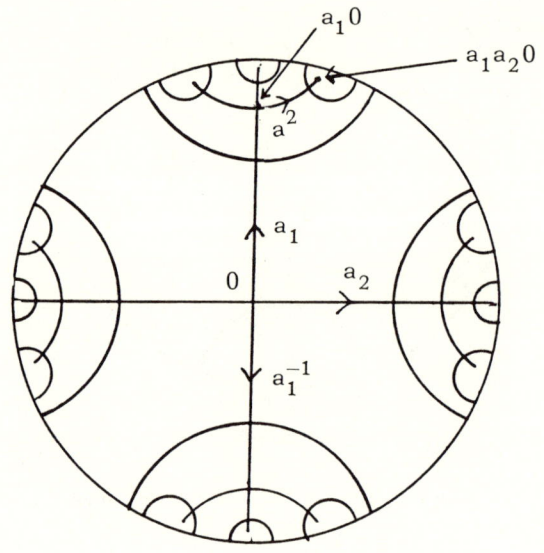

Figure 4

Note that, if $h \epsilon G$, then h takes each labelled arc $[g0, ga_i 0]$ to an arc $[hg0, hga_i 0]$ with the same label. Conversely, given arcs $[g0, ga_i 0], [f0, fa_i 0]$ with the same label a_i, there exists $h \epsilon G$ taking the first to the second (viz. $h = fg^{-1}$).

To compare lines in Γ with hyperbolic lines we need the

Definition. Let L_1 and L_2 be topological lines in Int D^2 which are the interiors of properly embedded arcs $\overline{L}_1, \overline{L}_2$ in D^2. (Such lines are called *proper lines* and we denote $\partial L = \bar{L} - L$.) We say that L_1 and L_2 are *parallel at infinity* — written $L_1 \| L_2$ — iff $\partial L_1 = \partial L_2$.

Note that there exists exactly one hyperbolic line L_2 which is parallel at infinity to a given proper line L_1. The case where $L_2 =$ Axis X for some $X \in G$ occurs in our context as follows. Let $a: I \to M^2$ be a loop in M^2 with lift $a_0: I \to \tilde{M}$. Let a_1 be the lift with $a_1(0) = a_0(1)$ and, in general, for $n \in Z$ let a_n be the lift with $a_n(0) = a_{n-1}(1)$. Then, if $L_1 = \underset{n}{\cup} a_n$ is an arc and $X \in G$ is the element with $X(a_n(0)) = a_n(1)$ then $L_2 = $ Axis X satisfies $L_1 \| L_2$.

To every vertex Q of Γ and every cyclically reduced word W in the generators $a_1, a_1^{-1}, \cdots, a_q, a_q^{-1}$ we associate two proper lines, an edge-path $E(Q,W)$ and a hyperbolic line $Ax(Q,W)$:

(1) $E(Q,W) \equiv$ the bi-infinite edge-path in Γ which starts at Q and reads $WWW\cdots$ in one direction and $W^{-1}W^{-1}\cdots$ in the other direction.

(2) $Ax(Q,W) \equiv$ the unique hyperbolic line with $Ax(Q,W) \| E(Q,W)$.

(2) specifies the hyperbolic line in a way which we can also use in Part II, where edge paths are a bit more complicated. But we can be more explicit here: Note that, if $Q = g0$, then

$$\partial E(Q,W) = \lim_{n \to \pm\infty} gW^n(0) = \lim_{n \to \pm\infty} (gWg^{-1})^n(0) = \partial(\text{Axis } gWg^{-1}).$$

Thus, by (2), $Ax(Q,W) = Axis(gWg^{-1})$ and $p(Ax(Q,W)) = \gamma(W)$.

Here are the properties of $Ax(Q,W)$ and $E(Q,W)$ which (except for P5) will be used in the proof of the Main Theorem. We formalize these properties so that, once their analogues are verified (for slightly more complicated edge paths) in Part II, the proof of the corresponding theorem won't have to be repeated.

THE BASIC PROPERTIES

P1: a) $(f \in G) \Longrightarrow f(E(Q,W)) = E(fQ,W)$
    b) $(f \in G) \Longrightarrow f(Ax(Q,W)) = Ax(fQ,W)$
    c) $E(Q,W) = E(Q',W') \Longrightarrow Ax(Q,W) = Ax(Q',W')$.

P2: If Q' is a vertex of $E(Q,W)$ then $E(Q,W) = E(Q',W_k)$ for some cyclic permutation W_k of W.

P3: If $X \in G$ there is a unique edge path $E(X)$ in Γ which is a proper line and satisfies $E(X) \| $ Axis X. Indeed, if W is a cyclically reduced word representing the free homotopy class determined by X then $E(X) = E(Q,W)$ for some vertex Q and Axis $X = Ax(Q,W)$.

P4: If $X, Y \in G$ and $E(X) \cap E(Y) \neq \emptyset$ then $E(X) \cap E(Y)$ is either an arc or a point.

P5: If $X \in G$ then $E(X)$ and Axis X traverse exactly the same sequence of fundamental domains in $\widetilde{M}$.

REMARK. While P5 is not used in the proof of the Main Theorem it explains (with P2 and P3) the intuitive content of the theorem: An equivalence class $(j_1, k_1) \sim \cdots \sim (j_s, k_s)$ corresponds to a maximal chain of fundamental domains through which two axes which cover $\gamma(V)$ and $\gamma(W)$ run together. In one of these they cross, and this crossing projects to an element of $\gamma(V) \cap \gamma(W)$.

Verification of properties P1-P5

    P1: Since f preserves labels of arcs, $fE(Q,W)$ is the bi-infinite edge-path starting at fQ and reading $WWW \cdots$ and $W^{-1}W^{-1} \cdots$ in opposite directions. Thus $fE(Q,W) = E(fQ,W)$, proving *P1a)*. Assertions *P1b)* and *P1c)* follow from the defining property (2) of $Ax(Q,W)$.

    P2 is immediate from the definition of $E(Q,W)$.

    P3: Suppose $E_1 \|$ Axis X and $E_2 \|$ Axis X; so $E_1 \| E_2$. If Q is a vertex of E_1 then Q must be a vertex of E_2. (For otherwise one of the ends of E_1 would be separated in the tree Γ from all of E_2. The limit of this end would then not be a limit point of E_2, since ends of the tree Γ correspond to distinct points of S^1. Hence we would not have $E_1 \| E_2$.) Thus {vertices of E_1} = {vertices of E_2}. Since there exists a unique arc between any two vertices of Γ, we have $E_1 = E_2$ proving uniqueness. Now $X = gWg^{-1}$ for W cyclically reduced, $g \in G$. Then $\partial E(g0,W) = \lim_{n \to \pm\infty} gW^n g^{-1}(0) = \text{Fix}(X)$. So $E(X) = E(g0,W)$ satisfies $E(X) \| \text{Axis } X$.

P4 is clear since Γ is a tree and $E(X)$ and $E(Y)$ are proper lines in Γ.

P5: Since Axis X meets each fundamental domain gF in a hyperbolic arc, it enters and leaves through different faces. (Axis X can only meet an arc $g\alpha_i$ (or similarly $g\bar{\alpha}_i$) transversely, since otherwise the arc α_i is invariant under $g^{-1}Xg$. This contradicts the fact that the arcs α_i project to arcs and not loops in M because the only identifications on F induced by G are those induced by $a_1, \cdots, a_q$ [Fe, p. 128].) Construct the edge-path $E_1(X)$ which contains the geodesic arc $[g0,h0]$ whenever Axis X crosses from gF to hF. Clearly $E_1(X) \| \text{Axis } X$. Thus, by P3, $E(X) = E_1(X)$. □

§4. *Proof of the theorem*

We use Properties P1-P5 of the previous section to prove our Main Theorem. Suppose $V = v_1 \cdots v_n$ and $W = w_1 \cdots w_m$ are cyclically reduced words which are not proper powers.

In the axiomatic setting of P1-P5 we define a pair (j,k) to be a *linking pair* $(i \leq j \leq n, 1 \leq k \leq m)$ if, for any vertex Q of Γ we have

$$\partial E(Q,V_j) \text{ links } \partial E(Q,W_k) \text{ in } \partial D^2.$$

This is independent of Q by P1a). (This is the "graphically" given definition of §1. It is equivalent to that given "hyperbolically" since it's equivalent to $Ax(Q,V_j)$ and $Ax(Q,W_k)$ meeting transversely, where $Ax(Q,V_j) = g(\text{Axis } V_j)$ and $Ax(Q,W_k) = g(\text{Axis } W_k)$ if $Q = g0$.)

Proof of (A) when V and W are not conjugate

The intersection numbers we seek are realized by geodesics: $|V \cap W| = |\gamma(V) \cap \gamma(W)|$ and $|V| = |\gamma(V)|$. (The proof of this, extending the argument of [P, p. 468], is omitted since it is simpler but similar to the proof that $|V \cap V| \geq 2|V|$ which we give in the proposition at the end of this section. Alternatively see [F-H-S].) Thus our problem reduces to counting $|\gamma(V) \cap \gamma(W)|$ and $|\gamma(V)|$. This reduction is only valid because the words are not proper powers and are not conjugate, whence $\gamma(V)$ and $\gamma(W)$ are transverse, self-transverse maps. (For example, if $V = a_1^2$ then $|V \cap V| = 0 \neq 2 = 2|V|$.)

We denote

$$\hat{S} = \{(j,k) | (j,k) \text{ is a linking pair, } 1 \leq j \leq n, 1 \leq k \leq m\}$$

$$S = \{[j,k] | (j,k) \text{ is a linking pair, } 1 \leq j \leq n, 1 \leq k \leq m\}$$

$$= \hat{S}/\sim, \text{ where "} \sim \text{" is defined in the Main Theorem.}$$

Define $\hat{\phi} : \hat{S} \to \gamma(V) \cap \gamma(W)$ by

$$\hat{\phi}(j,k) = p(Ax(Q,V_j)) \cap Ax(Q,W_k)) .$$

This is independent of Q, by P1b). Clearly $\hat{\phi}(j,k) = p(\text{Axis } V_j \cap \text{Axis } W_k) \subset \gamma(V) \cap \gamma(W)$, since V_j and W_k are conjugate to V and W respectively.

Define $\phi : S \to \gamma(V) \cap \gamma(W)$ by

$$\phi[j,k] = \hat{\phi}(j,k) .$$

We shall show that ϕ is a bijection.

ϕ *is well-defined:* Let $[Q,P]$ be an edge labelled $v_j (= w_k^{-1})$. Then the edge $[Q,P]$ lies on both $E(Q,V_j)$ and $E(Q,W_{k+1})$. (Figure 5.) Hence $E(Q,V_j) = E(P,V_{j+1})$ and $E(P,W_k) = E(Q,W_{k+1})$. Thus, by P1c), $\hat{\phi}(j,k+1) = \hat{\phi}(j+1,k)$. [The case $v_j = w_k$ is similar but easier.] Therefore equivalent pairs go to the same element of $\gamma(V) \cap \gamma(W)$ under $\hat{\phi}$ and ϕ is well defined.

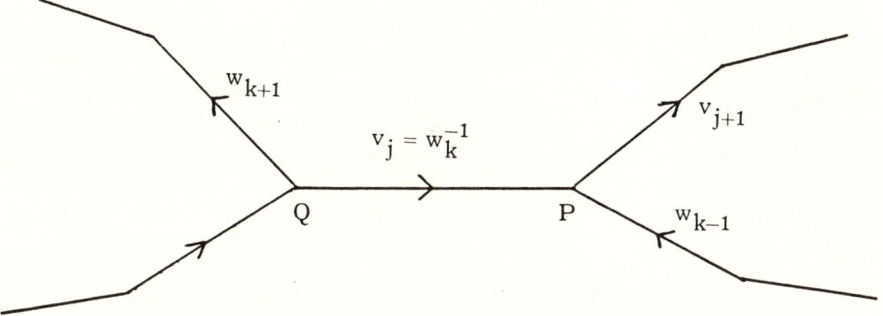

Figure 5

ϕ *is onto:* (Reinhart [R]; Birman-Series [B-S 1]). Let an intersection of $\gamma(V)$ and $\gamma(W)$ be given by two arcs crossing at $x_0 \in M^2$. Let $z_0 \in p^{-1}(x_0)$ and let Axis X and Axis Y be the transverse geodesics through z_0 whose images contain these arcs. Now $E(X) \| \text{Axis X}$ and $E(Y) \| \text{Axis Y}$ *(P3)*; so $\partial E(X)$ links $\partial E(Y)$. Then there is a vertex $Q \in E(X) \cap E(Y)$, and, by P3 and P2, $E(X) = E(Q,V_j)$, $E(Y) = E(Q,W_k)$ for some (j,k). Thus $\hat{\phi}(j,k) = p(Ax(Q,V_j) \cap Ax(Q,W_k)) = p(\text{Axis X} \cap \text{Axis Y}) = x_0$.

REMARK. One can use P5 to make this more precise: If M^2 is viewed as a disk F with handles and if the intersection x_0 is reached (starting in F) after $\gamma(V)$ has crossed over j handles and $\gamma(W)$ has crossed over k handles then $x_0 = \hat{\phi}(j,k)$.

ϕ *is one-one:* Suppose $x_0 = \phi[j_1,k_1] = \phi[j_2,k_2]$. Fix any vertex Q of Γ. Thus $x_0 = p(z_i)$ where $z_i = Ax(Q,V_{j_i}) \cap Ax(Q,W_{k_i})$, $i = 1,2$. Since the pair $Ax(Q,V_{j_i}), Ax(Q,W_{k_i})$ meets a neighborhood of z_i in a pair of arcs covering the two given arcs through x_0, there is an element $f \in G$ with

$$f(z_2) = z_1, \; f(Ax(Q,V_{j_2})) = Ax(Q,V_{j_1}), \; f(Ax(Q,W_{k_2})) = Ax(Q,W_{k_1}).$$

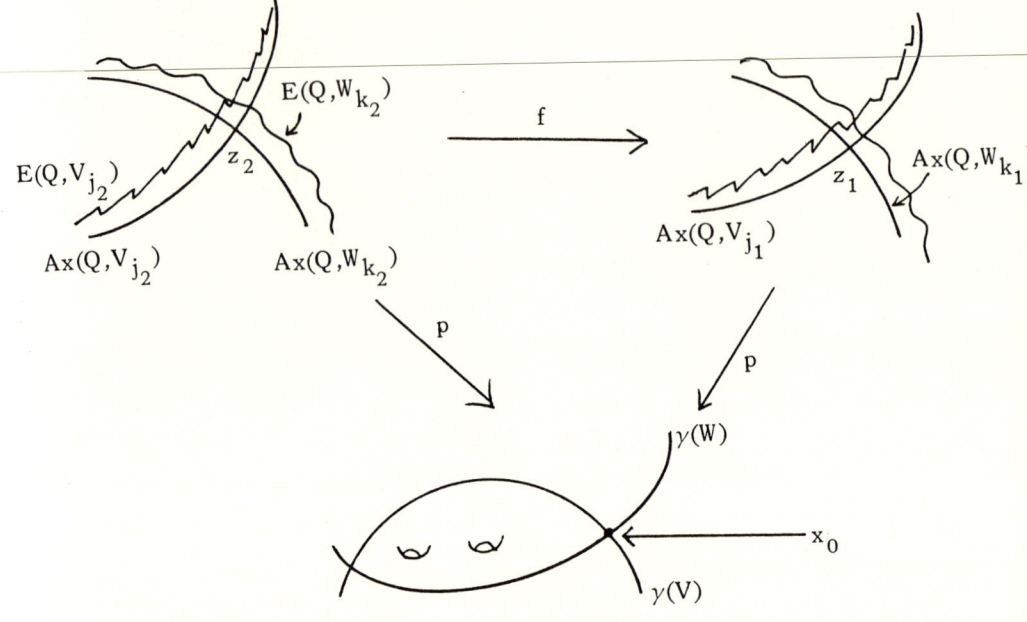

Figure 6

From Properties P1a) and P3 we also have (see Figure 6)

$$E(fQ, V_{j_2}) = fE(Q, V_{j_2}) = E(Q, V_{j_1})$$
$$E(fQ, W_{k_2}) = fE(Q, W_{k_2}) = E(Q, W_{k_1}) .$$

Now Q and $f(Q)$ are vertices of $E(Q, V_{j_1}) \cap E(Q, W_{k_1})$. This intersection is a (possibly degenerate) arc, by P4. As an arc in $E(Q, V_{j_1}) = E(fQ, V_{j_2})$ it is of the form shown in Figure 7(i), assuming (WLG) that the arc is directed from Q to $f(Q)$. Similarly, as an arc of $E(Q, W_{k_1}) = E(fQ, W_{k_2})$, this arc has labelling (ii) or (iii) of Figure 7.

In case (ii): $v_{j_1} = w_{k_1}, v_{j_1+1} = w_{k_1+1}, \ldots, v_{j_2-1} = w_{k_2-1}$

Thus $\qquad (j_1, k_1) \sim (j_1+1, k_1+1) \sim \cdots \sim (j_2, k_2) .$

In case (iii): $v_{j_1} = w_{k_1-1}^{-1}$, $v_{j_1+1} = w_{k_1-2}^{-1}, \ldots, v_{j_2-1} = w_{k_2}^{-1}$.

Thus $(j_1, k_1) \sim (j_1+1, k_1-1) \sim (j_1+2, k_1-2) \sim \cdots \sim (j_2-1, k_2+1) \sim (j_2, k_2)$.

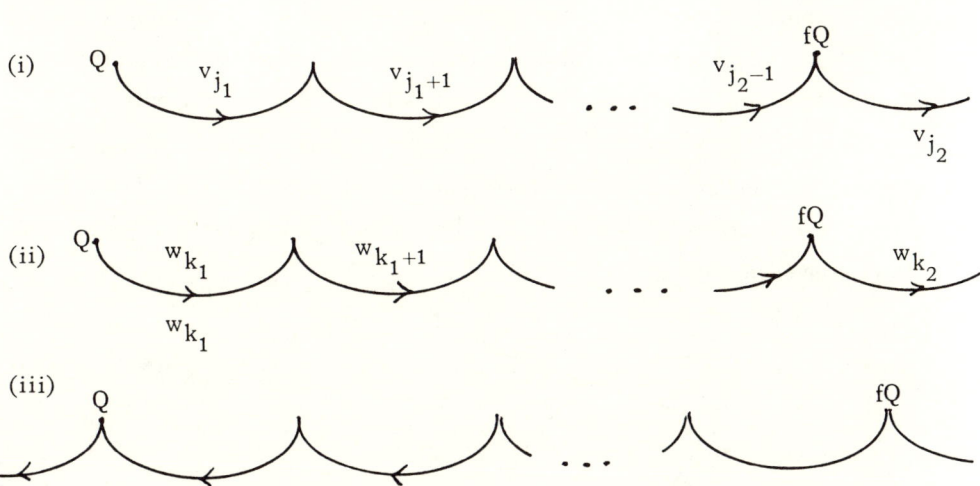

Figure 7

Therefore ϕ is one-one and (A) is proved in the case where V and W are not conjugate.

Proof of (B) and of (A) when V is conjugate to W

PROPOSITION. *If V is not a proper power, let $\gamma = \gamma(V)$ be the geodesic representing V and let γ_1 be a curve in the tubular neighborhood of γ which is ε-parallel to γ. Then, for sufficiently small $\varepsilon > 0$, $|V \cap V| = |\gamma \cap \gamma_1| = 2|V|$.*

Sufficiency of the proposition. If V is conjugate to W, their representative curves are freely homotopic; $\gamma(V) = \gamma(W)$ and the preceding proof does not apply, since $\gamma(V)$ and $\gamma(W)$ are not in general position. But by the Proposition, $|V \cap W| = |V \cap V| = |\gamma \cap \gamma_1|$. The infinitely iterated lifts of γ_1 —denoted $\{L_g(\gamma_1)\}$ —are not hyperbolic lines but they are parallel to the hyperbolic lines $\{\text{Axis } gWg^{-1}\}$ covering γ, and the argument when V,W are not conjugate will go through with $\{L_g(\gamma_1)\}$ playing the role of $\{\text{Axis } gWg^{-1}\}$. This proves (A) when V and W are conjugate.

To prove (B) we note (Figure 8) that, because ε is sufficiently small, each self-intersection of γ leads to precisely two intersections —corresponding to $[j,k]$ and $[k,j]$ —of $\gamma \cap \gamma_1$. By the proposition $|\gamma \cap \gamma_1| = 2|V|$, so there are no other intersections. A pair (j,j) is never a linking pair (since $(\text{Axis } W_j \| L_g(\gamma_1)$ if $gWg^{-1} = W_j$). Thus $|V|$ is in 1-1 correspondence with $\{[j,k]|(j,k)$ is a linking pair, $1 \leq j < k \leq n$ as claimed in (B).

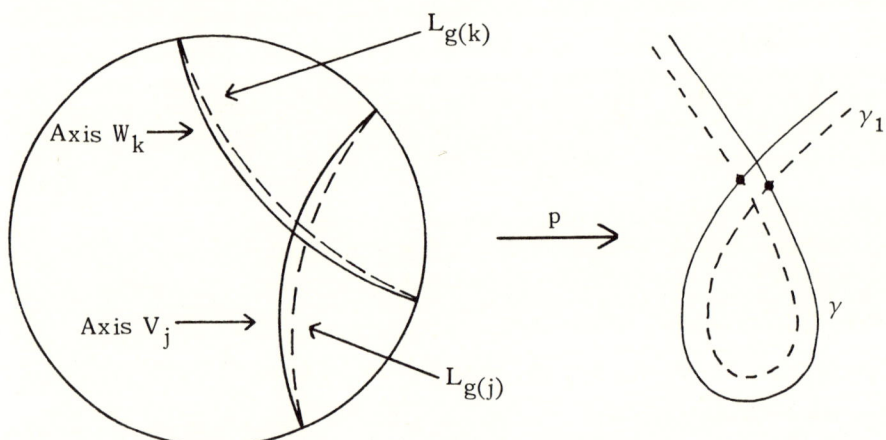

Figure 8

Proof of the proposition. Clearly, for sufficiently small $\varepsilon > 0$, the ε-parallel path is a closed path homotopic to γ (since M^2 is orientable). γ and γ_1 are transverse and self-transverse with $|\gamma \cap \gamma_1| = 2|\gamma|$. (See Figure 8.) Thus, since $|V| = |\gamma|$, we have $|V \cap V| \leq 2|V|$. To show that $|V \cap V| \geq 2|V|$ we take arbitrary loops α, β in general position realizing $|V \cap V|$ and show that $|\alpha \cap \beta| \geq 2|\gamma|$.

Given a self-intersection x_0 of γ —that is a pair of small subarcs meeting at x_0 —choose a lift $z_0 \in p^{-1}(x_0)$, and hyperbolic lines Axis gVg^{-1}, Axis hVh^{-1} through z_0 which in a small neighborhood of z_0 project to these arcs. Then a homotopy $\alpha \simeq \gamma$ yields lifts of α which (iterated) give proper paths $\widetilde{\alpha}_g \| \text{Axis}(gVg^{-1})$ and $\widetilde{\alpha}_h \| \text{Axis}(hVh^{-1})$. (We use the notation of §3 even though $\widetilde{\alpha}_g, \widetilde{\alpha}_h$ may not be arcs.) Similarly we get paths $\widetilde{\beta}_g, \widetilde{\beta}_h$ (Figure 9). Then $(\widetilde{\beta}_g \cap \widetilde{\alpha}_h) \neq \emptyset \neq (\widetilde{\beta}_h \cap \widetilde{\alpha}_g)$ since Fix(gVg^{-1}) links Fix(hVh^{-1}). Choose an intersection (necessarily transverse since α and β are in general position) in each case; let z,w be these intersections. The small arcs intersecting at z and w cannot project to the same pair of intersecting arcs in M^2. [For suppose they did. Because α, β are self-transverse the covering transformation f with $f(z) = w$ satisfies $f(\widetilde{\alpha}_h) = \widetilde{\alpha}_g$ and $f(\widetilde{\beta}_g) = \widetilde{\beta}_h$. But then $f(\text{Fix } gVg^{-1}) = \text{Fix}(hVh^{-1})$ and $f(\text{Fix } hVh^{-1}) = \text{Fix}(gVg^{-1})$; so f interchanges the axes of gVg^{-1} and hVh^{-1}. Therefore f fixes their intersection point z_0. Thus f = identity and these axes are equal! This contradicts the fact that Axis(gVg^{-1}) and Axis(hVh^{-1}) are transverse.] Thus the self-intersection x_0 of γ determines two of $\alpha \cap \beta$.

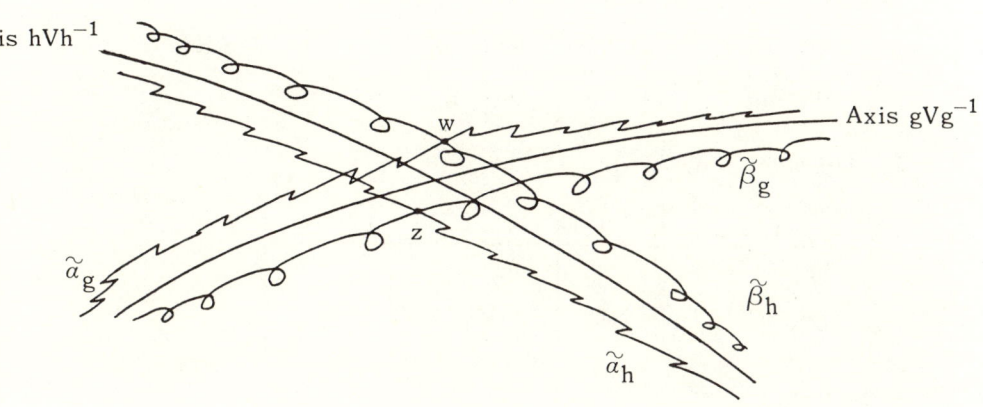

Figure 9

A different self-intersection y_0 of γ would determine axes of conjugates $g_1 V g_1^{-1}$ and $h_1 V h_1^{-1}$ such that $\{Fix(g_1 V g_1^{-1}), Fix(h_1 V h_1^{-1})\}$ is not equivalent to $\{Fix(gVg^{-1}), Fix(hVh^{-1})\}$ by a covering transformation. In this case no element of $\tilde{\alpha}_{g_1} \cap \tilde{\beta}_{h_1}$ or $\tilde{\alpha}_{h_1} \cap \tilde{\beta}_{g_1}$ can project to either of the two elements of $\alpha \cap \beta$ chosen using x_0, since if it did a covering map would have to take one pair of fixed point sets to the other. Thus distinct self-intersections of γ determine disjoint pairs of elements of $\alpha \cap \beta$. Hence $|\alpha \cap \beta| \geq 2|\gamma|$. $\square$

§5. Some intersection numbers

A. *Alphabet:* a, b, a^{-1}, b^{-1}

V	W	|V|	|W|	|V ∩ W|	
$ab^2a^{-1}b^{-3}$	$ba^2b^{-1}a^{-3}$	3	3	13	
a^3b^7	a^5b^{13}	12	48	54	See Remark 1
a^6b^{15}	$a^{10}b^8$	70	63	172	

B. *Alphabet:* $a, b, a^{-1}, b^{-1}, c, d, c^{-1}, d^{-1}$, {Let $X = b^{-1}aba^{-1}$, $Y = cd^{-1}c^{-1}d$}

V	W	|V|	|W|	|V ∩ W|	
cXc^2d^2Yc	cXc^2d^2Xc	11	9	20	
$a^{70}b^{-3}c^{17}$		292			See Remark 2
$a^3ba^{-1}c^2$	b^2c	2	1	6	

C. *Alphabet:* $a, b, c, a^{-1}, b^{-1}, c^{-1}$

V	W	|V|	|W|	|V ∩ W|
$a^5ba^{-1}c^{28}b^{-9}$		408		
$a^3ba^{-1}c^2$	b^2c	8	0	11

REMARKS.

1) Computations like this have led to a proof of the proposition:
   In the a, b, a^{-1}, b^{-1} alphabet let $V = a^p b^q$, $W = a^r b^s$.

Then $|V \cap W| = (p-1)(s-1) + (q-1)(r-1) + \Delta$

where $\Delta = \begin{cases} p-r+s-q & \text{if } p > r, s \geq q \\ p-r+q-s-2 & \text{if } p > r, q > s. \end{cases}$

2) In the closed manifold given by these identifications on an octagon the basic relation read in the graph of the group is $b^{-1}aba^{-1}d^{-1}cdc^{-1} = 1$; i.e., $X = Y$. Thus V and W are equal in the closed manifold while they have different self-intersection numbers in the punctured manifold.

3) An exercise for the reader: How can it be on the punctured torus that $|a^5 b^{13}| = 48$, when everybody knows that, on a closed torus, $a^p b^q$ with p,q relatively prime is represented by a simple closed curve? Give the path of a simple closed curve on the torus realizing the homotopy class $a^5 b^{13}$.

MARSHALL COHEN
DEPARTMENT OF MATHEMATICS
CORNELL UNIVERSITY
ITHACA, NEW YORK 14853

MARTIN LUSTIG
DEPARTMENT OF MATHEMATICS
M.I.T.
CAMBRIDGE, MASSACHUSETTS 02139

BIBLIOGRAPHY

[Be] A. F. Beardon, *The geometry of discrete groups*, GTM 91, Springer-Verlag New York (1983).

[B-S 1] J. S. Birman and C. Series, *An algorithm for simple curves on surfaces*, J. London Math Soc. 29 (1984), 331-342.

[B-S 2] _____, *Dehn's algorithm revisited, with applications to simple curves on surfaces*, preprint.

[B-S 3] _____, *Geodesics with multiple self-intersections and symmetries on Riemann surfaces*, preprint.

[Ca] G. Călugăreanu, *Sur les courbes fermées simples tracées sur une surface fermée orientable*, Mathematica Vol. 8 (31) 1, 1966, 29-38.

[Ch 1] D. R. J. Chillingsworth, *Simple closed curves on surfaces*, Bull. London Math. Soc. 1 (1969), 310-314.

[Ch 2] _____, *An algorithm for families of disjoint simple closed curves on surfaces*, Bull. London Math. Soc. 3 (1971), 23-26.

[Ch 3]   D. R. J. Chillingsworth, *Winding numbers on Surfaces, II*, Math. Ann. 199 (1972), 131-153.

[Fe]   R. A. Fenn, *Techniques of geometric topology*, L.M.S. Lecture Note Series 57, Cambridge University Press 1983.

[F-H-S]   M. H. Freedman, J. Hass and P. Scott, *Closed geodesics on surfaces*, Bull. London Math. Soc. 14 (1982), 385-391.

[Fr]   C. Frohman, *An algorithm for computing geometric intersection numbers*, preprint.

[H-S]   J. Hass and P. Scott, *Intersections of curves on surfaces*, Israel J. Math 51 (1985), 90-120.

[He]   J. Hempel, *Traces, lengths and simplicity for loops on surfaces*, Topology Appl. 18 (1984), 153-161.

[L-S]   J. Lehner and M. Sheingorn, *Computing self-intersections of closed geodesics on finite-sheeted covers of the modular surface*, preprint (submitted for publication).

[P]   H. Poincaré, *Cinquième complement à l'analysis situs*, Oevres de Henri Poincaré, vol. VI, 435-498.

[R]   B. L. Reinhart, *Algorithms for Jordan curves on compact surfaces*, Annals of Math. 75 (1962), 209-222.

[T]   V. G. Turaev, *Intersections of loops in two-dimensional manifolds*, Mat. Sb. 106 (148) (1978), 566-588; English transl. in Math. USSR Sb. 35 (1979), no. 2.

[T-V]   V. G. Turaev and O. Ya. Viro, *Intersections of loops in two-dimensional manifolds, II. Free loops*, Mat. Sb. 121 (163) (1983), English transl. in Math. USSR Sb. 49 (1984), no. 2, 357-366.

[Z 1]   H. Zieschang, *Algorithmen für einfache Kurven auf Flächen*, Math. Scand. 17 (1965), 17-40.

[Z 2]   _____, *II.*, Math. Scand. 25 (1969), 49-58.

PATHS OF GEODESICS AND GEOMETRIC INTERSECTION NUMBERS: II

Martin Lustig[*]

§0. *Introduction*

This paper extends the work described in Part I ([5]), the previous paper in this volume. We assume M^2 to be a closed orientable surface provided with a hyperbolic structure. The latter is given by a rotationally symmetric tesselation graph N of the open unit disc, corresponding to some "geometric" presentation of $\pi_1 M^2$.

The main results (given in §4 and §5 respectively) are algorithms which determine, for given words $V, W \in \pi_1 M^2$:

1. the path of a closed geodesic $\gamma(W)$ representing the free homotopy class [W]. It is described in terms of the sequence of intersections of $\gamma(W)$ with the image of the tesselation graph N in M^2.
2. the geometric intersection number $|V \cap W|$ as defined in Part I. (Actually there is a stronger result which may have applications for computational purposes concerning Heegard splittings of 3-manifolds, see Problem 5.5.)

The algorithms are (as in Part I) purely combinatorial, involving only the presentation of $\pi_1 M^2$. However we want to stress the fact that the underlying ideas are geometric and are more easy accessible than their combinatorial analogues.

It is well known that in each free homotopy class [W] of closed curves on M^2 there is exactly one geodesic representative $\gamma(W)$. In contrast to the case of bounded surfaces there is no algebraically well

---

[*] partially supported by the Studienstiftung des Deutschen Volkes.

distinguished word in the generators of $\pi_1 M^2$ which describes the path of the geodesic $\gamma(W)$; in general even shortest words are far from being unique within their free homotopy class (see [3]). The following example might help to illustrate the typical ambiguity (see Figure 0 for a schematic representation):

Using $<a,b,c,d\,|\,ab\overline{ab}cd\overline{cd}>$ (here and in the following $\bar{x}$ means x^{-1}) as presentation for the fundamental group of a sphere with two handles, the words $W_1 = c\overline{abab}cab\bar{a}^2\overline{bc}^3$, $W_2 = c\bar{a}^2dc\overline{d}ab\bar{a}^2\overline{bc}^3$, $W_3 = c\bar{a}^2dc^2\overline{dc}\overline{b}ab\overline{c}^3$, $W_4 = c\bar{a}^2dc^2\overline{dc}adcd\overline{c}^2$ represent the same free homotopy class $[W_i]$. They are related to each other through replacements of a subword equal to half of the relator $ab\overline{ab}cd\overline{cd}$ by its complementary half.

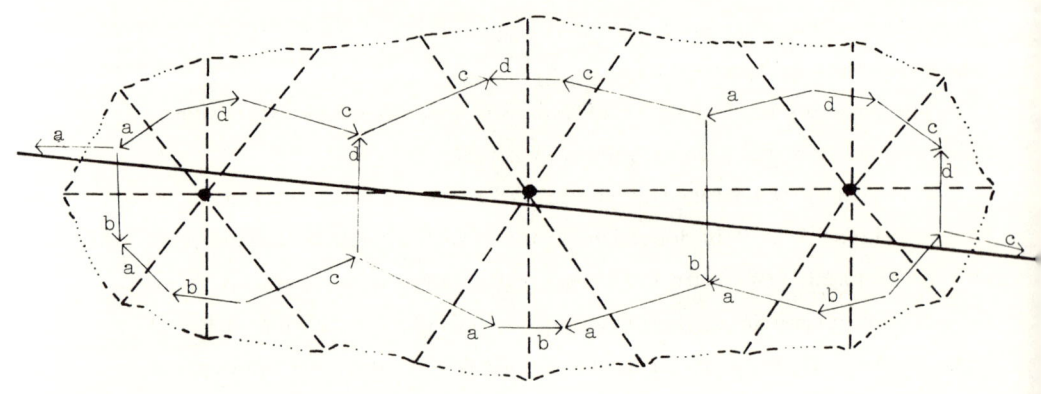

- - - - tesselation graph N
———→ dually embedded group graph
——— hyperbolic line γ (lift of $\gamma(W_i)$)

Figure 0

For each such possible replacement the geodesic $\gamma(W_i)$ has the "choice" of passing to the left or to the right of the tesselation vertex (or crossing over it). Though neither of the possible choices seems to have an immediate preference it turns out (§4) that a lift γ of $\gamma(W_i)$ indeed crosses through the tesselation as indicated in Figure 0. More generally it follows (see Remark 4.5 for a detailed statement) that, in similar situations, the determination of the geodesic path depends on the remainder of the given word only for the middle vertex.

Sections 2 and 3 are entirely devoted to the development of some new techniques needed to overcome the difficulties described above. Since they might be of interest in their own right I summarize them briefly: Introducing additional vertices and edges we enlarge the graph N (§1). In the bigger graph thus obtained we can define "normal" edgepaths (§2) which approximate hyperbolic lines closely. It is shown (Lemma 2.3 and Theorem 2.9) that they are uniquely determined by their endpoints, thus eliminating ambiguities of the type described above. In §3 the combinatorics of the words read off from normal edgepaths are developed using a correspondingly enlarged alphabet. Theorem 3.5 gives a combinatorial characterization, and the algorithm in Theorem 3.7 assures the existence of normal edgepaths for any free homotopy class. This yields a 1-1 correspondence between geodesics on the closed surface M^2 and bi-infinite words in the enlarged alphabet which respect the given restrictions. (The extension to non-closed infinite geodesics is not given here but can be easily carried out.)

The methods and the corresponding results developed in this paper apply also to non-orientable surfaces; a more general attempt to carry them over to arbitrary Fuchsian groups should be successful at least for the case where the tesselation net is mapped onto itself by the reflection through any of its vertices and where the fundamental domain has more than four sides.

I would like to thank Marshall M. Cohen for his constant encouragement and for his advice on the presentation of the material and on the use of the English language. I am also indebted to Joan S. Birman for her helpful comments. Last but not least I want to express my gratitude to Isabella Malara for the time and energy she spent on the drawings.

§1. Preliminaries

Throughout this paper we assume M^2 to be a closed orientable surface of genus $q/2 \geq 2$. Let

$$<a_1, a_2, \cdots, a_q | R>$$

be some *geometric presentation* of $\pi_1 M^2$; i.e. the corresponding standard one vertex 2-complex is homeomorphic to M^2.

A partial word of length s of some cyclic permutation of the relator R (or its inverse) is called a *positive* (or *negative*) s-cycle. We want to stress the fact that for geometric presentations of $\pi_1 M^2$ a cycle is well defined given its length, its sign and its first or last symbol.

Corresponding to the given presentation of $\pi_1 M^2$ we provide (1-3 below) M^2 with an explicit hyperbolic structure:

(1) Tesselate the interior of the unit disc D^2 by hyperbolically isometric regular 2q-polygons, obtaining a net N that consists of hyperbolic lines which are disjoint or have as intersection angle a multiple of π/q.

(2) Embed the group graph Γ of $\pi_1 M^2$ corresponding to the generators $a_1, \cdots, a_q$ into D^2, using the centers of the 2q-polygons of N as vertices and geodesic segments as edges (which then are orthogonal to those of N). Every vertex of N is then surrounded by a cycle of edges in Γ that reads off the relator R in counterclockwise order (WLG). The 2q-polygon of N containing the origin of Γ will be called F.

(3) The natural operation of $\pi_1 M^2$ on Γ extends to a hyperbolic action on D^2 (thus identifying Int D^2 with the universal cover $\widetilde{M}^2$) with fundamental domain F (or equivalently gF for all $g \in \pi_1 M^2$); the obtained pairing of the sides of F corresponds to the set of generators $a_1, \cdots, a_q$ (given by the labels of the perpendicular edges in Γ).

We label and direct the edges of N by symbols $\alpha_1, \cdots, \alpha_q$ such that an edge $a_i \in \Gamma$ is mapped onto $\alpha_i \in N$ by a counterclockwise rotation of $\frac{\pi}{2}$ around its center (see Figure 1.1). We adapt the convention that one might simultaneously reverse the direction of an edge and invert its label. The author would like to warn the inexperienced reader: in general the word in the a_i's read off from the boundary of F is quite different from the word in the a_i's given by the relator R.

We introduce further edges (i.e. geodesic segments) joining a vertex P of Γ with a neighboring vertex Q of N, oriented from P to Q and labelled h_{xy} with $x,y \in \{a_i, \bar{a}_i\}$ such that x, h_{xy}, y are successive edges radiating from P in counterclockwise order. Note that each of the two subscripts of h_{xy} determines the other; thus we denote the edge sometimes by h_{x*} or h_{*y} (or even h_{**} if the subscripts are irrelevant).

Let $\Delta = \Gamma \cup N \cup \{h_{xy}\}$ be the graph consisting of all the edges and vertices introduced above. The intersection points of edges a_i and a_j are not in Δ. We shall call them *virtual vertices* of Δ. Two distinct edges of Δ (and also their labels) are called *diagonally opposed* if they have a common final vertex and lie on the same hyperbolic line.

A drawing of a finite part of Δ is given schematically in Figure 1.2. Here and wherever a particular presentation for $\pi_1 M^2$ is illustrated we use the surface of genus 2 and the presentation

$$<a,b,c,d \mid ab\bar{a}\bar{b}cd\bar{c}\bar{d}> .$$

However we want to remark that for the less usual presentations

$$<a_1, a_2, \cdots, a_q \mid a_1 a_2 \cdots a_q \bar{a}_1 \bar{a}_2 \cdots \bar{a}_q>$$

the graph Δ is more tractable since all edges lying on a common hyperbolic line have the same label. The representation of Δ in Figure 1.2 is not metrically accurate since otherwise parts of the drawing would necessarily turn out too small to give a clear, intuitive idea of the geometric pattern. This will be true for all drawings throughout this paper. We also adopt the conventions that edges of Γ are represented by continuous lines, edges of N are dashed, and edges labelled h_{**} are dotted.

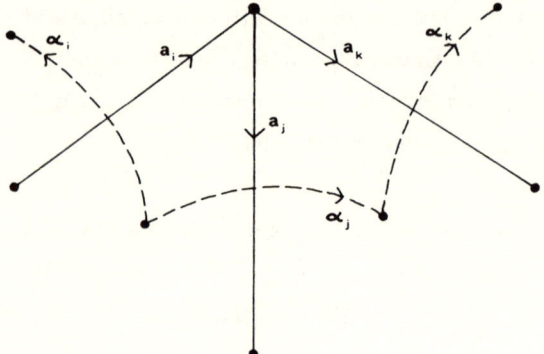

Figure 1.1

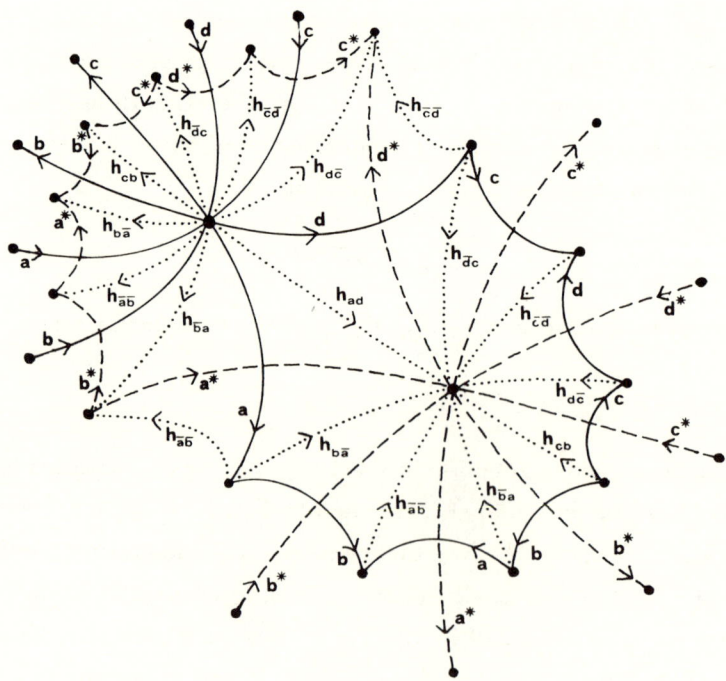

Figure 1.2

An *edge path* E in Δ is a connected subgraph of Δ homeomorphic to an immersed topological line segment. It can be finite, infinite in only one direction or bi-infinite; we assume that it is always provided with a direction. The word W(E) in symbols $a_i, \bar{a}_i, a_i, \bar{a}_i$, $h_{xy}, \bar{h}_{xy}$ (in the sequel called the *enlarged alphabet*) is the sequence of labels of the directed edges read off when running along E in the given direction. Especially for any directed edge e of Δ we let W(e) denote its label.

In general a *word* W in the enlarged alphabet is any sequence of symbols such that an edge path $E \subset \Delta$ exists with $W = W(E)$.

For notational convenience we distinguish the symbols corresponding to the different edge types in the following way: by latin symbols u, v, w, x, y, z (with or without subscripts) we always mean the original generators $a_1, \cdots, a_q$ of $\pi_1 M^2$ or their inverses. If $x = a_i$ (or $x = \bar{a}_i$) we let x* denote the symbol a_i (or $\bar{a}_i$ respectively). Symbols which are not yet specified are given by the greek letters ω, ν.

We will freely use classical facts about hyperbolic geometry; as references see [1], [7] and Part I.

§2. Normal edge paths

In this section we introduce a class of edge paths which satisfy certain geometric properties. We prove that they approximate hyperbolic line segments and we deduce some uniqueness results.

DEFINITION 2.1. For each directed edge e of Δ we define what will be called an *admissible region* $H(e) \subset D^2$. For some of the edge types several different choices for H(e) are possible. H(e) is always the intersection of some closed or open hyperbolic half planes; the explicit information about the shape of the H(e) is given in the Figures 2.1 to 2.4, where H(e) is represented by the shaded areas.

$W(e) = x \in \{a_i, \bar{a}_i\}$

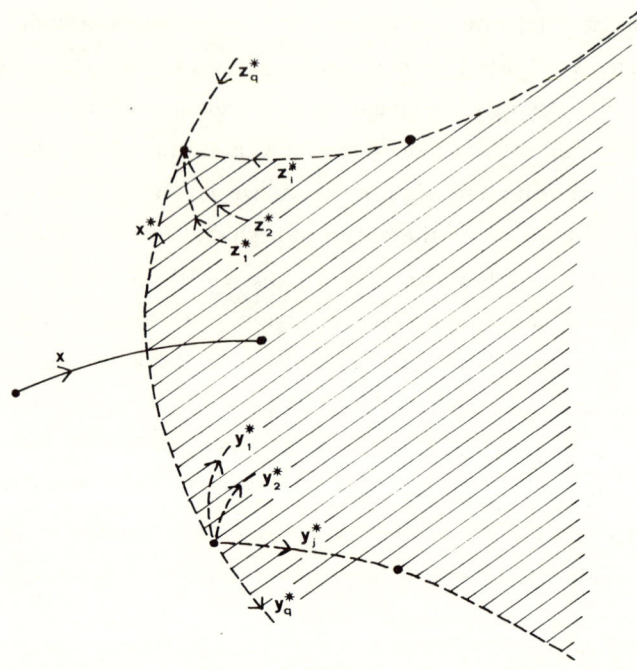

Figure 2.1

$H(x)$ is the open region bounded by the edge x^* and two half lines one containing some z_i^* and the other some y_j^* such that $xz_1 \cdots z_i \cdots z_q$ $(1 \leq i \leq q-1)$ is a positive and $xy_1 \cdots y_j \cdots y_q$ $(1 \leq j \leq q-1)$ is a negative $(q+1)$-cycle.

$W(e) = x^* \in \{a_i, \bar{a}_i\}$

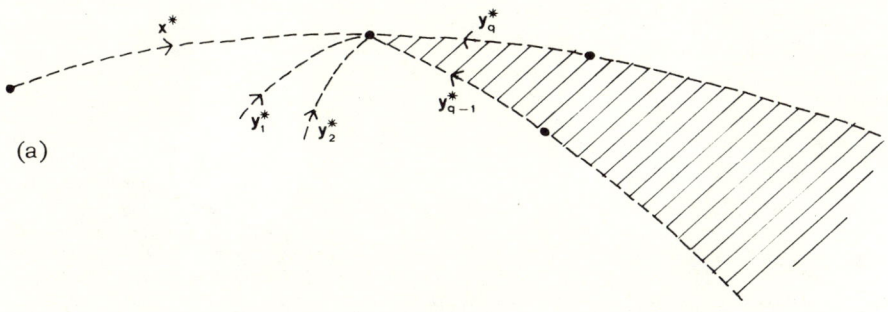

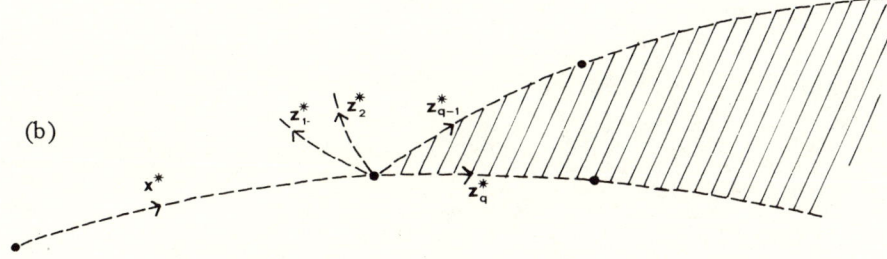

Figure 2.2

$H(x^*)$ is (a) the closed region bounded by the two half lines containing y_{q-1}^* and y_q^* where $xy_1 \cdots y_q$ is a positive $(q+1)$-cycle or (b) the closed region bounded by the two half lines containing z_{q-1}^* and z_q^* where $\bar{x}z_1 \cdots z_q$ is a negative $(q+1)$-cycle.

$W(e) = h_{xy}$

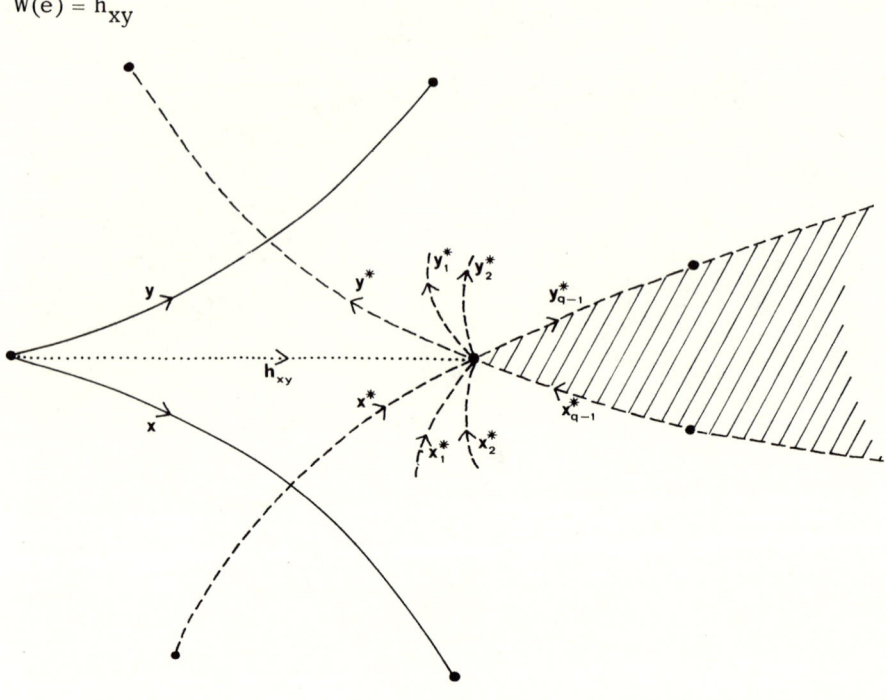

Figure 2.3

$H(h_{xy})$ is the closed region bounded by the two half lines containing the edges x^*_{q-1} and y^*_{q-1}, where $xx_1 \cdots x_{q-1}$ is a positive and $yy_1 \cdots y_{q-1}$ is a negative half cycle (= q-cycle).

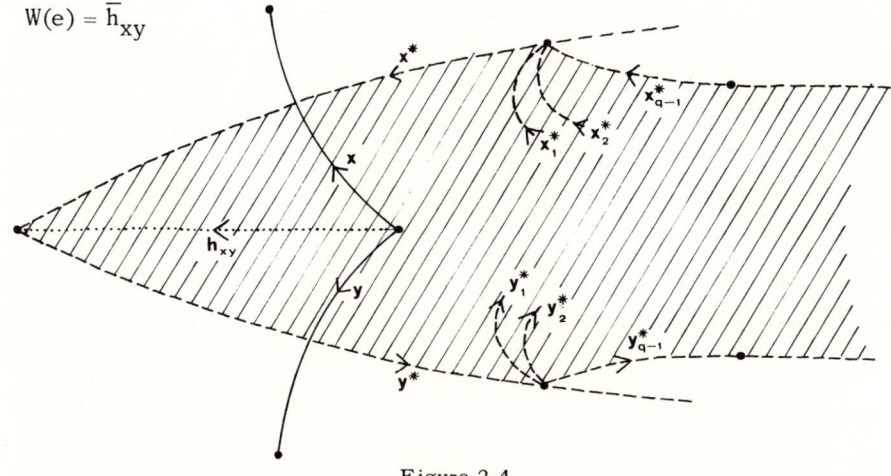

Figure 2.4

$H(\bar{h}_{xy})$ is the open region bounded by the edges x^* and y^* and the two half lines containing the edges x^*_{q-1} and y^*_{q-1}, where $\bar{x}x_1\cdots x_{q-1}$ is a positive and $\bar{y}y_1\cdots y_{q-1}$ is a negative half cycle (the reader should notice that the two half lines do not intersect since $q \geq 4$; this can be seen easily by choosing the initial vertex of h_{xy} to coincide with the center of D^2).

DEFINITION 2.2. A (directed) edge path $E = \cdots e_{i-1} e_i e_{i+1} \cdots$ in Δ is said to be *normal* if one can assign to each edge e_i of E (with the induced orientation) an admissible region $H(e_i)$ such that for each pair of successive edges we have

$$H(e_i) \supset H(e_{i+1}).$$

Such an assignment $e_i \to H(e_i)$ will be called an *admissible assignment*. A normal edge path with normal inverse will be called *binormal*.

It is immediate from this definition that any subpath of a normal edge path is again normal.

LEMMA 2.3. *If P and Q are vertices of Δ (possibly virtual) then there exists at most one normal path with initial vertex P and final vertex Q (with P or Q lying on the first or last edge of the path if they are virtual vertices).*

Proof. For any normal path $e_1 e_2 \cdots e_r$ running from P to Q we have

$$H(e_1) \supset H(e_2) \supset \cdots \supset H(e_r) \ni Q.$$

But for any two distinct directed edges with common initial vertex (or common virtual vertex) the possible admissible regions are always disjoint; thus P and Q determine uniquely the normal path $e_1 \cdots e_r$. □

The interested reader might derive the existence of a normal path running from P to Q from the fact that for each edge e of Δ every admissible region H(e) splits into a disjoint union of admissible regions $H(e_1), H(e_2), \cdots, H(e_t)$ with $e_1, e_2, \cdots, e_t$ being edges of Δ which have the final vertex of e as initial vertex. For $P, Q \in \Gamma$ the existence statement is given in the next section (Theorem 3.7).

The next proposition describes the crucial property of normal edge paths; its proof is immediate from the last lemma.

THEOREM 2.4. *Two binormal edge paths are disjoint or intersect in a common possibly virtual vertex or a common arc (possibly infinite).*

We now establish a first close connection between normal edge paths and hyperbolic lines.

THEOREM 2.5 (Approximation Theorem). *Given vertices $P \neq Q$ of Δ and a finite normal edge path E running from P to Q, let $g_1 F$ and $g_2 F$ be fundamental domains containing P and Q respectively and γ any hyperbolic line segment which intersects both $g_1 F$ and $g_2 F$. Then the union $B(\gamma)$ of all fundamental domains gF that have nonempty intersection with γ contains the edge path E.*

Proof. By induction on the length of the path E it is enough to prove that the first edge e of E is contained in $B(\gamma)$. We have to distinguish several cases according to the label of this edge. Except for the case $W(e) = h_{xy}$, which is trivial (since $e \in g_1 F$ holds), we always proceed in the same way: In Figures 2.5-2.7 hyperbolically convex regions $R(e)$ are exhibited for the remaining cases $W(e) \in \{a_i, \bar{a}_i\}$, $W(e) \in \{a_i, \bar{a}_i\}$ and $W(e) = \bar{h}_{xy}$. Each of these regions splits into two connected components when one removes the union $U(e)$ of some fundamental domains gF, all of them containing the final vertex of e ($U(e)$ is given explicitly in the drawings). One of the two components contains all translates of F that contain P (with exception of all points contained already in $U(e)$) and thus $g_1 F - U(e)$. The other one contains all possible admissible $H(e)$ in its interior (again excluding $U(e)$) and thus it contains $g_2 F - U(e)$ (by the normality of E). Therefore the line segment γ intersects both components or $U(e)$ and hence ($R(e)$ is convex!) it must intersect at least one $gF \in U(e)$, which contains the final vertex of e. But with its two endpoints e too is contained in $B(\gamma)$ and the proof is completed. □

$W(e) = x \in \{a_i, \bar{a}_i\}$

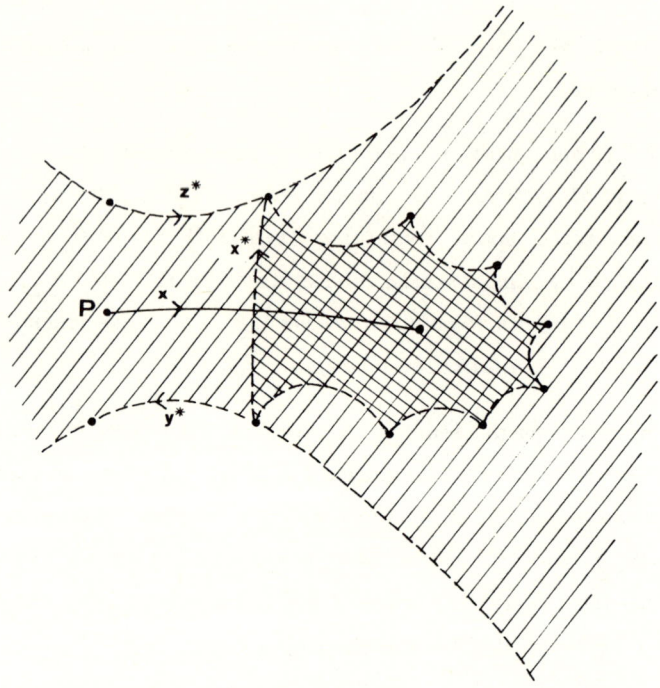

Figure 2.5

R(e) is the closed region (shaded) bounded by the two hyperbolic lines containing the edges y^* and z^*, where $\overline{xy}$ and zx are positive 2-cycles.

U(e) is the translate of the fundamental region F (twice shaded) that contains the final vertex of the edge x.

$W(e) = x^* \epsilon \{a_i, \overline{a}_i\}$

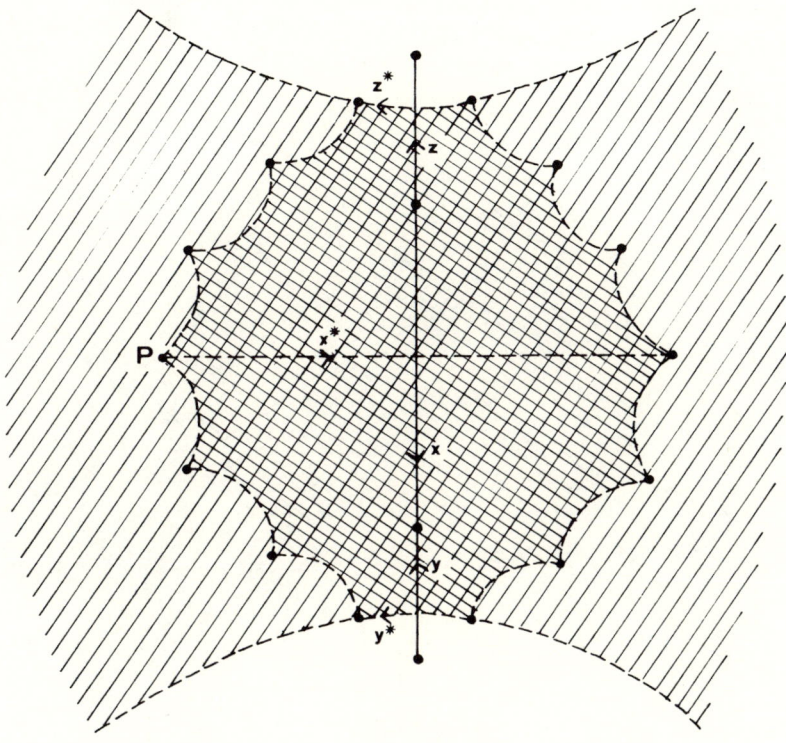

Figure 2.6

R(e) is the closed region (shaded) bounded by the two lines containing the edges y^* and z^* where $\overline{x}, \overline{z}$ and x, y are diagonally opposed pairs of edges.

U(e) consists of the two translates of F (twice shaded) adjacent to the edge x^*.

$W(e) = \bar{h}_{xy}$

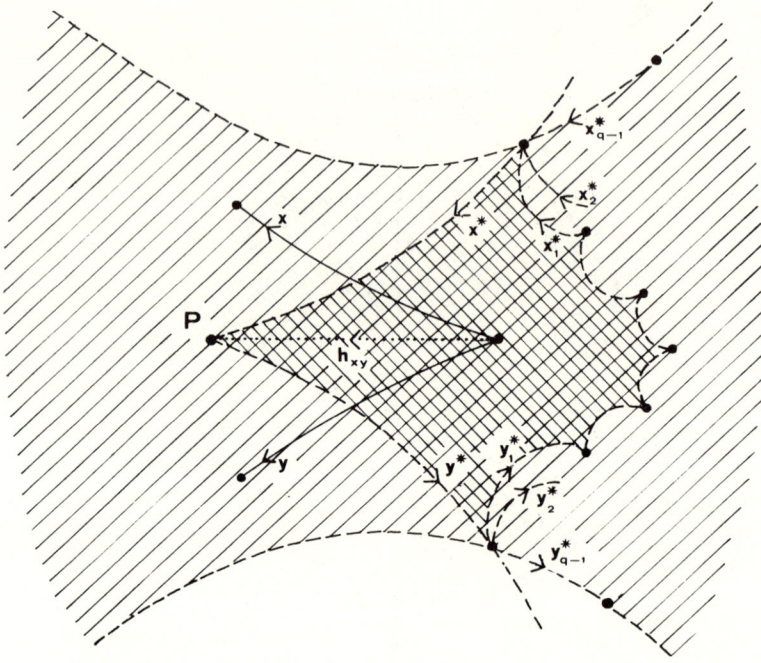

Figure 2.7

R(e) is the closed region (shaded) bounded by the two lines containing the edges x^*_{q-1} and y^*_{q-1}, where $\bar{x}x_1 x_2 \cdots x_{q-1}$ is a positive and $\bar{y}y_1 y_2 \cdots y_{q-1}$ is a negative half cycle.

U(e) consists of the translate of the fundamental domain F (twice shaded) that contains the initial vertex of h_{xy}.

The last proof remains valid if, in the theorem, one replaces the hypothesis $\gamma \cap g_2 F = \emptyset$ by the condition that γ is a hyperbolic half line with limit point $S \subset \partial D^2 \cap \overline{H(e_r)}$, where $E = e_1 \cdots e_r$. Thus an infinite normal path $E = e_1 e_2 \cdots$ is contained in $B(\gamma)$ if γ joins $g_1 F$ with a point $S \in \bigcap_{i \in 1}^{\infty} \overline{H(e_i)}$. This implies that $\bigcap_{i=1}^{\infty} \overline{H(e_i)}$ consists of exactly one point of ∂D^2, since a normal path can not have accumulation points in Int D^2.

COROLLARY 2.6. *An infinite normal edge path* $E = e_1 e_2 \cdots$ *has a well defined final limit point on* ∂D^2. □

COROLLARY 2.7. *A bi-infinite binormal edge path* $E = \cdots e_{i-1} e_i e_{i+1} \cdots$ *is contained in the neighborhood* $B(\gamma)$ *(defined in 2.5) of the hyperbolic line* γ *that joins the two limit points of* E.

Proof. If e_n intersects $g_n F$ and γ_n joins the final limit point S of E with $g_n F$ then $B(\gamma_n)$ contains $E_n = e_n e_{n+1} \cdots$. Thus for each vertex P of E the intersection of the closed region $\cup \{gF | P \in gF\}$ with γ_n for large enough n and hence with $\gamma = \lim_{n \to -\infty} \gamma_n$ is nonempty. □

With the Approximation Theorem 2.5 we are now able to extend the uniqueness result of Lemma 2.3 to the case of infinite edge paths. For this purpose we need to slightly restrict the class of normal edge paths in a natural manner:

DEFINITION 2.8. An infinite normal edge path $E = \cdots e_{i-1} e_i e_{i+1} \cdots$ is said to be ∞-*normal* if its final limit point on ∂D^2 is contained in all $H(e_i)$ for some admissible assignment $e_i \to H(e_i)$. An ∞-normal path E with ∞-normal inverse is called ∞-*binormal*.

THEOREM 2.9. *Two* ∞-*binormal edge paths* E, E' *with the same pair of limit points are identical.*

An existence result for periodic ∞-normal edge paths will be given in the next section (Theorem 3.7). For the proof of the above stated theorem we need two (technical) lemmas:

LEMMA 2.10. *For a binormal edge path* $E = \cdots e_{i-1} e_i e_{i+1} \cdots$ *there exist admissible assignments* $e_i \to H(e_i), \bar{e}_i \to H(\bar{e}_i)$ *such that, if* $e_j \in N$ *(so* $W(e_j) \in \{a_i, \bar{a}_i\}$ *),* $H(e_j)$ *and* $H(\bar{e}_j)$ *lie on different sides of the hyperbolic line* γ *containing the edge* e_j.

Proof. If E coincides with the hyperbolic line $\gamma \subset N$ containing e_j the proposition is obvious. If not, there exists for any $e_j \subset N$ (after possibly inverting the direction of E) some first edge e_{j+r} not contained in γ, thus determining $H(e_{j+r-1})$ and hence $H(e_{j+r-2})$, $H(e_{j+r-3}), \cdots, H(e_j)$ (see Figure 2.8). But for $e_{j+r-1} \subset \gamma$ the region $H(e_{j+r-1})$ contains only two edges with initial vertex equal to the final vertex of e_{j+r-1} and not lying on γ. The edge e_{j+r} has to coincide with one of them, and in both cases all admissible regions for $H(\bar{e}_{j+r})$ are situated on the other side of γ from e_{j+r}. By the normality of the inverse of E the same has to be true for $H(\bar{e}_{j+r-1}), H(\bar{e}_{j+r-2}), \cdots, H(\bar{e}_j)$. □

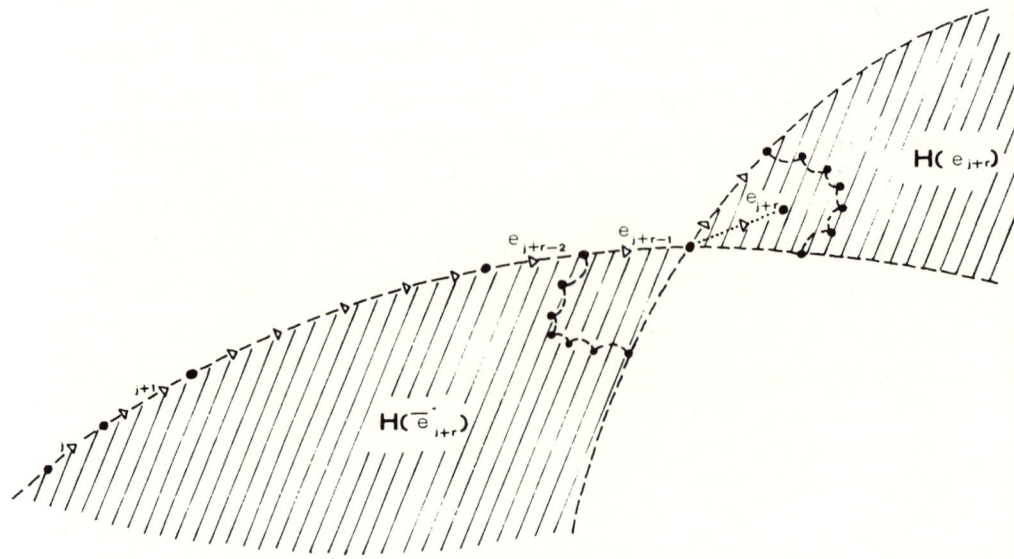

Figure 2.8

LEMMA 2.11. *Assume given (1) an edge* e *with* $W(e) \in \{a_i, \bar{a}_i\}$ *or* $W(e) = \bar{h}_{xy}$ *(2) the union* $U(e)$ *of all fundamental domains* gF *containing* e *(given in Figure 2.6 and 2.7) and (3) some finite binormal edge path* $E = e_1 e_2 \cdots e_r$ *passing through* $U(e)$ *(i.e. intersecting* $U(e)$ *but not starting or ending in* Int $U(e)$ *). If there exist admissible assignments* $e_i \to H(e_i)$, $\bar{e}_i \to H(\bar{e}_i)$ *such that both* $H(e) \cap H(e_r)$ *and* $H(\bar{e}) \cap H(\bar{e}_1)$ *are nonempty (where we assume that* $H(e)$, $H(\bar{e})$ *are situated as given by the previous lemma if* $W(e) \in \{a_i, \bar{a}_i\}$ *) then* E *contains the edge* e.

Proof. Every path passing through $U(e)$ contains a minimal path passing through $U(e)$. There are only finitely many such minimal paths which are binormal; it is immediate from Definition 2.1 that none of those which avoid the edge e can possibly have nonempty $H(e) \cap H(e_r)$ and $H(\bar{e}) \cap H(\bar{e}_1)$ simultaneously. □

Proof of Theorem 2.9. We are given two points S, T on ∂D^2. If there exists a vertex of Δ which is contained in every ∞-binormal edge path E joining these two points then a double application of the proof of Lemma 2.3 shows that there is at most one ∞-binormal edge path joining S and T. To find such a vertex we distinguish the following two cases (in both assuming γ to be the line which joins S, T):

1.) Suppose E contains some edge e with $W(e) \in \{a_i, \bar{a}_i\}$ or $W(e) = \bar{h}_{xy}$. Using the definitions for the regions $R(e)$ and $U(e)$ exhibited in the proof of Theorem 2.5 (see Figures 2.6 and 2.7) one may check that γ is contained in Int $R(e)$ (since the two limit points of E are contained in $H(e)$ and $H(\bar{e})$ respectively). Thus $B(\gamma) \subset R(e)$ holds, and by 2.7 another ∞-binormal edge path E′ with same limit points S, T as E has to lie in $R(e)$ and hence passes through $U(e)$. By Lemmas 2.10 and 2.11 (the admissible regions of E and E′ intersect since each contains a limit point S or T) the path E′ then contains $e \subset E$.

2.) Suppose $E \subset \Gamma$. We may assume $\gamma \not\subset N$; otherwise we apply case 1 to an alternative edge path E' running along γ. Let gF be some fundamental domain with $\gamma \cap \text{Int } gF \neq \emptyset$. If γ has a limit point contained in $H(e)$ for some edge $e = h_{xy}$ with the center Q of gF as initial vertex, then the same arguments as in case 1 apply in order to show that E contains one of the edges h_{xy}, x^* or $\bar{y}^*$ (all of them having the same final vertex) according to where the other limit point of γ is located. Thus we may assume in this case 2 that the two limit points of γ are contained in some admissible $H(e')$ and $H(e'')$ where e' and e'' are distinct edges of Γ with initial vertex Q. Using again the definition given in the proof of Theorem 2.5 we define $R_0(e')$, $R_0(e'')$ to be the component of $R(e') - gF$ and $R(e'') - gF$ respectively that contains the corresponding limit point of γ. For the closed region $R = R_0(e') \cup R_0(e'') \cup gF$ we obtain $\gamma \subset \text{Int } R$ and thus $B(\gamma) \subset R$; hence $E \ (\subset B(\gamma))$ has to pass through gF (since $R_0(e') \cap R_0(e'') = \emptyset$) and by the hypothesis $E \subset \Gamma$ it passes furthermore through the vertex Q. □

§3. Words in canonical form

After having introduced the geometric pattern for a normal edge path E in the preceding section, we now study the combinatorics of the word $W(E)$ read off from the labels of E. In terms of $W(E)$ we give, in the second half of this section, an algorithm that transforms an arbitrary finite or periodic edge path $E \subset \Gamma$ into a normal or ∞-normal path with the same pair of endpoints.

In order to motivate our main definition (3.1) we look at normal edge paths $E = \cdots e_{i-1} e_i e_{i+1} \cdots$ which are completely contained in N (so $W(e_i) \in \{a_i, \bar{a}_i\}$). Let $e_i \to H(e_i)$ be an admissible assignment. We associate to each edge $e_i = x^*$ a pair $P(e_i) = (p_1(e_i), p_2(e_i))$ which is of the form $(h_{\bar{x}*}, h_{x*})$ or $(h_{*x}, h_{*\bar{x}})$ according to the rule that the edge labelled $p_2(e_i)$ with the same final vertex as e_i satisfies $H(p_2(e_i)) = H(e_i)$ (see Figure 3.1). The edge with label $p_1(e_i)$ and the same final vertex as $\bar{e}_i$ then lies on the other side of the hyperbolic line γ which contains e_i. By the normality of E we have $H(e_{i-1}) \supset H(e_i)$; thus $p_2(e_i)$ and $p_2(e_{i-1})$ lie on the same side of γ and so $p_1(e_i)$ and $p_2(e_{i-1})$ are diagonally opposed. This is true for all pairs e_{i-1}, e_i of edges of the path $E \subset N$. A typical situation is illustrated in Figure 3.1.

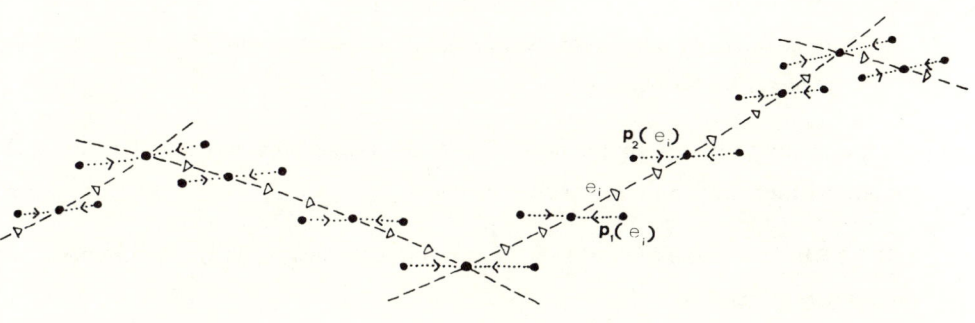

Figure 3.1

The following definition will turn out to be the combinatorial analogue of the geometric definition (2.2) of a normal edgepath.

DEFINITION 3.1. A word $W = \cdots \omega_{i-1} \omega_i \omega_{i+1} \cdots$ in the enlarged alphabet (i.e. $\omega_i \in \{a_i, \bar{a}_i, a_i, \bar{a}_i, h_{xy}, \bar{h}_{xy}\}$) is said to be in *canonical form* if the following conditions are satisfied:

1.) W contains no s-cycle with $s \geq q$.
2.) To each symbol $\omega_i = x^* \in \{a_i, \bar{a}_i\}$ we can associate a pair $P(\omega_i) = (p_1(\omega_i), p_2(\omega_i))$, which equals $(h_{\bar{x}*}, h_{x*})$ or $(h_{*x}, h_{*\bar{x}})$ such that:
   a) If $\omega_i = x^*$, $\omega_{i+1} = y^*$ then $p_2(\omega_i)$ diagonally opposes $p_1(\omega_{i+1})$.
   b) If $\omega_i = h_{**}$, $\omega_{i+1} = y^*$ then ω_i diagonally opposes $p_1(\omega_{i+1})$.
   c) If $\omega_i = x^*$, $\omega_{i+1} = \bar{h}_{**}$ then $p_2(\omega_i)$ diagonally opposes $\bar{\omega}_{i+1}$.
   d) If $\omega_i = h_{**}$, $\omega_{i+1} = \bar{h}_{**}$ then ω_i diagonally opposes $\bar{\omega}_{i+1}$.

3.) None of the following sequences may occur as partial words of W:

a) $xh_{\bar{x}*}$; $xh_{*\bar{x}}$; $\bar{h}_{\bar{x}*}$; $h_{*\bar{x}}x$

b) $\bar{h}_{*x}h_{x*}$; $\bar{h}_{x*}h_{*x}$

c) $\bar{h}_{xy}z_1z_2\cdots z_{q-1}$ for $\bar{x}z_1\cdots z_{q-1}$ a positive or $\bar{y}z_1\cdots z_{q-1}$ a negative halfcycle;

$z_1z_2\cdots z_{q-1}h_{xy}$ for $z_1\cdots z_{q-1}y$ a positive or $z_1\cdots z_{q-1}x$ a negative halfcycle.

d) $\bar{h}_{xy}z_1z_2\cdots z_{q-2}h_{uv}$ for $\bar{x}z_1\cdots z_{q-2}v$ a positive or $\bar{y}z_1\cdots z_{q-2}u$ a negative halfcycle.

It is easy to check that a normal path E reads off a word $W(E)$ in canonical form. We now prove the converse:

THEOREM 3.2. *An edge path $\hat{E}$ that reads off a word $W(\hat{E})$ in canonical form is normal.*

REMARK 3.3. By the symmetry of Definition 3.1 it follows from this theorem that the inverse of a normal path is normal.

Proof of Theorem 3.2. We define below the forms of special words in the enlarged alphabet which we call *elementary words*:

a) $W(a_i, \bar{a}_i)$ bi-infinite [$W(a_i, \bar{a}_i)$ means an arbitrary word using only symbols a_i and $\bar{a}_i$].

b) $W(a_i, \bar{a}_i)\bar{h}_{**}$ infinite

c) $h_{**}W(a_i, \bar{a}_i)$ infinite

d) $h_{**}W(a_i, a_i)\bar{h}_{**}$

e) $\bar{h}_{xy}z_1\cdots z_s$ with $\bar{x}z_1\cdots z_s$ a positive or $\bar{y}z_1\cdots z_s$ a negative $(s+1)$-cycle.

f) $z_1\cdots z_sh_{xy}$ with $z_1\cdots z_sy$ a positive or $z_1\cdots z_sx$ a negative $(s+1)$-cycle.

g) $\bar{h}_{xy}z_1\cdots z_sh_{uv}$ with $\bar{x}z_1\cdots z_sv$ a positive or $\bar{y}z_1\cdots z_s$ a negative $(s+2)$-cycle.

h) $z_1\cdots z_s$, a positive or negative s-cycle.

i) $\bar{h}_{xy}z$ with neither $\bar{x}z$ nor $\bar{y}z$ a cycle.
j) zh_{xy} with neither zx nor zy a cycle.
k) xy where xy is not a cycle.
l) $\bar{h}_{xy}h_{uv}$ with $x \neq v$ and $y \neq u$.

If an edge path $E = e_{i-1}e_i e_{i+1} \cdots$ reads off an elementary word $W(E)$ in canonical form, then there exists an admissible assignment $e_i \to H(e_i)$, satisfying $H(e_i) \supset H(e_{i+1})$. For the words a)-d) this follows immediately from condition 2) of Definition 3.1 (see the discussion preceding 3.1). For all other elementary words this can be checked easily from the definition of the admissible regions (Definition 2.1). For example we check the (hardest) case g) in Figure 3.2. For the cases e), f), h), i), j) and k), where various possibilities for the admissible assignment exist, we impose the further condition that the admissible region for the first edge is chosen minimally and for the last edge maximally among the different possibilities.

Any two elementary words $W = \cdots \omega_{i-2}\omega_{i-1}\omega_i$ and $V = \nu_j \nu_{j+1} \cdots$ with $\omega_{i-s} = \nu_j$, $\omega_{i-s+1} = \nu_{j+1}, \cdots, \omega_i = \nu_{j+1}$ always give rise, for $s > 0$, to a larger elementary word $U = \cdots \omega_{i-2}\omega_{i-1}\omega_i \nu_{j+s+1} \nu_{j+s+2} \cdots$ For $s = 0$ this might not be true; in this case let E be an edgepath reading off U. If W and V are in canonical form, then the above described admissible assignments for the subpaths corresponding to W and V induce in all cases an admissible assignment for E.

Now let $W(\hat{E}) = \cdots \omega_{i-1}\omega_i \omega_{i+1} \cdots$ be the word in canonical form read off from the given edgepath $\hat{E}$. Each symbol ω_i which is not the first symbol of $W(\hat{E})$ is contained as other than the first symbol in a (possibly infinite or bi-infinite) maximal partial word of $W(\hat{E})$ which is equal to one of the given elementary words. We see from the above statements that $W(\hat{E})$ is covered by such maximal partial words, the last letter of any one of them being the first letter of another. Thus the above defined admissible assignments for the corresponding subpaths of E induce an admissible assignment for the whole path E which is hence normal by Definition 2.2. □

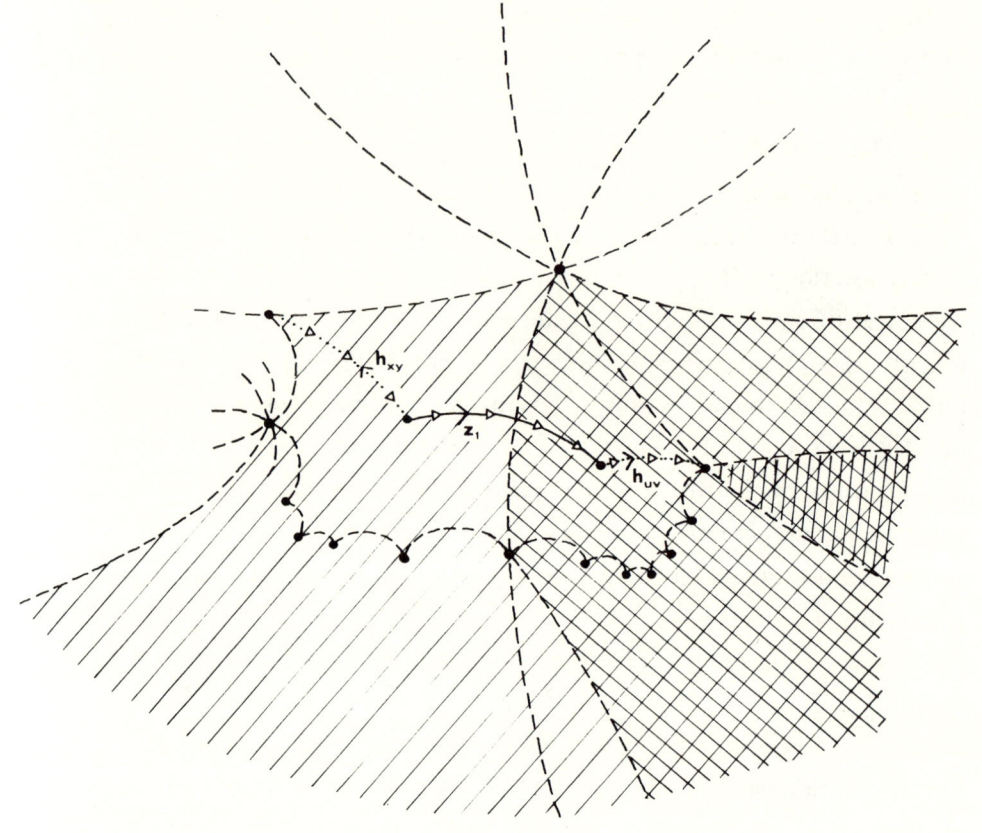

Figure 3.2

The following construction illustrates the fact that an edge path E reading off a word $W(E)$ in canonical form does not necessarily need to be ∞-normal (an example is given in Figure 3.3):

Let W be a product

$$W = w_{1,1} w_{1,2} \cdots w_{1,q-1} w_{2,1} w_{2,2} \cdots w_{2,q-1} \cdots w_{s,1} w_{s,2} \cdots w_{s,q-1}$$

of sequences $w_{i,1}w_{i,2}\cdots w_{i,q-1}$ where all are positive or all are negative (q–1)-cycles and where the corresponding (q+1)-cycles $w_{i,0}w_{i,1}w_{i,2}\cdots w_{i,q-1}w_{i,q}$ further satisfy the conditions $w_{i,q} = \overline{w}_{i+1,0}$ (for $1 \le i \le s-1$) and $w_{s,q} = \overline{w}_{1,0}$.

Let $\hat{W}$ be the word $w_{1,0}^* w_{2,0}^* \cdots w_{s,0}^*$ for the case that the $w_{i,1}w_{i,2}\cdots w_{i,q-1}$ are all positive (q–1)-cycles and set $\hat{W} = \overline{w}_{1,0}^* \overline{w}_{2,0}^* \cdots \overline{w}_{s,0}^*$ if they are negative (q–1)-cycles.

A bi-infinite edge path E periodically reading off W has the same limit points as the hyperbolic line γ contained in the net N which reads off, when understood as edge path E_γ, the word $\hat{W}$ (periodically). But whereas E_γ is clearly ∞-normal it is easy to see that E is not.

DEFINITION 3.4. A periodic word W as given in the above construction (possibly after a suitable permutation) is called *exceptional*. A word in canonical form which is not exceptional is said to be in *restricted canonical form* (abbreviated in the sequel "res. can. form").*

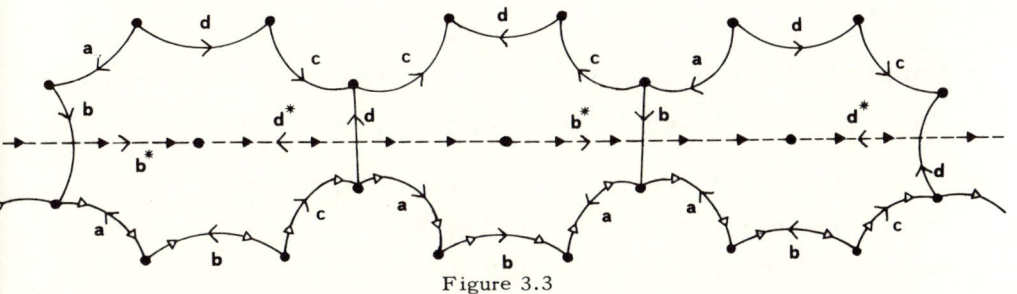

Figure 3.3

–▷–▷–▷– the edge path E periodically reading off $W = ab\overline{a}^2bc$

–▶–▶–▶– the edge path E_γ periodically reading off $\hat{W} = \overline{d}^*b^*$.

---

*For the extension of the theorems following Definition 3.4 to the case of non-periodic bi-infinite edge paths one has to carry over Definition 3.4 to the slightly generalized situation where the edge path reading off the ∞-biinfinite word $\hat{W}$ is contained in the union of two intersecting hyperbolic lines contained in N.

It is clear that for a given geometric presentation of $\pi_1 M^2$ only finitely many exceptional words exist which are not proper powers. If we exclude them we obtain a stronger version of Theorem 3.2.

THEOREM 3.5. *An edge path* E *which periodically reads off a cyclic word* W(E) *in res. can. form is* ∞-*normal.*

Proof. The path $E = \cdots e_{i-1} e_i e_{i+1} \cdots$ is normal by Theorem 3.2. Thus the final point $S \in \partial D^2$ of E is contained in every $\overline{H(e_i)}$. If $S \in H(e_i)$ holds for some e_i, the assertion follows from the periodicity of $W = W(E)$. Otherwise we have $E \subset \Gamma$ (since all $H(e_i)$ have to be open). For some e_i we fix the hyperbolic line γ which bounds $H(e_i)$ and has $S \in \overline{H(e_i)} - H(e_i)$ as limit point. Then γ has to bound all $H(e_j)$ for $j \geq i$ (by the shape of the admissible regions defined in 2.1) and thus all $H(e_j)$ for arbitrary $j \in Z$ since W is periodic. But this exactly characterizes an edge path which reads off an exceptional word W; the line γ understood as edge path E_γ reads off the corresponding word $\hat{W}$. □

COROLLARY 3.6. *In each free homotopy class there is at most one loop covered by some edgepath which reads off a word in res. can. form.*

Proof. Otherwise the infinite lift of the free homotopy between two such loops would (by 3.5) give two different ∞-normal edge paths with same limit points, contradicting Theorem 2.9. □

The algorithm

We now describe an algorithm that transforms a given (cyclic) word in the generators $a_i, \bar{a}_i$ of $\pi_1 M^2$ into a word in res. can. form, without changing the (free) homotopy class of the corresponding path in M^2. Thus we obtain the existence statements promised in §2. Furthermore the algorithm (Theorem 3.7), together with Corollary 3.6, constitutes an independent solution of the conjugacy problem (and also of the word problem by Theorem 3.2 and Lemma 2.3) in $\pi_1 M^2$ for an arbitrarily given geometric presentation.*

Let $W = w_1 w_2 \cdots w_r$ be a word or a cyclic word in the symbols $a_i, \bar{a}_i$. We treat both cases simultaneously; for the case of a cyclic word all indices are understood modulo the word length r.

We now specify the six steps I-VI out of which the algorithm is built. It is shown below (Theorem 3.7) that the algorithm is finite and that the resultant word $\widehat{W}$ is in (restricted) canonical form.

In each of the steps the user has to check for the occurrence of a certain type of subword and to replace each such subword by a given alternative word. The result of the algorithm doesn't depend on the order in which the checks and replacements within one step are performed (by Lemma 2.3 and Theorem 2.9). We let $W^{(i)}$ denote the word obtained from W after having performed the first i steps of the algorithm.

For each substitution we give an example using

$$<a,b,c,d \mid ab\bar{a}\bar{b}cd\bar{c}\bar{d}>$$

as presentation of $\pi_1 M^2$ (see §1). The corresponding edge paths (Figures 3.4 - 3.10) are represented

⟶▷—▷—▷— (before substitution)

and

⟶▶—▶—▶— (after substitution).

I) Find all positive or negative s-cycles $x_1 \cdots x_s$ with $s > q$ occurring in W and replace them by the complementary $(2q-s)$-cycle $y_1 \cdots y_{2q-s}$ (i.e. $x_1 \cdots x_s \bar{y}_{2q-s} \cdots \bar{y}_1$ equals the relator R up to cyclic permutation or inversion).

---

*The reader should notice that we do not need the assumption that the presentation is "alternating" in the terminology of [3].

Example: (see Figure 3.4)

$$a^2 \bar{b} \bar{a} dcb \longrightarrow a \bar{b} cdb$$

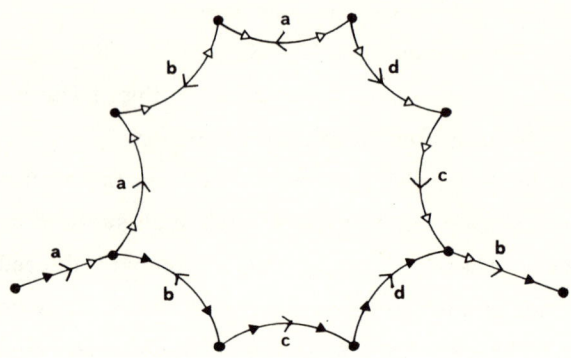

Figure 3.4

II) Check if $W^{(1)}$ contains a halfcycle $w_i w_{i+1} \cdots w_{i+q-1}$. If so, replace it by its complementary halfcycle $v_i v_{i+1} \cdots v_{i+q-1}$ if the resulting sequence $v_{i+q-1} w_{i+q} w_{i+q+1} \cdots w_{i+2q-2}$ is again a halfcycle. Repeat the procedure with this new halfcycle and so on. (These successive replacements can go on only finitely many times, by an argument given in the proof of Theorem 3.7 below.) If in the course of these substitutions a cycle of length $> q$ occurs, replace it (as in step I) by its complementary cycle (of strictly shorter length).

Go back to the beginning of step II and repeat the whole procedure until all halfcycles occurring in $W^{(1)}$ have been checked and possibly replaced.

An exceptional case for this step II occurs if $W^{(1)}$ equals the halfcycle $w_i w_{i+1} \cdots w_{i+q-1}$. Here the replacement by the complementary halfcycle $v_i v_{i+1} \cdots v_{i+q-1}$ takes place if and only if $v_1 = \bar{v}_{i+q-1}$ such that the obtained word $W^{(2)}$ is of strictly shorter length

Example: (see Figure 3.5)

$$c\bar{b}cd\bar{c}^2\bar{d}a^2 \longrightarrow ca\bar{b}ad\bar{c}da^2$$

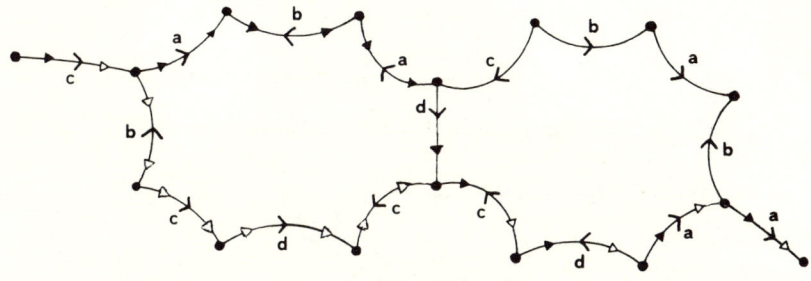

Figure 3.5

III) Find all positive (or negative) halfcycles $xz_1 \cdots z_{q-2}y$ occurring in $W^{(2)}$ and replace them by the corresponding word $h_{x*}\bar{h}_{*y}$ ($h_{*x}\bar{h}_{\bar{y}*}$).

Example: (see Figure 3.6)

$$c\bar{b}cd\bar{c}a \longrightarrow ch^-_{\bar{b}a}\bar{h}^-_{\bar{d}c}a$$

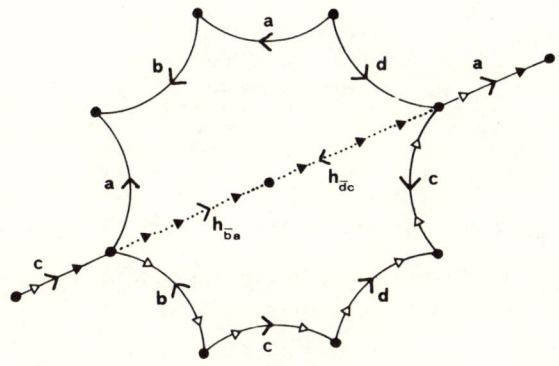

Figure 3.6

IV) In $W^{(3)}$ find all segments of the type

$$\underbrace{\cdots \bar{h}_{xy} z_1 z_2 \cdots z_{q-2} v \cdots}_{A}$$

and then:

a) If $\bar{y} z_1 \cdots z_{q-2} v$ is a negative halfcycle, replace A by $y * \bar{h}_{\bar{v}*}$

b) If $\bar{x} z_1 \cdots z_{q-2} v$ is a positive halfcycle, replace A by $\bar{x} * \bar{h}_{*\bar{v}}$

Analogously for all segments of type

$$\underbrace{\cdots v z_1 \cdots z_{q-2} h_{xy} \cdots}_{A}$$

c) If $v z_1 \cdots z_{q-2} x$ is a negative halfcycle, replace A by $h_{*v} x*$

d) If $v z_1 \cdots z_{q-2} y$ is a positive halfcycle, replace A by $h_{v*} \bar{y}*$

Example: (see Figure 3.7)

$$ch_{\overline{ba}} \bar{h}_{\overline{dc}} \overline{cda}^2 \longrightarrow ch_{\overline{ba}} d * \bar{h}_{b\bar{a}} a$$

V) In $W^{(4)}$ find all segments of type

$$\underbrace{\cdots \bar{h}_{xy} z_1 \cdots z_{q-2} h_{vw} \cdots}_{A}$$

and then:

a) For $\bar{x} z_1 \cdots z_{q-2} w$ a positive halfcycle, replace A by $\bar{x} * \bar{w} *$

b) For $\bar{y} z_1 \cdots z_{q-2} v$ a negative halfcycle, replace A by $y * v *$

INTERSECTION NUMBERS: II   531

Example: (see Figure 3.8)

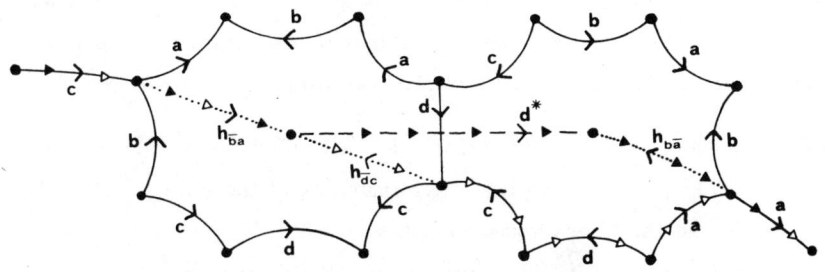

Figure 3.7

Figure 3.8

VI) If W is a cyclic word, check if $W^{(5)}$ is exceptional. If so replace it by the corresponding word $\widehat{W}^{(5)}$ in res. can. form (see Definition 3.3 and the preceding discussion).

THEOREM 3.7. *The algorithm consisting of the steps I through VI transforms a given (cyclic) word W in the generators $a_i, \bar{a}_i$ into a word $\widehat{W}$ in (restricted) canonical form. The algorithm is finite and does not change the (free) homotopy class of the corresponding path in M^2.*

Proof. It is clear from the definitions that all substitutions defined in the steps I - VI correspond to local deformations of the path in M^2; thus we never change the (free) homotopy class of W.

The algorithm is finite: Except for step II (in the case of a cyclic word W) this is immediate from the definitions. The only conceivable problem would occur if in step II the cyclic word $W^{(1)}$ gave rise to a sequence of replacements $v_i v_{i+1} \cdots v_{i+q-1}$ for $w_i w_{i+1} \cdots w_{i+q-1}$ which created a new halfcycle $v_{i+q-1} w_{i+q} w_{i+q+1} \cdots w_{i+2q-2}$ which in turn became the object of the next replacement, and so on. But such an infinite chain of substitutions cannot occur since all the (q–1)-cycles obtained by the substitutions have the same orientation (otherwise W would not have been a reduced word), whereas the (q–1)-cycle $w_i w_{i+1} \cdots w_{i+q-1}$ occurring before each substitution has the opposite sign. Thus a sequence of substitutions cannot proceed further than once through the cyclic word $W^{(1)}$.

We assert that the word W obtained from the algorithm is in (restricted) canonical form: It is immediate that $W^{(3)}$ satisfies the conditions 1), 2) and 3 a) of Definition 3.1. Going through steps IV and V of the algorithm doesn't affect conditions 1) and 3 a). Each symbol $\omega_i \in \{a_i, a_i\}$ introduced in IV or V takes place of some symbol h_{xy} (or its inverse); if we define the pair $P(\omega_i)$ using h_{xy} as $p_2(\omega_i)$ (or as $p_1(\omega_i)$ respectively), condition 2) turns out to be satisfied inductively after each substitution throughout steps IV and V. Thus it follows from the definitions of steps IV and V that the word $W^{(5)}$ satisfies all parts of Definition 3.1 with the possible exception of condition 3 b). But $W^{(5)}$ cannot be in contrast with 3 b) either; otherwise this would already be true for $W^{(4)}$. This would imply (as can be verified by checking all possibilities for words $W^{(3)}$ which give rise to a word $W^{(4)}$ contradicting 3 b)) that in $W^{(2)}$ we had either a cycle of length $> q$ or a sequence $w_i w_{i+1} \cdots w_{i+q-1}$ which was a halfcycle and for which the complementary halfcycle $v_i v_{i+1} \cdots v_{i+q-1}$ gave rise to a halfcycle $v_{i+q-1} w_{i+q} w_{i+q+1} \cdots w_{i+2q-2}$ (or an exceptional situation as discussed in the last paragraph of step II with $v_i = v_{i+q-1}$); both possibilities are excluded by the definition of step II.

Thus the word $W^{(5)}$ obtained via the algorithm is in canonical form, with $\hat{W} = W^{(6)}$ being the corresponding res. can. form (for W cyclic). □

§4. *The path of a geodesic*

In each free homotopy class $[W]$ of closed curves in M^2 there exists exactly one geodesic $\gamma(W)$. In this section we determine the path of $\gamma(W)$, i.e. the sequence of crossings of $\gamma(W)$ with the image of the net N in M^2.

Throughout this section let $\gamma(W) \subset M^2$ be a fixed geodesic which represents the given free homotopy class $[W]$ and let $\gamma \subset D^2$ be some lift of $\gamma(W)$. From Corollary 2.7 we know that the unique ∞-binormal edge path E with the same limit points as γ is already a very close approximation to γ. The exact information about the fundamental domains gF and the edges and vertices of the net N traversed by γ is derived from the path E in Lemmas 4.1-4.3. Theorem 4.4 then gives the periodic word W_γ (in terms of the unique word $\hat{W} \in [W]$ in res. can. form) which describes the sequence of crossings of the net N by γ. The complete picture is summarized in Remark 4.5.

The proof of the following lemma uses an idea similar to that of the "palindromic" words in [4]:

LEMMA 4.1. *The hyperbolic line γ passes through a vertex P of N if and only if the edgepath E passes through P and E is invariant with respect to a hyperbolic radial reflection of D^2 through P.*

Proof. a) "if": If E passes through P and is invariant under the reflection through P then γ is mapped by this reflection onto itself, and hence it passes through P.

b) "only if": γ is mapped by the reflection through P onto itself and thus E is mapped onto an edge path E' with the same pair of limit points. By Theorem 2.9 the paths E and E' are identical, and hence E passes through P. □

LEMMA 4.2. *If E intersects the interior of some gF then the same holds for γ.*

Proof. From Corollary 2.7 we know that γ intersects gF. If $E \cap gF$ doesn't contain any edge h_{xy} or $\bar{h}_{xy}$, the assertion follows from Lemma 4.1. Otherwise γ passes through Int gF or through the final vertex of h_{xy}; in the latter case the assertion follows from the shape of the admissible region $H(\bar{h}_{xy})$ as defined in 2.1. □

Finally consider the case where E passes through a vertex P of N but is not invariant under reflection through P. Let $N(P)$ be the union of all fundamental domains gF which contain P. By the normality of E the set $N(P) \cap E$ is topologically a closed line segment that separates $N(P)$ into two connected components. If $E = \cdots e_{i-1} e_i \cdots$ let e_i be the edge of E with final vertex P. Let $E' = \cdots e_{i-1} e_i e'_{i+1} e'_{i+2} \cdots$ be the path obtained by reflecting the subpath $\cdots e_{i-2} e_{i-1} e_i$ of E through P; E' is again ∞-binormal by Theorem 3.5.

LEMMA 4.3. *In this case the hyperbolic line γ intersects all nonboundary edges of N which are contained in that component of $N(P) - E$ which lies on the other side of E from the final limit point of E'.*

Proof. By Corollary 2.7 γ intersects some fundamental regions $g_1 F$ containing e_i and $g_2 F$ containing e_{i+1} (so $g_1 F$ and $g_2 F$ are contained in $N(P)$). On the other hand $N(P)$ is hyperbolically convex (since it is the intersection of half planes); thus any line segment $\gamma_0 \subset \gamma$ bounded by intersection points $P_1 \in \gamma \cap g_1 F$ and $P_2 \in \gamma \cap g_2 F$ is contained in $N(P)$. Since γ does not pass through P we can choose P_1 and P_2 so that γ_0 doesn't intersect E in its interior. Then all edges of N contained in the interior of the corresponding component of $N(P) - E$ are crossed by γ_0 and hence by γ.

The statement that the final limit point S of E' lies on the other side of E from γ_0 is equivalent to the assertion that P and S lie on the same side of γ. (In fact, S has to lie on the same side of E as of γ, since E and γ have equal limit points, and by definition γ_0 lies left of E if P lies right of γ and vice-versa.) But the hyperbolic line γ' that joins the two end points of E' passes (by the symmetry of E') through P. Since γ and γ' have a common limit point on D^2 they do not intersect in $\operatorname{Int} D^2$ and hence P and S lie on the same side of γ. □

By Lemma 4.1 γ does not pass through the end points of the edges of N contained in the component of $N(P) - E$ described in the last lemma. Therefore γ has to pass through the interior of the fundamental domains on which they lie. Together with Lemma 4.1 and Lemma 4.2 this gives the complete information about the fundamental domains gF and their edges and vertices traversed by the hyperbolic line γ with the same limit points as the ∞-binormal path E. Theorem 4.4 below is nothing else but a translation of this geometric knowledge into the corresponding combinatorial terminology.

Before we give Theorem 4.4 we want to describe a combinatorical test in terms of the words $W(E)$ and $W(E')$ that determines (in the situation of Lemma 4.3) the side of E on which the final limit point of E' is located.

We cyclically order the set of all $a_i, \bar{a}_i$ and h_{**} corresponding to the counterclockwise order of the labelled edges which have as initial point a fixed vertex P of Γ. For a geometric presentation $\langle a_1, \cdots, a_q | R \rangle$ and conventions as given in §1 this order is obtained from R by first defining the symbol succeeding $x \in \{a_i, \bar{a}_i\}$ to be the inverse of the left neighbor of the symbol x in the word R, and second by inserting the symbol h_{xy} between two succeeding $x, y \in \{a_i, \bar{a}_i\}$.

Analogously a cyclic order for the symbols $a_i, \bar{a}_i, \bar{h}_{**}$ (corresponding to the counterclockwise order of the edges radiating from some vertex $Q \in N$) is obtained from R by first defining the successor of $x^* \in \{a_i, \bar{a}_i\}$ to be $y^* \in \{a_i, \bar{a}_i\}$ where $\bar{y}$ is the right neighbor of $\bar{x}$ in the word R, and then inserting the symbol $\bar{h}_{\overline{yx}}$ between them. We refer to the two ordered sets as the *first* and the *second cyclic alphabets*. For the standard presentation $\langle a,b,c,d, | ab\overline{ab}cd\overline{cd} \rangle$ they are given by $a, h_{ad}, d, h_{d\bar{c}}, \bar{c}, h_{\overline{cd}}, d, h_{\overline{dc}}, c, h_{cb}, b, h_{ba}, \bar{a}, h_{\overline{ab}}, \bar{b}, h_{\overline{ba}}$ and $a^*, \bar{h}_{ba}, b^*, \bar{h}_{cb}, \bar{c}^*, \bar{h}_{d\bar{c}}, \bar{d}^*, \bar{h}_{\overline{cd}}, c^*, \bar{h}_{\overline{dc}}, d^*, \bar{h}_{ad}, \bar{a}^*, \bar{h}_{ba}, \bar{b}^*, \bar{h}_{\overline{ab}}$ respectively.

For any word $W = \cdots \omega_{i-2} \omega_{i-1} \omega_i$ we define the *reflected* word $W^+ = \omega_0^+ \omega_1^+ \omega_2^+ \cdots$ to be the word obtained recursively in the following way: First invert W and obtain $\overline{W} = \bar{\omega}_i \bar{\omega}_{i-1} \bar{\omega}_{i-2} \cdots$. Then define ω_0^+ to be the symbol diagonally opposed to $\bar{\omega}_i$ (i.e. the $(2q)^{th}$ symbol succeeding $\bar{\omega}_i$ in the corresponding cyclic alphabet) and ω_{n+1}^+ to be the pth symbol succeeding the symbol $\bar{\omega}_n^+$ in its cyclic alphabet, with p defined by the fact that $\bar{\omega}_{i-n-1}$ is the pth symbol succeeding ω_{i-n}.

It is immediate from the definitions that for any edge path $E = \cdots e_{i-2} e_{i-1} e_i$ the path E' obtained from reflecting E at the final vertex of e_i reads off the word $W^+ = W(E')$ obtained from reflecting the word $W = W(E)$.

Suppose now that $W^{(j)} = \omega_1^{(j)} \omega_2^{(j)} \omega_3^{(j)} \cdots$ is a set of words (none of them contained in another as initial subword) with first symbols $\omega_1^{(j)}$ all contained in the same cyclic alphabet. We define a cyclic lexicographic order of these words analogous to [2] as given in Part I, §2, where one switches between the two cyclic alphabets according to the actual symbol in question. (By our definition of "words" (§1) two symbols $\omega_n^{(j)}, \omega_n^{(j')}$ are in the same alphabet if $\omega_i^{(j)} = \omega_i^{(j')}$ for all $i < n$.) For words $W^{(j)}$ in canonical form this cyclic order corresponds to the counterclockwise order of the normal edge paths $E^{(j)}$ starting at some appropriate vertex of Δ and reading off $W(E^{(j)}) = W^{(j)}$.

Now let $W = \omega_1 \omega_2 \cdots \omega_n$ be a word in res. can. form. For any partial word $\omega_i \omega_{i+1}$ with $\bar{\omega}_i, \omega_{i+1}$ symbols of the second cyclic alphabet we define an *insertion sequence* $I(\omega_i, \omega_{i+1})$ that corresponds to the sequence of edges traversed by γ as given in Lemma 4.3: If for $W_{i+1} = \omega_{i+1} \omega_{i+2} \cdots \omega_n \omega_1 \omega_2 \cdots \omega_i$ we have $W_{i+1} = W_{i+1}^+$ then let $I(\omega_i, \omega_{i+1})$ be empty. Otherwise the lexicographic order defined above gives $\overline{W}_{i+1} < W_{i+1} < W_{i+1}^+$ or $\overline{W}_{i+1} < W_{i+1}^+ < W_{i+1}$ (up to cyclic permutation). In the first case we define $I(\omega_i, \omega_{i+1})$ as follows: Take the positive cycle $\bar{y}xz_1 \cdots z_r \bar{v}u$, where y and u are determined by the condition that $\omega_i \epsilon \{h_{xy}, \bar{y}*\}$ and $\omega_{i+1} \epsilon \{\bar{h}_{uv}, \bar{u}*\}$, define $I(\omega_i, \omega_{i+1}) = xz_1 \cdots z_r \bar{v}$. In the second case we define $I(\omega_i, \omega_{i+1}) = yz_1 \cdots z_r \bar{u}$, where y and u are determined by the conditions that $\omega_i \epsilon \{h_{xy}, x*\}$, $\omega_{i+1} \epsilon \{\bar{h}_{uv}, v*\}$ and $\bar{x}yz_1 \cdots z_r \bar{u}v$ is a negative cycle.

We can now construct the *geodesic word* W_γ that describes the path of the geodesic $\gamma(W) \epsilon [W]$ (see Theorem 4.4). Let $\hat{W} = \omega_1 \omega_2 \cdots \omega_r$ be the well-defined representative of W in res. can. form. If ω_i is diagonally opposed to $\bar{\omega}_{i+1}$ for all i (mod r) we define $W_\gamma = \hat{W}$. Otherwise we build a word W' from $\hat{W}$ by inserting between each pair of symbols $\omega_i \omega_{i+1}$ of W, with $\bar{\omega}_i, \omega_{i+1}$ elements of the second cyclic alphabet, the corresponding insertion sequence $I(\omega_i, \omega_{i+1})$. We now obtain W_γ from W' by operating on all symbols $\omega_i, \bar{\omega}_i$ and $h_{**}, \bar{h}_{**}$ in W' as follows:

(1) Assume $\omega_i = x* \epsilon \{a_i, \bar{a}_i\}$: If $I(\omega_{i-1}, \omega_i)$ and $I(\omega_i, \omega_{i+1})$ are both nonempty cycles with the same sign, then remove ω_i without replacing it. If $I(\omega_{i-1}, \omega_i)$ and $I(\omega_i, \omega_{i+1})$ have opposite sign, then $\omega_i = x*$ is replaced either by x if $I(\omega_{i-1}, \omega_i)$ is a positive cycle or by $\bar{x}$ otherwise.

If $I(\omega_i, \omega_{i+1})$ is empty then $\bar{\omega}_{i+1}$ is diagonally opposed to ω_i and neither $I(\omega_{i-1}, \omega_i)$ nor $I(\omega_{i+1}, \omega_{i+2})$ can be empty (otherwise all $\bar{\omega}_j$ and ω_{j+1} in W would have been diagonally opposed). We replace $\omega_i = x^*$ by $h_{*\bar{x}}$ if the cycle $I(\omega_{i-1}, \omega_i)$ is positive and by h_{x*} otherwise. Analogously ω_{i+1} is replaced by $\bar{h}_{\bar{y}*}$ if $I(\omega_{i+1}, \omega_{i+2})$ is positive and by $\bar{h}_{*y}$ otherwise.

(2) If $\omega_i = h_{**}$ (or $\omega_i = \bar{h}_{**}$) and $I(\omega_i, \omega_{i+1})$ ($I(\omega_{i-1}, \omega_i)$ resp.) is nonempty we cancel ω_i without replacement. If $I(\omega_i, \omega_{i+1})$ is empty then ω_{i+1} is diagonally opposed to ω_i (thus covering the case $\bar{h}_{**}$). In this case both symbols remain unchanged.

Having presented the construction of W_γ to reflect the geometry in Lemmas 4.1 - 4.3 we have now proved:

THEOREM 4.4. *Suppose γ is a lift of the geodesic $\gamma(W) \in [W]$. An edge path E that crosses the same open edges and the same vertices of N (or coincides with γ if γ is contained in N) will periodically read off the geodesic word W_γ, derived above from the representative $\hat{W} \in [W]$ in res. can. form.*

The following remark relates our results with the work done in [3] and pinpoints an interesting fact for the distinction of the geodesic path among all shortest ones:

REMARK 4.5. If the geodesic word W_γ of a given free homotopy class [W] uses only the original generators $a_i, \bar{a}_i$ then it is shortest among all such words in [W] (for a detailed discussion see [3]). In order to get some shortest word from an arbitrary representative of [W] (in the original generators) we may apply the steps I and II of the algorithm described in §3 (thus eliminating "long cycles" and "long chains" in the terminology of [3]). If the resultant word W (nonexceptional by the above hypothesis on W_γ) contains no half cycle then it coincides with W_γ which then is the unique shortest representative of [W]. Otherwise all shortest representatives of [W] are related to W by sequences of half-cycle switches. Their corresponding edge paths coincide with the well-defined unique ∞-binormal edge path $\hat{E}$ (representing $[W]$) in all edges of $\hat{E}$ contained in Γ (i.e. labelled $a_i, \bar{a}_i$). The possible half-cycle switches correspond exactly to the vertices of $\hat{E}$ contained in N: Shortest paths using only edges of Γ (corresponding to the shortest words using only the original generators) have the possibility of passing on the left or on the right side of these vertices.

An interesting consequence of the above investigations is the fact that for most vertices the determination of where the geodesic path will go does not involve the whole word W but depends only on the local circumstances: For a sequence of vertices all lying on the same geodesic γ_0 contained in the net N the geodesic path is well defined by the two neighboring edges of $\hat{E}$ not contained in γ_0 except for the middle vertex in the case their number is odd. An example is given in the introduction (§0); see Figure 0.

§5. *The geometric intersection number*

In this last section we prove the analogue of the Main Theorem in Part I that enables us to determine the geometric intersection number for two given free homotopy classes $[V]$ and $[W]$ of loops in the closed surface M^2. Furthermore we show that the same methods apply to give an algorithmic solution of a problem that arises in working computionally with Heegard splittings.

Throughout this section we assume all (cyclic) words to be in (restricted) canonical form and correspondingly all edge paths to be ∞-normal. If $W = \omega_1 \omega_2 \cdots \omega_r$ and Q is a vertex (or virtual vertex) of Δ which is the initial (resp. central) vertex of an edge labelled ω_1, we define $E(Q,W)$ (as in Part I, §3) to be the bi-infinite edge path which periodically reads off W with an edge ω_1 starting at Q (or crossing over Q if Q is a virtual vertex). We shall freely use the terminology introduced in Part I, §1 and §3, if the definitions remain unchanged or can be carried over unequivocally (e.g. parallel at infinity, Ax(Q,W), geometric intersection number, ···).

THEOREM 5.1. *Suppose* $V = \nu_1 \cdots \nu_n$ *and* $W = \omega_1 \cdots \omega_m$ *are cyclic words in res. can. form which are not proper powers of other words. Then*

(A) *the geometric intersection number* $|V \cap W|$ *is the cardinality of the set of equivalence classes*

$$\{[j,k] \mid (j,k) \text{ is a linking pair}, 1 \le j \le n, 1 \le k \le m\}$$

where the equivalence of pairs is the relation generated by

(1) $\qquad (j,k) \sim (j+1, k+1) \quad$ *if* $\quad \nu_j = \omega_k$

(2) $\qquad (j+1,k) \sim (j, k+1) \quad$ *if* $\quad \nu_j = \overline{\omega}_k$

with j+1 *and* k+1 *taken modulo* n *and* m *respectively.*

(B) $|V| = (1/2)|V \cap V| = \text{card}\{[j,k] \mid (j,k) \text{ is a linking pair}, 1 \le j < k \le n\}$
where equivalence of pairs is the relation generated by (1) and (2) with the interpretation that (a,b) *is replaced by* (b,a) *if* $a > b$.

Here (j,k) *is a linking pair if there exist an appropriate vertex* Q *in* Δ *such that* $\partial E(Q,V_j)$ *and* $\partial E(Q,W_k)$ *link in* ∂D^2 (Q *possibly virtual!*).

REMARK 5.2. A word W in res. can. form is a proper power if and only if the corresponding group element of $\pi_1 M^2$ is a proper power. This is a consequence of Corollary 3.6.

REMARK 5.3. From the definitions above and in §4 we see that (j,k) is a linking pair if and only if (1) ν_j and ω_k are dual to each other (i.e. if $\nu_j = x \in \{a_i, \overline{a}_i\}$ then $\omega_k = x^*$ or $\omega_k = \overline{x}^*$ and conversely) or (2) the cyclic lexicographic order defined in §4 yields $V_j^\infty < W_k^\infty < \overline{V}_j^\infty < \overline{W}_k^\infty$ or $V_j^\infty < \overline{W}_k^\infty < \overline{V}_j^\infty < W_k^\infty$ (both understood cyclically).

REMARK 5.4. The algorithm given in Part I, §2 for computing $|V \cap W|$ carries over immediately (with the obvious changes) to the present situation of a closed surface M^2 and given cyclic words V,W in res. can. form. For arbitrarily given words V,W in the generators $a_i, \bar{a}_i$ the first computational work is of course to transform them into their corresponding res. can. form, using the algorithm given in §3. However the author wants to emphasize that in most cases not all of the steps I-VI actually have to be performed: if for example a given word does not contain any s-cycle for $s \geq q$ one may pass directly over to step VI or even omit this one too; it is not hard to see that the exceptional words also always yield the correct intersection number.

Proof of Theorem 5.1. An edge path E(Q,W) maps onto a path freely homotopic to the well-defined geodesic $\gamma(W)$. The periodic lift of this free homotopy shows that the hyperbolic line Ax(Q,W) with the same limit points as E(Q,W) is a covering of $\gamma(W)$. Thus we can define a map
$$\phi : \{[j,k] \mid (j,k) \text{ is a linking pair}\} \longrightarrow \gamma(V) \cap \gamma(W)$$
as in Part I, §4. By the structure of the proof given there it suffices to show that the properties P1-P4 hold in the present situation too. For P1 and P2 this is obvious from the definitions. Property P4 is stated explicitly by Theorem 2.4. The uniqueness of the path E(X) in P3 is given by Theorem 2.9. Its existence follows from Theorem 3.7. The image of some edge path reading off the word W in res. can. form is freely homotopic to the closed geodesic $\gamma(W) \in [X]$. The periodic lift of the free homotopy that contains Axis X gives an edge path E(Q,W) with the same limit points as Axis X, where Q is an appropriate vertex of Δ. □

A possible application of Theorem 5.1 is to decide whether two sets of free homotopy classes of curves on a closed surface M^2 determine a Heegard splitting of some 3-manifold M^3. The methods developed in this paper actually allow us to solve the slightly harder Problem 5.5; the algorithm sketched below can be used, for example, to determine a presentation of $\pi_1 M^3$ corresponding to the Heegard splitting given in terms of generators $a_1, \cdots, a_q$ of $\pi_1 M^2$. Another application is that one may carry through the test described in [6], p. 149 for whether a given Heegard splitting defines the sphere S^3.

Problem 5.5. Suppose $W^{(1)}, W^{(2)}, \cdots, W^{(r)}$ is a set of cyclic words in res. can. form describing distinct oriented geodesics $\gamma(W^{(i)}) \subset M^2$ with no intersection or self-intersection. Let V be a cyclic word in res. can. form. Determine the sequence of crossings of the geodesic $\gamma(V)$ with the $\gamma(W^{(i)})$'s and their direction (i.e. if $\gamma(W^{(i)})$ crosses from the right or the left while running along $\gamma(V)$).

We know that each crossing $\gamma(V) \cap \gamma(W^{(i)})$ corresponds exactly to an equivalence class of linking pairs (j,k). For a linking pair (j,k) corresponding to a virtual intersection vertex the direction of the crossing is given by the two crossing edges with label ν_j and $\omega_k^{(i)}$; in all other cases we obtain it from the cyclic order of the end points of $E(Q, V_j)$ and $E(Q, W_k^{(i)})$ for some appropriate vertex $Q \subset \Delta$, computed from V and $W^{(i)}$ as described in Remark 5.3.

The cyclic order in which the crossings occur is induced by a linear order: Set $C = [j, k^{(i)}] < [j', k'^{(i')}] = C'$ if and only if

(a) $j < j'$ for some j' occurring in the equivalence class C' and all j in the equivalence class C

or (b) $j < j'$ for some j in the class C and all j' in the class C'

or (c) $j = j'$ for some j, j' occurring in the classes C and C' resp. and $\overline{V}_j^\infty < (W_k^{(i)})^{\pm\infty} < (W_{k'}^{(i')})^{\pm\infty} < V_j^\infty$ [$(j, k^{(i)}) \epsilon C, (j', k'^{(i')}) \epsilon C'$].

Note that while (c) overlaps with (a) and (b), it always gives the same linear order.

In fact this order corresponds to the order of the limit points of edge paths periodically reading off $W^{(i)}$ which cross a fixed edge path that periodically reads off V. Since the $\gamma(W^{(i)})$ have been assumed to be without intersection or self-intersection this gives exactly the sequence of crossings of the $\gamma(W^{(i)})$'s with $\gamma(V)$.

MATHEMATICS DEPARTMENT
M.I.T.
CAMBRIDGE, MASSACHUSETTS 02139

BIBLIOGRAPHY

[1] A. F. Beardon, *The geometry of discrete groups*, GTM 91, Springer-Verlag, New York (1983).

[2] J. S. Birman and C. Series, *An algorithm for simple curves on surfaces*, J. London Math Soc. 29 (1984), 331-342.

[3] _____, *Dehn's algorithm revisited, with applications to simple curves on surfaces*, this volume.

[4] _____, *Geodesics with multiple self-intersections and symmetries on Riemann surfaces*, preprint.

[5] M. M. Cohen and M. Lustig, *Paths of geodesics and geometric intersection numbers* I, this volume.

[6] W. Haken, *Various aspects of the three-dimensional Poincaré problem*, Proc. Univ. of Georgia, Markham Publishing Comp. Chicago (1970), 140-152.

[7] C. L. Siegel, *Topics in complex function theory* II, Wiley-Interscience, New York (1971).

SELECTED PROBLEMS

S. M. Gersten

This brief coda to the Proceedings consists of a selection of problems submitted by participants and one repentant nonparticipant. I am grateful to John Stallings for having aided me in making this selection and for helping me formulate some of these problems in a way we could understand. Of course I'm grateful to all who submitted problems, whether they were used or not. Any problem without a specific attribution can be assumed to be my own.

In problems 1-6, F denotes a finitely generated free group. In problem 13, F denotes R. J. Thompson's group, defined there.

Problem 1.

If $a \in \text{Aut } F$ and $\text{Fix}(a) = \{x \in F | a(x) = x\}$, then it is known that $\text{Fix}(a)$ is a finitely generated subgroup of F. It is still open in general whether rank $(\text{Fix}(a)) \leq$ rank F. See Stallings' article on automorphisms in this volume for a more detailed discussion. Let $R(a) = \bigcup_{n>0} \text{Fix}(a^n)$, the set of recurrent points of a. Is it true that $R(a)$ is finitely generated and if so can one give an effective bound on the rank of $R(a)$? Both questions have affirmative answers if a is geometric, by results of Jaco and Shalen.

Problem 2.

Does Aut F (respectively Out (F) if rank $F \geq 3$) admit a faithful linear representation? In fact it is even unknown whether or not these groups are arithmetic. The expected answers are "no" but there is no compelling evidence. In this connection, B. Weisfeiler has shown (unpublished) that Aut F is not a lattice in a simple Lie group. Serre has observed that if Γ = Out F were arithmetic then it would follow from results of Culler-Vogtmann and Gersten that $H^i(\Gamma, Z[\Gamma]) = 0$ for $i \neq 2n-3$ (n = rank F). Closely related is the question whether Γ is a duality group, in the sense of Bieri and Eckmann.

Problem 3.

Let N = Ker (Out F $\to$ Aut F_{ab}), the group of outer IA-automorphisms, in Bachmuth's terminology. Is N finitely presented? If rank F = 2, then N = {1} (Nielsen) and in general N is always finitely generated by a result of Magnus.

Problem 4 (the Hanna Neumann Conjecture).

Let A and B be finitely generated subgroups of F and let C = A $\cap$ B. Let the ranks of A, B and C be a, b and c respectively. Is $c-1 \leq (a-1)(b-1)$ if $c \neq 0$? Hanna Neumann proved that $c-1 \leq 2(a-1)(b-1)$ if $c \neq 0$. Partial results have been obtained by R. G. Burns and by Gersten.

Problem 5.

Is an extension G of F by Z coherent? Recall that a group G is called coherent if the finitely generated subgroups of G are finitely presentable. If $G = F \rtimes_\alpha Z$, where $\alpha \in$ Aut F is geometric, then G. P. Scott has shown that G is coherent. Scott has also asked whether an extension G of F by Z is LERF (a group G is called LERF if each finitely generated subgroup of G is the intersection of subgroups of finite index).

Problem 6.

Let w be a (cyclic) word in F and let $\ell(w)$ denote the length of a cyclically reduced word conjugate to w in F. Let $\alpha \in$ Out F and let $\lambda = \limsup_{n\to\infty} \ell(\alpha^n(w))^{1/n}$. Is λ an algebraic integer? Is the set of such λ as w varies, for fixed α, a finite set? If α is geometric then both questions have affirmative answers, by results of Thurston.

Problem 7.

Let G be a finite group and let G<t> be the free product with the infinite cyclic group <t>. If w ∈ G<t> is such that the exponent sum $\exp_t(w)$ of t in w is not zero, then the equation "w = 1" can be solved in an overgroup of G by a theorem of Gerstenhaber and Rothaus. Their argument cannot be given a constructive interpretation. Find a constructive proof of the Gerstenhaber-Rothaus theorem. It is interesting to observe in this context that Howie's result for locally indicable groups, can be interpreted constructively. The general question, whether "w = 1" can be solved in an overgroup of G when $\exp_t(w) \neq 0$, is one form of the Kervaire-Laudenbach conjecture which is discussed more fully in Howie's survey article.

Problem 8 (suggested by Joan Birman).

Is there an obstruction to a very strong version of the Nielsen Realization Theorem: namely, can one realize the entire mapping class group of a surface as a group of homeomorphisms? Equivalently, in the case when M is a closed surface, is there an obstruction to the existence of a section to the natural map $\text{Aut}(\pi_1(M)) \to \text{Out}(\pi_1(M))$? (Editorial note: no such section exists for open surface. If F is a free group of rank 2, then $\text{Out}(F)$ has an element of order 6, but $\text{Aut}(F)$ does not.) Steve Kerckhoff remarks that this question is related to a recent result of Morita, that you can't lift $\pi_0(\text{Diff } M_g^2)$ to $\text{Diff } M_g^2$ for sufficiently high genus g. This result hinges on the existence of cohomology classes in $H^*(\pi_0(\text{Diff}))$ which are of sufficiently high dimension that they must vanish in $H^*(\text{Diff}^\delta)$ (where the superscript δ refers to the discrete topology) by the Bott Vanishing Theorem. This latter result holds only for Diff and not for $\text{Homeo}(M_g^2)$. Nevertheless, Kerckhoff admits he would be surprised if Birman's conjectured section existed.

Problem 9 (Joan Birman).

Let M_g be an orientable surface of negative Euler characteristic, closed or once-punctured, g = genus, and let $h: M_g \to M_g$ be a pseudo-Anosov map; in particular h fixes no conjugacy class in $\pi_1 M_g$, for any $n \neq 0$. Let $\lambda = \lambda(h)$ be the stretch factor for h. It is known that λ is real and bigger than 1. Let $\lambda_g = \inf.(\lambda(h))$, for all maps $h: M_g \to M_g$ as above.

Problem.

Study the sequence $\{\lambda_g; g = 1,2,3,\cdots\}$.

REMARK. It is "known" that λ_g is achieved, and that $\lim_{g \to \infty} \lambda_g = 1$, but proofs of both would be a contribution. The real numbers λ_g are algebraic integers. It is not known which algebraic integers occur as $\lambda(h)$ so it might be necessary to study that question first. For fixed g, arbitrarily large λ can occur, since if h is pseudo-Anosov, so is h^n for every $n > 1$, and $\{\lambda(h^n), n = 1,2,3,\cdots\}$ is unbounded.

Problem 10 (Leo Comerford).

Given an equation $W(x_1, x_2, \cdots, a_1, a_2, \cdots) = 1$ which has no solution in a free group $F = \langle a_1, a_2, \cdots \rangle$, is there a finite quotient of F in which w = 1 has no solution? This property might be called "equational separability." For equations $U(a_1, a_2, \cdots) = 1$ it is residual finiteness, and for equations $x^{-1} U(a_1, a_2, \cdots) x = V(a_1, a_2, \cdots)$ it is conjugacy separability.

EDITORIAL NOTE. The problem may be rephrased in terms of the profinite completion $\hat{F}$ of F as follows. If the equation $W(x_1, x_2, \cdots; a_1, a_2, \cdots) = 1$ above has a solution in $\hat{F}$, then does it have a solution in F? The corresponding assertion for the completion of F in the p-lower central series topology (p prime) is false, by results due to Howie (Math Z. 187 (1984), 25-27). A special case of Comerford's problem is of particular interest to this editor. Suppose that $g_1, g_2, \cdots, g_r \in F$ are such that $1 \in \prod_{i=1}^{r} g_i^{F\hat{}}$, where $g_i^{F\hat{}}$ denotes the conjugacy class of g_i in $F\hat{}$. Then is $1 \in \prod_{i=1}^{r} g_i^{F}$?

Problem 11 (Malcolm Wicks).

Let F be a free group of rank n, k a positive integer. Define $Pr(n,k) = \#\{w \in F | \ell(w) = k$ and w is primitive$\}$, where $\ell(w)$ is the reduced length of w. Describe the growth of the function $Pr(n,k)$ with k, for fixed n.

Problem 12 (Gerhard Rosenberger).

Let $G = <a_1, b_1, \cdots, a_g, b_g | \sum_{i=1}^{g} [a_i, b_i] = 1>$ be the surface group of genus $g \geq 2$. Let $x_1, x_2, \cdots, x_q$, $q \geq 2g$, be a set of generators for G. Is there a free (Nielsen) transformation from $(x_1, x_2, \cdots, x_q)$ to the generating system $(a_1, b_1, \cdots, b_g, 1, \cdots, 1)$?

Problem 13 (due to Ross Geoghegan, proposed by Mathew Brin).

Let $F = <x,y | y^{xy} = y^{x^2}, y^{x^2 y} = y^{x^3}> \cong <x_0, x_1, \cdots | x_i^{-1} x_j x_i = x_{j+1}$ for all $i<j>$ be the group discovered by Thompson and studied by Freyd, Heller, Dydak, Minc, Geoghegan, Brown, Brin and Squier, i.a. Is the group F amenable? Recall that a group G is called amenable if there exists a function $\mu: P(G) \to [0, \infty]$, where $P(G)$ is the power set of G, so that

(1) μ is finitely additive (i.e., if $A_1, A_2, \cdots, A_n \subseteq G$ are pairwise disjoint subsets, then $\mu(A_1 \cup \cdots \cup A_n) = \sum_{i=1}^{n} \mu(A_i)$),

(2) μ is nontrivial and bounded (so one may as well assume $\mu(G) = 1$), and

(3) μ is left invariant (so $\mu(gA) = \mu(A)$ for all $g \in G$, $A \subseteq G$). Such a function μ is called a left invariant mean on G.

By way of background to this problem, Von Neumann conjectured that a finitely generated group G is amenable iff every free subgroup of G

is cyclic. There exist finitely generated but not finitely presented counterexamples to von Neumann's conjecture due to Adian and Ol'shanskii. Geoghegan asks whether Thompson's group F is a finitely presented counterexample to von Neumann's conjecture. As evidence, one knows that every free subgroup of F is cyclic (Freyd, Brin-Squier) and F satisfies no law (Freyd-Heller, Brin-Squier). In particular, for each number $n > 0$ there is a homomorphism $h_n : F_2 \to F$ (where F_2 is the free group of rank 2) so that h_n is injective on the set of words of length no greater than n.

Problem 14 (Andrew Nicas).

Characterize all groups which occur as the fundamental group of a closed aspherical manifold.

Problem 15 (Tim Cochran).

Is every finitely generated perfect group the normal closure of one of its elements (a group G such that $G = <<w>>_G$ for some $w \in G$ is said to be of weight 1)? This question is of interest since a homology 3-sphere Σ can arise from ± 1 Dehn surgery on a knot in S^3 only if $\pi_1(\Sigma)$ has weight 1. It has been conjectured that no nontrivial free product G*G can be the fundamental group of such a surgered manifold Σ. Note however that if A and B are perfect groups which are normally generated by elements a, b respectively of relatively prime order, then A*B has weight 1.

Problem 16 (suggested by Tim Cochran, a problem of Gilbert Baumslag's).

A group P is called parafree if there is a homomorphism $F \to P$ with F free inducing isomorphisms $F/\gamma_n(F) \xrightarrow{\cong} P/\gamma_n(P)$ of the quotient of the finite terms of the lower central series and if $\gamma_\omega(P) = 1$ where $\gamma_\omega(P) = \bigcap_{n \geq 1} \gamma_n(P)$. If P is a finitely generated parafree group, then is $H_2(P; Z) = 0$? The corresponding assertion if P is not finitely generated is false (for background, see M. Gutierrez, J. Algebra 51 (1978),

354-366; MR 58 (1979) #5972). What can one say about the higher integral homology of P? The cited article of Gutierrez purports to prove that $H_2(P;Z) = 0$ if P is finitely generated and parafree. However, there are difficulties in the argument which are not suggested in any review of this paper and which suggest that Baumslag's problem should be considered open.

Problem 17 (Tim Cochran).

Let $\mathcal{H}$ denote the class of finitely presented groups such that $H_1(G) \cong Z^m$ and $H_2(G) = 0$ $(m \geq 1)$. Can every group G in $\mathcal{H}$ be mapped to another group P of $\mathcal{H}$ such that the induced map $H_1(G) \to H_1(P)$ is an isomorphism and such that $H_n(P) = 0$ for $n \geq 2$? This question relates to a fundamental open question in higher dimensional knot theory: "Is every link of m codimension 2 spheres in S^n $(n \geq 4)$ concordant to a boundary link?" (see Cappell and Shaneson, Link Cobordism, Comm. Math. Helv. 55 (1980), 20-49 for the relevant definitions). The relevant point here is that a link (with fundamental group G of the complement) is a boundary link only if P can be chosen to be free of rank m. For the concordance question it is only necessary that P have the homology of a free group (T. Cochran, TAMS 285 (1984), 389-401).

DEPARTMENT OF MATHEMATICS
UNIVERSITY OF UTAH
SALT LAKE CITY, UTAH 84112

Library of Congress Cataloging-in-Publication Data

Combinatorial group theory and topology.

(Annals of mathematics studies ; no. 111)
Presentations from a conference held at Alta Lodge, Utah, July 15-18, 1984.
Bibliography: p.
Includes index.
1. Combinatorial group theory—Congresses.
2. Topology—Congresses. I. Gersten, S. M., 1940-
II. Stallings, John R. (John Robert), 1935-
III. Series.
QA171.C6784 1986   512'.22   85-43283
ISBN 0-691-08409-2   ISBN 0-691-08410-6 (pbk.)

Stephen M. Gersten is Professor of Mathematics at the University of Utah. John R. Stallings is Professor of Mathematics at the University of California at Berkeley

RAYMOND H. FOGLER LIBRARY
DATE DUE

BOOKS ARE SUBJECT TO
RECALL AFTER TWO WEEKS